Nutrition Labeling Handbook

FOOD SCIENCE AND TECHNOLOGY

A Series of Monographs, Textbooks, and Reference Books

1. Flavor Research: Principles and Techniques, *R. Teranishi, I. Hornstein, P. Issenberg, and E. L. Wick*
2. Principles of Enzymology for the Food Sciences, *John R. Whitaker*
3. Low-Temperature Preservation of Foods and Living Matter, *Owen R. Fennema, William D. Powrie, and Elmer H. Marth*
4. Principles of Food Science
   Part I: Food Chemistry, *edited by Owen R. Fennema*
   Part II: Physical Methods of Food Preservation, *Marcus Karel, Owen R. Fennema, and Daryl B. Lund*
5. Food Emulsions, *edited by Stig E. Friberg*
6. Nutritional and Safety Aspects of Food Processing, *edited by Steven R. Tannenbaum*
7. Flavor Research: Recent Advances, *edited by R. Teranishi, Robert A. Flath, and Hiroshi Sugisawa*
8. Computer-Aided Techniques in Food Technology, *edited by Israel Saguy*
9. Handbook of Tropical Foods, *edited by Harvey T. Chan*
10. Antimicrobials in Foods, *edited by Alfred Larry Branen and P. Michael Davidson*
11. Food Constituents and Food Residues: Their Chromatographic Determination, *edited by James F. Lawrence*
12. Aspartame: Physiology and Biochemistry, *edited by Lewis D. Stegink and L. J. Filer, Jr.*
13. Handbook of Vitamins: Nutritional, Biochemical, and Clinical Aspects, *edited by Lawrence J. Machlin*
14. Starch Conversion Technology, *edited by G. M. A. van Beynum and J. A. Roels*
15. Food Chemistry: Second Edition, Revised and Expanded, *edited by Owen R. Fennema*

16. Sensory Evaluation of Food: Statistical Methods and Procedures, *Michael O'Mahony*
17. Alternative Sweetners, *edited by Lyn O'Brien Nabors and Robert C. Gelardi*
18. Citrus Fruits and Their Products: Analysis and Technology, *S. V. Ting and Russell L. Rouseff*
19. Engineering Properties of Foods, *edited by M. A. Rao and S. S. H. Rizvi*
20. Umami: A Basic Taste, *edited by Yojiro Kawamura and Morley R. Kare*
21. Food Biotechnology, *edited by Dietrich Knorr*
22. Food Texture: Instrumental and Sensory Measurement, *edited by Howard R. Moskowitz*
23. Seafoods and Fish Oils in Human Health and Disease, *John E. Kinsella*
24. Postharvest Physiology of Vegetables, *edited by J. Weichmann*
25. Handbook of Dietary Fiber: An Applied Approach, *Mark L. Dreher*
26. Food Toxicology, Parts A and B, *Jose M. Concon*
27. Modern Carbohydrate Chemistry, *Roger W. Binkley*
28. Trace Minerals in Foods, *edited by Kenneth T. Smith*
29. Protein Quality and the Effects of Processing, *edited by R. Dixon Phillips and John W. Finley*
30. Adulteration of Fruit Juice Beverages, *edited by Steven Nagy, John A. Attaway, and Martha E. Rhodes*
31. Foodborne Bacterial Pathogens, *edited by Michael P. Doyle*
32. Legumes: Chemistry, Technology, and Human Nutrition, *edited by Ruth H. Matthews*
33. Industrialization of Indigenous Fermented Foods, *edited by Keith H. Steinkraus*
34. International Food Regulation Handbook: Policy · Science · Law, *edited by Roger D. Middlekauff and Philippe Shubik*
35. Food Additives, *edited by A. Larry Branen, P. Michael Davidson, and Seppo Salminen*
36. Safety of Irradiated Foods, *J. F. Diehl*
37. Omega-3 Fatty Acids in Health and Disease, *edited by Robert S. Lees and Marcus Karel*
38. Food Emulsions: Second Edition, Revised and Expanded, *edited by Kåre Larsson and Stig E. Friberg*
39. Seafood: Effects of Technology on Nutrition, *George M. Pigott and Barbee W. Tucker*
40. Handbook of Vitamins: Second Edition, Revised and Expanded, *edited by Lawrence J. Machlin*
41. Handbook of Cereal Science and Technology, *Klaus J. Lorenz and Karel Kulp*
42. Food Processing Operations and Scale-Up, *Kenneth J. Valentas, Leon Levine, and J. Peter Clark*
43. Fish Quality Control by Computer Vision, *edited by L. F. Pau and R. Olafsson*
44. Volatile Compounds in Foods and Beverages, *edited by Henk Maarse*

45. Instrumental Methods for Quality Assurance in Foods, *edited by Daniel Y. C. Fung and Richard F. Matthews*
46. Listeria, Listeriosis, and Food Safety, *Elliot T. Ryser and Elmer H. Marth*
47. Acesulfame-K, *edited by D. G. Mayer and F. H. Kemper*
48. Alternative Sweeteners: Second Edition, Revised and Expanded, *edited by Lyn O'Brien Nabors and Robert C. Gelardi*
49. Food Extrusion Science and Technology, *edited by Jozef L. Kokini, Chi-Tang Ho, and Mukund V. Karwe*
50. Surimi Technology, *edited by Tyre C. Lanier and Chong M. Lee*
51. Handbook of Food Engineering, *edited by Dennis R. Heldman and Daryl B. Lund*
52. Food Analysis by HPLC, *edited by Leo M. L. Nollet*
53. Fatty Acids in Foods and Their Health Implications, *edited by Ching Kuang Chow*
54. *Clostridium botulinum*: Ecology and Control in Foods, *edited by Andreas H. W. Hauschild and Karen L. Dodds*
55. Cereals in Breadmaking: A Molecular Colloidal Approach, *Ann-Charlotte Eliasson and Kåre Larsson*
56. Low-Calorie Foods Handbook, *edited by Aaron M. Altschul*
57. Antimicrobials in Foods: Second Edition, Revised and Expanded, *edited by P. Michael Davidson and Alfred Larry Branen*
58. Lactic Acid Bacteria, *edited by Seppo Salminen and Atte von Wright*
59. Rice Science and Technology, *edited by Wayne E. Marshall and James I. Wadsworth*
60. Food Biosensor Analysis, *edited by Gabriele Wagner and George G. Guilbault*
61. Principles of Enzymology for the Food Sciences: Second Edition, *John R. Whitaker*
62. Carbohydrate Polyesters as Fat Substitutes, *edited by Casimir C. Akoh and Barry G. Swanson*
63. Engineering Properties of Foods: Second Edition, Revised and Expanded, *edited by M. A. Rao and S. S. H. Rizvi*
64. Handbook of Brewing, *edited by William A. Hardwick*
65. Analyzing Food for Nutrition Labeling and Hazardous Contaminants, *edited by Ike J. Jeon and William G. Ikins*
66. Ingredient Interactions: Effects on Food Quality, *edited by Anilkumar G. Gaonkar*
67. Food Polysaccharides and Their Applications, *edited by Alistair M. Stephen*
68. Safety of Irradiated Foods: Second Edition, Revised and Expanded, *J. F. Diehl*
69. Nutrition Labeling Handbook, *edited by Ralph Shapiro*
70. Handbook of Fruit Science and Technology: Production, Composition, Storage, and Processing, *edited by D. K. Salunkhe and S. S. Kadam*
71. Freezing Effects on Food Quality, *edited by Lester E. Jeremiah*
72. Food Antioxidants: Technological, Toxicological, and Health Perspectives, *edited by D. L. Madhavi, S. S. Deshpande, and D. K. Salunkhe*

Additional Volumes in Preparation

Handbook of Indigenous Fermented Foods: Second Edition, Revised and Expanded, *edited by Keith H. Steinkraus*

Carbohydrates in Food, *edited by Ann-Charlotte Eliasson*

Handbook of Food Analysis: Volume 1, *edited by Leo M. L. Nollet*

Nutrition Labeling Handbook

edited by

Ralph Shapiro

President
Nutrition International
Dayton, New Jersey

Marcel Dekker, Inc. **New York • Basel • Hong Kong**

Library of Congress Cataloging-in-Publication Data

Nutrition labeling handbook / edited by Ralph Shapiro
p. cm.— (Food science and technology ; 69)
Includes index.
ISBN 0-8247-9285-8 (hardcover : alk. paper)
1. Food—Labeling—United States—Handbooks, manuals, etc.
2. Food—Labeling—Law and legislation—United States—Handbooks, manuals, etc. 3. Nutrition—United States—Handbooks, manuals, etc.
I. Shapiro, Ralph. II. Series: Food science and technology (Marcel Dekker, Inc.) ; 69.
TX551.N79 1995
363.19'2'0973—dc20 95-21838
CIP

The publisher offers discounts on this book when ordered in bulk quantities. For more information, write to Special Sales/Professional Marketing at the address below.

This book is printed on acid-free paper.

Marcel Dekker, Inc.
270 Madison Avenue, New York, New York 10016

Current printing (last digit):
10 9 8 7 6 5 4 3 2 1

PRINTED IN THE UNITED STATES OF AMERICA

Foreword

Never before in my almost half a century in the food industry, serving in various research, development, and management capacities, have I been aware of an issue that has generated as much unanimity of interest as has the Nutrition Labeling and Education Act (NLEA) passed by Congress in 1990. Without question in my mind, the food industry is for NLEA, nutritionists are for it, consumer activists are for it, and, of course, the politicians are for it or they would not have passed the law. Congress has not acted directly on many food issues in our times other than the original establishment of the USDA and the FDA and the several modifications to these laws. A particularly interesting dimension of the current law is that it is not only a technical content and labeling bill, but is also a food and nutrition *education* act. And, as I said, the most remarkable dimension is that the majority of the stakeholders, from whatever sector of society they come, appear to be behind it. Not that they're all looking to it for the same reasons or outcomes, but they clearly all view it as a move in the right direction.

How did all of this come about? Simply put, it is the direct result of the highest level ever of consumer interest in food and in health and in the connection between the two. The current generation in the United States, and increasingly elsewhere in the developed world, is very much focused on healthful living, involving lifestyle, exercise, diet, and nutrition. This in itself is a consumer food "revolution" of unprecedented dimensions. I'm not aware of any period of time in the past where people more firmly believed that "you are what you eat" and recognized the potentially healthful dimensions of proper diet beyond the simple necessity of sustenance. There have been fringe food groups, food faddists, and even health food movements in the past, but they have always been minority activities essentially ignored by the vast bulk of the population. This is not true today. Today's food consumers may not fully understand what they are dealing with when it comes to the healthful dimensions of proper diet, but they do believe that there is a connection between the two, and, more importantly, in increasing numbers they are acting accordingly.

Like all new things, NLEA has a way to go yet in terms of being as effective as all of the stakeholders would like to see it become. In this book, you will find most of the unsolved issues delineated. These issues lie in the regulatory arena that is charged with implementing NLEA. They lie in industry desires to be free to promote their products. They lie in consumer activists' concerns over possible abuse and/or misinformation. They lie in nutritionist concerns over how to insure the proper *educational* dimensions get across to the consumer which result in right nutritional behavior. They lie in the technical research needs for proper resources to be invested in new health and nutrition knowledge.

For these reasons and others, this book is needed. There is need for a concise and authoritative delineation of the issues and expert perspectives on how to deal with them. This book can serve as a seminal text for guiding direction and adjustments to NLEA in the future, for much has yet to be done to make NLEA the "magic bullet" of a more healthful society that all of the stakeholders would like to see it become. You will find in this text experts of unmatched experience and knowledge in their given fields. They all have something to say that needs to be considered very carefully.

It is rare in our society that so many different perspectives can view with optimism a new direction such as NLEA. While there are problems to be resolved, the vision of NLEA leading to a healthier food supply and a more nutritionally informed consumer will be achieved with additional compromise and mutual understanding by all of the interested parties. Once again, the consumer is the driving innovation in food and now it falls on all of the players involved to make NLEA work.

Al Clausi
Past President,
National Institute of Food Technologists

Preface

This book is a primary, easy-to-read reference to the 1990 Nutrition Labeling and Education Act (NLEA) regulations. It is intended for food industry professionals, educators, and consultants who want an overview of the regulations as well as specifics for their implementation. It will be useful to food scientists involved in product development; sales and marketing professionals; importers and distributors; food service professionals; dieticians; nutritionists; college and university teachers; consumer interest groups; and physicians and others in the medical community concerned with nutrition and disease prevention.

The book is unique in that under one cover it reviews a broad range of issues related to nutrition labeling. It examines the goals of the NLEA and its impact on the food industry in general as well as on specific segments of the industry (e.g., raw produce and fish, calorie control products, meat and poultry, frozen foods, dairy, food service, functional foods). It discusses industrial, government, and consumer perspectives on the NLEA; provides a comprehensive review of the regulations with practical guidelines for compliance, including label formats; assesses the status of international harmonization; and addresses issues related to advertising, communication, promotion, and consumer education. In addition, the appendices provide FDA's list of reference amounts for determining the serving size of foods; FDA's list of protein digestibility coefficients for determining protein quality of common foods; FTC's Enforcement Policy Statement for Food Advertising; a list of AOAC official and recommended analytical methods with references for determining the nutrient content of foods; data base resources; a list of USDA food composition publications; and a list of FDA and USDA experts (with telephone numbers) who can be contacted for detailed information about NLEA regulations.

Each chapter is written by a preeminent leader in the food arena. Peter Barton Hutt, Former Chief Counsel to FDA, traces the history and evolution of FDA regulation of food labeling with respect to nutrient content. F. Edward Scarbrough, Director, Office of Food Labeling, FDA; John McCutcheon, Deputy Administrator, Regulatory Programs, USDA/FSIS; and Deborah Owen, Former Commissioner of FTC, provide insights into the impact of NLEA on the industries that they regulate.

Several chapters deal with the practical implementation of the regulations. One reduces approximately 1,000 pages of cumbersome and complicated regulatory language from the Code of Federal Regulations into concise, comprehensible form, providing a comprehensive review of the most salient features of the regulations including: "mandatory" and "voluntary" listing of nutrients, daily values, permissible nutrient content claims, standardization and determination of serving sizes, ingredient listing, health claims, and labeling of juices and rules for label formats.

Another chapter provides practical guidelines for compliance, discussing sampling, analytical, and other data base programs to obtain the required nutrient data for labeling; conversion of this "raw data" into acceptable label values; and procedures for statistical evaluation to assess their reliability and the risk of being out of compliance.

A review of official analytical procedures accepted by FDA and USDA with appropriate references is presented in a subsequent chapter. This chapter also discusses the use of data bases for obtaining nutrient values and each agency's position with respect to their acceptance. The reliability, practicality, and economics of each system for obtaining nutrient data are also evaluated.

Strategic planning and management of package design and modification to comply with the NLEA regulations are discussed in two separate chapters. Instructions with graphic examples are provided.

In the final chapters, FDA experts discuss the means by which the goals of the NLEA, namely consumer education and improved dietary patterns, can be achieved. Insights are provided that reveal anticipated consumer understanding of the new regulations, their attitudes, and expected use of the new labels. Strategies are suggested whereby the food industry, government, consumer groups, the medical profession, and academics can individually and collectively, through accurate and effective marketing, advertising, and communication, simultaneously promote healthier diets and better nutrition in the United States using the NLEA as the stimulus.

Overall, the book will familiarize readers with virtually all aspects of the NLEA, will provide them with the information necessary for developing accurate nutrition information panels that comply with the labeling regulations, and will enable them to provide advice and effectively answer questions generated internally or externally by customers, consumers, and others.

I would like to express my deep appreciation to all of the contributing authors who took the time from their very busy schedules to provide their unique insights and expertise on the NLEA.

I would also like to thank my family for their forbearance, understanding, and encouragement during all aspects of this endeavor.

Lastly, a special "thank you" to Sophie Holbrook, whose continuous focus, organization, and energy enabled this book to come to fruition.

Ralph Shapiro

Contents

Contributors

Thomas Balmer Scientific Director, International Dairy Foods Association, Washington, D.C.

Mary M. Bender Department of Health and Human Services, Food and Drug Administration, Washington, D.C.

Aaron L. Brody President, Rubbright·Brody Inc., Devon, Pennsylvania

Paul Bruns Director of Technical Regulatory Affairs, Nabisco Brands, Inc., East Hanover, New Jersey

Cheryl Callen Nabisco Brands, Inc., East Hanover, New Jersey

Kimberlee J. Carson International Dairy Foods Association, Washington, D.C.

Eve M. Caudill Business Administration, University of Illinois at Urbana-Champaign, Urbana, Illinois

Brenda M. Derby Statistician, Consumer Studies Branch, Center for Food Safety and Applied Nutrition, Department of Health and Human Services, Food and Drug Administration, Washington, D.C.

S. Kathleen Egan TAS, Inc., Washington, D.C.

Sara Beck Fein Department of Health and Human Services, Food and Drug Administration, Washington, D.C.

Stephen Gardner Former Assistant Attorney General, Texas and Southern Methodist University, Dallas, Texas

Clare M. Hasler Director, Functional Foods for Health Program, University of Illinois, Chicago and Urbana, Illinois

James T. Heimbach Chief Operating Officer, TAS, Inc., Washington, D.C.

Paul F. Hopper* General Foods Corp., White Plains, New York

Linda R. Horton Director, International Policy Staff, Department of Health and Human Services, Food and Drug Administration, Rockville, Maryland

Richard L. Huston Office of the Vice Chancellor for Research, University of Illinois, Chicago, Illinois

Peter Barton Hutt Former Chief Counsel, Food and Drug Administration; Partner, Covington & Burling, Inc., Washington, D.C.

Guy H. Johnson Director of Nutrition, The Pillsbury Company, Minneapolis, Minnesota

Naomi Kulakow Coordinator, Food Labeling Education, Office of Food Labeling, Department of Health and Human Services, Food and Drug Administration, Washington, D.C.

Jung Keun Lee Department of Health and Human Services, Food and Drug Administration, Washington, D.C.

Guy E. Livingston President, Food Science Associates, Inc., Dobbs Ferry, New York

James L. Marsden Vice President, Scientific Affairs, American Meat Institute, Washington, D.C.

John W. McCutcheon Deputy Administrator, Regulatory Programs, Food Safety and Inspection Service, United States Department of Agriculture, Washington, D.C.

Lyn O'Brien Nabors Executive Director, Calorie Control Council, Atlanta, Georgia

**Present affiliation:* Department of Food Science, Cornell University, Ithaca, New York.

Deborah K. Owen* Former Commissioner, Federal Trade Commission, Washington, D.C.

J. Roy Parcels Partner, Dixon & Parcels Associates, Inc., New York, New York

Jean A. T. Pennington Nutritionist—Data Base Coordinator, Department of Health and Human Services, Food and Drug Administration, Washington, D.C.

Joyce E. Plyler Attorney, Office of General Counsel, Federal Trade Commission, Washington, D.C.

Leslie G. Sarasin Executive Vice President and Staff Counsel, American Frozen Food Institute, McLean, Virginia

F. Edward Scarbrough Director, Office of Food Labeling, Department of Health and Human Services, Food and Drug Administration, Washington, D.C.

Ralph Shapiro President, Nutrition International, Dayton, New Jersey

Lauren Swann President, Concept Nutrition, Inc., Bensalem, Pennsylvania

Virginia L. Wilkening Food and Drug Administration, Washington, D.C.

**Present affiliation:* Arent, Fox, Kintner, Plotkin and Kahn, Washington, D.C.

Nutrition Labeling Handbook

1

A Brief History of FDA Regulation Relating to the Nutrient Content of Food

Peter Barton Hutt

Covington & Burling, Inc., Washington, D.C.

I. INTRODUCTION

Concern about the relationship between diet and health extends back more than twenty centuries. In the fourth century B.C., Hippocrates argued that "differences of diseases depend upon nutriment."[1] Hippocrates concluded that either too little or too much food "harms the man just as much."[2] The renowned Roman physician, Galen, who lived in the second century A.D., similarly advocated moderation in dietary habits to achieve sound nutrition and good health:

> And in the nature of eating and drinking, in quantity, quality and faculty, the objective again there also is moderation, so as to take neither too much nor too little, but as much as, digested and distributed and nourishing the body well, if need be will supply symmetry to the still growing parts, and leave nothing superfluous or lacking[3].

These ancient physicians believed that there were four basic elements (air, fire, water, and earth) reflecting a combination of four fundamental powers or qualities (hot, dry, wet, and cold). Health existed when there was a natural balance or "coction" of the four bodily "humors" (blood, phlegm, yellow bile, and black bile) which originated in four specific parts of the body (the heart, brain, liver, and spleen respectively)[4]. Disease, in turn, existed when there was an unnatural imbalance among these four humors. Treatment of disease consisted primarily of selecting the foods with the correct powers or qualities to bring the four bodily humors back into a natural balance.

This humoral theory of medicine on which Hippocrates, Galen, and their followers related diet to health prevailed for more than 2,000 years. In spite of its lack of scientific support, it was still being taught in medical schools throughout the world as recent as the late nineteenth century. Of course, the humoral theory of medicine has now been shown to be without any medical or nutritional support. Just at the time that it was being displaced by modern scientific medicine, the field of scientific nutrition was also in its infancy.

Government regulation of the quality of the food supply has existed throughout recorded history[5]. Concern about adulteration has lead every civilized society to impose standards and requirements to preserve the integrity of the food supply. Although this early regulation was designed simply to make certain that food remained "wholesome"[6] in general, it also undoubtedly had the unintended effect of preserving the basic nutritional quality of the food supply as well. The regulatory standards for bread which existed in ancient Rome[7], and for at least 600 years in England from the end of the Dark Ages through the mid-1800s[8], had the same fortuitous effect. Only in the past sixty years have food standards explicitly addressed the nutrient content of food.

Food labeling requirements also originated in ancient times. Weights and measures were standardized and primitive forms of food labeling (e.g., writing the name of the food producer on every loaf of bread) were required as well[9]. Misbranding of food—any form of false or misleading labeling—has long been prohibited. Not until the past sixty years, however, have more sophisticated forms of food labeling relating to the nutrient content of food been explored and affirmatively required on food packages.

This chapter provides a brief overview of the history of regulation relating to the nutrition content of food by the Food and Drug Administration (FDA) in the United States. It is intended to provide the reader with a broad perspective on the progress that has been made over the past century, not to focus on specific details.

II. THE DEPARTMENT OF AGRICULTURE ACT OF 1862

When the Department of Agriculture (USDA) was created by Congress in 1862[10], it included the chemical laboratory that previously had been established in the Patent Office. From that chemical laboratory came the successive governmental agencies that ultimately became our current FDA[11].

Throughout the nineteenth century, the food supply was regulated by city, county, and state laws and regulations[12]. The first bill to establish federal legislation to regulate food adulteration and misbranding was introduced in Congress in 1879, but this legislation was not seriously considered at that time because of the prevailing constitutional doctrine that this type of governmental activity was properly within the jurisdiction of state and local governments and thus outside the authority of Congress. Even without a national food law, however, FDA's predecessor agencies within USDA launched a long series of investigations into the quality of the American food supply. Particularly under the leadership of Dr. Harvey W. Wiley, who served as the USDA Chief Chemist from 1883 to 1912, extensive food adulteration was documented and publicized. As a result of its investigations into food adulteration, USDA established some 200 informal standards for unadulterated food products even before federal legislation was enacted. These informal standards formed the basis for much of the enforcement work undertaken by state and local authorities.

III. THE PURE FOOD AND DRUG ACT OF 1906

With the growth of a national economy, the publicity given to food adulteration by Dr. Wiley and his staff at USDA, publication of *The Jungle* by Upton Sinclair documenting gross manufacturing conditions in the meat industry, and the development of constitutional theories compatible with federal regulation of a burgeoning food industry, Congress enacted the Pure Food and Drug Act of 1906[13]. At the same time, Congress also enacted the Federal Meat Inspection Act of 1906[14].

The Pure Food and Drug Act of 1906 was relatively short and, by modern standards, might be regarded as antiquated. At the time, however, it represented a revolutionary federal initiative that had an enormous impact on the quality of the American food supply. The statute broadly prohibited all adulteration and misbranding of food in interstate commerce. A food was deemed to be adulterated if any substance had been added or substituted to reduce its quality or conceal inferiority (economic adulteration) or if it contained any added poisonous or deleterious ingredient which may render the food injurious to health (safety adulteration). A food was deemed to be misbranded if labeled so as to deceive or mislead the purchaser. Based upon these provisions, literally thousands upon thousands of regulatory actions were taken against adulterated and misbranded food products, leading the United States into an unparalleled era of public confidence in the food supply[15].

The 1906 Act contained no provision requiring the affirmative labeling of food products with product information that consumers might regard as important to their purchasing decisions. The Gould Amendment of 1913 for the first time required that the net quantity of contents be labeled on food packages[16].

No other amendments to the 1906 Act, however, required additional mandatory disclosure of product information in food labeling.

Because the prohibitions against adulteration and misbranding were phrased in broad and comprehensive terms, FDA had sufficient statutory authority to handle virtually all issues that related to the quality of food products marketed in the United States. Exaggerated health claims for food products brought prompt FDA regulatory action[17]. The record of FDA regulatory accomplishments during this era is nothing short of remarkable.

Although the 1906 Act contained no provision authorizing the establishment of standards of purity or identity for food, the food standards work that was begun under Dr. Wiley before enactment of the statute was continued[18]. Throughout the history of the 1906 Act, FDA maintained informal food standards, both as guidance for its own regulatory action against economic adulteration and to assist state and local officials.

While this regulatory work was going forward, substantial limitations were identified in the 1906 Act. The 1917 USDA Annual Report, for example, cited the lack of legal standards for food, authority to inspect food establishments, and jurisdiction over false or misleading claims not made on food labels[19]. In 1933, following the election of President Roosevelt, the FDA Commissioner successfully persuaded the new administration to undertake a complete revision of the 1906 Act. The resulting legislation was ultimately enacted into law five years later, in 1938.

One of the principal reasons why FDA concluded that the 1906 law needed revision was the rapid development of the science of nutrition. As knowledge about nutrition developed, entirely new and more sophisticated forms of health representations for food products began to appear in the marketplace.

Vitamins were first characterized and found to be essential to human life early in this century. The annual reports of USDA and FDA document the gradual increase in interest both by the public and by the food industry in vitamins, and the resulting problems of regulating health claims related to nutrition. The first USDA Annual Report mentioning scientific investigation of vitamins was in 1916[20]. From then on, numerous reports related both scientific study and regulatory action relating to specific nutrients and nutrition in general.

The earliest form of dietary supplement was cod liver oil labeled with claims for the vitamin D content, beginning in the 1920s. FDA devoted considerable regulatory time and effort to assure that vitamin claims for these products were accurate and truthful[21]. By 1932, moreover, some 400 other types of dietary supplement products were already on the market, and FDA was forced to establish a new vitamin unit to maintain adequate regulatory surveillance[22].

Closely behind the development of dietary supplement products came the use of vitamins to fortify staple food products[23]. In the early 1930s, the American Medical Association (AMA) Council on Food and Nutrition endorsed the addition of vitamin D to milk and of iodine to table salt. In March 1936 and December 1938, the AMA Council announced the first general policies on food fortification in this country, urging that fortification should be reserved for exceptional cases where there is convincing evidence of a need for an enhanced amount of a vita- min or mineral in the general food supply and the food to be fortified is suitable for fortification, or where the restoration of a vitamin or mineral is needed to maintain the nutrient at an appropriate level in processed food. Throughout this time, the AMA Council identified particular foods for which fortification was justified. In mid-1940, the National Academy of Sciences (NAS) established a Subcommittee on Medical Nutrition, which later became the Food and Nutrition Board, which also strongly supported appropriate food fortification.

FDA was at the same time confronted by a number of obesity and weight reduction products. The Agency publicly stated that the limited definition of the term "drug" made it difficult to take appropriate action against diet products, but nonetheless brought extensive regulatory proceedings to control the claims for these products[24].

IV. THE FEDERAL FOOD, DRUG AND COSMETIC ACT OF 1938

FDA regulation relating to the nutrient content of food under the Federal Food, Drug and Cosmetic Act of 1938[25] (FD&C Act) falls into three distinct eras: 1938–1970, 1970–1990, and 1990 to the present. The provisions of the FD&C Act relating to nutrition remained unchanged through the first two of these three eras. The enactment of the Nutrition Labeling and Education Act of 1990[26] (NL&E Act) begins the third era. Following a general description of the provisions enacted in the 1938 Act, the remainder of this chapter traces the development of FDA regulatory policy regarding the nutrient content of food throughout these three eras.

A. The Provisions of the FD&C Act

From the initial introduction of the new legislation in 1933 through to its final enactment in 1938, a major purpose of the FD&C Act was to broaden FDA authority with regard to the nutrient content of food. First, it authorized the establishment of mandatory food standards. Second, it continued to prohibit any false or misleading statement in food labels and expanded this prohibition to all food labeling. Third, it required any imitation food to be labeled as such. Fourth, it required affirmative labeling of food with particular information specified in the statute, including the name of the food and a statement of ingredients. Fifth, it authorized FDA to require additional affirmative label information for food for special dietary use. Sixth, it required the labeling affirmatively to reveal all facts material in the light of any other representations for the product. This represented a broad array of legal authorities for FDA to embark upon a new era of nutrition regulation.

B. 1938–1970

Section 401 of the FD&C Act authorized FDA to promulgate definitions and standards of identity for any food product in order to "promote honesty and fair dealing in the interest of consumers." In August 1940, FDA announced public hearings to establish a standard of identity for flour[27]. Neither FDA nor the food industry proposed to include consideration of nutrient fortification of flour, but the American Medical Association and the National Academy of Sciences urged FDA to include a standard for enriched flour. The industry ultimately supported this proposal and FDA also agreed. The enriched flour standard became effective on January 1, 1942. This marked the beginning of what became, and remains to this day, a major FDA program of using food standards to assure the nutritional quality of staple food products.

Even at this early date, however, FDA was concerned about overfortification of food. FDA issued a statement in July 1943[28], taking the position that fortification constitutes an implicit representation that the fortified food contained sufficient vitamins and minerals to make a substantial contrition to the nutritional welfare of persons eating the food in customary amounts. FDA stated that the kinds and quantities of nutrients to be added to any particular food must be determined in light of specific nutrient deficiencies in the diet of both the general population and of significant population groups, the place occupied by the food in such diets, and the suitability of the food as a carrier for those nutrients. FDA argued that most natural foods contain a wide variety of nutrients in significant amounts and that adequate nutrition is better assured through a choice of natural food than through reliance on food fortification. Nonetheless, FDA did endorse restoration of nutrients lost during processing and addition of specific nutrients needed to correct deficiencies in the ordinary diet. The 1943 FDA statement of policy remains an accurate statement of the Agency's position fifty years later and, in fact, has never been revoked.

Following World War II, food fortification accelerated. FDA also became more concerned about overfortification and, after devoting a large portion of its enforcement resources to an increasingly difficult effort to police the fortification of food[29] and a major loss in the courts against a fortified sugar product[30], ultimately concluded that explicit regulatory restrictions should be placed on the types of nutrients that could be used in dietary supplements and fortified food and the types of food to which these nutrients could be added. These

policies were incorporated into proposed regulations[31] that were the subject of formal public hearings conducted by FDA during 1968–1969.

Immediately following enactment of the 1938 Act, FDA also sought to implement the statutory provision authorizing additional label information for special dietary food. Section 403(j) of the Act permitted FDA to prescribe, by regulation, appropriate labeling information for any food represented for "special dietary uses." Following a public hearing, FDA promulgated regulations in November 1941, governing dietary supplements, fortified food, infant food, hypoallergenic food, and food used in weight control[32]. The regulations governing dietary supplements and fortified food required the label declaration of the percent of the "minimum daily requirement" (MDR) for a vitamin or mineral for which a specific representation was made, when consumed in a specified quantity during a period of one day. Infant food was required to bear specific information about both the micronutrient and macronutrient content. Food used to control body weight was required to declare the percent by weight of protein, fat, carbohydrates, and calories. Food represented to regulate the intake of sodium was required to bear the number of milligrams of sodium in 100 g of the food and in an average serving of the food. Thus, the concept of nutrition labeling was created by FDA in the first few years under the 1938 Act.

Throughout this period, FDA continued to take strong regulatory action against false or misleading nutrition claims for dietary supplement products and general purpose food products. For the same reasons that the Agency ultimately concluded in the 1960s that case-by-case enforcement against overfortification of food was inadequate, however, the Agency determined that it would be more efficient to ban representative misleading claims by regulation rather than to continue to use the traditional court enforcement approach. The Agency therefore included in the 1968–1969 public hearings a number of provisions designed to ban types of false or misleading nutrition claims against which it had been taking court enforcement action for the past twenty years[33].

In December 1969, however, President Nixon convened the White House Conference on Food, Nutrition and Health, with Jean Mayer, M.D., as the Chairman. The Report of the White House Conference[34] rejected the highly restrictive approach to food standards, imitation labeling, dietary supplements, and nutrition claims that had been the subject of the formal public hearing conducted by FDA during 1968–1969. The Report emphasized the need for sound nutrition, the capability of modern food technology to provide products to fill that need, the use of increased public information about nutrition, and the prevalence of hunger in America, rather than overfortification of food.

At just the time of the White House Conference, moreover, the top three FDA officials were removed and new individuals were recruited from outside to take their place. Within eighteen months, five individuals who had helped shape the White House Conference policy and prepare the Report had left their former positions to join FDA and were committed to implement the Report's regulatory recommendations[35].

C. 1970–1990

The foundation for all FDA policy regarding the nutrient content of food during this twenty-year period was nutrition labeling. Although the concept of nutrition labeling was set firmly in place with promulgation of the special dietary food regulations in November 1941, the approach taken in those regulations was inadequate to implement the White House Conference Report. The November 1941 regulations applied only where a specific claim was made about a particular nutrient, and the required label declaration related only to that nutrient. The November 1941 regulations were, moreover, directed only to special dietary food, not to general purpose food. It was therefore necessary to construct a new approach.

To construct that approach, FDA turned to Section 201(n) of the Act, which stated that, in determining whether labeling is misleading, there shall be taken into account the extent to which the labeling "fails to reveal facts material in the light of such representations" as are made elsewhere in the labeling. FDA took the position that, where the manufacturer added a nutrient to a food or made any representation about the nutrient content of the food in advertising or labeling, full nutrition labeling was required in order to present all of the facts material to that representation. Those material facts, moreover, included not just the nutrients that were added or about which claims were made, but rather extended to representative micronutrients and macronutrients that would provide a balanced nutrient profile of the food. Whereas the November 1941 regulations focused only on the positive nutrient attributes of the food, the new nutrition labeling regulations required disclosure of any negative nutrient attributes as well. Thus, food marketers were for the first time required to present the negative as well as the positive nutrition aspects of their products[36].

The 1973 nutrition labeling requirements included a uniform format[37]. Under the overall heading of "Nutrition Information," data were required for specified micronutrients and macronutrients. The data on the micronutrients (vitamins and minerals) were required to be presented in terms of the percentage of the United States recommended daily allowances (U.S. RDA), a level derived by FDA from the recommended dietary allowances established periodically by the Food and Nutrition Board of the National Academy of Sciences. Thus, these values were stated in percentages, not in international units or metric weight (grams and milligrams). In contrast, the data on macronutrients were presented solely in terms of metric weight and provided no percentage information. Unlike the November 1941 regulations, which provided the percent of the minimum daily requirement of micronutrients per day, the 1973 regulations required declaration of the percent of the U.S. RDA per serving.

The initial 1973 nutrition labeling regulations did not specify either a generalized standard for determining how to calculate a serving of food for purposes of the new regulations or set particular serving sizes for specific foods. When labeling began to appear that specified differing serving sizes for identical food products, FDA proposed a regulation to standardize serving sizes[38]. The Agency failed to carry through with that approach, however, and withdrew that proposal some years later[39].

Following promulgation of the final nutrition labeling regulations in January 1973, FDA provided an additional thirty days for public comment on technical changes[40]. The Agency was properly concerned that, in a rulemaking of this complexity, errors might inadvertently have been made. Because FDA wanted to respond to all the technical comments immediately, in order to permit industry as much time as possible to comply with the new labeling requirements, the Agency reviewed all the comments, made a number of important changes in the regulations, and promulgated final revisions only twenty days after the close of the comment period[41].

The final regulations initially provided that all labeling ordered after December 31, 1973 (except where FDA extended that date for good cause shown) and all labeling for food shipped in interstate commerce after December 31, 1974, must be in compliance. Faced with industry information demonstrating that this would require the destruction of large inventories of already-printed packaging, FDA allowed additional extensions for up to December 31, 1975, for specific packaging where the manufacturer could demonstrate a particular need, or other relevant proposed regulations were pending, or for other sound reasons[42].

There was widespread industry concern that, because of the complexity and lack of guidance for every aspect of the new regulations, inadvertent errors might lead to stringent FDA enforcement. FDA officials therefore reassured the food industry that the Agency would approach the matter on a cooperative, rather than a regulatory enforcement, basis. When FDA found violations of the new regulations, it informed the company, answered questions about how to make changes that would bring the product into compliance, and provided time to use up the old labeling and introduce the new.

To make nutrition labeling work, FDA at the same time promulgated additional related labeling requirements. Nutrition labeling was required to appear in a standardized format, with a minimum type size, in a uniform place on the package[43]. This meant that consumers could find information more readily and understand it more easily.

Other regulations were promulgated at the same time to back up the new approach to nutrition labeling. Since 1938, Section 403(c) of the FD&C Act had required that any "imitation" food be labeled as such. FDA for the first time defined the term "imitation" to mean nutritional inferiority[44], thus prompting food manufacturers to formulate new substitute and processed foods in a way that would assure nutritional equivalence to traditional foods and to use nutrition labeling on these foods. New foods that were specially formulated to improve their nutrition profile (e.g., low fat or low sodium) were no longer regarded as illegal violations of existing food standards, but were permitted to bear their own "common or usual name" to distinguish them from the traditional product[45] and to bear nutrition labeling. FDA began to promulgate nutrition quality guidelines[46] and issued a general statement of fortification policy[47] to establish criteria under which vitamins and minerals should be added to food.

In promulgating regulations for food used in weight control, following the 1968–1969 public hearing, FDA for the first time included definitions for nutrient descriptors[48]. The new regulations defined such terms as low calorie, reduced calorie, sugar free, and related claims. This approach established the FDA policy that was later to be used for definitions of all nutrient descriptors.

Although FDA attempted to establish a new policy with regard to the formulation of dietary supplements of vitamins and minerals[49], the Agency was humiliated by the dietary supplement industry when this approach was overwhelmingly rejected both in the courts[50] and in Congress[51], and it was thus abandoned in 1979[52]. Even an attempt to promulgate regulations limiting levels of vitamins A and D[53] resulted in rare losses for FDA in the courts[54], causing FDA to abandon that effort as well[55]. FDA therefore reverted to its earlier policy of 1938–1970, issuing regulatory letters and bringing legal action against specific dietary supplement products for misleading labeling on a case-by-case approach.

The 1970s ended with a series of four public hearings on food labeling conducted jointly throughout the country by FDA, the FTC, and USDA during 1978 and 1979[56]. These hearings culminated in publication by FDA in December 1979 of a major analysis of the food labeling issues that must be addressed by the Agency in the near future[57]. That analysis identified such issues as expansion of nutrition labeling to include additional macronutrients, further definitions for widely-used nutrient descriptors, and permission for disease prevention claims in food labeling in order to address emerging new information on the relation between diet and health.

Throughout the 1980s, however, FDA lagged behind congressional and public expectations in addressing these issues. The Agency did define nutrient descriptors for sodium[58], and proposed to define some nutrient descriptors for cholesterol[59], but it did not address this area systematically. Only after industry forced the issue in 1984 by beginning to use disease prevention claims in food labeling did FDA attempt to address that area[60]. No consideration was given to a substantial review of the content of nutrition labeling, to reflect new information obtained on diet and health since nutrition labeling was developed in the early 1970s. As a result, FDA came under increasing criticism from Congress and the public, and Congress ultimately took the matter into their own hands in 1990.

D. 1990–The Present

The Nutrition Labeling and Education Act (NL&E Act) was signed into law by President Bush on November 8, 1990[61]. This date was extremely important because all of the statutory time deadlines flowed from it. The statute dealt with six separate matters, all of which involved, directly or indirectly, FDA regulation relating to the nutrient content of food.

First, Section 403(q) was added to the FD&C Act to require nutrition labeling for virtually all food products. This was not a controversial issue. FDA had belatedly proposed to do this by regulation in 1989[62] and the industry had supported it. The statute required a shift in emphasis from micronutrients to macronutrients and authorized FDA to set standards for serving sizes. Because FDA could do this under existing law however, this represented little change.

Second, FDA was required to define the nutrient descriptors commonly in use throughout the food industry—high fiber, low fat, reduced cholesterol, and so forth. FDA had begun this task in the 1970s, but had not completed it. The Agency could do this under existing law, and thus the new authority represented no additional statutory power for FDA.

Third, FDA was required to review specified disease prevention claims to determine whether they were appropriate for labeling. The Agency asserted that it could do this under existing law, and thus it represented no new statutory power. Even though FDA had twice proposed regulations[63] to govern these matters prior to enactment of the NL&E Act, this provision was quite controversial because it reflected some of the same criteria that had been proposed by FDA and strongly criticized by the food industry. It was correctly anticipated that this provision would ultimately produce the most serious controversy.

Fourth, the NL&E Act contained several new food labeling and food standards provisions. Vegetable and fruit juice beverages were required to bear the percent of each juice on the information panel. All ingredients in standardized food products were required to be included in the label statement of ingredients. All certified color additives were required to be included by name in the label statement of ingredients. In the future, food standards (except for dairy products and maple sirup) were permitted to be promulgated, amended, and repealed using informal notice-and-comment rulemaking, rather than the formal trial-type administrative hearing procedure that prevailed in the past.

Fifth, all of the labeling requirements under the FD&C Act were explicitly made subject to national uniformity (federal preemption) except for the general prohibition against false or misleading labeling, health and safety warnings, special dietary food regulation, and such local matters as unit pricing and open date labeling. Thus, by requiring national uniformity in nutrition labeling, nutrient descriptors, disease prevention claims, food standards, and the other food labeling provisions involved, the NL&E Act effectively removed state and local government from establishing regulatory requirements relating to the nutrient content of food. This represented a major change in national food policy, one that had been sought by the food industry for a century.

Sixth, in return for national uniformity, states were for the first time explicitly authorized to enforce the food labeling provisions of the FD&C Act, with the same exceptions that apply to national uniformity, in federal courts, rather than state courts, after first informing FDA and giving FDA the opportunity to take action. Because state officials can enforce directly in state courts the identical requirements under state law, however, without any of the procedural requirements established in the NL&E Act, it is doubtful that this provision has any practical impact except perhaps in a rare situation where a state has not yet adopted the federal requirements.

The NL&E Act established a rigid timetable for FDA implementation. Proposed regulations were required to be promulgated within one year (i.e., by November 8, 1991). FDA announced the proposed regulations on time but inexplicably actually published them on November 27, 1991, thus missing this deadline by nineteen days[64]. After time for public comment, the NL&E Act required the final regulations to be promulgated within an additional year (i.e. by November 8, 1992). If FDA missed this deadline, the proposed regulations became final under the "hammer" provision of the statute. FDA again missed this deadline, for reasons discussed subsequently, this time by two months[65]. Finally, the regulations were required to become effective six months later (i.e., by May 8, 1993), with the discretionary authority in FDA to extend this deadline by up to an additional year (i.e., up to May 8, 1994). Once the regulations became effective, all food labeled after that date was required to comply.

The proposed regulations that were published in November 1991 reflected serious mismanagement of the issues within FDA. They did not contain a proposed format for nutrition labeling, which was in fact not proposed until eight months later[66], less than four months before the final regulations were required to be promulgated. They contained no information on the ultimate effective date. The proposed nutrient descriptor regulation was widely understood to be unduly restrictive. Although a revised version was agreed to among FDA, OMB, and the Council on Competitiveness, was publicly released in Federal Register format, and was signed by the Commissioner, in fact it was never published in the Federal Register for public comment because FDA was concerned that it would become the final regulation in the event that the statutory hammer provision were invoked. Although the FDA Commissioner personally championed the use in labeling of additional nutrition information specifically designed for children, the proposed regulations themselves explicitly prohibited any such approach. The Agency also included non-NL&E Act issues in the proposed regulations, dealing with such matters as the definition of "fresh"[67] and with various ingredient labeling issues that were not raised by the new statute[68], thus unnecessarily complicating an already complex rulemaking.

Perhaps more important, the proposed regulations greatly expanded the scope of the statutory provisions themselves. Instead of setting standards for serving sizes, FDA proposed to set the serving sizes themselves[69]. The concept of a nutrient descriptor was expanded to include all nutrient content statements[70]. The category of disease prevention claims was expanded to include all dietary advice[71]. Very stringent standards were proposed both for nutrient descriptors and for disease prevention claims, thus providing a disincentive for the food industry to improve the nutrition profile of many foods or to conduct research on the relationship of diet to health that would justify additional disease prevention claims.

The NL&E Act inadvertently required the new ingredient labeling amendments to go into effect before the nutrition related provisions of the law. FDA could have used its administrative discretion to solve this problem, but it refused to do so. As a result, Congress was forced to amend the NL&E Act in August 1991 to change the effective date[72].

When the initial nutrition labeling regulations were promulgated in 1973, they included fresh fruits and vegetables. Faced with a lawsuit challenging FDA's authority to do this, the Agency withdrew that portion of the regulations and said that it would address this issue in a separate rulemaking[73]. The NL&E Act provided voluntary, not mandatory, nutrition labeling for fresh fruits and vegetables as well as for seafood. FDA promulgated the regulations needed to implement this approach promptly[74], and this approach has in fact proved to be successful[75]. Nonetheless, FDA proposed in mid-1994 to revise the guidelines for this voluntary approach, thus changing the labeling that must be used in order to assure compliance by retailers[76].

The NL&E Act covered dietary supplements as well as general purpose food. It contained a provision explicitly authorizing FDA to establish separate requirements for disease prevention claims for dietary supplements, using a different procedure and standard than for general purpose food, if that was appropriate. Not only did FDA choose to ignore this provision, but reflecting its decades-old war with the dietary supplement industry it proposed even more stringent requirements for dietary supplements than for general purpose food and proposed to reduce the United States recommended daily allowances (U.S. RDAs), now renamed "reference daily intakes" (RDIs), for essential vitamins and minerals[77]. The result was another humiliating defeat for FDA in Congress, which enacted a statute to overrule the FDA RDI levels, to require FDA to conduct new rulemaking, and to grant the dietary supplement industry an additional one-year extension[78].

Although meat and poultry are food within the meaning of the FD&C Act, they are regulated separately by USDA under their own statutes. The NL&E Act specifically provided that this jurisdictional line was to be respected and thus the proposed FDA regulations did not include meat or poultry. Nonetheless, because USDA Secretary Edward R. Madigan had been the ranking minority member of the House Subcommittee that had produced the NL&E Act, and had personally supported and worked on the legislation, he made the decision that USDA would propose its own nutrition labeling requirements. The USDA proposals were published on the same day as the FDA proposals[79] and USDA developed its final regulations in parallel with the FDA work on its own final regulations.

This parallel development of regulations relating to the nutrient content of food within two separate departments inevitably lead to disagreement. The offices responsible for resolving such matters—the Council on Competitiveness within the White House and the Office of Information and Regulatory Affairs (OIRA) within the Office of Management and Budget—were unable to resolve the dispute, because FDA refused to bring them into its process, to provide early drafts of regulations for review, or to consider ways to compromise.

FDA's strategy was to stall as long as it could, discuss controversial provisions but not compromise on them, and rely on the hammer provision—that automatically adopted the proposed regulations as final if FDA missed the November 8 statutory deadline—to force USDA, OMB, and the White House to accept its approach. The Agency seriously miscalculated its own power and the resolve of USDA Secretary Madigan and others in the White House. USDA and the White House concluded that FDA itself recognized that its proposed regulations were badly in need of change and that the invocation of the hammer would embarrass FDA more than anyone else. They did not want to let FDA isolate just a few remaining issues. OMB and the White House therefore refused to allow FDA to publish any of its final regulations—including those not in dispute—until the disagreement between the two agencies was resolved.

Normally, the matter could have been resolved by the President immediately. Throughout the fall of 1992, however, President Bush was preoccupied with his bid for reelection, and on November 3 President Bush lost the national election and he and his top officials left for vacations. The statutory deadline of November 8 therefore passed without resolution of the issues and FDA determined that all of the proposed regulations except those not required by the NL&E Act automatically became final under the hammer provision[80].

The precise details of the way in which the disputes about the final regulations were resolved is difficult to reconstruct. Based upon discussion with a number of the participants, however, it appears that the events unfolded in the following way.

The White House considered this the type of dispute on which a single policy—covering all food—was important and therefore rejected the concept of different policies for FDA and USDA. Such disputes were routinely resolved by OIRA and the Council on Competitiveness in other instances. As the dispute between FDA and USDA deepened during the fall of 1992, however, OIRA sided almost completely with USDA and the effectiveness of the Council on Competitiveness was weakened by continuing media attacks. OIRA also raised a number of important issues relating to the draft regulations, independent of the USDA concerns. When FDA adamantly refused any compromise with either USDA or OIRA, it became clear that OIRA would be unable to resolve the matter.

As the election approached, the FDA Commissioner became more insistent on his position and made it known that, if he did not get his way, he would take it to the media in a way calculated to hurt the President just prior to the election. To avoid further media criticism of the already controversial Council on Competitiveness, responsibility for the matter was assigned to a person who served both as an Associate Director of OMB and as a Deputy Assistant to the President, who attempted to find a middle ground. By then, however, there was such intense disagreement between FDA and USDA that compromise was hopeless on both sides.

Throughout the FDA/USDA interagency dispute, FDA insisted that it was the expert on interpretation of the statute and that USDA had no similar expertise. In fact, however, no one in FDA or HHS had been significantly involved in development of the statute itself, whereas USDA Secretary Madigan and another top USDA official had actively participated in the drafting and enactment of the statute and thus had far greater expertise than FDA. The USDA staff had greater practical experience with nutrition labeling than the FDA staff because of the requirement for USDA premarket approval of nutrition labeling for each individual meat and poultry product. Thus, the two agencies grew further apart and became even more hardened in their positions.

Realizing that compromise was unachievable, the White House sought to prevent a media disaster by continuing the talks between the two agencies until after the election. During this period, numerous position papers were prepared by FDA, USDA, and OIRA on a wide variety of issues. As time went on, some matters were resolved and others were discarded, but a number of important issues remained. Although the FDA Commissioner repeatedly threatened media exposure, in fact none occurred until after the election.

Following the election, the matter was further complicated by the fact that the President lost. It was apparent to all of those involved that the FDA Commissioner was maneuvering to keep his job under the new Clinton administration and would pursue whatever position would be most likely to help achieve that objective. This angered many who remained loyal to President Bush and were attempting to find a reasonable middle ground to resolve the remaining issues.

The Associate Director of OMB/Deputy Assistant to the President who had been brought into the matter during the fall continued in that capacity following the election. It was he, in consultation with the Associate Counsel to the President and others in the White House who decided that no regulations of any kind would be published in the Federal Register until all of the outstanding issues were resolved. For some three weeks after the election, meetings were held between the White House and the agencies individually, and with both of the agencies present. A number of the issues raised by the individual agencies and by OIRA were resolved or discarded in the course of these meetings.

It is not feasible to recount the dozens of issues that USDA and OIRA raised with FDA during nonstop meetings throughout the fall of 1992—both before and after the election—or how or why they were resolved or discarded. Some of these included the type of statements that would be regarded as disease prevention claims, whether the NL&E Act applies to factually accurate statements about the current status of nutrition research on a particular nutrient, the criteria for particular nutrient descriptors (e.g., the criteria for a "light" sodium claim), the use of nutrient amounts in lieu of descriptor terms, the use of a nutrient density requirement to prevent the application of descriptors to foods consumed in small quantities, whether and how to define the term "healthy," whether serving sizes should be declared per one ounce of food in order to facilitate simple comparisons among different foods, details of nutrition labeling, and the one year extension of the effective date. As time went on, the list of remaining issues was inevitably narrowed.

During late November, HHS Secretary Sullivan and USDA Secretary Madigan and their top staff members met in a final attempt to reach a mutually agreeable compromise. FDA refused to compromise and USDA refused to give in. It was agreed that only the White House could resolve the matter.

By early December, three events had occurred that required a further change in approach. First, the Associate Director of OMB/Deputy Assistant to the President who had been responsible for the matter had begun to discuss a possible position with a Washington law firm that had food industry clients, and thus concluded that he must recuse himself from the matter in order to avoid the appearance of a conflict of interest. Second, the Chief of Staff of the White House was occupying himself with his return to private life, leaving responsibility for most matters with the Deputy Chief of Staff. Third, after the meeting between Secretary Sullivan and Secretary Madigan in late November resulted in no progress, the Deputy Chief of Staff chaired a meeting in the White House with Secretary Madigan and Secretary Sullivan in another attempt to break the impasse. It was apparent that there was outright intransigence on both sides, however, and that the matter could only be resolved by the President personally[81].

Throughout November, a briefing book had been prepared for the President under the direction of the Associate Director of OMB/Deputy Assistant to the President, with the contributions of both FDA and USDA and with the review of OIRA and the Associate Counsel to the President. By early December, the issues that had to be resolved by the President appeared to have been reduced to two: the format for nutrition labeling and whether restaurant menus must meet the same standards as other restaurant labeling in order to use nutrient descriptors and disease prevention claims for the daily entrees on their menus. At this point, the Deputy Chief of Staff turned the matter over to the Acting General Counsel of OMB for preparation of a final decision memorandum for the President. The Acting OMB General Counsel talked to both agencies as well as the White House staff and made a strong attempt to have both sides represented fairly and adequately in the final memorandum.

That memorandum has not been made public and thus the precise issues presented to the President remain uncertain. It is the consensus opinion of those involved, however, that the primary issue was the nutrition labeling format and the secondary issue was the rules for nutrient descriptors and disease prevention claims for restaurant menus.

The details of the differences between the nutrition labeling formats advocated by FDA and USDA have not been made public. A newspaper article published in mid-November[82] purporting to show the differences was later described as reflecting the FDA version correctly but not reflecting the USDA version correctly. It appears that FDA was arguing for an approach based on an "average person" who consumed 2,000 calories, contending that consumers could calculate their own appropriate intake based on the information provided on the label about the amount of the nutrient in a serving and the percentage of the daily value (DV) that this amount represented based upon the 2,000 calorie diet. FDA placed greater emphasis on the importance of the percentage of DV than on the amount of the nutrient. USDA contended that use of an average person and an average caloric intake, together with a fixed percent of the DV, was misleading, that consumers would be unable to make the mathematical calculations necessary to translate the label information to the level of calories that they actually consumed, and therefore that the amount of nutrients and the actual DV, but no percentages, would be far more useful and meaningful. USDA strongly attacked the FDA approach as a "one size fits all" policy that assumed consumers were unintelligent and incompetent and therefore sacrificed accuracy to achieve simplification.

With these two divergent approaches as guidelines, the White House staff proceeded to consider what options should be presented to the President. Both the Associate Director of OMB/Deputy Assistant to the President who handled the matter through November and the Deputy Chief of Staff who took it over from there came up with several hybrid options and mock labels that included but went beyond what either FDA or USDA proposed, thus providing a wider range of options for the President to consider.

At the request of the Secretary of HHS, the President met with USDA Secretary Madigan and HHS Secretary Sullivan in early December to hear both sides of the remaining issues. Vice President Quayle, Chief of Staff James Baker, and Press Secretary Marlin Fitzwater were also reportedly present. The President made no decisions at that meeting[83].

Vice President Quayle felt strongly that the nutrition labeling issues should be resolved at a level below the President, and he declined to participate in any of the other meetings on this matter. As they were waiting to meet with the President, he commented to the two Secretaries that he could not believe that the President was actually being asked to decide these issues.

By mid-December, the matter had been fully briefed. The Deputy Chief of Staff met with the President to present the matter, with no one else present, in the President's hideaway office just off the Oval Office. After reviewing all the options, the President made his decision at the end of that meeting. The President chose one of the options that had been formulated by the White House staff, and rejected both the FDA and the USDA approach.

President Bush's decision was clearly intended as a compromise between the two agencies. The result was labeling that is more cumbersome and confusing than either agency had urged. As a result of the President's decision, FDA added what is now the bottom third of the nutrition label, showing both a 2,000 and a 2,500 caloric approach. Thus, the label was made more lengthy and less intelligible than the FDA or USDA proposals.

The only other issue in contention between FDA and USDA that was presented to the President related to the rules to be followed by restaurants in determining compliance with nutrient descriptors and disease prevention claims. USDA and OMB contended that restaurant menus should be completely exempt from these requirements. FDA took the position that menus should be required to comply with the same rules as any other restaurant labeling and pointed out that the final regulations would only require restaurant labeling to make a good faith effort to comply (because of the impossibility of subjecting every item on the menu to a nutrient analysis each day). FDA argued that its historical approach to the definition of "labeling" would be compromised and that, in any event, there was no statutory authority to exempt restaurant menus under the NL&E Act. The Department of Justice was brought into the dispute, and disagreed with FDA on this matter.

This issue was resolved by the President in favor of the USDA and OMB menu exemption[84]. The President and his staff felt strongly that the FDA approach would place a disproportionate burden on small restaurants, discourage many restaurants from even attempting to provide useful information, impose compliance standards that would be exceedingly difficult for restaurants to prove that they have met, and leave restaurants open to invidious state and local enforcement actions. As soon as the Clinton administration came into office, however, FDA proposed to reverse that decision and to impose its own position[85].

The final regulations were published in early January 1993[86], approximately two months after the statutory deadline. They explicitly superseded the proposed regulations that had automatically become final under the statutory hammer provision[87]. Virtually all of the changes receded from the unrealistic provisions contained in the proposed regulations. A number of problems remained, but the revisions that had been made by the Agency resulted in much more workable regulations. USDA promulgated its own final regulations at the same time[88].

As in 1973, FDA again provided thirty days for technical comment[89]. This time, however, it took the Agency more than seven months to publish the final technical amendments[90]. Even then, it admitted that it was unable to deal with all of the comments that were submitted, thus leaving a number of important issues unresolved. In the interim, the food industry was unable to begin to change its labeling, because FDA had not made the final requirements available. Still more "clarifications" were made available in the months that followed, providing industry with continuing uncertainty about the new requirements[91].

To add to this difficulty, another new statute was enacted in 1992, requiring the food industry to adopt metric labeling for the net contents declaration[92]. FDA failed to expedite the implementing regulations[93], which more than two years later were still not yet promulgated. The food industry thus had to comply with the final effective date for the NL&E Act regulations without final regulations governing metric labeling. This further exacerbated the industry problems in attempting to change its labeling to comply with all new requirements at one time.

The NL&E Act contained a small business exemption, but it was so narrow that it applied to very few food manufacturers. In August 1993, Congress acted to expand the exemption to cover a much larger number of small businesses[94].

The final regulations governing nutrition labeling and nutrient descriptors, although still presenting problems, were promptly implemented by the food industry in anticipation of the final deadline of May 8, 1994. The nutrition labeling format was gradually made more flexible in the final regulations and the subsequent revisions and interpretations, and thus could be adapted to virtually any type of package. Basically, it adopted the same format that FDA had developed in the early 1970s, changed only by the new heading of *Nutrition Facts*, the increased emphasis on macronutrients and decreased emphasis on micronutrients, the extension of a percent format to macronutrients through the development of specific "daily values" for macronutrients based on a 2,000 calorie diet, and a larger type size. It thus represented a clear improvement in, although not a fundamental change from, the nutrition labeling established in 1973.

When FDA developed nutrition labeling in the early 1970s, the Agency determined that compliance would be evaluated by direct analysis of a representative sample of a food product. FDA concluded that the industry-wide data on the nutrient composition of food that had been accumulated over the years in USDA Handbook No. 8[95] did not represent a statistically significant sampling of the products involved and thus were insufficient to support the declaration of particular nutrient levels as part of nutrition labeling. In the intervening years, only a few food industries—primarily the dairy industry—conducted studies adequate to develop a comprehensive data base to support uniform nutrition labeling throughout the country. Recognizing that this could be important for the future, in 1992 FDA made available a draft manual on how adequate databases could be developed to support nutrition labeling in the future[96].

The nutrient descriptor regulations were clearly needed to complete the job begun by FDA in the 1970s to define the various nutrient descriptors that were so prevalent in food labeling. Although there continued to be controversy about some details of the specific definitions for particular nutrients, the basic concept was well accepted and implemented.

As part of its initial rulemaking under the NL&E Act, FDA considered the possibility of promulgating regulations governing such terms as "natural" and "healthy."[97] FDA subsequently abandoned any definition of "natural,"[98] but proceeded to define the term "healthy" both when used as an implied nutrient descriptor and when used generally to describe food[99]. USDA issued parallel regulations[100]. The definition of "healthy" was strongly criticized by the food industry.

Most of the controversy relating to the final regulations focused on disease prevention claims. Forced to admit that the proposed disease prevention claim regulations were far too restrictive and that they would have prevented virtually any nutrition education materials from being used in labeling, FDA conceded in the preamble to the final regulations that three quite limited types of claims would still be permitted in food labeling as outside the definition of a disease prevention claim: (1) general dietary guidance that refers either to a specific food substance or to a disease or health related condition, but not to both[101], (2) structure and function (physiological) claims that refer to the effect of a substance on normal human physiology but do not refer to any dysfunction of or damage to a bodily part or to a change in a biological parameter that is associated with a disease or health-related condition[102], and (3) diseases resulting from essential nutrient deficiencies[103]. Nonetheless, these exceptions were very limited. The Agency's criteria for disease prevention claims were so narrow that only seven claims were permitted[104]. Five were explicitly rejected[105] and an untold number were never even considered. Even when the Centers for Disease Control announced to the world the relationship between folic acid and neural tube defects, FDA said that the data were insufficient and dragged its heels on the matter until political and public pressure forced it to concede[106].

The Agency's very broad definition of what constitutes a disease prevention "claim," together with its very restrictive criteria for approving such claims, has provoked widespread criticism. This criticism has come both from those who believe the FDA policy violates the First Amendment and those who strongly support public education by the food industry on important nutrition issues. A major effort has therefore been undertaken to force the Agency to permit labeling that does no more than state the views of other government health agencies and to allow nutrition educational materials that are accurate and not misleading but are not approved by FDA. The ultimate resolution of this matter remains uncertain at this time.

Following the end of the one-year moratorium enacted by Congress for dietary supplements, FDA moved quickly to impose the same requirements that it would have imposed without the moratorium[107]. This added additional fuel to an already-inflamed controversy. Responding to widespread public criticism, Congress passed the Dietary Supplement Health and Education Act of 1994[108], which overruled not only the new nutrition labeling regulations that FDA had promulgated under the NL&E Act for dietary supplements but went far beyond and imposed substantial restrictions on other existing FDA authority over dietary supplement products as well. The new law defined dietary supplements very broadly, exempted them from the drug definition (unless therapeutic claims are made), exempted them from regulation as food additives (thus also exempting them from the Delaney Clause), imposed on FDA the burden of proving lack of safety for any dietary supplement, limited FDA authority over dietary supplement labeling, and specified new ingredient and nutrition labeling requirements for these products. Two new government entities were created, outside FDA, to advise FDA on regulation of dietary supplements. FDA was given explicit authority to promulgate regulations governing good manufacturing practices (GMP) for dietary supplements, but even this provision was more limited than the existing food FDA GMP authority that had previously been upheld in the courts. The 1994 Act thus represents the single largest legislative rebuke and defeat suffered by FDA in its history.

Just as the final implementation date of May 8, 1994, was approaching, yet another controversy erupted. FDA had initially informed industry that packages labeled prior to the effective date would remain lawful. Later, FDA reversed itself and said this would be true only if the packages were filled with food. When representatives of the food packaging industry took the matter to the courts, the judge determined that FDA had sent "confused signals" to the industry and enjoined FDA from changing its determination[109]. Congress was forced to enact yet another statute to correct FDA's error, extending the effective date an additional three months (i.e, to August 8, 1994[110]).

The contrast between the way in which nutrition labeling was developed in the early 1970s and the way that the NL&E Act was implemented in the early 1990s is both substantial and puzzling. It undoubtedly reflects the different management styles and regulatory policies of those in FDA who were responsible for these virtually identical tasks.

In the early 1970s, FDA sought to implement the recommendations of the White House Conference in a very open and cooperative way with both consumer advocates and industry representatives. The regulations represented a compromise among competing views, with ample time for compliance. FDA employees worked day and night, using a team approach to meet self-imposed short deadlines. The entire process took less than a year, and all issues were resolved.

In contrast, in the early 1990s, FDA pursued implementation of the NL&E Act through a closed door and arms length approach both with its coordinate government agencies and with the food industry. It failed to meet a single statutory deadline, and by its obstinate and arrogant insistence on its own views it forced the President to overrule some of its approaches and the Congress to enact four separate statutes to overrule other approaches. The process took more than two years, and even then FDA has been unable to resolve a number of important issues, including the entire subject of metric labeling.

In spite of the difficulty created by FDA during this process, the ultimate result of the NL&E Act is improved labeling for the American consumer and for the food industry in two respects. Nutrition labeling is improved, by presenting more information on macronutrients and using a percentage presentation that can be understood by the consumer. The definitions for nutrient descriptors assure consistent terminology that can be understood by consumers and that provides a level playing field for industry. National uniformity in these approaches prevents the states from adopting additional or different approaches and therefore assures consistent food labeling throughout the country. FDA's current policy of strict censorship over disease prevention claims inhibits public education and harms the public health, but once the issues surrounding disease prevention claims are resolved it can anticipated that greater information will be made available to the public on the relationship of diet to health.

V. CONCLUSION

As one looks back over the past sixty years, the progression of FDA regulation relating to the nutrient content of food is readily apparent. The concept of nutrition labeling was begun in 1941, developed to its present form in 1973, and refined to its new content in 1993. The approach of defining nutrient descriptors was initiated in the 1970s and expanded to include the full range of descriptors in the early 1990s. Although disease prevention claims were prohibited from labeling by FDA until the mid-1980s, and thus could be made only in advertising until then, recognition of the importance of the relationship between diet and health has lead to a gradual reduction in the barriers to these claims and ultimately will force FDA to change its current restrictive policy. This will inevitably lead to greater opportunity for educating the American public about these important matters.

Thus, the history of FDA regulation relating to the nutrient content of food reflects the need for greater information being made available to the public on the nutrient content of food and the relationship of food to health. It comes full circle back to the philosophical view prevalent in ancient Greece and Rome: that control of the daily diet is an important part of assuring sound health. Although we have come a long way since then in understanding greater detail about the relationship between diet and health, even today we are still at a very primitive stage in the development of scientific knowledge about nutrition and health. It is essential, for the future progress of this scientific knowledge, to replace the current disincentives against private research in this area resulting from FDA's policies in implementing the NL&E Act with important incentives that will encourage private investment in nutrition research and thus facilitate development of new knowledge that will have an important bearing on public and private health.

VI. ENDNOTES

1. 1 *Hippocrates* 351 (W.H.S. Jones, ed., 1923).
2. *Id.* at 27.
3. Galen, *Hygiene* 162–163 (R.M. Green ed. 1951).
4. Hutt & Hutt, *A History of Government Regulation of Adulteration and Misbranding of Food*, 39 Food Drug Cosmetic Law Journal 2, 9 (1984).
5. *Id.* at 2.
6. E.g., 51 Hen. III, Stat. 6 (1266), reprinted in D. Pickering, 1 *The Statutes at Large From Magna Charta to the End of the Eleventh Parliament of Great Britain* 47 (1762).
7. C. Pharr, *The Theodotian Code* 418 (1952).
8. Hutt & Hutt, note 4 *supra*, at 14.
9. *Id.* at 4–14.
10. 12 Stat. 387 (1862).

11. For a history of FDA and its predecessor agencies, see Hutt, *A Historical Introduction*, 45 Food Drug Cosmetic Law Journal 17 (1990).
12. Hutt & Hutt, note 4 *supra*, at 47–52.
13. 34 Stat. 768 (1906).
14. 34 Stat. 669, 674 (1906). This law was reenacted a year later, 34 Stat. 1256, 1260 (1907), with minor change, because the 1906 law had been enacted as part of appropriations legislation and not as part of the substantive law.
15. Hutt & Hutt, note 4 *supra*, at 55–61.
16. 37 Stat. 732 (1913).
17. Hutt, *Government Regulation of Health Claims in Food Labeling and Advertising*, 41 Food Drug Cosmetic Law Journal 3, 6 (1986).
18. Hutt & Hutt, note 4 *supra*, at 60.
19. Bureau of Chemistry, *Annual Report* 16 (1917), in Food Law Institute, *Federal Food, Drug, and Cosmetic Act Law Administrative Reports 1907–1949* 355, 370 (1951).
20. Bureau of Chemistry, *Annual Report* 1 (1916), reprinted in Food Law Institute, note 19 *supra*, at 341.
21. Hutt, note 17 *supra*, at 8.
22. Food and Drug Administration, *Annual Report* 10–11 (1932), reprinted in Food Law Institute, note 19 *supra*, at 778–779.
23. Hutt, *Government Regulation of the Integrity of the Food Supply*, in Darby, ed., *Annual Review of Nutrition* 1, 8–9 (1984).
24. Hutt, note 17 *supra*, at 8–9.
25. 52 Stat. 111 (1938), 21 U.S.C. § 201 et seq.
26. 104 Stat. 2353 (1990).
27. Hutt, note 23 *supra*, at 9.
28. 8 *Federal Register* 9170 (July 3, 1943).
29. Hutt, note 17 *supra*, at 57–58.
30. *United States* v. *119 Cases. . ." New Dextra Brand Fortified Cane Sugar*, 231 F. Supp. 551 (S.D. Fla. 1963), aff'd, 334 F.2d 238 (5th Cir. 1964).
31. Hutt, note 17 *supra*, at 60–61.
32. 5 *Federal Register* 3565 (September 5, 1940); 6 *Federal Register* 3304 (July 8, 1941); 6 *Federal Register* 5921 (November 22, 1941).
33. Hutt, note 17 *supra*, at 61.
34. *Final Report of the White House Conference on Food, Nutrition and Health* (1970).
35. Hutt, *Regulating the Misbranding of Food*, 43 Food Technology, No. 9, at 288, 292 (September 1989).
36. Wodicka, *The 1970s: The Decade of Regulation*, 45 Food Drug Cosmetic Law Journal 59 (1990).
37. 37 *Federal Register* 6493 (March 30, 1972); 38 *Federal Register* 2125 (January 19, 1973).
38. 39 *Federal Register* 20887 (June 14, 1974).

39. 51 *Federal Register* 15653 (April 25, 1986).
40. 38 *Federal Register* 2163 (January 19, 1973).
41. 38 *Federal Register* 6951 (March 14, 1973).
42. 39 *Federal Register* 3650 (October 10, 1974).
43. 37 *Federal Register* 6493 (March 30, 1972); 38 *Federal Register* 2124 (January 19, 1973). Austern, *The Regulatory Gospel According to St. Peter*, 29 Food Drug Cosmetic Law Journal 316 (June 1974).
44. 38 *Federal Register* 2138 (January 19, 1973); 38 *Federal Register* 20702 (August 2, 1973).
45. 37 *Federal Register* 12327 (June 22, 1972); 38 *Federal Register* 6964 (March 14, 1973).
46. 36 *Federal Register* 24822 (December 23, 1971); 38 *Federal Register* 6969 (March 14, 1973).
47. 39 *Federal Register* 20900 (June 14, 1974); 45 *Federal Register* 6314 (January 25, 1980).
48. 42 *Federal Register* 37166 (July 19, 1977); 43 *Federal Register* 43248 (September 22, 1978).
49. 38 *Federal Register* 2143, 2152 (January 17, 1973); 38 *Federal Register* 20708, 20730 (August 2, 1973).
50. *National Nutritional Foods Association* v. *FDA*, 504 F.2d 761 (2d Cir. 1974); *National Nutritional Foods Association* v. *Kennedy*, 572 F.2d 377 (2d Cir. 1978).
51. 90 Stat. 401, 410 (1976).
52. 44 *Federal Register* 16005 (March 16, 1979).
53. 37 *Federal Register* 26618 (December 14, 1972); 38 *Federal Register* 20723 (August 2, 1973).
54. *National Nutritional Foods Ass'n* v. *Weinberger*, 512 F.2d 688 (2d Cir. 1975); *National Nutritional Foods Ass'n* v. *Mathews*, 557 F.2d 325 (2d Cir. 1977).
55. 43 *Federal Register* 10551 (March 14, 1978).
56. 43 *Federal Register* 25296 (June 9, 1978).
57. 44 *Federal Register* 75990 (December 21, 1979).
58. 47 *Federal Register* 26580 (June 18, 1982); 49 *Federal Register* 15510 (April 18, 1984).
59. 51 *Federal Register* 42584 (November 25, 1986).
60. 52 *Federal Register* 28843 (August 4, 1987); 55 *Federal Register* 5176 (February 13, 1990).
61. 104 Stat. 2353 (1990).
62. 55 *Federal Register* 29487 (July 19, 1989).
63. Note 60 *supra*.
64. 56 *Federal Register* 60366–60878 (November 27, 1991).
65. 58 *Federal Register* 2066–2964 (January 6, 1993). Because FDA failed to meet the first statutory deadline for the proposed regulations, the hammer provision technically was inapplicable. Because final regulations ultimately were promulgated by FDA, however, this technicality became moot.

66. 57 *Federal Register* 33283 (July 28, 1992).
67. 56 *Federal Register* 60421, 60462–60467 (November 27, 1991).
68. 56 *Federal Register* 28592 (June 21, 1991).
69. 56 *Federal Register* 60394 (November 27, 1991).
70. 55 *Federal Register* 60421 (November 27, 1991).
71. 56 *Federal Register* 60537 (November 27, 1991).
72. 105 Stat. 549 (1991); 56 *Federal Register* 60877 (November 27, 1991).
73. 38 *Federal Register* 32786 (November 28, 1973); 40 *Federal Register* 8214 (February 26, 1975); 48 *Federal Register* 27266 (June 14, 1983).
74. 56 *Federal Register* 30468 (July 2, 1991); 56 *Federal Register* 60880 (November 27, 1991).
75. 58 *Federal Register* 28985 (May 18, 1993); 59 *Federal Register* 36379 (July 18, 1994).
76. 59 *Federal Register* 36379 (July 18, 1994).
77. 56 *Federal Register* 60366 (November 27, 1991).
78. 106 Stat. 4491, 4500 (1992).
79. 56 *Federal Register* 60366 (November 27, 1991).
80. 57 *Federal Register* 56347 (November 27, 1992).
81. The only other time that a president has had to make a decision on an FDA matter was in early 1908 when President Theodore Roosevelt overruled Harvey W. Wiley's decision to ban saccharin from food. H.W. Wiley, *An Autobiography* 240–241 (1930); H.W. Wiley, *The History of a Crime Against the Food Law* 160–163 (1929).
82. Gladwell, *Interagency Fight Puts Overhaul of Food Label Rules on Bush's Plate*, Washington Post, November 15, 1992, page A1.
83. During this meeting, Brent Scowcroft, Assistant to the President for National Security Affairs, interrupted to inform the President that Boris N. Yeltsin was telephoning him from Russia. President Bush said that he would call Yeltsin back later, because he had to deal with the nutrition labeling issues first.
84. 58 *Federal Register* 2302, 2478 (January 6, 1993).
85. 58 *Federal Register* 33055 (June 15, 1993).
86. Note 65 *supra*.
87. 58 *Federal Register* 2066 (January 6, 1993).
88. 58 *Federal Register* 632 (January 6, 1993); 59 *Federal Register* 45189 (September 1, 1994). 60 *Federal Register* 174 (January 3, 1995).
89. 58 *Federal Register* 2066 (January 6, 1993).
90. 58 *Federal Register* 44020–44090, (August 18, 1993). Pertinent questions and answers were also made available in *id.* 44096 and subsequently.
91. 58 *Federal Register* 44091 (August 18, 1993); 59 *Federal Register* 15049 (March 31, 1994). FDA has also issued a large amount of informal explanatory information.
92. 106 Stat. 7, 13 (1992); 106 Stat. 847 (1992).

93. 58 *Federal Register* 29716 (May 21, 1993); 58 *Federal Register* 67444 (December 21, 1993).
94. 107 Stat. 773 (1993).
95. USDA, *Composition of Foods: Raw, Processed, Prepared,* Agriculture Handbook No. 8 (revised yearly).
96. 57 *Federal Register* 32796 (July 23, 1992); 59 *Federal Register* 36379 (July 18, 1994).
97. 56 *Federal Register* 60421, 60423, 60466 (November 27, 1991).
98. 58 *Federal Register* 2302, 2407 (January 6, 1993).
99. 58 *Federal Register* 2944 (January 6, 1993); 59 *Federal Register* 24232 (May 10, 1994).
100. 58 *Federal Register* 688 (January 6, 1993); 59 *Federal Register* 24220 (May 10, 1994).
101. 58 *Federal Register* 2478, 2479, 2482, 2487–2488 (January 6, 1993); 59 *Federal Register* 395, 409 (January 4, 1994).
102. 58 *Federal Register* 2478, 2482, (January 6, 1993).
103. 58 *Federal Register* 2478, 2481 (January 6, 1993); 21 C.F.R. § 101.14(a)(6).
104. (1) Calcium and osteoporosis, (2) dietary lipids and cancer, (3) sodium and hypertension, (4) dietary saturated fat and cholesterol and risk of coronary heart disease, (5) fiber–containing grain products, fruits, and vegetables and cancer, (6) fruits, vegetables, and grain products, particularly soluble fiber, and risk of coronary heart disease, and (7) fruits and vegetables and cancer. 59 *Federal Register* 436 (January 4, 1994).
105. (1) Dietary fiber and cancer, (2) dietary fiber and cardiovascular disease, (3) antioxidant vitamins and cancer, (4) omega-3 fatty acids and coronary heart disease, and (5) zinc and immune deficiency in the elderly. 59 *Federal Register* 436 (January 4, 1994).
106. 56 *Federal Register* 60610 (November 27, 1991); 57 *Federal Register* 32751 (July 23, 1992); 57 *Federal Register* 52781 (November 5, 1992); 58 *Federal Register* 2606 (January 6, 1993); 58 *Federal Register* 53254 (October 14, 1993); 59 *Federal Register* 433 (January 4, 1994).
107. 58 *Federal Register* 33690–33751 (June 18, 1993); 59 *Federal Register* 350–426 (January 4, 1994).
108. E.g., S. 784, 103d Cong., 1st Sess. (1993); 140 Cong. Rec. S11705–11715 (August 13, 1994) (daily edition).
109. *Can Manufacturers Institute* v. *FDA*, CCH Food Drug Cosmetic Law Reporter ¶38,354 (D.D.C. 1994).
110. P.L. 103–261, 108 Stat. 705 (1994); 59 *Federal Register* 28480 (June 2, 1994).

2

Perspectives on Nutrition Labeling and Education Act

F. Edward Scarbrough

Food and Drug Administration, Washington, D.C.

I. INTRODUCTION

The food label has many roles; much is expected from such a small vehicle. For the United States Food and Drug Administration (FDA), the label has long served as a regulatory tool. The Agency uses the label to contact the responsible persons whenever a product needs to be recalled from the market. Also, the FDA compares contents of packages with information on the label about the identity of the product, the quantity of contents, and the ingredients to ensure that consumers are not being misled or cheated. For the food manufacturer, the label is a major vehicle for promoting products and product lines. The consumer is told about the taste, convenience, and nutritional advantages of the product, about recipes and alternative uses of the food, and about other products that the manufacturer is promoting. But for consumers, the label and associated labeling have become the most widely available source of nutrition and food content information used today. The food label, because of its proximity during the purchase, preparation, and even consumption of food, is considered by many consumers, as well as by food processors and health professionals, as an ideal vehicle for conveying dietary guidance and information about the nutritional attributes of food. However, because of the limited size of most food labels and the manufacturer's vested interest in the product, there is a real potential for information on a food label to be confusing or even false or misleading. Therefore, governments are inevitably and rightly placed in the challenging position of regulating the marketplace and the label in such a way that the information which appears on food labels makes positive contributions to the enhancement of public health.

Whenever the food label is considered in its evolving consumer information role, there are at least two major considerations that must be kept in mind. First, the label represents a potent force for disseminating dietary information and improving public understanding of proper dietary choices. Such a potent force should be used. However, it *must* be used responsibly and carefully. Second, the food label, as a source of information, has a unique degree of credibility. Consumers expect the label to be a much more factual piece of information than, for example, food advertising as seen on television (1). The food label is much like the owner's manual found in the glove compartment of the new car. While the consumer or owner is willing to grant a certain amount of creative license to the television or print advertisement, he or she expects the information in the owner's manual or food label to be factual and helpful. It is therefore important that the food label retain the trust that has been carefully developed over the years by the FDA and USDA and by responsible industry.

In November, 1990, the United States Congress passed, and former President Bush signed, the Nutrition Labeling and Education Act of 1990 (NLEA) (2), the most significant food labeling legislation in the United States in at least fifty years. The NLEA requires nutrition information to be on virtually all packaged foods on the American market. The NLEA also strictly controls claims that may be made on food labels. However, it is important to remember that the NLEA amended the Federal Food, Drug and Cosmetic Act (3), the statute that FDA administers. Therefore, the NLEA applies only to the labels of food products regulated by the FDA. While the FDA has regulatory authority over the labels of most foods, the labels of meat and poultry products are controlled by the Department of Agriculture (USDA) and alcoholic beverage labeling by the Bureau of Alcohol, Tobacco, and Firearms (BATF). It is also important to remember that all advertising, including food advertising, is regulated by the Federal Trade Commission (FTC). Fortunately, FDA has traditionally enjoyed close working relations with its sister agencies. The USDA has voluntarily embarked on a food label reform effort consistent with the NLEA requirements[1]. FTC is also working to achieve the greatest degree of harmonization possible between the requirements for food advertising and food labeling, while recognizing that advertising and labeling have different expectations and limitations[2].

The major goals of FDA's food label reform initiative are threefold:

- to clear up the consumer confusion about information and claims on food labels,
- to aid consumers in making food choices leading to healthier dietary patterns,

- to encourage manufacturers to produce innovative products that are truly healthier

However, the food label reform is not simply about a better label on food packages. It is about Americans living longer, with better quality lives and with lower health care costs. The success of the food label reform initiatives will not be measured by how many packages are correctly labeled or by how many Americans read and understand the new label. It will be measured by whether dietary habits actually change, by the types of new food products that are marketed, and, ultimately, by whether Americans lead healthier lives.

II. HISTORY[3]

Understanding the current food label reform activities requires a brief look at the historical background. Certain elements of the label have been around essentially since 1906 when the first Pure Food and Drug Act was passed (4). The 1906 act authorized the federal government to regulate the safety and quality of food and certain label requirements were a direct outgrowth of this regulatory authority, subsequent amendments to that act, and precedent-setting court cases.

In 1938, the Federal Food, Drug and Cosmetic Act (FD&C Act) replaced the 1906 Pure Food and Drug Act and specifically required processed packaged food labels to contain such elements as a statement of identity of the product, net quantity of contents statements, and the manufacturer's name and address. Also, in the 1930s an amendment to the 1906 act [the McNary-Mapes Act (5)] introduced food standards for canned foods. The FD&C Act also incorporated and broadened the concept of food standards, which are essentially legally established recipes for common staple food products. For example, cheddar cheese cannot be called cheddar cheese unless it has 50% milk fat on a solids basis. Therefore standards have profound implications for how foods are described on labels. A list of ingredients was also required by the FD&C Act on most food products (the exceptions being mandatory ingredients in standardized foods).

The White House Conference on Food, Nutrition, and Health (6), convened by President Nixon in 1969, addressed deficiencies in the U.S. diet. One of the many recommendations from the Conference was that the federal government consider use of the food label for identifying the nutritional qualities of food. In response to this recommendation, FDA, in 1973, issued proposed regulations requiring nutrition labeling (the listing of macronutrients, vitamins, and minerals on the information panel) on foods for which the label, labeling, or advertising made claims about the nutritional properties of the food and on foods with one or more added nutrients (7). Further, when manufacturers voluntarily provided nutrition labeling, the content and format provisions of the regulations were applicable. The final regulations based on this proposal became fully effective in 1975 and established the nutrient content and format of nutrition labeling which was to remain essentially unchanged until 1993.

Also, in 1978, FDA's final regulations for the use of "low calorie" and "reduced calorie" claims (8) started a program of regulations defining nutrient content claims or descriptors, terms that can be used on food labels to describe foods. In 1984, FDA continued this program by defining such terms as "sodium free," "very low sodium," "low sodium," and "reduced sodium" when used on the food label to make sodium content claims (9). These regulations also added sodium to the list of nutrients required to be listed in nutrition labeling.

And in 1984, the Kellogg Company kicked off the modern era of health claims on food labels, by using the label, as well as advertising, to connect the consumption of their bran-containing cereals with possible reduction in the risk of certain kinds of cancer. Health claims are the explicit discussion of diet and disease on the food label.

III. DRIVING FORCES

In 1979, the FDA, USDA, and FTC conducted a series of food label reform hearings across the country to determine how consumers and the food industry believed that the food label could be improved. In December 1979, the three agencies published the findings of the hearings and discussed a number of potential food labeling changes that were contemplated (10). Yet, almost a decade later, virtually none of the changes had been accomplished, even though most of the issues raised were exactly those that have been addressed by the current food label reform initiative. Why then did the 1979 food label reform movements fail to bring about noticeable change, whereas in 1993 the food label has been fundamentally and almost completely revamped? The success of the current initiative, I believe, represents a convergence of a number of driving forces. These forces include advances in science, dietary guidance, consumer interest, food industry use of the label, continued Congressional involvement, and the entry of State governments into the food labeling arena.

Scientific advances: Underlying the movement for change in the food label is the fact that the scientific data base linking diet to chronic degenerative diseases has become progressively stronger over the past decade. There is now scientific consensus about many specific diet and disease relationships, such as the role of dietary fats in heart disease and the role of dietary calcium in osteoporosis.

Dietary guidance: The scientific evidence has reached such a point that august bodies such as the National Academy of Sciences now feel confident in making rather explicit dietary recommendations to the American public, as evidenced by their 1989 report on diet and health (11). Perhaps more importantly from the perspective of the FDA, explicit dietary recommendations have also been made by the Public Health Service, in the form of the Surgeon General's Report on Nutrition and Health (12). This report, along with the Dietary Guidelines for Americans (13) issued jointly by the Departments of Health and Human Services (DHHS) and Agriculture and the USDA Food Guide Pyramid (14), constitute the backbone of the federal government's dietary advice to the American consumer.

Other public health service agencies, such as the National Heart, Lung, and Blood Institute through its National Cholesterol Education Program (15) and the National Cancer Institute through activities such as the five-a-day program (16), are also issuing dietary advice that is consistent with the Dietary Guidelines.

And many groups outside the federal government, particularly biomedical professional groups, such as the American Heart Association (17), American Society of Clinical Nutrition (18), and the American Dietetic Association (19), are issuing dietary advice and are looking at the food label as a vehicle for spreading their particular dietary messages.

Consumer interest: Consumers are more interested in diet and health than ever before. They are aware of the dietary recommendations that are being made. Large segments of the population have expressed interest in following these recommendations, and they have looked to the food label for the information needed to apply this advice. The FDA as well as industry-sponsored surveys confirm that the food label is one of the most important sources of nutrition and food composition information for the American public today (e.g., (20,21)). However, consumers expressed concern because they have found the food label lacking much of the information needed to follow dietary recommendations. They were asking for more information to be placed on the label. At the same time, they often expressed the somewhat contradictory desire for simpler, more easily understood labels. But above all, in the late 1980s, consumers often stated that they were confused by, and somewhat skeptical of, the information that did appear on the label.

Industry use of the label: The food industry knows its consumers and, therefore, was well attuned to the growing interests in diet and health in the 1980s. This led to a situation in which, in a manner much more intense and widespread than ever before, the industry increasingly used the food label to tout the nutritional qualities of its products. Much of this label promotion of products was done in largely unregulated ways, which contributed to consumer confusion. And while the vast majority of the food industry acted responsibly, there were those few who crossed the line and made label claims that were misleading and occasionally even false. Yet throughout the early 1980s, because of Administration decisions not to devote FDA resources to issues that were not of acute public health significance and that were viewed as being largely economic in nature, there was little FDA enforcement action against these misleading claims.

Congress: Congress, which is attuned to the concerns of its constituencies and which has had a long-standing interest in the food label, was certainly aware of the consumer and industry interest in labeling. For example, at the direction of Congress, the Government Accounting Office (GAO) issued two reports, one in 1982 and one in 1988, with the common theme that there was a need to improve communication to the public about food through use of the food label. Also, the Senate Committee report accompanying the 1989 appropriations bill directed the FDA to initiate food labeling rulemaking. Congressional activity reached a crescendo in the 101st Congress with over a dozen bills on food labeling being introduced. Clearly, Congress was telling the Executive Branch to act, or the Legislative Branch would. And, of course, Congressional activity culminated with the passage of the NLEA in November 1990.

States: In the late 1980s several state legislatures concerned about the period of regulatory inactivity by the federal government introduced or began to draft food labeling laws that would apply to products sold within their states. The food industry was particularly concerned that, given the modern food distribution systems, it would be unable to produce and control distribution of foods labeled for fifty different states.

International: One force that received little attention in the early phases of the current food label reform debates, but which in retrospect should have been more prominently factored into the equation is international harmonization[4]. In fact, there is no evidence that the U.S. Congress considered international implications during the debate that led to the passage of NLEA. There are a number of international pressures that are being brought to bear on the FDA. For products to move freely across international boundaries there must be consistency in labeling requirements. With the ratification of the US/Canada Free Trade Agreement and now the passage of the North American Free Trade Agreement, these concerns are even greater. Also, consideration must be given to activities of the Codex Alimentarius Commission, a joint undertaking of the Food and Agriculture Organization and the World Health Organization of the United Nations. The USDA and FDA are active participants in this body which was formed to reduce barriers to the movement of food products across international boundaries. As the food supply becomes more international in scope and with the era of expanded international trade agreements, particularly under GATT (General Agreement on Tariffs and Trade), this lack of attention to international forces will complicate the future of food label reform efforts.

IV. PROCESS

With this ever-increasing interest in the food label, in the summer of 1989, Dr. Louis Sullivan, then Secretary of Health and Human Services, directed FDA to undertake a major food labeling initiative to bring about the needed reforms.

As the first publicly visible step in this process, on August 8, 1989, FDA published an advance notice of proposed rulemaking (ANPRM) in the *Federal Register*, calling for comment on all aspects of the food label (22). The Agency received about 2000 comments in response to the ANPRM from a wide diversity of interests—ranging from consumers, to health professionals, to the food industry.

Almost concurrent with the ANPRM, the DHHS and USDA sponsored a study by the Food and Nutrition Board, Institute of Medicine, National Academy of Sciences (NAS), which called for the NAS to have an expert panel consider how food labels could be improved to help consumers adopt or adhere to healthy diets. The report from that study, *Nutrition Labeling: Issues and Directions for the 1990s* (23), was released in November, 1990, and was used as support for many of the subsequent food labeling policy decisions.

Throughout the fall and early winter of 1989, the FDA, jointly with the USDA, carried out a series of public meetings to obtain further comment on food label issues. There were four major public hearings across the country, each focusing on different topics (Chicago, nutrition labeling content; San Antonio, ingredient labeling, food standards, and food descriptors; Seattle, health claims; and Atlanta, nutrition label format) (24). At these hearings, approximately 200 people, representing a wide diversity of interests, from the food industry to individual consumers, presented testimony. Also, over 500 people attended the hearings as observers.

The FDA also held over 50 consumer exchange meetings in various locations across the country. Many of these meetings were targeted toward specific populations, such as urban or rural areas or minorities. For example, three meetings were held in Spanish and one meeting in Chinese. In total, over 1500 people participated in the consumer meetings.

On July 19, 1990, the FDA published proposed rules on nutrition labeling, label reference values, and nutrient content claims dealing with cholesterol, fat, and fatty acids (25). The comment period on these proposals, scheduled to run until November 18, was just ending when Congress passed, and President Bush signed, the NLEA.

The NLEA required FDA to establish proposed implementing regulations within one year (by November 8, 1991) and final regulations within 2 years (by November 8, 1992). (In a curious provision of the NLEA, fondly known as the "hammer," Congress dictated that if final regulations were not in place by the deadline, the proposed regulations would become final.) Considering the pace at which regulation development typically moves and the fact that food label reform had been debated for years, with a lack of consensus on many issues, many thought that the Congressionally mandated timelines were virtually impossible for the Agency to adhere to. Yet, through the hard work of a dedicated staff at the Center for Food Safety and Applied Nutrition, of General Counsel, and of the Commissioner and his policy staffs, the FDA met the deadlines by releasing proposed regulations on November 6, 1991 (published in the *Federal Register* on November 27, 1991 (26)) and final regulations on December 6, 1992 (published in the *Federal Register* on January 6, 1993 (27)).

V. MAJOR FOOD LABELING ISSUES

For the purposes of the debate on the food label, particularly starting with the 1989 ANPRM, the FDA had divided the issues into six topic areas. These areas were:

- NUTRITION LABEL, APPLICABILITY, and CONTENT
- NUTRITION LABEL, FORMAT
- INGREDIENT LABELING
- NUTRIENT CONTENT CLAIMS or DESCRIPTORS
- FOOD STANDARDS
- HEALTH CLAIMS

During the 1989–90 debates, a seventh area was repeatedly raised, especially by the food industry—NATIONAL UNIFORMITY.

For the remainder of this chapter, I briefly look at a few of the major issues and concerns in each of these topic areas that were raised to the FDA, at the hearings, from the comments to the ANPRM, and during the Congressional debate. And in each topic area, I would like to consider how the NLEA and the implementing regulations responded to these issues.

A. Nutrition Label, Applicability and Content[5]

Issue: Consumers stated that nutrition labeling should be mandatory on all foods. More comments were received on this than any other single issue. By 1990, over half of the processed, packaged foods regulated by FDA bore nutrition labeling (23). Nutrition labeling was required if a nutrition claim was made for the food product or if nutrients were added to the food. Otherwise, nutrition labeling was provided voluntarily, and over half of the nutrition labeling seen on foods at that time was there because the manufacturer provided it voluntarily. Many segments of the food industry generally supported the concept of providing nutrition labeling on all foods. However, many in industry linked their support to the condition that they be assured of national uniformity on all aspects of food labeling.

Response: The NLEA settled this issue. The final regulations require nutrition labeling on most processed packaged food labels. There are a few exemptions, including bulk foods for further manufacture, foods specifically for the food service sector, and foods of no nutritional significance (e.g., unsweetened coffee and tea and most spices, colorings, and flavorings).

Perhaps the most far-reaching exemption is that granted to small businesses. NLEA, as passed in 1990, established an exemption for firms based on the total dollar volume of sales for that firm (e.g., firms with less than $500,000 total sales were exempt from nutrition labeling). Many small firms pointed out that, while this exemption might be appropriate for firms that produced only one or two products, small firms producing 50 or 100 products could easily exceed the dollar sales limit, yet the cost of developing so many different labels would be such a significant portion of their profits that some product lines would be abandoned and businesses would be jeopardized. These firms further pointed out that most of the products in question were gourmet specialty items that made insignificant contributions to the total diet. In August 1993, Congress responded to these concerns and changed the small business exemption to focus on products rather than firms and on units per year volumes rather than on dollar values of sales (28). Now a product is exempt if fewer than 100,000 units of the product are sold in a year and it is produced by a firm with fewer than 100 employees. This exemption will be, in fact, phased in over a period of several years with, for example, during the first year (May 1994 to May 1995) a product with sales of fewer than 600,000 units, produced by a firm with fewer than 300 employees, will be given an additional year to provide nutrition labeling. The FDA supports this approach to small business exemptions because the overall nutritional impact of these exempted products on the American diet is relatively small and the delayed effective dates grant relief to those firms that were having the greatest difficulty meeting the statutory deadlines. The FDA also believes that competitive marketplace pressures will cause many of these products to be nutrition labeled, even though they are exempt.

In addition to exemptions, alternative means of providing nutrition information was granted to certain products. For example, very small packages may provide a telephone number or address where nutrient content information may be obtained in lieu of nutrition labeling. Also foods with limited nutritional value (e.g., soft drinks) may use a simplified form of nutrition labeling. The net result is that consumers will now know the nutrient content of the foods that make up the bulk of their diets, will be able to compare products at the point of purchase and, hopefully, will be motivated to make healthier dietary choices.

Issue: Consumers also often expressed the idea that nutrition labeling should be extended to restaurant foods, because it is soon expected that over half of the food dollar will be spent on, and approximately a third of the diet will come from, food eaten away from home (29). The restaurant industry generally opposed the requirement of nutrition labeling, citing the variability of menus, recipes, and portion sizes from day to day.

Response: The FDA is also concerned about the impact of food eaten away from home on the overall healthy diet. If consumers ignore, or are unable to determine, nutritional attributes of food away from home, the efforts to improve dietary patterns are severely compromised. However, after careful consideration, the Agency determined that it had neither sufficient experience with providing nutrition information in the restaurant sector nor adequate resources for effective enforcement to require nutrition labeling of all food service foods. However, the FDA was extremely concerned that more and more claims for foods were being made in restaurant settings (e.g., "low calorie meals" or "low fat entrees") when in some cases there appeared to be little basis in fact to support these claims. Therefore, the final regulations require that when a claim is made for food service foods, the claim must adhere to the FDA regulatory definitions and the food service operator must, on request, provide to government inspectors or to consumers a reasonable basis for making the claim. Consumers deserve some assurance that claims made in a restaurant have meaning[6].

Issue: Consumers also expressed a desire for nutrition labeling on traditionally non-packaged foods, such as fresh produce and fish. The industry, particularly the trade associations representing the supermarket sector, pointed out the many problems associated with the ever-changing wide variety of produce and fish and the logistical challenges of providing non-label information in the produce and fish sections. The industry generally advocated a voluntary program.

Response: As directed by the NLEA, the FDA has established a voluntary program for food retailers to provide nutrition information, at the point of purchase, for the 20 most commonly consumed fruits, vegetables, and fish (30). The FDA will conduct a survey every two years to determine if there is substantial compliance by the retail food sector in providing nutrition information for these 60 items. If there is substantial compliance, the voluntary program will continue for another two years. However, if retailers fail to provide information at a level that meets the Agency's definition of substantial compliance, the NLEA directs the FDA to make the program mandatory, at which point it could be expanded beyond the 20 most commonly consumed items to encompass more of the fresh produce and fish market. The voluntary approach gives consumers information on the fruits, vegetables, and fish that contribute most to the overall American diet. At the same time, the approach as devised by Congress gives retailers a very real incentive to provide the information at point of purchase[7].

Issue: Consumers and health professionals often stated that there was a need to change some of the nutrients required in nutrition labeling, to focus on those nutrients related to chronic disease, such as saturated fat, cholesterol, and fiber, and to de-emphasize those nutrients, such as niacin, thiamin, and riboflavin, that are related to deficiency diseases that are no longer prevalent in the United States.

Response: The FDA recognized that many of the nutrients that had been required to be on nutrition labeling since 1973 were those that the scientific community had focused on following the confirmation of widespread vitamin deficiencies in parts of the United States at the time recruits were being screened for World War II. However, scientific concerns and public health focus have now shifted to diseases related to over-consumption rather than under-nutrition. Therefore the new food label (Figure 2.1) emphasizes nutrients related to chronic diseases, such as cancer, heart disease, and even obesity, by placing nutrients such as calories, fat, saturated fat, cholesterol, sodium, etc. in bold-face type at the top of the label. The vitamins and minerals are de-emphasized by being non-bolded at the bottom of the label and by not being required to be listed one after another. Additionally, certain nutrients, such as niacin, riboflavin, and thiamin, are no longer required to be listed in nutrition labeling, although they may be voluntarily provided. Also, the order of nutrients in the label was selected to place those of greater public health concern nearer the top of the lists. The FDA believes that this new label will be an important public health education tool.

Issue: Perhaps the biggest complaint lodged by consumers throughout the listening process was that some in the industry appeared to be improperly manipulating serving sizes to make products appear to be of greater nutritional value than they were. Serving size was, and is, the basis for nutrient content values in nutrition labeling. Consumers brought numerous products to the hearings with what they considered inappropriate serving sizes. For example, consumers questioned how there could be 1.4 servings in an 8 oz. box of fruit juice that obviously could not be closed once opened and with only one straw attached, indicating that the juice was not intended to be easily shared.

Response: The NLEA directed the FDA to establish standards for serving sizes. The law also specified that servings were to be based on amounts customarily consumed, not on amounts recommended for consumption or on the manufacturers' desired serving size. For a number of reasons, the serving size requirements were, perhaps, the most difficult portions of the NLEA for the FDA to implement. They required the FDA to create serving sizes for the great diversity of products in the market and to derive those serving sizes from data, such as nationwide food consumption surveys, that were not necessarily designed to measure amounts customarily consumed. In the final rules, the FDA established reference amounts for 139 categories of foods. Categories grouped foods that are generally interchangeable in the diet so that those foods would have essentially the same labeled serving sizes, thus, facilitating consumer

Nutrition Facts

Serving Size 1/2 cup (114g)
Servings Per Container 4

Amount Per Serving

Calories 260 Calories from Fat 120

	% Daily Value*
Total Fat 13g	**20%**
Saturated Fat 5g	**25%**
Cholesterol 30mg	**10%**
Sodium 660mg	**28%**
Total Carbohydrate 31g	**11%**
Dietary Fiber 0g	**0%**
Sugars 5g	
Protein 5g	

Vitamin A 4%	•	Vitamin C 2%
Calcium 15%	•	Iron 4%

* Percent Daily Values are based on a 2,000 calorie diet. Your daily values may be higher or lower depending on your calorie needs:

	Calories:	2,000	2,500
Total Fat	Less than	65g	80g
Sat Fat	Less than	20g	25g
Cholesterol	Less than	300mg	300mg
Sodium	Less than	2,400mg	2,400mg
Total Carbohydrate		300g	375g
Dietary Fiber		25g	30g

Calories per gram:
Fat 9 • Carbohydrate 4 • Protein 4

Figure 2.1 The new food label.

comparisons. The regulations further provide specific guidelines for converting reference amounts to labeled serving sizes, which, as required by law, are stated in common household units. The FDA believes that the new regulations faithfully implement the requirements of the statute. However, in this very complicated area there are many minor technical details that may require further refinement. It may take several years and some additional rulemaking to smooth down all the rough edges of serving size. Yet, the Agency believes that the consumer now has serving sizes on food labels that are more realistic (more in line with what people actually eat) and that will promote product comparisons.

B. Nutrition Labeling, Format[8]

Issue: The nutrition label format previously required was tabular. Nutrient values were given in a mixture of metric measurements and percentages. The FDA repeatedly heard from consumers and from industry that this label was not a very good communication tool. There was a wide variety of ideas expressed on how to improve the label format. These included pie charts and bar graphs and the use of adjectives instead of numbers. But it was clear that there was no consensus on what the best format would be for all foods. In fact, many commenters, particularly from industry, told the Agency not to throw out the old label for a new one until it was certain that the new label was a better communicator.

Response: To determine which type of label format would be a better communication device, additional research was needed. The FDA and a number of food industries and trade associations (31–33) conducted an impressive variety of consumer tests on a wide array of different label formats. In July, 1992, the FDA published the results of its testing in the *Federal Register* (34). While no one format was perfect for fulfilling all the possible uses of nutrition labeling, virtually all the research demonstrated that graphical presentations (e.g., pie charts, bar graphs) were not suited for conveying the diversity and amount of information required on nutrition labels. In the end, the FDA adopted a format based more on consumers' abilities to use and comprehend, rather than on stated consumer preference. The format adopted (Figure 7.1) features a column of nutrient values expressed as a percent of a label reference value called the "Daily Value." Research showed that this percent presentation aided consumers in quickly determining whether a food contained a little or a lot of a nutrient. This is particularly true when compared to the old format's listing of nutrients by quantitative amounts and consumer confusion about whether 50 milligrams of sodium or 10 grams of fat represented high or low values. The percentage listing also facilitates comparison of different foods at the point of purchase, by putting all values on the same scale. The FDA has specified certain graphical and size requirements for the new format that causes the "Nutrition Facts" panel to be extremely visible and easier to read on food packages. All in all, the Agency believes that the new label is a significant step forward in aiding Americans to select healthier diets.

C. Ingredient Labeling

For many years, FDA regulations have required that ingredients be listed on the label, by their common or usual name, in descending order of predominance by weight. The FD&C Act permitted spices, flavors, and colors to be designated by the generic heading and not specifically identified. Also, the FD&C Act stated that mandatory ingredients of standardized foods need not be declared in ingredient lists.

Issue: The number one ingredient labeling issue raised by consumers was an opposition to the so-called "and/or" labeling of fats and oils. Under this regulation, a manufacturer may list all fats and oils that may be used in a product and not necessarily those that are present. For example, the label may read "Vegetable oil (soy, corn, and/or coconut)" if all of the named oils, alone or in blends, are used in the product from time to time. The argument most often put forward by consumers was that, because of the varying fatty acid profiles of different oils, the public health significance of the oils may be so different that consumers need to know the precise oil(s) being used. Industry, on the other hand, indicated strong opposition to changing "and/or" labeling, because they saw this as an important economic issue, in that "and/or" labeling allows manufacturers to switch oils as market conditions dictate.

Response: The NLEA regulations resolve this issue in an indirect fashion. Nutrition labeling is mandatory on virtually all foods, and total fat and saturated fat content is required to be declared. Therefore, consumers now have the important public health relevant information needed to construct healthier diets regardless of the oil used. On the other hand, this approach allows a continuation of "and/or" labeling for fats and oils, thus giving manufacturers a degree of flexibility in selecting oils for economic reasons. It does, however, constrain the substitution of oils somewhat in that manufacturers will have to select oils or blends of oils with similar fatty acid profiles, *or* change the label each time oils are changed, *or* always declare nutrient values as if the oil with the greatest saturated fat were used, thus in many cases overstating the saturated fat content of, and consequently reflecting badly on, their product.

Issues: There were a number of other ingredient labeling issues raised. These included:

- labeling of all ingredients on standardized foods
- percentage declaration of major ingredients
- grouping of sweeteners at one location in the ingredient list

- specific identification of flavors, spices, and colors

None of these issues had the level of interest of "and/or" fat labeling.

Responses: The NLEA and the FDA's final regulations address all of these issues to some extent.

- NLEA changed the FD&C Act to require full ingredient listing on all products, including the mandatory ingredients of standardized foods. This change will have little impact on the market, because the trend in the industry has been to provide a complete listing on all foods and the FDA has for several years been carrying out a program to specify that most ingredients in standardized foods are optional and therefore must be listed in the ingredient statement.
- Although consumers expressed interest in percent declaration of ingredients, industry opposed such requirements because of their concern that trade secret formulations would be divulged. The FDA determined that, because all foods must now bear nutrition labeling, percent ingredient labeling would not provide additional information of public health significance. Therefore, although the final regulations provide manufacturers with a format if they choose to declare the percentage of ingredients, such declarations will not be required.
- Consumers raised the sweeteners issue because of their belief that some very highly sweetened products obscured that fact by using a number of sweeteners. Such a hypothetical product might list sucrose, high fructose corn syrup, and honey as the second, fourth, and sixth ingredient in the ingredient list, whereas if all these ingredients, which were all added for the same purpose, were grouped and listed as "sweeteners" (sucrose, high fructose corn syrup, honey), the collective name would be the first item in the ingredient list. The FDA believes that because now "sugars" content declaration is required in nutrition labeling, consumers will have the important public health relevant information, without requiring that sweeteners be grouped in the ingredient list.
- The NLEA amended the FD&C Act to require that certified colors be declared by the common or usual name. Accordingly, the final regulations require that colors such as FD&C Blue No. 2 be listed by name in the ingredient list. Non-certified colors, such as paprika or beet juice, may continue to be listed by the collective term, "colors." Flavors and spices may also continue to be listed by the collective terms. In many processed foods a very large number of flavoring compounds are added, each at very low levels. Requiring that all compounds be declared by name would create an extraordinarily long ingredient list full of complex chemical names that would be confusing to the average consumer without providing information of public health relevance.

D. Nutrient Content Claims or Descriptors

By 1989, FDA had, by regulation, defined a number of nutrient content claims. These included such terms as "Low Calorie," "Reduced Calorie," "Sodium Free," "Very Low Sodium," "Low Sodium," and "Reduced Sodium." The Agency had also proposed a regulation that would define terms related to cholesterol, such as "Reduced Cholesterol," "Low Cholesterol" and "Cholesterol Free" (35). Throughout the food labeling debates, the Agency heard of a number of other terms for which definitions would be desirable. Many of these terms pertained to fat content, such as "low fat," and "low in saturated fat." For what are considered "desirable" nutrients, definitions were sought for terms such as "high in calcium" and "good source of fiber." Additionally there were a number of non-specific terms such as "lite," "natural," "fresh," "organic," etc. that were being used with no regulatory definition, often in very confusing ways.

Issue: Although there were disputes about the individual definitions, there appeared to be a general consensus that the Federal Government should provide definitions for descriptive terms used on food labels. Consumers wanted the definitions so that when the terms were used they had a consistent meaning.

Industry generally supported the concept of definitions because of their desire for a "level playing field."

Response: The NLEA settled this issue by amending the FD&C Act to require that the FDA define terms related to nutrient content before they could be used on food labels. The NLEA also listed certain terms for which the FDA was specifically instructed to provide definitions. In the final regulations the FDA selected a number of core terms (or descriptors), such as "free," "low," "reduced" or "less," "good source," and "high in," and established definitions based on dietary guidance and nutrition principles. As instructed by the NLEA, the FDA also listed what the Agency considered to be acceptable synonyms for each of the core terms. The FDA then applied the descriptors to the range of nutrients under consideration, taking into account the distribution of the nutrient in the food supply and specific dietary guidelines, to create the criteria for each specific nutrient content claim, such as "fat free," "low sodium," "reduced saturated fat," "good source of calcium," "high in fiber," etc. The FDA also adopted the USDA definitions for "lean" and "extra lean" to be applied to seafood and game meat products, the labeling of which the FDA regulates. The NLEA also required the FDA to define the popular marketing term, "light" (or "lite"). The Agency originally proposed a definition that focused on a calorie reduction in products bearing the claim. However, comments to the proposal convinced the FDA that consumers also saw the term "light" as implying a reduction in fat content in some cases, and the final rule focuses on both calorie and fat reduction criteria for use of the term on food labels. This illustrates that public comment does have major impact on regulations. The FDA believes that one of the major gains for consumers from the NLEA is that now they can believe claims related to nutrient content on the front panel of food labels, because these claims have meanings grounded in dietary guidance and are used consistently throughout the food supply.

E. Food Standards

Food standards are essentially recipes that specify the types and often levels of ingredients in familiar staple food products. For example, peanut butter must be 90% peanuts to be labeled as "peanut butter." Food standards were established in the 1930's to prevent consumer deception by inferior products.

Issue: The FDA heard a variety of positions expressed. Certain manufacturers, for example, the makers of traditional cheddar cheese, advocated keeping the standards just as they were. On the other hand, other manufacturers, for example, the makers of the so-called "low fat cheddar cheese" (that is, cheddar cheese with less than 50% fat on a solids basis as required by the food standard, and thus technically not permitted to be called "cheddar cheese") wanted the standards modified to accommodate their products. Consumers and health professionals, in general, did not focus on the food standard issues. The Office of Management and Budget (OMB) perceived food standards as anti-competitive, preventing products such as the low fat cheeses from reaching the marketplace, and they advocated eliminating food standards altogether and allowing marketplace forces determine which products would survive. This OMB position was generally unacceptable to industry.

The FDA had reached the tentative conclusion that the major problem with food standards was their lack of flexibility. They were inflexible because the administrative requirements in the FD&C Act relating to how standards were modified or established made them extremely time consuming and resource intensive to change. Thus, the Agency found the standards locked in place while food technology advanced and made new and different products possible. More importantly, as dietary advice changed and focused less on nutrient deficiencies and more on the diseases of over-consumption, fat, which was once considered a valued ingredient and the marker of an economically superior product, had become a dietary pariah. Yet, food standards interfered with the marketing of reduced fat products using the standardized name. The FDA believed that until there was some legislative action to make the standards easier to amend, its limited resources could best be spent elsewhere.

Response: The issue of food standards was addressed in two different ways. First, the NLEA amended the FD&C Act to move regulations establishing, or subsequently amending, standards from formal to informal rulemaking. This means that food standards can be changed through notice and comment rulemaking, rather than by the cumbersome and time-consuming process of not only notice and comment procedures, but also objections and requests for formal stays of regulations followed by evidentiary hearings. A curiosity of the NLEA occurred during the last minutes of passage of the law, when the senators from Wisconsin and Vermont intervened and moved dairy and maple syrup standards back into the formal rulemaking process. These glaring anomalies in the law demonstrate the legislative process in action.

The second response to the food standard issue is found in the FDA's final regulations. The Agency established a generic regulation, §130.10, that will make almost all food standards much more flexible. This regulation will allow standardized products that have been reduced in calories, or fat, or sodium, etc. and still retain the organoleptic characteristics of the standardized food to be marketed using the standardized name amended by the appropriate nutrient content claim. Therefore, product statements of identity, such as "low fat ice cream" or "reduced fat cheddar cheese" are now permitted by this regulation. The FDA believes that this is a significant step forward in healthier foods for the American public.

F. Health Claims

Before 1984, an explicit discussion of disease and its possible connection to a food product was prohibited on the label. Such labeling caused the product to become a drug, a result that compelled the manufacturer to submit a new drug application for the product with data supporting the health claim. Following the Kellogg labeling and advertisement campaign mentioned earlier, the FDA had been struggling to find a way to permit appropriate health claims on food labels without opening the door to misleading, or outright fraudulent, claims. The Agency often stated that it had no intention to deny to the American public information on the role of diet in good health. At the time of passage of the NLEA, however, no final regulation had been published on this topic. In 1987, the FDA published a proposed regulation on health claims (36). The proposal provoked sharp controversy, both within and outside of the government (37). Furthermore, during the comment period and debate on this proposal, it became apparent that the proposed standard was too vague and difficult to enforce to be viable. In fact, judicial interpretations of this proposal were inhibiting the FDA actions against types of products, outside the mainstream of the food supply, which had traditionally been regarded as fraudulent (38). On February 13, 1990, the FDA reproposed the regulation on health claims, with a more explicit scientific standard for determining the validity of a claim (39). The standard called for claims to be supported by the totality of available scientific information and that there be general agreement among scientists that the statements on the food label were in fact supported by science. Final action on the reproposal had not been taken when the NLEA was adopted.

Issue: During the food labeling debates, the FDA again heard the entire spectrum of opinions on health claims. Some consumers, most health professionals, and representatives of state government supported a total ban on health claims, contending that they are inherently confusing and blur the borderline between food and drugs. Others, notably nutrition professionals and consumer advocates, appeared to have accepted the inevitability of health claims on food labels and called for the FDA to strictly control them. Some industry representatives either endorsed the 1987 proposal or advocated even less stringent regulation by the FDA. The only point of agreement seemed to be that this was an issue that the FDA must resolve.

Response: Again, the NLEA directly addressed this issue by adopting into law many of the provisions of the FDA 1990 proposal. The law requires that any health claim on the label of a conventional food product be supported by the totality of publicly available information and that there be general agreement among scientists qualified by training and experience that the data support the claim. Applying this standard to available data, the FDA, in the final regulations, approved health claims relating: 1) dietary fat, especially saturated fat and cholesterol, to heart disease; 2) fat to cancer; 3) calcium to osteoporosis; 4) sodium to hypertension; 5) diets rich in fruits, vegetables, and grain products, which are foods high in fiber, to cancer; 6) diets rich in fruits and vegetables, food high in antioxidant vitamins, to cancer; and 7) diets high in fruits, vegetables, and grain products, which are foods high in fiber, particularly soluble fiber, to heart disease. The FDA has also proposed a claim relating the consumption of folic acid to a reduction in the risk of neural tube defect births (40). These final regulations have laid the groundwork for providing the American public much more explicit, but substantiated, dietary guidance on the labels of food products.

G. Uniformity

Issue: As indicated earlier, a seventh issue was added to the debate. A number of states were enacting, or proposing to enact, food labeling requirements that went beyond or ran counter to the federal requirements. For example, in 1990, 109 bills related to food labeling, ranging from extensive nutrition labeling regulations in California to definitions of "lite cheese" in Wisconsin, were introduced into state legislatures and 11 were enacted (41).

However, when consumers and industry talked about national uniformity they had different goals. Industry found the prospect of fifty different state food labeling laws to be unacceptable with today's nationwide delivery systems for food products and wanted federal pre-emption for all aspects of the food label, including food safety issues. Consumers on the other hand acknowledged the need for uniformity on nutrition labeling issues but believed the states should retain the right to impose additional food safety labeling requirements and take action in cases of deceptive practices.

Response: A primary feature of the NLEA is that most state regulations in the area of food labeling are now preempted by federal regulations. Therefore, the FDA believes that Congress has clearly spoken that no state law regarding food labeling should interfere with interstate commerce in food. However, Congress recognized the importance of states and state regulatory bodies and incorporated into the new law provisions that allow states to enforce federal food labeling regulations. Therefore, the NLEA has created the unique situation where the states and FDA are partners in the food label reform implementation. The FDA is working closely with state regulatory bodies to ensure that all partners have a common understanding of the new regulations and that the regulations are being enforced even-handedly throughout the United States.

VI. FOOD LABEL EDUCATION[9]

I can not close this chapter without saying a few words about what has now become a priority focus for the FDA—food label education. After Congress passed the NLEA, the FDA, because of resource constraints and because of the monumental task facing the Agency, was forced to concentrate on the L in NLEA—the labeling regulations. The FDA responded by producing proposed regulations and final regulations on a schedule that the rest of the world thought was impossible. However it is now time to concentrate more on the E part of NLEA and to turn some of this manpower and intellectual resource toward the education task that lies ahead. The regulations are virtually meaningless and a revised food label serves little purpose, unless consumers understand and use the information that they are being given. Education is a continuing commitment that may in fact be more important in the long run than the regulations FDA has produced.

The FDA, jointly with the USDA, has launched a campaign that is both visionary and pragmatic, taking into account limitations in budget, resources, and expertise.

The goals of the campaign are threefold. First, the agencies want to promote awareness and accurate understanding of the new label. Second, the FDA the USDA will work hard to build partnerships and facilitate information exchange. And third, the many partners will conduct a multi-year campaign to increase label use.

This campaign faces a number of challenges. An obvious one is that there was an extended transition period from the old label to the new one. Some new labels showed up in early summer 1993, but because of the way the law is worded and the pipeline, it may be the beginning of 1995 before all products on the grocery shelf have completely shifted to the new label. These two sets of labels complicate the education program. Another challenge is that the FDA's staffing and funding are limited and the Agency, traditionally, does not have experience with this type of education. Finally, the potential audience—anyone who eats—is extremely diverse, from the information seekers who want more and more information to the non-label users, whether because of low-literacy, non-English speaking, or cultural indifference. Ways need to be found to reach all of these groups.

Because of budget constraints, a key component of the campaign will be collaboration and information sharing with a number of partners, including health professional associations, consumer groups, trade associations, and individual industries. Also key will be relations with the print and broadcast media, because their power in reaching the public is unparalleled. The FDA also firmly believes that research is needed, particularly on how to reach some of the harder-to-reach groups. The outcome of all these efforts is intended to be a series of educational materials and programs, reaching all consumers.

VII. ENDNOTES

1 An overview of NLEA regulations from the USDA is presented in Chapter 3 by John McCutcheon.

2 The effects of NLEA on FTC advertising enforcement is discussed in Chapter 4 by Former Commissioner Deborah K. Owen and Joyce E. Plyler.

3 A brief history of FDA regulation relating to the nutrient content of food is presented in Chapter 1 by Peter Barton Hutt, formerly Chief Council to the FDA.

4 A review of international harmonization and NLEA compliance is presented in Chapter 5 by Linda Horton.

5 A comprehensive review of the NLEA regulations is presented in Chapter 7 by Ralph Shapiro.

6 The impact of NLEA on the food service industry is discussed in Chapter 20 by Guy Livingston.

7 FDA regulations for nutrition labeling of raw produce and fish is discussed in Chapter 8 by Jean A. T. Pennington et al.

8 A discussion of food packaging and label formats with illustrations is presented in Chapter 10 by Roy Parcels and Aaron Brody

9 Consumer education is discussed in Chapter 12 by Brenda Derby and Sarah Beck Fein and Chapter 11 by Naomi Kulakow.

VIII. REFERENCES

1. M. Venkatesan, W. Lancaster, and K.W. Kendall, An empirical study of alternate formats for nutritional information disclosure in advertising, *Journal of Public Policy and Marketing 5*:29–43 (1986).
2. Nutrition Labeling and Education Act of 1990, Public Law 101-535, 104 Stat. 2353 (1990).
3. Federal Food, Drug, and Cosmetic Act of 1938, Public Law 75-717, 52 Stat. 1040 (1938).
4. Food and Drugs Act of 1906, Public Law 59-384, 34 Stat. 768 (1906).
5. McNary-Mapes Amendments to the Food and Drugs Act, 46 Stat. 1019 (1930).
6. *White House Conference on Food, Nutrition, and Health: Final Report*, U.S. Government Printing Office, Washington, D.C., 1970, pp. 341
7. Food and Drug Administration, Regulations for the Enforcement of the Federal Food, Drug, and Cosmetic Act; Nutrition Labeling, *Federal Register 38*:2125–2132, U.S. Government Printing Office, Washington, D.C., January 19, 1973.
8. Food and Drug Administration, Special Dietary Foods; Label Statements, *Federal Register 43*:43248–43260. U.S. Government Printing Office, Washington, D.C., September 22, 1978.
9. Food and Drug Administration, Food Labeling: Declaration of Sodium Content of Foods and Label Claims for Foods on the Basis of Sodium Content, *Federal Register 49*:15510–15535, U.S. Government Printing Office, Washington, D.C., April 18, 1982.
10. U.S. Department of Health, Education, and Welfare, U.S. Department of Agriculture, and Federal Trade Commission, Advance Notice of Rulemaking, Tentative Position of Agencies, *Federal Register 44*:75990–76020, U.S. Government Printing Office, Washington, D.C., December 21, 1979.
11. Committee on Diet And Health, Food and Nutrition Board, Commission on Life Sciences, National Research Council, *Diet and Health: Implications for Reducing Chronic Disease Risk*, National Academy Press, Washington, D.C., 1989, pp. 749
12. U.S. Department of Health and Human Services, Public Health Service. *The Surgeon General's Report on Nutrition and Health*, DHHS(PHS) Publication No. 88-50210, U.S. Government Printing Office, Washington, D.C., 1988, pp. 727
13. U.S.Department of Agriculture and U.S. Department of Health and Human Services, *Nutrition and Your Health, Dietary Guidelines for Americans*, Home and Garden Bulletin No. 232, 3rd Edition, U.S. Government Printing Office, Washington, D.C., 1990.
14. U.S. Department of Agriculture, Human Nutrition Information Service, *Food Guide Pyramid*, Home and Garden Bulletin No. 249, U.S. Government Printing Office, Washington, D.C., 1992.
15. U.S. Department of Health and Human Services, National Heart, Lung, and Blood Institute; *National Cholesterol Education Program, Report of the Expert Panel on Population Strategies for Blood Cholesterol Reduction*, NIH Publication No. 90-3046, National Institutes of Health, Bethesda, MD, 1990.

16. National Cancer Institute, *Eat More Fruits and Vegetables—Five-a-Day for Better Health*, NIH Publication No. 92-3248, National Institutes of Health, Bethesda, MD, 1991.
17. American Heart Association, Committee on Nutrition, Dietary Guidelines for Healthy Adults, *Circulation 74*:1465a (1986).
18. American Institute of Nutrition and the American Society of Clinical Nutrition, Position Statement on Food Labeling, *Journal of Nutrition 120*:534–536 (1990).
19. American Dietetic Association, Position of the American Dietetic Association: Nutrition and Health Information on Food Labels, *Journal of the American Dietetic Association 90*:583–585 (1990).
20. A.S. Levy, O. Matthews, M. Stephenson, J.E. Tenny, and R.E. Schucker The impact of a nutrition information progam on food purchases, *Journal of Public Policy and Marketing 4*:1–3, 1985.
21. Opinion Research Corporation, *Food Labeling and Nutrition: What Americans Want, Survey conducted for the National Food Processors Association*, ORO, Washington, D.C., 1990 178 pp.
22. Food and Drug Administration, Food Labeling; Advance Notice of Proposed Rulemaking, *Federal Register 54*:32610–32615, U.S. Government Printing Office, Washington, D.C., August 8, 1989.
23. Committee on the Nutrition Components of Food Labeling, Food and Nutrition Board, Institute of Medicine, National Academy of Sciences, *Nutrition Labeling: Issues and Direction for the 1990s* (Porter, D.V. and Earl, R.O. editors), National Academy Press, Washington, D.C. 1990.
24. Food and Drug Administration, Food Labeling; Notice of Hearings and Extension of Comment Period, *Federal Register 54*:38806–38807, U.S. Government Printing Office, Washington, D.C., September 20, 1989.
25. Food and Drug Administration, Food Labeling; Reference Daily Intakes and Daily Reference Values; Mandatory Status of Nutrition Labeling and Nutrient Content Revision; Serving Size; Proposed Rules, *Federal Register 55*:29475–29533, U.S. Government Printing Office, Washington, D.C., July 10, 1990.
26. Food and Drug Administration, Food Labeling; Proposed Rules; *Federal Register 56*:60366–60878, U.S. Government Printing Office, Washington, D.C., November 27, 1991.
27. Food and Drug Administration, Food Labeling; Final Rules, *Federal Register 58*:2066–2941, U.S. Government Printing Office, Washington, D.C., January 6, 1993.
28. Nutrition Labeling and Education Act Amendments of 1993; Small Business Exemptions, Public Law 103-80, August 13, 1993.
29. National Restaurant Association, *Food Service Industry Forecast*, NRA, Washington, D.C., 1990, pp. 24
30. Food and Drug Administration, Food Labeling; Nutrition Labeling; Guidelines for Voluntary Nutrition Labeling; and Identification of the 20 Most Frequently Consumed Raw Fruit, Vegetables, and Fish; Definition of Substantial Compliance;

Final Rule, *Federal Register 56*:60880–60891, U.S. Government Printing Office, Washington, D.C., November 27, 1991.

31. A.S. Levy, S.B. Fein, and R.E. Schucker, "A Study of Nutrition Label Formats: Performance and Preference," Division of Consumer Studies, Center for Food Safety and Applied Nutrition, Food and Drug Administration, Washington, D.C., March 1991.
32. A.S. Levy, S.B. Fein, and R.E. Schucker. "Evaluation of Nutrition Label Formats: FDA Study 2," Division of Consumer Studies, Center for Food Safety and Applied Nutrition, Food and Drug Administration, Washington, D.C., March 1992.
33. T.S. Gardner and E.T. Guarino, GMA/NFPA Industry Study, Grocery Manufacturers of America, Letter to the Food and Drug Administration, June 3, 1992.
34. Food and Drug Administration, Food Labeling; Format for Nutrition Labeling; Proposed Rule, *Federal Register 57*:32063–32067. U.S. Government Printing Office, Washington, D.C., July 20, 1992.
35. Food and Drug Administration, Food Labeling; Definition of the Terms Cholesterol-Free, Low Cholesterol, and Reduced Cholesterol; Tentative Final Rule. *Federal Register 55*:29455–29473, U.S. Government Printing Office, Washington, D.C., July 19, 1990.
36. Food and Drug Administration, Food Labeling; Public Health Messages on Food Labels and Labeling, *Federal Register 52*:28843–28849, U.S. Government Printing Office, Washington, D.C., August 4, 1987.
37. Committee on Government Operations, House of Representatives, *FDA's Continuing Failure to Regulate Health Claims for Foods, Hearings before the Human Resources and Intergovernmental Relations Subcommittee*, October 31 and November 9, 1989. U.S Government Printing Office, Washington, D.C., 1990.
38. U.S. v. Health Club, Inc., United States District Court for the Southern District of New York, No. 87 Civ. 7779, Opinion dated July 11, 1989, in Food Drug Cosmetic Law Reports, (July 31, 1989), pp. 38, 265–38, 272, Commerce Clearing House, Inc.
39. Food and Drug Administration, Food Labeling; Health Messages and Label Statements; Reproposed Rule, *Federal Register 55*:5175–5192, U.S. Government Printing Office, Washington, D.C., February 13, 1990.
40. Food and Drug Administration, Food Labeling; Health Claims and Label Statements; Folate and Neural Tube Defects, *Federal Register 58*:53254–53295, U.S. Government Printing Office, Washington, D.C., October 14, 1993.
41. Food and Drug Administration. Annual Report on State Legislation for the Year 1990. State Programs Coordination Branch, Division of Federal–State Relations, Food and Drug Administration, Rockville, MD, 1991.

3

Nutrition Labeling of Meat and Poultry Products—An Overview of Regulations from the U.S. Department of Agriculture

John W. McCutcheon

Food Safety and Inspection Service, U.S. Department of Agriculture, Washington, D.C.

I. INTRODUCTION

The U.S. Department of Agriculture's Food Safety and Inspection Service (FSIS) is the agency with the authority for regulating the labeling of meat and poultry products. The Federal Meat Inspection Act and the Poultry Products Inspection Act authorize the Secretary of Agriculture to establish and maintain inspection programs designed to assure consumers that meat and poultry products distributed in commerce or within designated states are wholesome, not adulterated, and are properly marked, labeled, and packaged. The Food and Drug Administration (FDA) regulates labeling for foods other than meat and poultry.

Since 1973, the FSIS has required nutrition information on labels when nutrition related claims are made, for example, about fat or sodium content. Nutrition labeling guidelines for meat and poultry products had been issued through various policy memoranda, which are now superseded by the new, comprehensive regulations issued on January 6, 1993.

The issuance of updated nutrition labeling regulations was prompted by the realization that consumers are becoming increasingly aware of the relationship between nutrition and health. The USDA has been a leader in providing sound dietary advice to the public since the early 1900s, when the first food guide was issued.

While this advice has changed more rapidly in recent years, the nutrient information provided on meat and poultry labels has remained relatively static, limiting the ability of consumers to put this advice into practice. For instance, while dietary guidelines issued by the USDA and the Department of Health and Human Services (HHS) encourage consumers to avoid too much sodium, sodium information on meat and poultry labels is required only if a claim is made. In addition, the previous nutrition label focused heavily on nutrients, such as the vitamins thiamin and riboflavin, that historically were responsible for nutrient deficiency diseases but now are relatively plentiful in the American diet. Another concern was the trend towards nutrient content claims such as "lean" and "low fat" on meat and poultry products. The USDA was concerned that the use of such terms is subject to various interpretations and may mislead the consumer.

Recognizing the need to reform nutrition labeling to make it more relevant to contemporary health concerns and marketplace, the USDA and HHS jointly sponsored a study in 1989 by the National Academy of Sciences to provide options for improving food labeling. In November, 1991, the FSIS and FDA simultaneously proposed regulations on nutrition labeling. The FDA regulations for foods other than meat and poultry were designed to comply with the Nutrition Labeling and Education Act of 1990. The FSIS was not required by this legislation to issue regulations for meat and poultry. However, the FSIS chose to initiate new nutrition labeling requirements to help provide consistent regulations for all foods. After soliciting public input through joint hearings and comments in response to regulatory proposals, the FSIS and FDA issued final nutrition labeling regulations on January 6, 1993.

The FSIS's nutrition labeling requirements for meat and poultry became effective on July 6, 1994; the FDA effective date was May 8, 1994.

II. MANDATORY NUTRITION LABELING

The FSIS regulations for meat and poultry require nutrition labeling for most products other than single-ingredient, raw meat and poultry products available for purchase at the retail level. For instance, basted turkey, chicken franks, corned beef, and meat burritos are examples of products that are subject to mandatory nutrition labeling.

The label information is based on the products "as packaged." "As packaged" labeling must list information about the nutrients in a serving of food at the time it is purchased, without considering further consumer preparation. A label may, in addition, list information "as consumed," provided that it clearly states preparation and cooking instructions. Manufacturers of meat and poultry products for which nutrient content information changes significantly during preparation may find this dual declaration useful.

III. VOLUNTARY NUTRITION LABELING

Raw, single-ingredient meat and poultry products such as chicken breasts, ground beef, and whole unbasted turkeys will not be required to carry nutrition labeling, but manufacturers may provide it voluntarily. One reason the FSIS decided to include these products under the voluntary, rather than the mandatory program, is because nutrient information is easier to obtain for these products.

Manufacturers who choose to participate in the voluntary program must adhere to the same requirements of the mandatory program. The only exception is that nutrition information may be listed on the product "as packaged" *or* "as consumed." "As consumed" figures are based on meat and poultry products using common cooking methods that do not add any ingredients. For instance, a whole, raw chicken may be labeled with its nutrient content as is, without any preparation or it may be labeled with its nutrient content after it has been oven roasted, or prepared in another way indicated on the label.

The FSIS encourages firms to participate in the voluntary program and, in fact, will evaluate participation. If, by 1995, there is not significant participation, the FSIS will consider proposing regulations for a mandatory program for the products currently under the voluntary program. Significant participation is defined as 60 percent of stores evaluated providing information on 90 percent of the major cuts of meat and poultry listed in the regulation.

As an alternative to providing nutrition information on the label, manufacturers of meat and poultry products that fall in the voluntary nutrition labeling category may provide nutrition information through point-of-purchase materials, such as pamphlets, wall posters, and aisle hangings. The nutrition information may be supplemented by videos, live demonstrations, or other media.

IV. EXEMPTIONS

The FSIS has exempted certain products from the mandatory nutrition labeling provisions. They include:

- Products in small packages weighing less than 1/2 ounce
- Restaurant menus, since the FSIS does not believe they constitute labeling

- Small businesses with 500 or fewer employees will not be required to provide nutrition information on products produced in amounts of 100,000 pounds or less. The pounds of product limit will be phased in over a three year period. During the first year following the effective date of the regulations, the limit is 250,00 pounds. The limit is lowered to 175,000 pounds the second year and 100,000 pounds during the third and subsequent years. The FDA definition of a small business, which was mandated by the NLEA, differs from that of the FSIS. The FSIS chose to differ because of concern that the original criteria for exemption imposed by the NLEA would not provide adequate relief for small meat and poultry processors. Subsequently, the NLEA was amended to provide more adequate relief for small businesses under the FDA jurisdiction. (See Chapter 7.)
- Products not for sale to retail consumers, such as those intended for institutional users
- Products intended for further processing
- Products packaged, portioned, or processed at retail establishments, such as sliced bologna

Products are disqualified from these exemptions if a nutrient content claim is made.

V. NUTRIENT LABEL CONTENT

The FSIS and FDA require mandatory disclosure for the same 14 food components on the label: calories, calories from fat, total fat, saturated fat, cholesterol, sodium, total carbohydrate, dietary fiber, sugars, protein, vitamin A, vitamin C, calcium, and iron.

The following food components may be listed voluntarily: calories from saturated fat, polyunsaturated fat, monounsaturated fat, stearic acid, potassium, soluble fiber, insoluble fiber, sugar alcohols, other carbohydrates, and other vitamins and minerals.

One difference between the FDA and FSIS is that FSIS will permit the voluntary disclosure of "stearic acid" as a subcomponent of saturated fat. Although the definition of saturated fat included stearic acid, a near consensus has been reached that stearic acid does not have the same serum cholesterol-raising effect as the three other saturated fatty acids—myristic, palmitic, and lauric acids. Thus, FSIS is allowing the voluntary declaration of stearic acid to provide consumers with factual information concerning saturated fats that do not have a cholesterol-raising effect.

VI. FORMAT

The FSIS and FDA have the same format requirements for the full nutrition label. Nutrition information must be presented in a manner that is consistent with the regulations which include guidelines regarding type size, style, color and prominence.

The new format, titled "Nutrition Facts," must appear on all foods carrying nutrition labeling. A major feature of the new format is the use of the term "Daily Value." Daily Value is a label reference value that is the means of placing the nutrition information on the label in the context of a total daily diet. The daily values on the new nutrition label are based on a daily diet of 2,000 calories, which is appropriate for most women, including teenage girls and some sedentary men. The footnote provides daily values for a 2,500 calorie diet, which may be closer to what most men, teenage boys, and active women should consume.

To help consumers evaluate how a food fits into a healthful diet, the following nutrients must be expressed as a percent of the daily value on the label: total fat, saturated fat, cholesterol, sodium, total carbohydrate, dietary fiber, vitamins A and C, calcium, and iron.

Daily Value is actually a combination of two sets of dietary standards. The first, Daily Reference Values (DRVs) are set for total fat, saturated fat, cholesterol, total carbohydrate, dietary fiber, protein, potassium, and sodium.

The second, Reference Daily Intakes (RDIs), replaces the term "U.S. RDA," which was introduced in 1973 as a label reference value for vitamins, minerals, and protein. These two terms—Daily Reference Values and Reference Daily Intakes—will not appear on the label. To make label reading less confusing for consumers, the term Daily Value is used to represent these two sets of dietary standards.

There are several variations that can be made to the format depending on nutrient content and package size. Both the FSIS and FDA, however, require that the core nutrients (calories, total fat, total carbohydrate, protein, and sodium) be listed.

A simplified label format may be used for meat and poultry products when any of the required nutrients, other than core nutrients, are present in insignificant amounts. The "core" nutrients (calories, total fat, total carbohydrate, protein, and sodium) must be listed with their nutrient values on all labels covered under the mandatory program. However, non-core nutrients can be listed in a sentence within the nutrition label instead of in the tabular listing. An example would be "not a significant source of fiber and vitamin C."

The regulations also allow other alterations that give manufacturers additional flexibility. To accommodate different package shapes and sizes a format with the footnote to one side is permitted, a dual declaration format may be used when both "raw" and "cooked" data are being presented, and foods that have a total surface area available to bear labeling of 40 or fewer square inches (i.e., intermediate size packages) may modify the nutrition label by providing the information in a tabular or linear format if the package will not accommodate a column display on any panel. However, it should be noted that the linear format may only be used if the label will not accommodate a tabular display. There is also a format for toddler foods. Details on label formats with graphics are presented in Chapter 10.

VII. SERVING SIZES

Serving sizes are more standardized under the new regulations. In an effort to harmonize the serving sizes for all foods, the FDA worked with the FSIS to establish 139 food categories for non-meat or poultry products, 23 categories for meat products, and 22 categories for poultry products (the number of categories will change as the need for new categories is demonstrated). These food categories and their reference amounts are contained in the regulations. The joint effort provides consumers with a uniform food labeling system that categorizes similar foods by the same reference standard.

Meat and poultry cuts such as beef roasts and chicken breasts have a reference amount of 85 grams cooked, which amounts to a 3-ounce serving size. The 3-ounce serving size conforms with dietary recommendations, which will facilitate nutrition education and dietary counseling.

Items such as chili and spaghetti with sauce have a reference amount and serving size equal to one cup. For products in slices or pieces, such as luncheon meats or frankfurters, the serving size is the number of slices or pieces closest to the applicable reference amount—in this case, 55 grams. Meal type products such as TV dinners that meet the definition of a single serving container are considered to be a single serving.

These reference amounts will not appear on the label. However, they serve as the basis for determining serving sizes that will appear on the package. The regulations allow manufacturers some flexibility to determine the appropriate serving size for their specific products.

The serving size must be clearly labeled on the nutrition panel, in the appropriate household unit such as cups, pieces, or ounces, followed by the equivalent gram measure. When the common household measure is an ounce, the FSIS does not require the use of a visual unit of measure as does the FDA.

VIII. NUTRIENT CONTENT CLAIMS

FSIS has set specific requirements for using the following terms: high, good source of, free, low, light (lite), reduced/less/fewer, more/added, percent fat-free, lean and extra lean. The definitions match those set by the FDA. They are summarized as follows; more detailed definitions can be found in the regulations (see also Chapter 7).

- *High/Rich in/Excellent source of:* The product contains 20 percent or more of the daily value per reference amount for individual foods. Does not apply to meal-type products in their entirety but can identify a food within a meal-type product.
- *Good source/Contains/Provides:* The product contains 10-19 percent of the daily value per reference amount for individual foods. Does not apply to meal-type products in their entirety but can identify a food within a meal-type product.
- *Free/No/Zero/Without/Trivial source of/Negligible source of/Dietarily insignificant source of:* The product contains an insignificant amount of fat, saturated fat, cholesterol, sodium, sugar, or calories. For example, a "fat-free" product will contain less than 0.5 g. of fat per reference amount for individual foods.
- *Low/Few/Contains a small amount of/ Low source of/Low in:* The product is low in fat, saturated fat, cholesterol, sodium, or calories. For example, a "low-fat" product contains 3 grams or less of fat per reference amount for individual foods.
- *Light or Lite:* For individual foods that contain 50 percent or more of their calories from fat, the product must have at least 50 percent less fat per reference amount compared to an appropriate reference food. For products that derive less than 50 percent of their calories from fat, either a 50 percent reduction in fat or one-third fewer calories. "Light in sodium" means a product with 50 percent less sodium.
- *Reduced/Less/Fewer/Lower:* The product contains 25 percent less of a nutrient or calories per reference amount for individual foods. For example, a product may be labeled "25 percent less fat than our regular hot dogs."
- *More/Added/Fortified/Enriched:* An individual food must have at least 10 percent more of the daily value of protein, potassium, dietary fiber, or a vitamin or other mineral per reference amount compared to an appropriate reference food.
- *Percent Fat-Free:* May be used on products that meet the low fat definition. In addition, percent lean claims may be used for meat and poultry products that meet the low fat definition.

- *Lean:* The product contains less than 10 g fat, 4.5 g saturated fat, and 95 mg of cholesterol per reference amount and the 100 g for individual foods.
- *Extra lean:* The product has less than 5 g fat, 2 g of saturated fat, and 95 mg cholesterol per reference amount and per 100 g for individual foods.

Often, the brand name of a product will use a nutrient content claim—for example, "Lean Cuisine." The FSIS permits use of a defined nutrient content claim in a brand name if the product meets the definition of the term.

IX. HEALTH CLAIMS

The FDA has approved seven health claims and three more are under consideration. The FSIS has not published a proposal on health claims, but intends to harmonize with the FDA. For the time being, the FSIS will review health claims on meat and poultry products on a case-by-case basis.

In addition to health claims, both the FSIS and FDA have published proposed definitions for the term "healthy." Final rules have not yet been issued.

X. COMPLIANCE

The FSIS monitors labels in the marketplace as part of its regulatory mission. The agency will select products and then analyze the food to make sure the label information is accurate. If the nutrient values of the food do not match the values on the label, the FSIS will notify the company, which must either change the label or provide data that support the label information listed. Companies must keep records on their nutrient analyses and/or recipe calculations.

USDA has identified several ways that manufacturers can provide adequate documentation to support the nutrient values on the nutrition panel. The first is direct product analysis. While this method may be relatively expensive, many manufacturers will prefer to use direct products analysis—especially for complex, multi-component products or variable nutrients.

Although not required by the USDA, direct product analysis would make particular sense to support nutrient values for which claims are being made. Laboratory analyses might also be used to check database calculations. When laboratory results differ from calculations, comparisons are very valuable because they can be used to develop prediction algorithms or nutrient retention factors to account for such differences.

A second way manufacturers can provide adequate documentation is by using data base values. While the USDA does not plan to certify or approve data bases, the Department believes the USDA National Nutrient Data Bank is an appropriate data base to use for this purpose. Generally, nutrition labels using data base values, from the USDA data bank or from private data banks, will be subject to compliance review. The only exception is single-ingredient, raw meat and poultry products that are labeled using USDA's data bank. These data are based on considerable scientific research and adequately characterize the products. They have been screened and accepted by USDA's human nutrition information service, therefore, the USDA believes that it is appropriate to exclude products using these data for labeling purposes from compliance review if no claims are made. Furthermore, the nutrient content of these single-ingredient, raw foods is fairly predictable and does not require the same oversight as would be needed for more complex, processed products.

Third, manufacturers could use recipe calculations to determine nutrient values on labels. Data base vendors recommend using recipe analysis to calculate the nutrient content of multi-ingredient products as opposed to using data base values directly, which might only approximate the content of their specific products. Pizza is an example of a product for which a recipe calculation might make sense. Data base values for individual ingredients in the pizza could be used to calculate the nutrient contents for the label declaration.

Manufacturers can also use a combination of direct product analysis, data base values, and recipe calculations to arrive at a nutrition label. In fact, some combination of these may produce the most accurate and economical results in many instances.

The USDA believes that the use of data bases, especially computerized systems, offers a powerful tool for developing nutrient declarations. Significant cost savings and reduced turnaround times can be realized if accurate, calculated data can be used exclusively or to supplement limited analytical testing.

The efficiencies provided by data bases could result in savings to both large and small firms and ultimately to consumers. They also can lead to accelerated timetables for launching new and improved products and enhance market competitiveness. The USDA has published guidelines for the reliable use of databases to assist manufacturers.

XI. PRIOR LABEL APPROVAL

Meat and poultry product labels must be approved before they can be used on products in the marketplace. FSIS reviews about 170,000 label applications per year. That number may grow with the new nutrition labels. To deal with a steadily increasing workload and the concurrent impact on industry, the policy of approving every label in this way is being reviewed in order to make the current process more efficient, both in time and money costs. While a revised policy is being formulated via the regulatory process, FSIS will continue to examine and approve every label before the product can enter the marketplace. This includes all new nutrition labels.

XII. IMPACT ON INDUSTRY

Regulations requiring nutrition labeling for most processed meat and poultry products will obviously have an impact on industry. Essentially all manufacturers of processed meat and poultry products will need to revise currently approved labels to incorporate and/or amend nutrition information and to submit the revised labels to the USDA for approval. Manufacturers of single-ingredient, raw meat and poultry products are not required to include nutrition information on their labels. However, such manufacturers opting to include nutrition information on their labels also will be required to revise their labels to comply with requirements of the mandatory program.

The USDA no longer will review data beforehand to ensure that nutrition information is accurate. Manufacturers must assume responsibility for ensuring that nutrition information on their labels is accurate and must maintain records to support the nutrient values and any nutrition claims made on their labels. Such records will serve as the basis for compliance checks conducted by the USDA.

When appropriate and feasible, the USDA has tried to minimize the impact on industry. For example, the USDA developed the final nutrition labeling regulations with the goal of providing nutrition labeling on the maximum volume of packaged processed retail product while assuring that small businesses producing small volumes of specialty product for limited markets are not at risk of going out of business or having to discontinue producing such specialty products. The final regulations achieve this goal by providing an exemption for small businesses which alleviates much of the impact on industry.

The Federal Government continually tries to minimize the impact on industry of new regulations, but total elimination generally is not practical. The USDA recognizes that nutrition labeling will generate both costs and benefits and considered this in determining that nutrition labeling should be required. The USDA estimates that the regulations requiring nutrition labeling for most processed meat and poultry products will result in a net societal benefit of $1.5 billion during the first 20 years after the effective date of the regulations. After careful analysis of the factors, the USDA concluded that the impact on industry is reasonable and warranted when weighed against the benefits.

XIII. PUBLIC EDUCATION ACTIVITIES

The FSIS and FDA are coordinating a public and private sector "National Exchange for Food Labeling Education" (NEFLE) to help educate the American public about nutrition labeling. NEFLE includes representatives from government and health, consumer, industry, and educational groups. These groups bring varied resources to the massive public education campaign. The centerpiece of the exchange is the FDA/USDA Food Labeling Education Information Center, located in the Food and Nutrition Information Center of USDA's National Agricultural Library.

4

Effects of the Nutrition Labeling and Education Act on FTC Advertising Enforcement Policy

Deborah K. Owen and Joyce E. Plyler*

Federal Trade Commission, Washington, D.C.

After the Nutrition Labeling and Education Act (NLEA)[1] was enacted, attention swiftly focused on the Federal Trade Commission (FTC): what impact would the NLEA, and related rules promulgated by the Food and Drug Administration (FDA), have on advertising, where the Act did not directly apply? Since 1954, the FTC and the FDA have shared a Memorandum of Understanding (MOU) that allocates primary enforcement responsibility for food advertising to the Commission and primary responsibility for food labeling to the FDA[2]. The MOU has successfully contributed to a close, coordinated working relationship between the two agencies, a relationship that may be expected to continue as FDA implements the NLEA. But there is no doubt that the NLEA introduces a host of complex new issues that the Commission must address in the effort to vigorously enforce the laws against deception and unfairness, without unnecessarily interfering with legitimate free speech and the flow of useful information to consumers.

---

* Deborah K. Owen, a Commisioner of the Federal Trade Commission at the time this was written, is now a partner in the law firm of Arent, Fox, Kintner, Plotkin and Kahn. Joyce E. Plyler, formerly an advisor to Commisioner Owen, is an attorney in the FTC's Office of General Counsel. The views expressed herein are those of the authors and do not necessarily reflect the views of the Commission or any commissioner.

The Commission addresses a number of these questions in its "Enforcement Policy Statement on Food Advertising," issued in May, 1994, which is reprinted in the appendix of this book. The Policy Statement provides guidance concerning how the Commission is likely to respond to claims in food advertising that do or do not comply with FDA regulations.

It is a valuable starting point for learning the Commission's general policy on many of the major issues in this area; but because FDA's rules are so vast, the nature of advertising is dynamic, and consumer perceptions everchanging, the Policy Statement does not answer every question that may arise. A complete understanding of how the FTC is likely to respond to particular advertising claims is best obtained from review of the underlying FTC law. This includes the FTC's enabling statute, Commission interpretations, and legal precedent in food advertising and other cases.

This chapter will review these and other sources of guidance and highlight some of the more difficult issues that will face the Commission as it attempts to incorporate NLEA principles into its enforcement program. It will explain how statutory differences, and practical differences between labeling and advertising, make absolute uniformity unlikely (and even undesirable), although the prospects for consistency remain high. Key points from the Policy Statement will also be noted.

I. STATUTORY FRAMEWORK

The statutory mandates bestowed by the Congress on the FTC and the FDA are complementary, but different. The FTC is authorized under Section 5 of the Federal Trade Commission Act to prevent unfair or deceptive trade practices in or affecting commerce[3]. Section 5 and Sections 12 and 15 of the FTC Act, pertaining to false advertisements, are the primary statutory provisions governing how the agency can enforce the law with respect to food advertising[4]. In addition, the Commission is sensitive to the First Amendment's protection of free speech, including truthful, nondeceptive, commercial speech[5].

While the FTC is limited to prohibiting claims that are unfair or deceptive, the FDA operates under a considerably broader statutory scheme. Like the FTC, the FDA may prohibit false or misleading claims on food labels, but the NLEA also gives it the authority to broadly require the affirmative disclosure of information, beyond those disclosures necessary to avoid material omissions. This consumer education function has played an important role in food marketing over the years and has been greatly expanded by the NLEA. In contrast, the FTC's authority to affirmatively require disclosures is limited to situations where disclosure is necessary to remedy or prevent unfairness or deception. In short, the Commission's authority to mandate optimal speech is much more circumscribed than that of the FDA.

The Commission has issued three general policy statements regarding deception, unfairness, and substantiation for claims. These three policy statements identify specific criteria that the Commission follows when determining whether a law violation has occurred. Briefly, they provide as follows: the deception policy statement states that deception occurs when there is a representation or omission that is likely to mislead consumers acting reasonably under the circumstances, to the consumers' detriment[6]. Unfairness exists where a practice causes substantial injury that is not outweighed by countervailing benefits, and that cannot be reasonably avoided[7]. Finally, under the advertising substantiation policy statement, advertisers must have a reasonable basis for objective product claims before the claim is made. Whether a reasonable basis exists depends on a variety of factors, including the type of claim or product and the amount of substantiation experts in the field believe is reasonable[8].

II. FUNCTIONS OF ADVERTISING AND LABELING

As both competitive tools and communication devices, advertising and labeling are the primary means by which marketers convey product attributes to consumers. Their ability to effectively communicate such information is significant. A landmark study by the Commission's Bureau of Economics found that advertising and labeling claims for ready-to-eat cereals increased consumer awareness of the potential nutritional benefits of fiber in the diet[9]. In fact, they found that for some segments of the population, advertising and labeling claims did a better job of communicating beneficial information than did other information sources, such as public service campaigns by government and private groups.

Nonetheless, as the Commission[10], its staff[11], and FDA Commissioner David Kessler[12] have all noted in the past, there are important differences between advertising and labeling that warrant somewhat different, although congruent, approaches to preventing deceptive claims. In advertisements, visual techniques and the order and positioning of information can be critical to the messages that consumers receive. While these factors are also present in the labeling context, their impact may be diminished because consumers have more time to closely examine the entire label.

Another difference relates to the costs involved in disseminating advertisements and creating labels, and the time that is either available or that consumers tend to devote to the task of reading and absorbing the information presented. For example, it would be virtually impossible to communicate coherently in a 15- or 30-second television ad all of the information that could be placed on a fixed label or in a package insert. Small print may be incomprehensible on a TV ad, but is more readable, and more likely to be read, on a label[13].

A third significant factor concerns the frame of reference that consumers use to interpret ads and labels. Consumers generally do not expect advertisements to contain the same degree of information as labels. While they may expect labels to contain more detailed nutritional information, they do not necessarily expect advertising to have the same detail, as long as any omissions are not material and therefore deceptive[14].

Of course, the fact that advertising and labeling serve some different functions does not mean that false claims or deceptive half-truths should be tolerated in either medium. It does, however, indicate that all of the same information need not appear in both media for consumers to be given accurate and beneficial information, and that it may be inappropriate to apply identical rules to differing media.

III. GENERAL POLICY ISSUES

In fact, applying absolutely identical rules to both labeling and advertising would be unworkable. Some requirements, such as disclosing on food labels the content of a list of specified nutrients and related information (the "Nutrition Facts" panel)[15], would simply be impracticable and of questionable benefit in advertisements. On the other hand, an advertising policy that did not take into account the impact of FDA's labeling rules on consumer understanding would ignore a major factor affecting whether an advertising claim was unfair or deceptive. Thus, the FTC has committed to harmonizing its policies with those of FDA to the extent possible under existing law[16]. The following sections give an overview of the more significant of FDA's labeling regulations, and explain the issues the FTC can be expected to face in applying them to advertising.

A. FDA Primary Labeling Restrictions and Prohibitions

The NLEA regulations most relevant to FTC advertising enforcement policy address two broad classes of claims—nutrient content claims and health claims. Nutrient content claims are those which expressly or implicitly characterize the level of nutrients in foods[17], such as fat, cholesterol, fiber and sodium. They fall into two categories—absolute nutrient content claims (e.g., "low fat," "high fiber") and relative nutrient content claims (e.g., "less fat"). Health claims are those which characterize the relationship of any substance to a disease or health-related condition[18].

1. Nutrient Content Claims

The NLEA and FDA's rules closely control the manner in which nutrients may be characterized on food labels. Several of these restrictions raise significant issues in the advertising context. Labels may not include a nutrient content claim unless it has been authorized by regulation[19]. Only descriptive terms defined by FDA, or approved synonyms, may be used[20]. All nutrient content claims must be accompanied by a "referral statement" (e.g., "See side panel for nutrition information.) [21] The statement must refer specifically to total fat, saturated fat, cholesterol and sodium if the amount of those nutrients in the food exceeds a certain level. For example, if a product contains fat, the statement must say, "see side panel for information about total fat and other nutrients" [22].

Some claims trigger additional required claims. For example, all fiber claims trigger a disclosure of the level of total fat, unless the food is "low fat" [23]. In addition, some claims for allowed nutrients are impermissible if the food contains certain levels of other nutrients. For example, claims characterizing the level of cholesterol are not authorized unless the food contains less than 2 grams of saturated fat[24].

Comparative claims may be made only if the differences in the nutrient content between the labeled food and the reference food meet a specified threshold. For example, to say "lower in calories," the labeled food must contain at least 25% fewer calories than the reference food[25]. The foods that may be referenced by relative claims are also restricted. References may be made only to "similar" foods, e.g., potato chips to potato chips, for the descriptors, "light," "reduced," "added," "fortified," and "enriched" [26]. Claims of "less" and "more" can reference similar foods, or dissimilar foods within the same product category, e.g., potato chips to pretzels[27]. Foods bearing relative claims must also state the percentage of the nutrient by which the labeled food differs from the reference food (e.g., "50% less fat than brand X") as well as quantitative information comparing the amount of the nutrient in the labeled food with that in the reference food[28].

2. Health Claims

Health claims may be made only if explicitly permitted by FDA[29]. The NLEA requires that approved claims be supported by significant scientific agreement among qualified experts[30]. Labels on foods that contain "disqualifying levels" of certain nutrients, as defined by FDA, may not make any health claims[31]. In addition, health claims cannot be made for foods that do not contain at least 10% of the Daily Value of at least one of six specified nutrients prior to nutrient addition[32]. Additional requirements apply to particular approved claims. For example, a fiber/cancer claim may be made only if the food is: 1) at least a "good source" of fiber; 2) a grain, fruit, or vegetable product; and 3) low in fat[33]. Authorized health claims may be made only in the context of specified disclosures, which may be given through model statements, written by the FDA, or which may use wording different from the model statements as long as all of the required information is stated[34].

B. Application to Advertising

Although much of FDA's labeling rules easily can be accommodated within the FTC's advertising enforcement authority, three general types of regulation present special concerns: 1) outright prohibitions on certain claims; 2) claims that are allowed only if approved language is used; and 3) disclosure requirements when permitted claims are made.

1. Outright Prohibitions

As noted above, under the NLEA, labels may not make any unapproved health claim. In addition, even approved health claims are prohibited if the food does not meet certain standards, i.e., the food must contain required levels of certain "good" nutrients, and must not contain disqualifying levels of "bad" nutrients. For relative nutrient content claims, the nutrient content difference between the labeled and reference foods must meet a specified percentage.

Statutory and constitutional limitations would appear to make the wholesale application of these prohibitions to advertising problematic. Because the FTC's enforcement powers are limited to instances where it finds that advertising is unfair, deceptive, or unsubstantiated, it is difficult to conceive of how the FTC could justify a rule prohibiting claims across the board to the same degree as FDA. On the other hand, in individual cases, depending on the context, analysis of FTC law could very well lead to results that prohibit the same claim in advertising as would be prohibited by the FDA on labels. The FTC Policy Statement indicates that, while the FTC would prohibit many of the same claims as the FDA, it would not take enforcement action in some instances of very carefully qualified claims, as explained below[35].

Certain of FDA's prohibitions are based on substantiation issues, such as that FDA must first rule on whether a health claim meets the NLEA's substantiation standard before the claim can be made. The FTC also imposes a substantiation standard on all advertising claims (though it does not require prior approval). In fact, many of the FTC's recent food advertising cases have alleged that the respondents did not have adequate substantiation for their express or implied health claims. Three cases involved health benefit claims for cooking oils. The Commission charged that one company did not have a reasonable basis for its claims that eating olive oil reduces blood pressure and blood sugar, and that olive oil reduces cholesterol more than other home cooking oils and is healthier for the heart[36]. Another company was alleged to have made unsubstantiated claims about olive oil's superiority to vegetable oil[37], and the manufacturer of a leading corn oil and margarine was charged with unsubstantiated claims that adding its products to the diet, without other dietary changes, would reduce serum cholesterol levels[38]. In a similar type of case, the Commission charged the maker of a rice bran cereal with failing to have a reasonable basis for claims that its cereal reduced cholesterol and would improve the ratio of HDL to LDL cholesterol in the blood, and that the cereal helps reduce the risk of heart disease[39].

Questions have arisen over whether the FTC's substantiation standard is as stringent as the NLEA standard. The FTC's substantiation requirement is rigorous. In cases involving food and health claims, the FTC has required substantiation to consist of competent and reliable scientific evidence[40]. The Commission has noted that the two standards have many similarities:

> Both standards require a high level of scientific support and have many common elements. Both standards look to the claim that is made in assessing whether there is adequate scientific support for that claim. Both standards require the use of the accepted methodologies of the relevent professions in assessing the reliability and accuracy of the evidence supporting the claim. Both standards require that the determination be made in the context of the entire body of available evidence.
>
> The recent actions by the Commission in the Pompeian, Bertolli and Pacific Rice Products cases demonstrate that advertisements cannot overstate available support for health claims and that "preliminary" or "inconclusive" science is simply not sufficient to substantiate claims under the FTC standard[41].

The Policy Statement also notes that the FTC and FDA standards share many elements. The Commission further clarified in the Statement that it does not require scientific "consensus" to support claims (nor does the FDA), and that it considers the FDA's "significant scientific agreement" standard "to be the principal guide to what experts in the field of diet-disease relationships would consider reasonable substantiation" for unqualified health claims[42].

It is important to note that, when questions of scientific substantiation arise, the Commission has traditionally given great weight to the findings of experts at FDA, as well as other agencies with scientific expertise, regarding the state of science in a particular area. This reliance has resulted in a great deal of consistency historically between FDA and the FTC.

The FTC staff constantly consults with FDA scientists concerning the evidence available to support claims made in advertising. There are numerous examples of cases in which the FTC has accorded substantial weight to even tentative or interim FDA scientific determinations. For example, in Thompson Medical[43], one of the Commission's most significant advertising cases, the FTC relied in part on FDA's interim determination that insufficient evidence existed to classify the product in question as safe and effective. In Removatron[44], the Commission gave substantial weight to an FDA panel's conclusion that the device marketed by the company for permanent hair removal was not proven effective. The Commission also challenged claims made by Sterling Drug for two separate analgesic ingredients that were the subject of FDA review procedures[45].

These cases illustrate the high value that the FTC places on ensuring consistency with FDA's substantive or scientific determinations. Moreover, they illustrate that the FTC can also move against allegedly unsubstantiated claims on a case-by-case basis even where the FDA has not completed its own regulatory process.

The FTC's reliance on, and deference to, FDA scientific determinations is also reflected in a number of recent FTC consent orders involving food as well as other products. For example, the Bertolli order provides that any claims that are specifically approved for labeling by the FDA pursuant to the NLEA will be deemed to have a reasonable basis for purposes of that order[46]. The Nestle order contains a "safe harbor" for all representations that are specifically permitted in labeling regulations promulgated by the FDA pursuant to the Federal Food, Drug and Cosmetic Act[47]. Similar order provisions are in other recent cases involving both health and nutrient content claims[48]. In addition, the Commission issued a final consent order against Revlon, Inc., involving sun protection claims for a sunscreen product[49]. The order specifies that if a tentative final or a final standard promulgated by FDA establishes that a representation covered by the order is supported by scientific evidence acceptable to the FDA, then this will also constitute adequate substantiation under the FTC order.

These cases, and the Policy Statement, indicate two things. First, if a claim has been approved by FDA, and complies with FDA's regulations, it is likely that such a claim would also be acceptable to the FTC, if used in the same context as that approved by FDA. Second, if a health claim has not been approved by FDA, it is likely that the FTC would scrutinize with great caution the scientific evidence surrounding such a claim, and would certainly utilize the expertise of FDA's scientists, before determining whether such a claim should be allowed in advertising. The Policy Statement recognizes that the Commission is likely to reach the same conclusions as FDA as to whether there is adequate scientific support for unqualified health claims. However, the Policy Statement also notes that in "certain limited instances" carefully qualified health claims may be permitted under FTC law even though they are as yet unapproved by the FDA[50].

Some of the FDA's absolute prohibitions appear to be based on the belief that certain claims, even if qualified, would not meet the objectives of the NLEA. Thus, for example, a "no cholesterol" claim may not be made for foods with specified levels of saturated fat. FTC cases similarly reflect a concern that even truthful representations about nutrients may lead to deceptive impressions. In a recent case, the FTC charged the maker of Klondike frozen dessert bars with making false and unsubstantiated claims for its Klondike Lite bar, which were represented in advertisements to be "low in cholesterol" [51]. The FTC charged that, among other implied false claims, the ads represented that the bar would cause little or no increase in serum cholesterol levels. However, according to the complaint, because the bar contained a substantial amount of saturated fat, it would in many cases cause a substantial increase in serum cholesterol. This case illustrates that the FTC may impose restrictions similar to those of FDA when presented with deceptive implied representations arising from truthful express statements.

One continuing issue in such cases, however, is whether the FTC will allow an advertisement to make claims that would be prohibited on labels, if the claim is adequately qualified in the advertisement. In a 1975 case involving advertising statements about the scientific evidence surrounding the effects of egg consumption on cholesterol and health, *FTC. v. National Commission on Egg Nutrition (NCEN)*[52], an appeals court limited the breadth of a preliminary injunction provision that would have prohibited NCEN from presenting its side of the debate over the scientific evidence surrounding the link between dietary cholesterol and certain health risks. While the court ruled that NCEN's claims were deceptive, and upheld prohibitions of certain claims, the court also determined that the Commission should not prohibit NCEN from espousing its side of the scientific debate, as long as NCEN disclosed that there was substantial contrary evidence[53]. NCEN thus reflects the principle that certain claims may be deceptive standing alone, but, with proper qualification, may be permissible[54].

A similar issue could arise in the following context. The FDA would prohibit a "less fat" claim for a food that was not at least 25% lower in fat than the reference food. But what if an advertisement includes the claim, "20% less fat than the other leading brand"? From the FTC's traditional perspective, whether this claim is deceptive or unfair would depend on the circumstances of the case and the context of the claim. Does the ad imply a benefit greater than is justified?[55] Is the qualification conspicuous enough so that consumers will read and understand it?[56] Do other aspects of the ad imply a difference greater than 20%? Again, though it may be problematic, from a legal standpoint, for the FTC to impose a policy that mirrors FDA's, many individual cases may present facts leading to similar results. The FTC's critical challenge, even if FDA's rules are not duplicated exactly, will be to ensure that advertising claims are truthful, substantiated, and not deceptive and that qualifying information necessary to avoid deception is adequately conveyed.

The Policy Statement highlights several principles that the Commission is likely to apply when considering qualified claims. For example, carefully qualified health claims may be permitted if they "convey clearly and fully the extent of the scientific support [for the claims]." The Policy Statement also cautions that the Commission will be "especially vigilant" in ensuring that claims are presented in such a way that consumers understand not only the extent of support but also "the existence of any significant contrary view within the scientific community" [57]. The Policy Statement observes, however, that the Commission is likely to find misleading even carefully qualified health claims that are inconsistent with the larger body of evidence. The Policy Statement also addresses qualifications for comparative nutrient content claims: such claims should make clear the basis for the comparison, should not overstate the significance of a nutrient difference, and should not mislead consumers about the absolute level of a nutrient[58].

2. Claims Allowed Using Only Approved Language

A second category of FDA labeling regulation concerns claims that are permitted, but only through the use of approved language. The best example is FDA's requirement that only defined nutrient descriptors may be used for nutrient content claims. Concomitantly, terms that have not been defined, including unapproved synonyms of defined terms, are prohibited. As explained in this section, the Commission's intent with respect to terms that FDA has defined is clear, but the issues surrounding unapproved terms, much like the issues with respect to unapproved health claims, are more complex. Even prior to FDA's definitions, the Commission has been active in challenging certain terms and nutrient content claims as deceptive, as will be seen in a few sample cases discussed at the end of the section.

a. Applying FDA's Nutrient Descriptor Definitions

The Commission stated in its Policy Statement that it will apply FDA's definitions for absolute nutrient content terms when they are used in the same context in advertising[59]. One rationale for this determination is that when a term has been defined by a government agency, consumers may come to rely on that definition whenever the term is used; different definitions, therefore, may lead to consumer confusion and potential deception. *FTC v. Brown & Williamson Tobacco Corp.*[60] focused on an analogous scenario: the FTC's own rating system for tar and nicotine content in cigarettes that had been used voluntarily for some time by cigarette manufacturers as the basis for advertised tar and nicotine ratings. The appeals court concluded that consumers had come to rely on the ratings achieved by the FTC method. It also concluded that consumers had no innate ability to assess a statement of tar content alone, a figure that was significant only as a measure of comparing the tar content in different brands. These conclusions supported the court's affirmance of the lower court's finding of deception, based on three factual premises: "1) consumers rely on tar ratings to make comparative assessments of the health effects of cigarettes;

2) Barclay cigarettes deliver disproportionately more tar than other similarly rated cigarettes; and 3) disclaimers explaining the difference will prove ineffective" [61]. While the facts in Brown & Williamson were unique and somewhat different than might be presented in a case involving food labeling and advertising issues under the NLEA scheme, it nevertheless illustrates judicial recognition that consumer perceptions can be influenced by a government regulatory scheme.

In December, 1991, the FTC filed a complaint in federal district court seeking a preliminary injunction against the maker of Sporicidin, a disinfectant used in medical applications[62]. Under a comprehensive regulatory scheme, the Environmental Protection Agency regulates the labeling of chemical germicides such as Sporicidin. In its complaint, the Commission recognized that EPA had registered Sporicidin as a "sterilant" in its undiluted form and in a 1:8 solution of one part Sporicidin to 7 parts water. However, it also pointed out that EPA had not registered Sporicidin as a sterilant in a 1:16 dilution of one part Sporicidin to 15 parts water. Nevertheless, according to the complaint, The Sporicidin Company had claimed in advertising that Sporicidin was an effective sterilant in a 1:16 solution. The matter was eventually resolved by consent agreement.

The Sporicidin case illustrates how the FTC has looked to the definitions and regulations of another agency as a factor in determining whether to allege deception under the FTC Act. In addition to relying on the EPA's criteria for registration as a "sterilant" or "sporicide," the FTC examined guidelines established by the Centers for Disease Control of the U.S. Public Health Service, which the Commission's complaint recognized as those "most universally consulted by the medical profession." Those guidelines referenced EPA's registration scheme in establishing recommended procedures for certain levels of disinfection.

b. Use of Unapproved Synonyms

The Commission's Policy Statement does not automatically preclude the use of unapproved synonyms for approved terms, such as "just a pinch" of sodium. However, it does caution that the Commission "will examine advertising to ensure that claims that characterize the level of a nutrient, including those using synonyms that are not provided for in FDA's regulations, are consistent with FDA definitions" [63]. In scrutinizing claims using synonyms, the Commission would consider such questions as how consumers interpret those synonyms, and whether any messages conveyed were influenced by FDA's comprehensive regulatory scheme and consumer understanding of it.

The FTC may hesitate to subscribe to a preapproved list of terms based on its experience in a recent proceeding involving tire advertising. The Commission was confronted with the issue of whether advertisers of used tires should be allowed to use only certain preapproved terms in their advertising to clearly distinguish used tires from new tires. FTC guidelines issued in 1967 specified that all advertisements for used tires should designate the tires as "retreads" or "retreaded."[64] No other terms were allowed. Responding to public comment that alternate terms could convey that advertised tires were used and not new, the Commission amended the guidelines to drop the statement that such tires must be designated as retreads. Explaining its decision, the Commission expressed the belief that it was preferable to avoid limiting the industry to only two terms because such limitations "may have the effect of inhibiting innovation by limiting the ability of manufacturers to distinguish their products in the marketplace in a nondeceptive manner" [65]. The Commission considered, but rejected, another amendment that would have expanded the list of acceptable terms, concluding that the record did not show that it was necessary to impose a restricted list of terms to avoid consumer deception or to inform consumers adequately of the nature of their purchase. The Commission concluded that manufacturers should have the flexibility to use any non-deceptive terms that clearly and conspicuously disclosed the fact that the advertised tires were not new.

Certainly, issues surrounding the advertising of used tires are different from those involving nutrient and health claims for foods. Nevertheless, the Commission's experience with the 1967 tire guidelines shows that a list of terms believed to be sufficient as descriptors today may not be adequate tomorrow[66].

c. FTC Cases Involving Nutrient Content Claims

It is clear that the Commission will continue to closely scrutinize all nutrient content claims in their advertising contexts for potential unfairness or deception. This commitment is reflected in a number of cases involving nutrient content claims, of which the following are merely a few examples. In Kraft, Inc.[67], the Commission considered advertisements stating that Kraft Singles cheese slices were made from five ounces of milk, and specifically mentioning the nutritional benefits of calcium. After full adjudication, the Commission found that Kraft had misrepresented that each Singles slice contains the same amount of calcium as five ounces of milk. The Commission also found that, in one set of advertisements, Kraft had misrepresented that Kraft Singles contain more calcium than most imitation slices.

Nestle Food Company, (NFC), formerly Carnation Company, settled charges that it falsely advertised that Carnation Coffee-Mate was "low-fat" when consumed in amounts used on cereal, over fruit, or in cooking[68]. The complaint noted that Coffee-Mate is not a low-fat product when consumed in those amounts and that, in fact, a half-cup serving has nearly twice the fat in an identical serving of whole milk. The consent order prohibits the company from misrepresenting the amount of total fat, saturated fat, or cholesterol in any milk product or non-dairy substitute relative to the serving size depicted in its ads or promotional material.

In an adjudicated case decided in September, 1994, the Commision found that Stouffer Foods Corp.[69], had made deceptive "low sodium" claims for its "Lean Cuisine" frozen entrees. The Commision concluded that several elements of the advertisements conveyed a low sodium claim, including the headlines, the language used, and the footnotes, which failed to disclose adequately that one gram, the measure of sodium used in the ads, is equivalent to 1000 milligrams, the commonly used unit of measurement for sodium.[70]

Significantly, two provisos of the Commision's final order against Souffer are hinged on FDA regulations. First, if Stouffer makes any nutrient content claim defined by the FDA, compliance with the FTC order will be determined by the FDA rule's qualifying amount for that term. Second, the order allows Stouffer to make in advertisements any content representation that is specifically permitted in labelling by the FDA, for the advertised serving size. The order applies to claims for any frozen food product.

Adjudicated cases, such as Kraft and Stouffer, provide considerable guidance to the public regarding the Commission's enforcement stance on the issues raised therein, because they often result in a written Commission opinion, which may explain in more detail the Commission's rationale for its conclusions that a violation occurred and why a particular remedy is warranted[71]. Cases resolved by consent agreement, such as Nestle and others cited within this chapter, are also important sources of guidance because they illustrate the types of claims the Commission is likely to challenge and the remedies it is likely to seek. While the FTC's Food Advertising Policy Statement is a valuable source of general guidance on how the FTC will apply certain of FDA's rules to advertising, individual cases will continue to be the most important source for information about the FTC's response in specific circumstances.

3. Required Disclosures

The third challenging category of labeling regulation involves triggered disclosures, that is, disclosures that must be included if permitted claims are made. Examples within this category are the requirements that referral statements accompany nutrient content claims; that certain nutrient content claims trigger disclosure of other nutrients (e.g., unless a product is "low fat" a fiber claim triggers a disclosure of the level of total fat); and that health claims must be made in the context of the information included in model statements.

Under FTC law, affirmative disclosure requirements are limited to those which are necessary to prevent a practice or representation from being deceptive or unfair. Thus, if a claim is likely to be interpreted by consumers in a deceptive manner, the Commission has required disclosures that would correct any misleading impressions. A recent example in the context of food advertising is the FTC's consent order against Campbell Soup[72]. In its complaint, the FTC charged that Campbell had represented that most of Campbell's soups are low in fat and cholesterol and, as part of a diet low in fat and cholesterol, may help reduce the risk of some forms of heart disease. The complaint further alleged that Campbell's failure to disclose that its soups are high in sodium was deceptive, given that diets high in sodium may increase the risk of heart disease. Under the consent order, Campbell must disclose the sodium content of its soups and the recommended maximum daily intake of sodium in any advertisement making a connection between heart disease and a soup that contains more than 500 milligrams of sodium per eight-ounce serving, or more than the amount set by FDA's final labeling regulations. The order also requires Campbell to have competent and reliable scientific evidence to substantiate any representation about a connection between its soups and a reduction in the risk of heart disease.

The Policy Statement recognizes that some of the disclosures required by FDA are necessary to fulfill the NLEA's educational goals, but would not necessarily be required under the FTC Act to prevent deception. These would include both the requirement that nutrient content claims be accompanied by a referral to the nutrition panel, and that health claims be presented in the context of the information contained in model statements. However, the Policy Statement also cautions that disclosures may be required if necessary to prevent consumers from being misled about the significance of diet as one of the factors in preventing a referenced condition, in the context of health claims. In addition, the Policy Statement notes that disclosures of the presence of nutrients at levels considered by FDA to increase the risk of disease may be required in ads making nutrient content claims if the context of the ad as a whole conveys the impression that the food makes only a positive contribution to diet, or that the food contains no risk-increasing levels of nutrients. The Policy Statement further observes that consumers often link nutrients to each other, for example, cholesterol and saturated fat, such that a claim about the one is likely to give an impression about the other. Misimpressions that arise in such circumstances would be deceptive if not corrected through adequate disclosures. In this category, the Commission highlighted cholesterol, saturated fat and fiber for special scrutiny.

IV. CONCLUSION

The NLEA regulations are based on statutory authority and practical considerations that look directly to labeling, not advertising. Even if the wholesale application of the NLEA rules to advertising were permissible by statute, such a result would be difficult, if not impossible, to accomplish practically. FDA is authorized not only to prohibit false and deceptive claims, but it is charged with implementing a Congressionally mandated policy of promoting consumer education of diet and health relationships through food labeling. The FTC's authority under the FTC Act and the Constitution, as reflected in its policy statements and case law, is limited to prohibiting claims that are unsubstantiated, deceptive, or unfair. This includes the concept that additional information must be disclosed if its absence renders the primary claim deceptive or unfair[73]. In short, the FTC Act authorizes the FTC to prohibit unfair and deceptive advertising, not necessarily to mandate optimal advertising[74].

While these fundamental principles have the potential to lead to some limited, and legitimate, differences in the way labels and advertisements are regulated, they do not necessarily lead to contradictory results. In fact, the FTC's application of its deception and substantiation policies have already prompted cases based on principles that are consistent with the NLEA[75]. Moreover, FTC determinations of unfairness and deception can be greatly influenced by the findings of agencies with special expertise, such as the FDA, as well as by the impact of such agencies' regulations on consumer perceptions. Because their roles in the health claims field are so interrelated, the FTC has always maintained a close working relationship with the FDA. This relationship can be expected to continue, and to continue with increased vigilance as the FTC attempts to harmonize to the greatest degree possible the rules FDA has promulgated under the NLEA. The FTC's recent Policy Statement on Food Advertising reflects this commitment.

V. ENDNOTES

1 Pub. L. No. 101-535, 104 Stat. 2353 (1990).

2 Working Agreement Between FTC and Food and Drug Administration, 4 Trade Reg. Rep. (CCH) ¶ 9,850.01 (1971).

3 15 U.S.C. § 45.

4 The FTC enforces a number of specific statutes other than the FTC Act, some relating to its antitrust mission and others that have primarily a consumer protection objective. However, the Commission's authority in the area of food advertising stems from the FTC Act alone.

5 The FDA also must consider First Amendment issues. While the FTC has not indicated an opinion one way or another, the FDA at least appears to consider labeling a "distinct category of communications" that differs from advertising for purposes of First Amendment analysis. The FDA has suggested that, "[u]nlike 'advertising pure and simple' labeling does not fall clearly within the bounds of commercial speech." Federal Register Notice Announcing FDA Final Rules, 58 F.R. 520, 2525 (Jan. 6, 1993) (citation omitted). FDA believes that labeling may have only "an incidental effect on speech in a narrowly defined area, food labeling." *Id.* Although it does not concede that labeling should be analyzed as commercial speech, which advertising clearly is, the FDA has concluded that, even if

labeling is considered commercial speech, its regulations are constitutional. FDA referred extensively to Congressional intent underlying the NLEA in reaching this conclusion. *See id.* at 2526. As noted in the text, the NLEA gives broader regulatory authority to the FDA with respect to regulation of speech concerning food labeling than the FTC Act gives to the FTC.

6 Statement of Policy on Deception, Letter from the Chairman of the Federal Trade Commission to Chairman John D. Dingell (Oct. 14, 1983), *Cliffdale Associates, Inc.*, 103 F.T.C. 110, 174, *appeal dismissed sub nom., Koven v. FTC*, No. 84-5337 (11th Cir. Oct. 10, 1984).

7 Statement of Policy on the Scope of the Consumer Unfairness Jurisdiction, Letter from the Federal Trade Commission to Senators Ford and Danforth (Dec. 17, 1980), *International Harvester Co.*, 104 F.T.C. 949, 1070 (1984).

8 Policy Statement Regarding Advertising Substantiation, *Thompson Medical Co.*, 104 F.T.C. 648, 839 (1984), *aff'd*, 791 F.2d 189 (D.C. Cir. 1986), *cert. denied*, 479 U.S. 1086 (1987).

9 Ippolito, P. and Mathios, A., *Health Claims in Advertising and Labeling: A Study of the Cereal Market*, Bureau of Economics Staff Report, Federal Trade Commission, August 1989.

10 Letter from the Federal Trade Commission to the Honorable Slade Gorton (September 25, 1991).

11 Comments of the Bureaus of Consumer Protection and Economics of the Federal Trade Commission, Submitted to the Food and Drug Administration Department of Health and Human Services in Response to a Request for Comments on its Proposal to Amend the Rules Governing Food Labeling; Health Messages and Label Statements; Reproposed Rule (May 18, 1990).

12 *See*, Kessler, *The Federal Regulation of Food Labeling*, 321 The New England Journal of Medicine 717, 723 (1989); Address by David A. Kessler, M.D., 20th Anniversary Conference of the Center for Science in the Public Interest, p. 5 (June 6, 1991).

13 The issue of whether mandated broadcast disclosures are comprehensible to consumers is not new to the Commission. Most recently, this issue prompted a partial dissent to consent order requirements in several FTC cases involving weight-loss claims. Responding to complaint allegations of deceptive weight loss maintenance claims, the Commission required in consent orders that future advertisements making maintenance claims include a series of complicated disclosures regarding the percent of weight loss maintained, the duration of that loss, and the representativeness of the triggering claim. Citing scholarly research questioning the effectiveness of complex broadcast disclosures, Commissioner Owen expressed concern that the numerically intricate disclosures required by the orders would not communicate the desired information. Alternatively, Commissioner Owen would have required that broadcast advertisements include a more concise general disclosure and alert consumers that additional information was available at the point of sale, coupled with a mandate that respondents give more detailed information in writing to consumers prior to execution of a contract. *See*, e.g., Statement of Commissioner Deborah K. Owen, Concurring in Part and Dissenting in Part, *Jason Pharmaceu-*

ticals, Inc., C-3392 (Sept. 16, 1992); *National Center for Nutrition*, C-3393 (Aug. 10, 1992); *Sandoz Nutrition Corp.*, C-3394 (Aug. 10, 1992).

14 *See* "The Federal Trade Commission's Consumer Protection Priorities," Remarks by Commissioner Deborah K. Owen Before the Grocery Manufacturers of America, Annual Executive Conference, June 17, 1991.

15 21 C.F.R. § 101.9 (1993).

16 Enforcement Policy Statement on Food Advertising, May 1994, at 3.

17 21 C.F.R. § 101.13(b).

18 21 C.F.R. § 101.14(a)(1).

19 (17).

20 (17).

21 21 C.F.R. § 101.13(g).

22 21 C.F.R. § 101.13(h).

23 21 C.F.R. § 101.54(d).

24 21 C.F.R. § 101.62(d).

25 21 C.F.R. § 101.60(b)(4)(i).

26 21 C.F.R. § 101.13(j)(1)(i)(B).

27 21 C.F.R. § 101.13(j)(1)(i)(A).

28 21 C.F.R. § 101.13(j)(2)(i) and (iv).

29 21 U.S.C. § 343(r)(1)(B).

30 The NLEA requires that claims are to be authorized "only if the Secretary determines, based on the totality of publicly available scientific evidence (including evidence from well-designed studies conducted in a manner which is consistent with generally recognized scientific procedures and principles), that there is significant scientific agreement, among experts qualified by scientific training and experience to evaluate such claims, that the claim is supported by such evidence." 21 U.S.C. § 343(r)(3)(B).

31 21 U.S.C. § 343(r)(3)(A)(ii); 21 C.F.R. § 101.14(a)(5).

32 21 C.F.R. § 101.14(e)(6).

33 21 C.F.R. § 101.76(c)(2)(ii).

34 *See* 21 C.F.R. § 101.14(d)(2) and 58 Fed. Reg. 2510.

35 See (57) and (58), *infra*, and accompanying text.

36 *Bertolli USA, Inc.*, C-3396 (Aug. 17, 1992).

37 *Pompeian, Inc.*, C-3402 (Oct. 27, 1992).

38 *CPC International, Inc.*, C-3321 (Jan. 2, 1991).

39 *Pacific Rice Products, Inc.*, C-3395 (Aug. 17, 1992).

40 "Competent and reliable evidence" is generally defined as "tests, analyses, research, studies or other evidence based on the expertise of professionals in the relevant area, that have been conducted and evaluated in an objective manner by persons qualified to do so, using procedures generally accepted in the profession to yield accurate and reliable results." *See*, e.g., *Gracewood Fruit Co.*, C-3470 (Oct. 26, 1993).

41 Prepared Statement of the Federal Trade Commission Before the Transportation and Hazardous Materials Subcommittee, Energy and Commerce Committee, U.S. House of Representatives, Nov. 21, 1991 (footnotes omitted).

42 (16) at 19.

43 *Thompson Medical Co.*, 104 F.T.C. 648 (1984), *aff'd*, 791 F.2d 189 (D.C. Cir. 1986), *cert. denied*, 479 U.S. 1086 (1987).

44 *Removatron International Corp.*, 111 F.T.C. 206 (1988), *aff'd*, 884 F.2d 1489 (1st Cir. 1989).

45 *United States v. Sterling Drug, Inc.*, No. CA90-1352 (D.D.C. June 21, 1990).

46 (36)

47 *Nestle Food Co.*, C-3365 (Jan. 21, 1992).

48 *See*, e.g., *Gracewood Fruit Company*, C-3470 (Oct. 26, 1993); *The Isaly Klondike Co.*, C-3412 (Jan. 28, 1993); *Campbell Soup Co.*, D-9223 (Aug. 18, 1992).

49 *Revlon, Inc.*, D-9231 (Nov. 17, 1993).

50 (16) at 19-20.

51 *The Isaly Klondike Co.*, C-3412 (Jan. 28, 1993).

52 517 F.2d 485 (7th Cir. 1975) *appeal after remand*, 570 F.2d 157 (7th Cir. 1977), *cert. denied*, 439 U.S. 821 (1978).

53 517 F.2d at 489-90. On appeal of the Commission's administrative order issued subsequent to the preliminary injunction, the Seventh Circuit upheld a number of order provisions that absolutely prohibited certain claims, which confirmed the Commission's authority to ban claims where necessary to prevent deception. However, noting that "[t]he First Amendment does not permit a remedy broader than that which is necessary to prevent deception," the court modified one order provision requiring an affirmative disclosure for one category of claims, though allowing the disclosure requirement for a more narrow category of claims. 570 F.2d. at 164 (citations omitted).

54 *See also*, *Beneficial Corp. v. FTC*, 542 F.2d. 611, 619 (1976) (Commission exceeded its remedial authority under Section 5 of the FTC Act in not allowing adequate qualifying language in lieu of absolute prohibition of offending language).

55 The Commission has found that advertising claims making product comparisons that imply a substantial difference of benefit to consumers, when in fact the difference is insignificant, may be deceptive. *See P. Lorillard Co. v. Federal Trade Commission*, 186 F.2d 52 (4th Cir. 1950).

56 As noted in the Commission's Deception Policy Statement, qualifications must be clear and conspicuous if they are to be adequate in preventing potential deception (6) at pp. 180-81. In *Kraft*, the Commission found that the company's attempt to qualify its claim of calcium equivalency was ineffective, both because of the qualification's appearance and its complicated wording. *Kraft, Inc.*, D-9208 (1991), slip op. at 10, *aff'd Kraft Inc. v. FTC*, 970 F.2d 311 (7th Cir. 1992), *cert. denied*, 113 S. Ct. 1254 (1993).

57 (16) at 20.

58 (16) at 10-11.

59 (16) at 7.

60 778 F.2d 35 (D.C. Cir. 1985).

61 (60) at 41-42.

62 *FTC v. The Sporicidin Company, et al.*, Civ. No. MJG 91-3543 (D. Md. Dec. 13, 1991). Commissioners Owen and Mary Azcuenaga, while concurring in the complaint filed by the Commission, dissented from the decision not to include in the requested preliminary injunction a safe harbor for claims approved by the Environmental Protection Agency under its germicide registration authority, for products not subject to a stop sale order.

63 (16) at 12.

64 Tire Advertising and Labeling Guides, 16 C.F.R. Part 228, § 228.9 (1993).

65 58 Fed. Reg. 64881 (Dec. 10, 1993).

66 While FDA's rules provide for a petition process whereby unapproved synonyms may be added to the list of approved terms, 21 C.F.R. § 101.69, it remains to be seen how this process will work in practice.

67 (56).

68 (47).

69 D-9250 (Complaint Issued Oct. 28, 1991).

70 The Administrative Law Judge found that the sodium content of Stouffer entrees exceeded public health guidelines for "low sodium," citing FDA's then-proposed regulations. Stouffer did not appeal the ALJ's finding that a low sodium claim was false and misleading.

71 The Commission serves both prosecutorial and judicial functions in such adjudicative matters. After it approves the filing of an administrative complaint—a prosecutorial decision—a full trial is held before an administrative law judge. Either the respondent or the FTC staff may appeal the ALJ's opinion to the Commission, which then acts in a judicial capacity to review the ALJ's decision *de novo*. Therefore, after the Commission approves a particular complaint allegation, it may or may not ultimately conclude that the allegation is supported by the trial record if the issue is presented to it on appeal of an ALJ's decision. *See* FTC Rules of Practice for Adjudicative Proceedings, 16 C.F.R. Part 3.

72 *Campbell Soup Co.*, D-9223 (Aug. 18, 1992).

73 *See* (6) and (7).

74 *See* "Federal Trade Commission Advertising and Promotion Enforcement Objectives," Remarks by Commissioner Deborah K. Owen before the Minnesota Institute of Legal Education, Aug. 10, 1993.

75 *See*, e.g., (47) and (51).

5

International Harmonization and Compliance

Linda R. Horton

Food and Drug Administration, Rockville, Maryland

I. INTRODUCTION

Food, a necessity for survival, is produced in all countries and traded among many of them. When food crosses the border from one country to another, it may be required to meet specified requirements not only for safety and quality, but also for labeling (1). Harmonizing requirements for food is a worthy but challenging objective for the world community and, as is discussed in this chapter, the challenge may be more difficult for food labeling than for most other aspects of food regulation. As an overview to the areas of difficulty and promise, this chapter describes:

- How NLEA affects the practical choices faced by those who sell food in the international market and must deal with divergent national laws
- How FDA's implementation of NLEA includes features responsive to international considerations
- How other countries likewise have particularized food labeling laws
- What harmonization opportunities exist through the Codex Alimentarius Commission and other mechanisms
- Why nutrition labeling is important for developing nations; and
- How recent trade agreements relate to countries' food labeling requirements

II. NLEA AND THE PRACTICAL CHOICES FACED BY GLOBAL FIRMS

A. NLEA's Specificity

As has been discussed in earlier chapters, before NLEA FDA's nutrition labeling rules had been developed under very general statutory authority that, due to its very generality, was more hospitable to administrative consideration of regulatory alternatives, to flexible interpretation, and thus to international harmonization (2). Furthermore, due to participation by FDA and other countries in the Codex Alimentarius Commission, the international body that develops international standards and guidelines for food, the Codex guidelines for food labeling are quite similar to FDA's pre-NLEA rules (3–5). Codex carries out the Food Standards Programme of the Food and Agriculture Organization of the United Nations (FAO) and the World Health Organization (WHO).

FDA believes in the NLEA label and enthusiastically endorses its use. The agency believes it represents a vitally important public health initiative, founded on sound science and policy, that will help people to live longer and healthier lives by enabling better dietary choices. The specificity of NLEA does, however, reduce FDA's latitude to accept food with labels that differ from what NLEA and FDA's implementing regulations require. (It should be remembered that FDA did not seek out such a prescriptive law: NLEA began as a Congressional initiative, paralleling FDA's efforts to achieve similar goals through rulemaking under the agency's general food labeling regulatory authority.)

FDA has, of course, fully embraced NLEA and is now wholeheartedly promoting compliance and understanding of the new food label that NLEA has brought forth. Although NLEA has moved FDA away from the Codex guidance and is a new complicating factor in striving toward international uniformity, NLEA does promote the Codex "principle that all consumers are entitled to safe, sound and wholesome food and to protection from unfair trade practices(1)."

B. The Regional Label

The practical reality is that international food companies that ship foods packaged for sale to the ultimate consumer confront an increasingly complex and divergent array of national norms for food labeling. NLEA is but one of many such laws.

For many exporters, the only practical choice is to design a region-specific label that meets the importing nation's requirements. After all, even if there were no disparities in rules on such matters as formats, nutrient declaration, descriptors, and claims, national laws requiring labels in one or more languages have the effect of requiring the printing of multiple labels. International standards and guidelines for food labeling in the Codex Alimentarius cannot overcome this linguistic hurdle and other specialized national requirements. As is discussed in Section V of this chapter, Codex itself acknowledges the limits of standards in the area of labeling. Other limitations are that Codex norms cannot cover all aspects of the label, and the Codex food labeling standards have received but limited acceptance (5).

Different populations have different nutritional needs and may therefore need to have different nutrients emphasized on the nutrition label (4). Codex recognizes that its Nutrient Reference Values for vitamin D, niacin, and iodine may not be applicable for countries where national nutrition policies or local conditions provide sufficient allowance to ensure that individual requirements are satisfied (4,6). In the United States, label declaration of niacin, riboflavin, and thiamine is now optional, as deficiencies of each are no longer considered to be of public health significance in the United States (7). As is discussed below in Section IV of this chapter, other major developed countries also have a similar, and quite understandable, tendency to meet their own domestic consumer protection goals when it comes to food labeling. Developing countries' populations have unique nutritional needs, as is discussed in Section IV of this chapter.

C. The Metric Question

The metric issue is another complication in the United States. The tenacity of avoirdupois (customary inch-pound) measurements in an increasingly metric world impedes harmonization efforts in the United States. In 1992 the Congress twice amended the U.S. Fair Packaging and Labeling Act (8). The first law required metric declarations in the quantity of contents declarations on labels of food and other products, then, in the face of industry concerns about adverse economic impact, the second law repealed that requirement and instead compelled use of both metric and avoirdupois declarations (9,10). In its implementation of the later amendment, FDA promoted international standardization by giving precedence to the metric declarations and in allowing use of a European mark on the label (9,10).

D. NLEA's Preemption and Food Standards Provisions

One benefit of NLEA for both domestic and foreign firms is the elimination, through preemption, of most disparate U.S. State requirements for food labeling. In recent years, states had increasingly enacted laws establishing additional food labeling requirements. The provisions in NLEA that require federal-state uniformity in many aspects of the food label were of sufficient value to the U.S. food industry to elicit its support of (or non-objection to) NLEA program. These uniformity provisions are likewise, for foreign firms, a harmonizing and trade enhancing feature.

Also, NLEA contains provisions that will facilitate the adoption by FDA of Codex Alimentarius food standards, if FDA concludes under the Federal Food Drug, and Cosmetic Act that it will promote honesty and fair dealing in the interest of consumers to do so (11). By simplifying the procedure for the issuance, amendment, and repeal of most food standards, as is discussed elsewhere in this book (Chapter 7), NLEA has made international harmonization of food standards more feasible for FDA. (It should be noted that standards of identity have been deemphasized by both the Codex Alimentarius Commission and FDA in recent years (6).)

E. Achieving Compliance

In sum, with NLEA, as with any other regulatory scheme, companies in this country and around the globe face the same set of three choices: to determine how to comply, to cease doing business in the affected market, or to challenge the law under domestic law or international trade law. Considering the drastic nature of the last two courses of action, compliance is, of course, the route most often taken. This is demonstrated by the enormous demand for information about NLEA, including copies of the proposed and final regulations, for seats at packed conferences on NLEA requirements, and for guides to compliance (such as this book). In fact, labels complying with NLEA began appearing in U.S. supermarkets many months before they were required.

To achieve compliance with NLEA, all firms, domestic and foreign, that do business in the U.S. food market must take steps, directly or through agents or consultants, to meet NLEA's requirements. One potentially useful option is one used by the German Democratic Republic. The German agricultural products board employs an official to keep track of food labeling and other U.S. requirements and to serve as a technical resource to German firms exporting food products for importation into the United States (12). This approach seems to be helpful in achieving compliance with U.S. food labeling requirements.

III. ASPECTS OF NLEA IMPLEMENTATION THAT ARE RESPONSIVE TO INTERNATIONAL CONSIDERATIONS

As discussed previously, the degree of prescriptive detail in NLEA gives FDA relatively little legal flexibility to allow food exporters in other countries latitude in meeting U.S. food labeling requirements. Moreover, given the high priority of nutrition labeling as a public health tool to help people live longer and healthier lives, as well as the comprehensiveness of NLEA label and the legitimate concerns that domestic producers would express about a "level playing field," FDA would be very unlikely, from a policy standpoint, to develop a laxer stance toward foreign firms' compliance with NLEA, even if it had the latitude to do so from a legal standpoint.

A. Technical Harmonization

Although international harmonization of label format and content is thus a stubbornly elusive goal, real progress has been made toward standardizing the technical underpinnings to the labels.

First, many Codex and FDA food standards are quite similar: standardizing what goes into the container makes it easier to standardize what goes onto the container (5). However, with the current trend away from foods with standardized compositions, labeling becomes increasingly important (6).

Second, in implementing NLEA, FDA either followed (or brought its definitions closer to) internationally recognized definitions for fatty acids, sugars, and other terms (13). FDA also adopted a number of internationally recognized methods of analysis, for example, on protein digestibility (13). Furthermore, it has been helpful that FDA scientists have worked closely for many years with the Association of Analytical Chemists International (AOAC). AOAC publishes methods that are internationally recognized and AOAC sends representatives to meetings of the Codex Alimentarius Commission. FDA's food regulations are replete with references to AOAC methods.

Third, the opportunity to establish or use databases on the nutrient values of food—rather than compelling individual companies to perform tests—will reduce costs for domestic and foreign firms alike (14). This effort builds upon 20 years of experience under FDA's original nutrition labeling regulations, which laid the groundwork for implementing the new requirements without undue costs.

Thus, while NLEA may at first seem to be a rather hegemonic labeling regime with respect to what must go on a label, the development of that required information is founded on a number of internationally recognized definitional and analytical building blocks.

Admittedly, FDA's efforts to harmonize these underlying technical building blocks have not been uniformly successful. Before NLEA, FDA had attempted to conform more closely to internationally recognized daily values for vitamins, minerals, and other nutrients (13). Nutrient Reference Values (NRV's) had been recommended two years earlier by the Joint FAO/WHO Expert Consultation on Recommended Allowances of Nutrients for Food Labelling Purposes, and the Codex Guidelines on Nutrition Labelling had been amended to incorporate these NRV's (4). FDA's discretion to adopt values identical or similar to these international guidelines has been circumscribed, in deference to the views of the U.S. dietary supplement industry (15).

Finally, it should be remembered that NLEA applies only to foods shipped to the United States in final packaging, a fact that reduces impact on countries supplying food components in bulk form for further processing or final packaging in the destination country (7,13).

B. Educational Activities

FDA has undertaken many educational and compliance activities, described in other sections of this book (Chapters 4, 11, and 23), that will aid domestic and foreign food firms in following NLEA. These activities promote understanding of the new label, compliance, and harmonized technical groundwork, for additional global uniformity.

Recognizing special difficulties faced by foreign suppliers, FDA took a number of special steps aimed at foreign audiences, including the following:

- FDA officials have participated in NLEA training programs in a number of countries, including Australia, Canada, the European Union, Indonesia, Mexico, Singapore, and Thailand.
- The agency provided copies of NLEA regulations to embassies and regulatory counterparts abroad.
- FDA provided special briefings in meetings of the Codex Committee on Food Labelling and the Canada-United States Free Trade Agreement (CUSFTA) Working Group on Packaging and Labelling.
- NLEA has been a prominent agenda topic in recent bilateral meetings with regulatory counterparts in Canada, the European Union, and Mexico.
- In February, 1994, FDA sent a letter to representatives of other governments reminding them of the impending effective date of the regulations implementing NLEA.

IV. FOOD LABELING INITIATIVES IN OTHER COUNTRIES

FDA officials and their counterparts in Europe, Canada, Mexico, Japan, and other countries exchange information and regulatory texts, in part to maximize opportunities for substantive regulatory agreement.

A. Europe

Single market activities of the European Union have made progress toward resolving some, but not all, of the differences among the 15 member states that have impeded trade in food products even among European countries. With the entry into force of the Maastricht Treaty on European Union on November 1, 1993, "the European Union" is now the generally preferred term for the 15 nations that make up the European Community (16). The term "European Community" or "European Communities" is still acceptable for discussions of the internal market policies of the Union. Several countries in the European Free Trade Area (EFTA) have signed an agreement with the EU to establish closer economic ties through the creation of a new European Economic Area that came into effect in 1993.

The European Union has issued a comprehensive set of directives on various aspects of food labeling (18). When nutrition labeling is provided, the following is required: as of October 1, 1993, information must be provided on energy value, protein, carbohydrate, and fat. As of October 1, 1995, with respect to any voluntary nutrition labeling or when a nutrition claim is made, information must also be given on sugars, saturated fat, fiber, and sodium (19).

Within the European Union, claims relating to the cure of diseases are forbidden except in the framework of the specific legislation on mineral waters and food for special dietary uses (6).

It should be noted that food labeling is an area of "mixed competency" in which the member countries retain some latitude with respect to national requirements. A recent report by the European Commission was critical of the current state of the law in Europe, especially with regard to challenges posed by providing consumer information in the many different languages (20). The report recommended a continuation of the current state of the law on food labeling (i.e., national legislatives may define which languages may be used on food sold in individual countries) (20).

Europe is also grappling with a number of controversial labeling issues, including "green laws" to identify on the label the biodegradability of packaging, and whether food products derived from biotechnology should bear special labeling (21).

B. Canada

Canada has comprehensive food labeling regulations that differ in many respects from FDA's new NLEA requirements. With the specificity of NLEA, coupled with Canadian regulatory requirements and French/English language requirements, firms outside the region find it increasingly difficult to attempt to design a "North American label" that would be lawful in both the Canadian and the U.S. markets (22). An effort to satisfy both countries' laws with one label would be confusing to consumers and might result in violations of the laws of each country (22). The two systems present large areas of similarity in basic requirements (such as ingredient listing, nutritional equivalency of foods, and standards), but set forth rules based on quite different views on fortification, health claims, serving sizes, and other matters (22,23). These differences impede United States-Canada harmonization.

The regular meetings of the CUSFTA Working Group on Packaging and Labelling have been a useful forum for exchange of information, developing comparisons of statutory and regulatory differences, and comparing and harmonizing scientific approaches. One recent success from a trade standpoint was a Canadian decision to eliminate food container size requirements that operated as a barrier to U.S. canned goods (22).

C. Mexico

Particularly with the ratification of the North American Free Trade Agreement (NAFTA), FDA has provided Mexican regulators with much information about the new U.S. requirements. For example, NLEA was discussed at the most recent bilateral meeting between FDA and its Mexican counterparts, held in December 1993. Also, FDA obtained assistance from the United States Agency for International Development (USAID) to provide technical assistance to Mexico on NLEA and has provided training. In 1994, Mexico proposed a food labeling rule that is so different from FDA's NLEA regulations that bilingual U.S.–Mexico food labels will be very difficult to achieve.

D. Japan

As in the United States and the European Union, Japan's labeling laws have imposed a number of unique requirements and the ordinary linguistic needs of the Japanese people dictate the creation of a market-specific label for food firms shipping products to Japan. Also, Japan has not adopted Codex standards for date marking of processed food products. Japanese retailers presently demand date of manufacturing rather than "best before" dates on food packaging (24).

V. HARMONIZATION OPPORTUNITIES THROUGH THE CODEX ALIMENTARIUS COMMISSION AND OTHER MECHANISMS

A. Codex Alimentarius Commission

Codex's Committee on Food Labelling and Committee on Nutrition and Foods for Special Dietary Uses have the responsibility to make recommendations in this area (25). Also, the Codex Food Standards Committees recommend labeling for the food categories within their areas of responsibility but must coordinate this work with the Codex Food Labelling Committee (25).

The Commission has made clear that the Codex Alimentarius is no substitute for, or alternative to, referring to national legislation: "Every country's laws and administrative procedures contain provisions which it is essential to understand and comply with. . . . The language which is required on labels will be a matter of national legislation. . . . (25)" The Commission goes on to encourage countries to minimize differences from the General Standard for the Labelling of Prepackaged information to be included on the labels of all foods:

> Governments are exhorted to use the General Standard as a basis for their national legislation and to keep differences to an absolute minimum especially those of detail or minutiae. Governments should . . . ensure that all compulsory provisions relating to presentation of information which are additional to, and different from, those in the standard should be notified [to the Commission]. Any other compulsory provisions in national legislation should also be notified if they are not provided for in the Codex standard. The labelling provisions in Codex standards include sections of the revised General Standard by reference. When accepting a Codex commodity standard, a country which has already accepted and responded to the General Standard can then refer to the terms of that acceptance in any subsequent responses. As much specific information as is relevant and helpful should be given. In particular, this should include the name and description relating to the food, the interpretation of any special requirements relating to the law or custom of the country, any additional details about presentation of the mandatory information and detailed differences if any in the labelling requirements, e.g., in relation to class names, declaration of added water, declaration of origin. It will be assumed that the language(s) in which the particulars should be given will be as indicated by national legislation or custom (25).

As an indication of the current work program of the Codex Food Labelling Committee, the following matters will be considered at its session held in October, 1994:

- Consideration of labeling provisions in Codex standards
- Proposed draft guidelines for the production, processing, labeling, and marketing of organically produced foods
- Proposed draft guidelines for the use of health and nutrition claims
- Recommendations for the labeling of potential allergens
- Proposed draft amendments to the Codex guidelines on claims on the use of the term "natural"
- Principles for the consideration of Nutrient Reference Values for labeling purposes; and
- Consideration of labeling as related to biotechnology (6)

As an example of an issue before the Codex Food Labelling Committee that has interesting international and public health ramifications, this section presents details about the question of how to improve the labeling of allergens in the declaration of ingredients in packaged food. The problem is a provision in the current General Standard for food labeling—and the law of many countries, but not the United States—that permits certain potentially allergenic food components to be present without declaration on the label, presenting a serious risk of injury or even death for certain consumers (3).

The so-called 25% rule, found in Section 4.2.1.3 of the General Standard, provides that, where an ingredient in a food is itself the product of two or more ingredients, the compound ingredient may be declared as such in the list of ingredients provided that it is immediately accompanied by a list in brackets of its ingredients in descending order of proportion. However, where a compound ingredient for which a name has been established in a Codex standard or in national legislation constitutes less than 25% of the food, the ingredients other than food additives which serve a technological function in the finished product need not be declared (3).

A report done for the Codex Committee on Food Labelling and discussed at its 1993 and 1994 meetings noted that this provision leads to the paradoxical situation that a substance added to a food in small amounts as an ingredient must be declared, whereas the same substance can be present in much larger amounts in another product if it is a component of a compound ingredient, without being declared (6). Although the 25% rule was intended to be applied to products subject to Codex or national standards—the theory being that the consumer allergic to certain ingredients would become acquainted with the contents of these standards—the report described several current trends in food policy that challenge the accuracy of this theory.

First, as was noted earlier, there is a trend away from vertical commodity standards, both internationally in Codex and nationally in the laws of individual countries. The trend is toward general horizontal legislation with more emphasis on informative labeling and less on recipe formulation (6).

Second, because of the trend away from standards, there are appearing on the market increasing numbers of products and ingredients without the standardized composition which is, according to the 25% rule, the prerequisite for being able to omit the declaration of the components of a compound ingredient. Aggravating the situation is the fact that some countries also apply the 25% rule to the labeling of traditional products (6).

Third, the trend toward increasing international trade and greater efforts to achieve international harmonization of food labeling regulations has resulted in a tendency for some countries to accept products which are in accordance with other countries' standards or traditions even if not strictly in accordance with the importing country's law (6). [The report paraphrased here was drafted by representatives of Nordic countries whose movement toward EU accession has meant decreased scrutiny of imports from EU countries.]

Thus, the risk of unwitting exposure to allergens increases when consumers in the importing country buy these imports without being acquainted with the food standards or customary foods of the country where produced. Obtaining the needed ingredient information from another country is difficult. The consequence is that consumers with allergens find they must severely restrict their purchases of processed food as they cannot trust food labels to describe what components are present (6).

The Codex Committee on Food Labelling is considering a recommendation to reduce the lower limit for specification of compound ingredients to 5%, rather than 25%, and to require that certain ingredients which can give rise to hypersensitivity must always be declared no matter the amount present. These ingredients would include specified grains, legumes, nuts, seafood, eggs, milk, and sulfites. Such a change would bring the Codex standard into closer harmony with the FDA rule under NLEA and with the public health need to protect sensitive consumers from allergens (6).

The work program set forth earlier shows how the Codex Committee on Food Labelling is targeting newer food labeling issues for harmonization activities, such as those involved in representations that food is "natural" or "organic," as well as the issue of whether food produced through biotechnology should be labeled as such (6,26). Clearly, despite the many challenges presented by such currently controversial issues, it is worthwhile to work toward harmony in new areas when countries are just beginning to develop laws.

Although an assignment was made at the October 1994 meeting to begin work on a nutrition label revision, it will probably take many years for the Codex Committee on Food Labelling to complete work on the issues presented by NLEA. It was in 1985 that the Commission adopted nutrition labeling guidelines that were similar to FDA's 1973 regulations. The Codex Committee is already grappling with the contentious issue of health claims, with the United States allowing a few such claims, where substantiated, while the European Union maintains a stance of prohibiting almost all health claims on food (6).

After a time, the revolutionary changes in food labeling occurring in the United States, Europe, and other countries will stabilize. It is now time for the Codex Alimentarius Commission's Committee on Food Labelling to consider which features of NLEA warrant adoption in Codex standards and guidelines. Indeed, the Commission, in its "Draft Medium Term Objectives," has included "harmonization of health and nutrition claims and warnings in food product labelling by 1995" as a priority for its Food Labelling Committee (27).

Thus, the U.S. government's delegates to the two Codex Committees interested in this field—the Committee on Food Labelling and the Committee on

Nutrition and Foods for Special Dietary Uses—will have the opportunity to persuade other countries of the virtues of NLEA—as well as to identify any shortcomings that U.S. experience shows to be unworthy of emulation by others.

B. Bilateral Opportunities

Other useful harmonization discussions are bilateral efforts mentioned earlier, particularly the continuing activities of the technical working groups under CUSFTA. Although, due to the ratification of NAFTA, the CUSFTA itself has been suspended, an earlier agreement between Canada and the United States continued these technical working groups for the time being (28).

New committees being established under NAFTA will provide similar structures for dialogue and work programs aimed at uniformity among regulators in Canada, Mexico, and the United States.

Without question, food regulatory authorities around the world need to exploit available opportunities to compare requirements, harmonize the methodology, definitions, and other technical work underlying labels, and identify areas of equivalence. Equivalence could entail not only requirements that are identical but ones that are in the range where regulators exercise enforcement discretion to refrain from regulatory action. Where differences exist, efforts first to understand the reasons, then to seek ways to narrow the differences, may be worthwhile. Although FDA is a public health agency, not a trade agency, this is one of many places where the agency's health aims and the nation's general interest in free trade can converge. What the United States needs to do here is to export its good ideas. The ideas to be exported are those of the Congress and of FDA, as expressed in NLEA and its regulations, at least to those countries whose nutritional profile resembles ours. Our country will not, of course, have a closed mind to good ideas from others about food labeling, even if it were to mean exploring the possibility of changes in our laws.

VI. THE IMPLICATIONS OF NUTRITION LABELING FOR DEVELOPING COUNTRIES

The citizens of developing countries have much to gain from attention to nutritional information. Although simply having enough food is a daily challenge for the poorest people, achieving a healthy diet is a universal challenge for all people. Nutrition labeling is dependent on a certain level of literacy and educational attainment, but can help meet that challenge. By accompanying the food, labeling is more likely to reach the ultimate consumer than is nutritional information in other forms. Thus, from a policy standpoint, nutrition labeling is a worthy objective for international observance, quite apart from the legal reality that it is mandated for imports into the United States and an increasing number of nations.

It should be noted that the information that needs to go on a food label in a developing country may be quite different from the information that needs to go on the food label in a developed country like the United States. NLEA and FDA's implementing regulations arose out of health concerns of a society in which overconsumption is a problem for more citizens than is malnutrition. In countries where many citizens suffer from malnutrition, there may well be a need for a different kind of nutrition label than what is being adopted in the United States.

Obviously, the first order of business in a developing country is to achieve the availability of enough food so that people do not starve or suffer adverse health effects from poor nutrition. Further, literacy must precede nutrition labeling before it can be helpful to people in choosing food. Thus, a requirement to label food as to its nutritional value requires some degree of development in a society's production and distribution of food. After all, in traditional societies, much of what a family eats is that which that family raised, and the food is unlabeled. In many local markets around the world, much of the traded food is unlabeled. Even NLEA gives special treatment to certain foods that, in developed as well as developing economies, are not prepackaged but are handpicked by consumers. Thus, the most popular fresh fruits, vegetables, and seafood are subject only to voluntary guidelines for nutrition labeling provided that there is substantial compliance with the guidelines (15).

The aims of early food labeling laws were to prevent fraud as to quantity and composition, not to control representations of the healthy properties of the food (29). Nutrition labeling came later not because it was unimportant, but because the necessary scientific knowledge underlying nutrition information—for example, the existence of vitamins—did not begin to exist until earlier in this century and is still being amassed (29).

Distribution of food with nutrition labeling in the language of the receiving country is a technique that can promote nutrition among poor countries. Therefore, multinational and national development assistance agencies that promote the availability of food to needy people can promote nutritionally balanced diets through labeling and a variety of other techniques as well, including education and information for consumers in the form of publications and educational aids. Thus, development assistance agencies could work with developing countries to learn what the governments and the people think are the needs for nutritional information, including label information, and to develop responsive programs and laws.

Nutrition labeling is thus not merely (for another country) related to a trade objective of complying with our country's NLEA requirements, nor merely (for the United States) an FDA compliance objective. Also, and more importantly, nutrition labeling is a very effective way of promoting public health goals for all people. People in countries at all stages of development might view nutrition labeling positively, as something of direct domestic benefit, not simply as a burdensome requirement to benefit affluent consumers in developed countries.

VII. HOW RECENT TRADE AGREEMENTS RELATE TO COUNTRIES' FOOD LABELING REQUIREMENTS

The final topic—left for last lest its legalisms lose readers—is the effect of recent trade agreements on countries' food labeling laws. The issue is whether another country wishing to help its food industry could attack NLEA under either the General Agreement on Tariffs and Trade (GATT) or NAFTA. These actions are a possibility, but are unlikely to occur due to many factors described earlier in this chapter such as the virtual inevitability, for other reasons, of manufacturers needing to print region-specific labeling. Among these reasons are not only NLEA, but *other* nations' quite prescriptive food labeling laws and different nutritional needs.

Even if another country were to initiate a trade action, it would probably be unsuccessful due to safeguards built into the trade agreements that protect the ability of a sovereign country to establish requirements that fulfill a legitimate objective, that are based on science, and that achieve the level of protection chosen by that country. The remainder of this chapter goes through the provisions in GATT and NAFTA that would be relevant in challenges to laws such as NLEA, for readers who are interested in this degree of technical legal detail.

A. GATT

GATT is a global trade agreement that entered into force on January 1, 1948, and has been amended several times following negotiations known as "rounds." The term "GATT" has also been used to describe the institutional body in Geneva that administers the agreement. As of January 1, 1995, this institution has been renamed the World Trade Organization (WTO) (30).

1. The Uruguay Round Agreement

The most recent GATT round, the Uruguay Round, was so-named because it began in Punta del Este, Uruguay in 1986. It was concluded in Geneva on December 15, 1993, and formally signed by the contracting parties at the Marrakesh Ministerial Meeting on April 15, 1994 (31).

Under U.S. law, a formally signed trade agreement enters into force only after both Houses of Congress have passed implementing legislation and the President has signed this legislation (32). The domestic ratification process for a trade agreement is thus somewhat different from that applicable to a treaty, which is ratified by a 2/3 vote of the Senate without need for House action or legislation presented to the President (32).

Congressional passage and Congressional endorsement of the legislation to implement the Uruguay Round GATT agreement occurred in late 1994. As with other recent trade agreements, including its 1993 ratification of NAFTA, Congress used a fast-track procedure during its consideration of the legislation to implement the Uruguay Round (32,33). This procedure limits debate, forbids amendment, includes expedited deadlines, and thus forces each member to simply vote up or down on the agreement. The purpose of the procedure is to protect the credibility of the President and his ambassador on these matters, the United States Trade Representative (USTR). Without an advance commitment by Congress to refrain from tampering with the carefully negotiated terms of trade agreements, the leaders of other countries would have no confidence in the ability of the President and the USTR to keep their word in follow-through on good faith trade negotiations (32).

Before turning to new provisions in GATT added by the Uruguay Round, we should review general long-standing GATT principles that remain relevant to the issue of whether a country can maintain its own unique food labeling regime.

2. National Treatment and Most Favored Nation

Article III, paragraph 4 of GATT has long required that "the products of the territory of any contracting party imported into the territory of any other contracting party shall be accorded treatment no less favorable than that accorded to like products of national origin in respect of all laws, regulations, and requirements affecting their internal sale, offering for sale, purchase, transportation, distribution, or use (34)."

Equality of treatment of domestic and imported goods is the key principle, closely related to Article I's assurance of "most favored nation" treatment among GATT contracting parties:

> With respect to . . . all rules and formalities in connection with importation and exportation, . . . any advantage, favor, privilege or immunity granted by any contracting party to any product originating in or destined for any other country shall be accorded immediately and unconditionally to the like product originating in or destined for the territories of all other contracting parties (34).

NLEA and FDA's implementing regulations clearly meet the national treatment and most favored nation tests. NLEA, like other U.S. labeling laws, is "trade blind," giving equal treatment to all foods of whatever origin. Because NLEA regulations thus create a level playing field, it should not be necessary for the United States, if called upon to defend NLEA, to invoke the provision of GATT, Article XX, that sets forth general exceptions to such general GATT principles as the national treatment test:

> Subject to the requirement that such measures are not applied in a manner which would constitute a means of arbitrary or unjustifiable discrimination between countries where the same conditions prevail, or a disguised restriction on international trade, nothing in this Agreement shall be construed to prevent the adoption or enforcement by any contracting parties of measures . . . necessary to protect human, animal or plant life or health [or] . . . necessary to secure compliance with laws or regulations which are not inconsistent with the provisions of this Agreement, including those relating to . . . prevention of deceptive practices. . . . (34)

NLEA and its implementing regulations are, in the view of the United States, necessary both to protect human health and to prevent deceptive practices and should pass muster under these criteria. The risk of life-threatening diseases such as cancer and heart disease can be reduced through diet, aided by nutrition labeling. In the unlikely event of a GATT dispute on NLEA, the case should not even reach the issue of the applicability of an exception, since the case ought to be resolved, under the national treatment standard and other disciplines discussed below, as a valid exercise of a sovereign country's right to establish a reasonable program to fulfill a legitimate objective.

No FDA regulation has ever been the subject of a GATT case. The agency is hopeful that the same level of inactivity will continue with the new GATT. Although concerns have been expressed about the significance of an adverse ruling against the U.S. mammal protection law in GATT tuna/dolphin case (35), that case involved concerns about national treatment, most-favored-nation treatment, and extraterritoriality, all issues not presented by FDA's food labeling laws. The USTR does not regard the case as presenting troublesome implications for either the new GATT or NAFTA (36). U.S. environmental and consumer groups disagree and have pressed their concerns to U.S. officials, the Congress, and the public (37–39).

To understand how a GATT challenge to NLEA would need to be framed, we turn to a discussion of the three relevant GATT agreements: the present Agreement on Technical Barriers to Trade (TBT), the new Agreement on TBT that is a part of the Uruguay Round, and the new Agreement on the Application of Sanitary and Phytosanitary Measures (SPS). The greatest emphasis is placed on the new TBT code, as it clearly would be applicable to technical regulations for products such as those under NLEA. The agreement on SPS Measures is discussed because, even though its applicability to NLEA regulations is unclear, it will undoubtedly be applicable to certain food labeling regulations (i.e., those directly related to food safety and those otherwise pertaining to protection of human life or health from risks arising from additives, contaminants, toxins or disease-causing organisms in foods, beverages or feedstuffs) (30).

3. The Present TBT Agreement

This agreement, sometimes called the standards code, entered into force on January 1, 1980, as part of the Tokyo Round of GATT concluded the preceding year. The 1980 TBT agreement was developed to address the fact that, as tariffs and other economic trade barriers fell as a result of earlier GATT rounds, trade still was hindered by national technical regulations and standards that had either the intent or the effect of excluding imports.

Although the 1980 TBT agreement will soon be replaced, its provisions are worth a brief summary, if for no other reason than as a reminder that the new GATT is not the first general trade agreement to target countries' use of regulations that have the intent or effect of unduly encumbering trade. It is interesting to note that, in the 14 years that this TBT code has been in effect, there have been no disputes resolved under it (40). Thus, there has as yet been no jurisprudence that foreshadows how issues will be resolved under the 1994 TBT code and it is possible that the new TBT code will likewise achieve success more as an "International APA" (Administrative Procedure Act) than as a license for international trade litigation.

In any event, the 1980 TBT agreement was, first and foremost, a transparency code designed to open up the rulemaking processes of trading partners with respect to mandatory technical regulations and voluntary standards. As discussed below, it also has encouraged adoption of international standards and requires contracting parties to notify each other of new technical regulations.

This notification requirement was satisfied by the U.S. government for NLEA regulations in 1991 (41). In response to the United States notification, in 1992 both the Japanese government and the European Commission filed comments with the U.S. Department of Commerce's National Institute of Standards and Technology, arguing that FDA's NLEA regulations violate Article 2.2 of the standards code (42). The Japanese government requested that the United States withdraw the proposed regulations and issue a new proposal based on Codex's international guidelines. The European Commission argued that the United States had "provided no convincing evidence of the necessity for such an immediate unilateral deviation from these [Codex] international standards." The Commission also argued that the financial burden imposed by such extensive regulations constituted an unnecessary obstacle to international trade (42). FDA believes that these arguments do not take into account the deference to national legislation both in the Codex (with respect to food labeling) or in GATT.

The 1980 TBT agreement contains several provisions—notably the prohibition against unnecessary obstacles to international trade in Article 2.1—that are pertinent to the present discussion:

> 2.1 Parties shall ensure that technical regulations and standards are not prepared, adopted or applied with a view to creating obstacles to international trade. Furthermore, products imported from the territory of any Party shall be accorded treatment no less favorable than that accorded to like products of national origin and to like products originating in any other country in relation to such technical regulations or standards. They shall likewise ensure that neither technical regulations nor standards themselves nor their application have the effect of creating unnecessary obstacles to international trade.

The U.S. implementing legislation clarified that,

> No standards-related activity of any private person, federal agency, or state agency shall be deemed to constitute an unnecessary obstacle to the foreign commerce of the United States if the demonstrable purpose of the standards-related activity is to achieve a legitimate domestic objective including, but not limited to, the protection of legitimate health or safety, essential security, environmental, or consumer interests and if such activity does not operate to exclude imported products which fully meet the objectives of such activity (33).

FDA firmly believes that NLEA regulations are consistent with these tests in the 1980 TBT code in that, under Article 2.1, the regulations neither are intended as an obstacle to trade, nor do they have the effect of creating unnecessary obstacles to trade. This issue is discussed in greater detail below with respect to corresponding language in the new TBT agreement.

4. New TBT Agreement

The new agreement largely carries over the existing TBT agreement, extending it beyond product-related technical regulations and voluntary standards to also cover conformity assessment procedures and product-related process and production methods (30).

The main provisions of the new TBT agreement are:

- Countries may promulgate technical regulations that are sufficient, but not more trade-restrictive than necessary, to fulfill a legitimate objective (e.g., protection of human health or safety, animal or plant life or health, or the environment) (Article 2).

- Technical regulations, when required, should be based on relevant international standards, whenever the international standards are appropriate and adequate to fulfill legitimate objectives (Article 2).
- Technical regulations in accordance with relevant international standards will be rebuttably presumed not to create an unnecessary obstacle to trade. Countries promulgating technical regulation not in accordance with relevant international standards must be able to justify the regulation (Article 2).
- Countries are encouraged to (1) participate in the development of international standards, and (2) accept an equivalent technical regulations of other countries (Article 2).
- Technical regulations should be based on product performance, rather than on product design on characteristics (Article 2).
- Countries preparing technical regulations must publish notices of their intentions and provide opportunities for other countries to comment on the proposed regulations, with provisions for taking more immediate actions when the technical regulations are needed to address urgent problems of safety, health, or environment (Article 2).
- National governments are responsible for local government and non-government bodies' conformance with the TBT agreement (Articles 3, 7, and 8).
- Countries should ensure that their standards bodies comply with the "Code of Good Practice for the Preparation, Adoption, and Application of Standards" (Article 4).
- Countries should apply prescribed procedures for assessing conformity with technical regulations (Article 5), and should accept, as equivalent to their own, conformity assessments of other countries (Article 6).
- Countries are encouraged to participate in international and regional systems for conformity assessments (Article 9).
- Countries should establish enquiry points for information about technical regulations, standards, and conformity assessment procedures (Article 10).
- Countries should assist other countries (especially developing ones) in the preparation of technical regulations (Article 11).
- Differential treatment should be accorded to developing countries (Article 12).
- The TBT agreement establishes The Committee on Technical Barriers to Trade (Article 13).
- Dispute settlements and consultations will take place under the auspices of the Dispute Settlement Body, following prescribed procedures (elaborated in the Dispute Settlement Understanding) (Article 14).

Of relevance to this chapter is the strengthening both of the language in the preamble recognizing a country's right to enact consumer protection laws and of the anti-trade barrier disciplines in Article 2.1 and 2.2. The language that follows shows how the new GATT TBT adds a phrase (*the italicized language*) to the 1980 preamble to clarify the ability of a country to take measures at the levels of protection it considers appropriate:

> No country should be prevented from taking measures necessary to ensure the quality of its exports, or for the protection of human, animal or plant health, of the environment, or for the prevention of deceptive practices, *at the levels it considers appropriate*, subject to the requirement that they are not applied in a manner which would constitute a means of arbitrary or unjustifiable discrimination between countries where the same conditions prevail or a disguised restriction on international trade . . . (30)

Part of the negotiating history for this language is a discussion in a letter dated December 15, 1993, from Peter D. Sutherland, the Director-General of GATT, to Ambassador John Schmidt, the Chief U.S. Negotiator:

> Representatives of the participants involved in the discussions agreed that each government is free to choose the levels it considers appropriate for the protection of human animal or plant life or health and of the environment and that no country should be prevented from taking measures necessary to achieve those levels subject to the requirement, inter alia, that the measures are in accordance with the provisions of the Agreement. This has been reflected in the fifth clause of the Preamble to the Agreement. It was clear from our consultations also that participants felt the Agreement could not be read, most particularly because of this preambular clause, to require the upward or downward harmonization of technical regulations or standards as a result of international harmonization activities.

New Article 2.2 carries over the 1980 requirement not to create unnecessary trade obstacles, then sets forth the central (and, arguably, the only) discipline in the new TBT, the "no more trade-restrictive than necessary" test italicized in this excerpt:

> 2.2 Members shall ensure that technical regulations are not prepared, adopted or applied with a view to or with the effect of creating unnecessary obstacles to international trade. *For this purpose, technical regulations shall not be more trade-restrictive than necessary to fulfil a legitimate objective taking account of the risks non fulfillment would create. Such legitimate objectives are inter alia* . . . the prevention or deceptive practices [or] protection of human health or safety. . . (34)

In the final days of the Uruguay Round negotiations, U.S. negotiators succeeded in securing the deletion of a footnote to Article 2.2 that might have proved problematic for countries defending regulations attacked as unduly trade-restrictive. The footnote, which read as follows, had been included in the 1991 draft version of GATT: "This provision is intended to ensure proportionality between regulations and the risks non-fulfillment of legitimate objectives would create."

This footnote had originally been favored by the United States as possibly helpful in shielding legitimate regulations from attack, but its so-called "proportionality approach" turned out to have some unwanted jurisprudential baggage—from European environmental law contexts—that could have been read to compel an economic balancing test even where inappropriate given the underlying legitimate objective at stake.

In the closing days of the Uruguay Round, the U.S. negotiators on the TBT text sought unsuccessfully to not merely delete the proportionality footnote to Article 2.2, but to replace it with a new footnote designed to clarify how Article 2.2 would operate in real life. This effort was unsuccessful because other countries' negotiators thought it inappropriate to reopen the TBT text to include an explanation considered unnecessary.

The December 15, 1993, letter from GATT's Peter Sutherland to U.S. Ambassador John Schmidt included the following helpful discussion of the TBT negotiations on this point:

> Representatives of the participants involved in the discussions also agreed that the removal of the existing footnote to Article 2.2 of the Agreement was an improvement to the text.
>
> It was not possible to achieve the necessary level of support for the U.S. proposal to add a new footnote to article 2.2 and 2.3 of the Agreement. The U.S. proposal for such a footnote was:
>
> For purposes of paragraphs 2 and 3 of Article 2, a technical regulation to protect human health or safety, animal or plant life or health, or the environment, or to prevent deceptive practices, is not more trade-restrictive than necessary, and the changed circumstances or objectives cannot be addressed in a less trade-restrictive manner, unless there is another technical regulation, including alternative versions of the technical regulation, reasonably available taking into account technical and economic feasibility, that fulfills the legitimate objective and is significantly less restrictive to trade.
>
> Nevertheless, it was clear from our consultations at the expert level that participants felt it was obvious from other provisions of the Agreement that the Agreement does not concern itself with insignificant trade effects nor could a measure be considered more trade restrictive than necessary in the absence of a reasonably available alternative.

The above negotiating history is also relevant to Article 2.3 of the TBT Agreement, a new provision that "technical regulations shall not be maintained if the circumstances or objectives can be addressed in a less trade-restrictive manner." This provision describes how regulations already in place will be judged under the new TBT (30).

A footnote of the type sought for the TBT had already, by a separate multilateral technical group in the final Uruguay Round discussions, been added to the draft text of the new SPS Agreement, at the same time the SPS language was revised to eliminate a statement compelling measures that were the "least restrictive to trade" in favor of language more in line with the test in Article 2.2 of TBT (and preferable from a consumer protection standpoint):

> 21. Without prejudice to paragraph 10 [according presumptive validity to measures which conform to international standards], when establishing or maintaining sanitary or phytosanitary measures to achieve the appropriate level of sanitary or phytosanitary protection, Members shall ensure that such measures are not more trade-restrictive than required to achieve their appropriate level of protection, taking into account technical and economic feasibility.

The new footnote accompanying paragraph 21 of the SPS reads:

> For purposes of paragraph 21, a measure is not more trade restrictive than required unless there is another measure, reasonably available taking into account technical and economic feasibility, that achieves the appropriate level of protection and is significantly less restrictive to trade (30).

It had been a U.S. negotiating objective in the concluding GATT negotiations to attach a footnote of this type not only to paragraph 21 of SPS but also to Article 2.2 of TBT. Although this objective was achieved for SPS but not TBT, the letter from Mr. Sutherland to Ambassador Schmidt is helpful.

It was also clear from the discussions in the final days of the Uruguay Round that no substantive difference was intended between "necessary" in Article 2.2 of the TBT Agreement ("not . . . more trade-restrictive than necessary") and "required" in paragraph 21 of the SPS ("not more trade restrictive than required"). The view was that the latter use of "required" referred to technical feasibility, not to legal compulsion.

Article 2.4 of the new TBT continues and strengthens the duty found in Article 2.2 of the 1980 TBT to consider international standards as a basis for national voluntary standards, as well as for mandatory technical regulations:

> 2.4 Where technical regulations are required and relevant international standards exist or their completion is imminent, Members

> shall use them, or the relevant parts of them, as a basis for their technical regulations except when such international standards or relevant parts would be an ineffective or inappropriate means for the fulfillment of the legitimate objectives pursued, for instance because of fundamental climatic or geographical factors or fundamental technological problems (30).

FDA believes that NLEA regulations have neither the intent nor the effect of creating unnecessary obstacles to international trade and are not more trade restrictive than necessary to fulfill their legitimate objective. The TBT preamble clearly states the right of a country to take measures necessary to protect human health and to prevent deceptive practices, at the level it considers appropriate (30). As NLEA regulations express the level of protection the United States considers appropriate, they are consistent with the TBT agreement.

A review of the legislative history of NLEA reveals the lack of any discriminatory intent: there is, in fact, no evidence that the U.S. Congress, in developing NLEA, even considered trade effects or international factors (43). Rather, its members had in mind the national objective of enhancing the availability to U.S. citizens of information needed for sound nutritional choices, a matter that falls squarely under permissible regulatory objectives under 2.2 (34,43). The encouragement to use international standards, such as those in Codex Alimentarius, merely requires that the parties use them as a basis for the parties' standards, except where inappropriate for the prevention of deceptive practices or protection of the public health (30). Moreover, standards are not static but need to change in response to changing societal needs (30). As discussed earlier in this chapter, Codex itself recognizes the need for national legislation on labeling (25).

5. New Agreement on the Application of Sanitary and Phytosanitary Measures (SPS)

As mentioned previously, some food labeling measures are subject to the SPS Agreement. These are those measures directly related to food safety and those otherwise pertaining to protection of human life or health from risks arising from additives, contaminants, toxins or disease-causing organisms in foods, beverages or feedstuffs (30).

Whether any given FDA food labeling regulation is challenged and judged under the TBT Agreement or the SPS Agreement, it should survive under the applicable standards. To summarize the provisions of the SPS agreement that would be relevant in a trade challenge to a food labeling law:

- Countries have a right to adopt and enforce measures necessary to protect human health that are non-discriminatory (preamble and Article 2).

- Countries shall ensure that any SPS measure is applied only to the extent necessary to protect human health, is based on scientific principles, and is not maintained without sufficient scientific evidence (Article 2).
- Countries shall base their SPS measures on international standards except that they may introduce or maintain SPS measures which result in a higher level of SPS protection if there is a scientific justification, or as a consequence of the level of protection a country determines to be appropriate (Article 3).
- There is a scientific justification to deviate from an international standard if, on the basis of an examination and evaluation of available scientific information in conformity with the relevant provisions of the SPS agreement, a country determines that the relevant international standards are not sufficient to achieve its appropriate level of protection (footnote to Article 3).
- Countries are required to accept the SPS measures of other countries as equivalent only if the exporting country objectively demonstrates to the importing country that its measures achieve the importing country's appropriate level of SPS protection (Article 4)(30).

FDA believes that its NLEA regulations would measure up to the standards laid out in the SPS text if it were to be considered to be applicable to any of these regulations. NLEA regulations are non-discriminatory, solidly grounded on science, and based on what the United States has chosen as the appropriate level of protection.

6. GATT Dispute Settlement

A detailed discussion of the new GATT dispute procedures is beyond the scope of this chapter. Suffice it to say that, if a trade challenge were to be brought against NLEA regulations—or any other nation's food labeling rules—under the new GATT and were to result in a final decision that the law being challenged is inconsistent with GATT (e.g., is more trade-restrictive than necessary to fulfil the objective at stake under the new TBT Agreement), there would be intense pressure on the country to re-examine the law. This pressure will be much greater than occurs at present when countries lose cases under the current GATT. The reason is the strengthened dispute resolution procedure. [In the following discussion, parenthetical citations refer to paragraphs in the Dispute Resolution Agreement of the new GATT (30).]

Although no country is ever compelled to change a law in response to an adverse GATT ruling, under the old or new procedures, trade sanctions in the form of compensation and the suspension of concessions or other obligations are available in the event that a country does not implement a GATT ruling within a reasonable time (Article 22) (30).

In the current GATT, a single member (even the losing party) can block adoption of a GATT decision; the new GATT dispute procedure requires that a GATT appellate report be adopted and unconditionally accepted by the parties to the dispute unless there is a consensus decision by the Dispute Settlement Body not to adopt the report within 30 days after its issuance to GATT member (Article 17.14). The new dispute settlement procedure also has deadlines and procedures for automatic appeals (as appeals are inevitable), designed to expedite resolution of trade disputes and prevent them from languishing (Article 20, Appendix 3).

It should be stressed that informal conciliatory processes precede dispute settlement. These steps include informal communications, consultations, conciliation, mediation, and arbitration (Articles 4, 5, 25). When a dispute goes to a panel for resolution, each party has the option of requesting the establishment of a technical expert group to assist in questions of a technical nature (Article 13.2, Appendix 4) (30).

As has been discussed elsewhere in this chapter, there is reason to hope that the factors that militate against formal trade cases in this area will spare food regulatory agencies the experience of having their labeling rules be a test case under these new procedures.

B. NAFTA

All three countries completed their respective domestic approval and implementation procedures in 1993 which allowed NAFTA to enter into force on January 1, 1994. The U.S. Congress passed implementing legislation in mid-November, and President Clinton signed the NAFTA implementing bill on December 8, 1993. In the case of Mexico, the Mexican Senate ratified NAFTA on November 22, 1993, and President Salinas sent numerous legislative changes to the Mexican Congress in November 1993 to bring Mexican laws into compliance with NAFTA obligations. The Canadian parliament passed NAFTA in June, 1993, and the Canadian Government proclaimed the legislation implementing NAFTA on December 30, 1993.

1. TBT and SPS

NAFTA's relevant provisions state that:

- Each party may adopt, maintain, or apply any standards-related measure, including any such measure relating to safety, the protection of human, animal or plant life or health, the environment or consumers, and any measure to ensure its enforcement or implementation (TBT Article 904.1, SPS Article 712.1).
- Each party may establish the levels or protection it considers appropriate (TBT Article 904.2, SPS Article 712.2).

- Measures must be non-discriminatory (TBT Article 904.3, 907.2; SPS Article 712.4).
- Measures may not have the intent or effect of creating an unnecessary obstacle to trade. An unnecessary obstacle to trade is not created where its purpose is to achieve a legitimate objective and it does not exclude a party's goods that meet that objective (TBT Article 904.4).
- International standards are to be used as the basis for national standards and are presumed to be consistent with NAFTA. However, parties can apply any standards-related measure that results in a higher level of protection (TBT Article 905, SPS Article 713).
- Parties shall work toward compatibility of measures and treat as equivalent a measure that an exporting party demonstrates to the satisfaction of the importing party will adequately fulfill the importing party's legitimate objectives (TBT Article 906, SPS Article 714).
- Each party may conduct risk assessment (TBT Article 907.1, SPS Article 715).
- Each party shall assure that its SPS measures are science-based (SPS Article 712.3).
- SPS measures shall be applied only to the extent necessary to achieve the appropriate level of protection, taking into account technical and economic feasibility (SPS Article 712.5).
- There is no language similar to that in 2.2 of the new GATT TBT Agreement or paragraph 21 of the SPS agreement to the effect that regulations must be no more trade restrictive than necessary (44).

2. NAFTA Dispute Settlement

As a general rule, disputes arising under both GATT and NAFTA may be settled in either forum "at the discretion of the complaining party" (Article 2005) (44). However, in any dispute arising under NAFTA TBT or NAFTA SPS, the responding party may request that the dispute be considered under NAFTA dispute resolution procedures. If the responding party so requests, the dispute is resolved under NAFTA. Thus, while NAFTA would not preclude Canada or Mexico from bringing a trade dispute against the United States under GATT, if the matter concerns a standard or an SPS measure, the United States has the option of having the matter resolved through NAFTA dispute resolution (44).

Disputes that arose between Canada and the United States under the CUSFTA are carried over to the counterpart procedures in NAFTA (28). There was one such dispute that has affected FDA. Canada challenged Puerto Rico's decision to exclude Ultra High Temperature milk from Quebec, resulting in a decision ruling that the United States had not violated the agreement but recommending a determination by the United States as to whether the milk in question was safe for importation into the United States (45). This decision set the stage for a study by the two countries of whether the regulatory systems are equivalent.

C. Practical Considerations Involving GATT and NAFTA

Practical considerations weigh against other countries bringing trade challenges against NLEA regulations. As was discussed previously, much world trade involves food in bulk form that will be further processed and then labeled. This food is not subject to laws like NLEA that apply to food as sold to the ultimate consumer (7,13). Rather, it is subject to much simpler identification requirements under customs and food laws.

Furthermore, it is commonplace for firms to have to design and use a variety of labels, to meet the various regulatory demands of importing countries—many of these driven by linguistic differences. Also, compliance costs for NLEA are initial start-up costs that occur in the first year and will not recur except as to new products. Thus, the fact that costs will flatten out after year one might dampen enthusiasm for longer-lived trade disputes. Additionally, some of the approaches that FDA has used, as discussed above—such as internationally recognized methods and definitions, and the opportunity to use nutritional databases that obviate costly product-specific analyses—assist foreign as well as domestic food firms in meeting the requirements of the regulations without undue costs.

If it is surprising that a comprehensive overhaul of U.S. food labeling laws could be enacted without consideration of global factors, it should be remembered that, although imported and exported goods occupy an increasing share of the U.S. economy—about 25%—this share is still small compared to the trade-dependency of other countries (46,47). Because of the sheer size of our internal national market and our unusual degree of self-sufficiency due to our food production sector, much of what we produce is labeled and sold here (46,47).

To say that international issues were not considered when NLEA was passed is not to say that NLEA is unfair to other countries. To reiterate, NLEA is an even-handed, perfectly valid exercise of a country's sovereign prerogative to pass a law that meets a legitimate objective, improving nutritional information to enhance health (43). Indeed, its principles are worthy of international consideration.

VIII. ACKNOWLEDGEMENTS

The views expressed are those of the author and not necessarily those of the FDA or the U.S. Department of Health and Human Services.

IX. REFERENCES

1. Food and Agriculture Organization of the United Nations (FAO)/World Health Organization (WHO), Code of Ethics for International Trade in Food (1985), *Codex Alimentarius* 2d Ed., 1992, Vol. 1, p. 20.
2. Federal Food, Drug, and Cosmetic Act, Sections 201 and 403 (21 USC 321, 343).
3. FAO/WHO, Codex General Standard for the Labelling of Prepackaged Foods (World-wide Standard), Codex Stan 1-1985 (Rev. 1-1991); Codex General Standard for the Labelling of and Claims for Prepackaged Foods for Special Dietary Uses (Codex Stan 146-1985).
4. FAO/WHO Codex Guidelines on Nutrition Labelling (CAC/GL 2-1985)[Draft Amendment on Nutrient Reference Values adopted by the 19th Session of the Commission]; Codex Guidelines on Claims (CAC/GL 1-1979 (Rev. 1-1991)).
5. FAO/WHO, *Codex Alimentarius* 2d Ed., 1992, Vol. 14.
6. FAO/WHO, Codex Alimentarius Commission, Twentieth Session, Geneva, 28 June-7 July 1993, Report of the Twenty-second Session of the Codex Committee on Food Labelling, Ottawa, Canada, 26-30 April 1993, pp. 6, 10, 15, 26, 35, 40, 45.
7. Food and Drug Administration, FDA Backgrounder, The New Food Label, December 10, 1992.
8. Pub. L. 102-329 repealed those provisions of Pub. L. 102-245 that amended the Fair Packaging and Labeling Act (15 USC 1451 *et seq.*).
9. Food and Drug Administration. Food Labeling; Metric Labeling Requirements; Proposed Rule. *Federal Register 58*: 29715-29726, May 21, 1993.
10. Food and Drug Administration, Metric Labeling; Quantity of Contents Labeling Requirements for Foods, Human and Animal Drugs, Animal Foods, Cosmetics, and Medical Devices. *Federal Register 58*: 67443-67464, December 21, 1993.
11. Federal Food, Drug, and Cosmetic Act, Section 401 (21 USC 341).
12. Speech at monthly seminar of the Washington, D.C. area chapter of the Institute of Food Technologists, German Embassy, Washington, D.C., November 4, 1993.
13. Food and Drug Administration, Food Labeling; Final Rules, *Federal Register 58*: 2066, 2081, 2087-89, 2097, 2099, 2103, 2104, 2128. Washington, D.C., January 6, 1993.
14. Food and Drug Administration, Draft Nutrition Labeling Manual; A Guide for Developing and Using Data Bases; Availability. *Federal Register 57*: 32796, July 23, 1992.
15. Nutrition Labeling and Education Act of 1990, Public Law 101-535, 104 Stat.2353 (1990).
16. Office of Press and Public Affairs, European Commission Delegation, Washington, D.C. "Post-Maastricht: EC Now Called European Union," *European Union News*, 61/93, December 13, 1993.
17. R. Watson and G. McIvor, "EU suitors face fresh snags as deadline nears," *The European*, No. 194, 28 January-3 February 1994.

18. Commission of the European Communities, Consolidated Instruments on Foodstuffs, June 14, 1993.
19. Directive 90/496/EEC (24 September 1990).
20. Commission of the European Communities, *Communication from the Commission to the Council and the European Parliament* Concerning Language Use in the Information of Consumers in the Community, COM 93-456, Brussels, 10 November 1993. "New Report [by European Consumers Organization to the European Parliament] Critiques Lack of Label Harmony in Europe," *Food Labeling News*, January 13, 1994, pp. 13-14.
21. "Greenlaws fall prey to subsidiarity," *The European*, No. 193, 17-23 December 1993. Joint Statement, Fifth Bilateral meeting between The Commission of the European Communities and The United States Food and Drug Administration, 16-17 November 1993.
22. Meeting Report, Canada-U.S. Free Trade Agreement Working Group on Packaging and Labelling, Washington, D.C., September 3, 1993.
23. Food Division, Consumer Products Branch, Consumer and Corporate Affairs Canada, Summary of Canadian, U.S. and Codex Requirements Pertaining to Ingredient Listing.
24. Japan Food Hygiene Association, Specifications and Standards for Foods, Their Containers, and Packages, Etc. [Based on the Food Sanitation Law], 1992.
25. FAO/WHO, *Codex Alimentarius Commission Procedural Manual*, Eighth Edition, 1993, pp. 47-48, 52-53, 131-132, 138-140.
26. Food and Drug Administration, Food Labeling; Foods Derived From New Plant Varieties. *Federal Register 58*: 25837-25841, April 28, 1993.
27. "Proposed Medium Term Plan for the Codex Alimentarius Commission 1993-1998," Joint FAO/WHO Food Standards Programme, Codex Alimentarius Commission, Twentieth Session, Geneva, 28 June-7 July 1993, Agenda Item 12.
28. Letter from C. Hills, then U.S. Trade Representative, to M. Wilson, then Canada's Minister of Industry, Science and Technology and Minister for International Trade, January 19, 1993.
29. R. Tannahill, *Food in History*, Stein and Day, New York, 1973, pp. 198-200, 375-377.
30. Office of the U.S. Trade Representative, *Final Act Embodying the Results of the Uruguay Round of Multilateral Trade Negotiations* (Marrakesh, 15 April 1994).
31. Office of the U.S. Trade Representative, *1994 Trade Policy Agenda and 1993 Annual Report of the President Under the Trade Agreements Program*, 1994, p. 19.
32. J. Jackson, *The World Trading System*, The MIT Press, Cambridge, 1992, pp. 62-74.
33. 19 USC 2191.
34. *The Text of the General Agreement on Tariffs and Trade*, Geneva, 1986, pp. 2, 6, 37-38.
35. United States Restrictions on Imports of Tuna, Panel Report [under GATT], September 3, 1991.

36. Letter from M. Kantor, U.S. Trade Representative, to Representative H. Waxman, September 7, 1993.
37. Letter from J. Ward, National Resource Defense Council [and four other environmental groups] to M. Kantor, December 3, 1993.
38. Testimony of Ralph Nader on the Uruguay Round of GATT, before the Trade Subcommittee of the House Ways and Means Committee, February 1, 1994.
39. Fair Trade Campaign of the U.S., "Will a Mad Rush to a False Deadline Lead to a GATT Failure in Congress?" [paid advertisement], *Financial Times*, December 9, 1993.
40. M. Rothschild, "Potential International Trade Challenges to the New FDA Food Labeling Requirements" [student paper], July 14, 1993, pp. 5-7.
41. TBT Notif. 91.330, December 11, 1991.
42. Letters to National Institute of Standards and Technology from Japanese government, dated February 24 and September 29, 1992, and from European Commission, dated February 25, 1992.
43. Nutrition Labeling and Education Act of 1990, H.Rept. No. 101-538, June 13, 1990; *Congressional Record*, July 30, 1990, pp. H5836-45.
44. North American Free Trade Agreement Between the Government of the United States of America, the Government of Canada, and the Government of the United Mexican States, 1992.
45. In the Matter of Puerto Rico Regulations on the Import, Distribution and Sale of U.H.T. Milk from Quebec, Final Report of the Panel [convened under Chapter 18 of the CUSFTA], June 3, 1993.
46. U.S. Imports and Exports by Principal Commodity Groupings, Office of Trade and Economic Analysis, U.S. Department of Commerce, 1992, reprinted in *The World Almanac 1994*, p. 184.
47. M. Cossolotto, *The Almanac of Transatlantic Politics, 1991-1992*, Brassey's, Washington, 1991, pp. 7, 21.

6

Food Labeling—The Agony and the Ecstasy: An Industry Perspective on Nutrition Labeling and Education Act

Paul F. Hopper*

General Foods Corp., White Plains, New York

I. INTRODUCTION

There can be little doubt that the Nutrition Labeling and Education Act of 1990 (NLEA) has brought about the most significant change in the way that nutrition information has been brought to the consumer's doorstep in the history of the food industry. In theory, it is a major step forward in providing the public with easier-to-use tools with which to make informed dietary choices. The implementing of the Food and Drug Administration regulations have brought a long sought after uniformity and consistency to the food label. On the other hand, NLEA has raised many issues of inflexibility, and even irrationality, that many view as impediments to the development of desirable new products, and to the communication of the healthful dimensions of both old and new foods.

This chapter will highlight some of the background events, and then examine NLEA as it is being implemented today, including some of the important issues. Finally, the education and communication needs will be discussed, along with a glimpse at NLEA tomorrow.

* *Present affiliation*: Department of Food Science, Cornell University, Ithaca, New York.

II. BACKGROUND

The historical background of the Nutrition Labeling and Education Act has been well covered in Chapter 1. These brief comments will serve to elucidate the role that industry has played.

Industry involvement over the last twenty years in developing and implementing regulations that would provide the consuming public with useful nutrition information has been enormous. Beginning in the early 1970s, scientists from industry joined with other experts in a study conducted by the National Academy of Sciences to provide recommendations to the Food and Drug Administration as to the kind of information that could be beneficial in making a more informed choice in the selection of foods that would meet individual dietary requirements. Having served on that Panel, I can testify as to the diversity of opinion that existed, and the difficulty of achieving consensus.

Once the report was forwarded to FDA, an equally laborious process took place in developing the format by which the chosen nutrition information would best be presented on the label of food products. Many of the same suggestions that have been considered in the current rule making for NLEA, were also a part of these earlier deliberations. The Technical Committee of the Grocery Manufacturers of America was very helpful in evaluating the various options. It was recognized from the beginning, by those knowledgeable in both the science and in consumer behavior, that this newly emerging source of nutrition information would be of only limited usefulness unless it were accompanied by a comprehensive and widespread educational program. The consuming public in general did not have the background to apply this information effectively to improve meal planning.

Shortly after the publication of the final 1973 voluntary labeling regulations, an ad hoc committee was formed to develop a booklet that would help fill the education gap for the consumer. I chaired that Committee which consisted of scientists from government, academia and industry. Chapters were assigned to experts in that field and a comprehensive peer review was conducted. The document entitled "Nutrition Labeling: How It Can Work for You" was published by the National Nutrition Consortium, Inc. with a grant from the General Foods Corp. Dr. Mark Hegsted of Harvard was the president of the Consortium at that time and he tried unsuccessfully to obtain government funding for a widespread distribution of the booklet.

This example is cited as a word of caution to the authorities responsible for the implementation of NLEA. The 1973 Nutrition Labeling initiative fell short of its objective in no small part because of a lack of a long term commitment, and the corresponding funding for a comprehensive program of promotion and education. The regulations implementing NLEA are voluminous and complicated. A major educational effort must accompany this program if it is to achieve its stated goals, all other issues aside. Later in this chapter this subject will be addressed, citing the very promising beginnings.

The food industry continued its cooperative efforts to help provide useful health related information to the consumer. In 1984, following considerable dialogue with industry, FDA published new regulations governing sodium labeling. The disappointment in this action however, was the refusal of FDA to permit label statements that cited the importance of reduced sodium intake to health.

In that same year, Kellogg drew a line in the sand and began making explicit label claims on the relation of high fiber diets in the reduction of the risk of certain kinds of cancer. Shortly thereafter, Associate Commissioner Paul Hile delivered a landmark speech announcing a change in FDA policy regarding health claims. Industry welcomed this change which no longer placed a food with certain health claims in the drug category.

III. NLEA TODAY

The relationship between diet and health rapidly became the new marketing strategy for many companies. Without appropriate guidelines, however, confusion reigned, and many individual interpretations of this policy change emerged. Food products increasingly touted their health benefits, and words like "lite," "low calorie," "no cholesterol," "high fiber," and other such statements, proliferated on label after label. "Horsepower" wars emerged, especially on the fiber-containing breakfast cereal category. In response, FDA proposed various draft regulations in order to restore some order to the cacaphony of practices. But progress was slow in finalizing these regulations. The first FDA proposal was issued in 1987, and the second proposal in 1990.

It could have been predicted that the patience of Congress would soon wear thin, but the speed with which NLEA was passed surprised many veterans of Capitol Hill debates. Suffice to say, as FDA began drafting the implementing regulations, it became apparent that the authorities were taking a very conservative approach. As the food industry looked to the future, concerns arose as to the viability of current and projected research and product development programs. This concern will be discussed later on.

As it stands today, "no expressed or implied health claim may be made on the label or in labeling" unless these requirements are met: 1) the claim must be "specifically provided for" in regulations issued by FDA; 2) the claim must conform to all general requirements established by FDA for health claims; 3) the food must not exceed "disqualifying levels" for fat, cholesterol, or sodium; 4) no substance may be present at an "inappropriate level"; 5) the food may not be represented for infants or toddlers without specific authorization; and 6) the food must contain at least 10% of the DRV for one of the following—vitamin A or C, iron, calcium, protein, or fiber.

While a key part of the NLEA regulations focuses on health claims, other provisions deal with nutrient content claims, the format for the Nutrition Facts panel, serving size, percent daily value and calories from fat. Details have been discussed in earlier chapters. NLEA specifically addressed food products governed by FDA's jurisdiction, but since its passage USDA has established similar requirements for processed meat and poultry.

Nutrition information for the 20 most commonly eaten raw fruits and vegetables, and the 20 most commonly eaten types of fresh seafood has been available for over a year on posters, shelf labels, or in brochures displayed in grocery stores. Certain food items such as coffee, tea, spices, and flavorings are exempt.

IV. CONTROVERSIAL PROVISIONS

Any document of the magnitude of FDA's new labeling regulations is bound to have its critics. NLEA is no exception. Without attempting to address each and every concern, the following list covers some of the more frequently voiced issues.

A. Percent Daily Value

The provision to require listing the Percent Daily Values for selected nutrients on the label is a totally new concept for the consumer to understand. It is further complicated by the explanatory section at the bottom of the Nutrition Facts panel on larger packages which lists the number of grams of fat, cholesterol, sodium, carbohydrate and dietary fiber that are considered ideal for a daily caloric intake of not only 2,000 calories but also for 2,500 calories.

FDA in one of its educational charts says: "Daily Values are also something new. Some are maximums, as with fat (65 grams or less); others are minimums, as with carbohydrates (300 grams or more)." As it states in the Nutrition Facts panel: "Your daily values may be higher or lower, depending on your caloric needs."

Not only can this be confusing to the reader, it encourages the tendency to categorize foods as "good" or "bad" when comparing packages. FDA in their instructional sheet "TIPS for Using the New Food Label" states, "So you can simply look at a single food and see which nutrients the food is high or low in, or you can use this column to compare two different foods." It goes on to say "You may want to pick out only one or two things to look at on the label.

Remember, you do not need to know everything on the label. Focus on what is important to you."

Clearly such instruction totally misses the mark of sound nutrition advice—which is Balance, Variety, and Moderation.

B. Claims

There are two types of claims which are authorized by NLEA. These are nutrient content claims and health claims. While there is general endorsement for the standardization of nutrient content claims such as "free," "low," "reduced/less," etc., and for relative or comparative claims such as "lite," "lean," "good source," and the like, the specific restrictions that accompany these definitions have proven to be very limiting. In certain instances, these limitations have required the reformulation of products of longstanding, resulting in an uncertain market acceptance in the future.

The health claim requirements raise a whole new set of issues. These will be discussed briefly in this section, and will be dealt with more fully later in this chapter, when we consider the NLEA of tomorrow.

The two principal concerns are the very limited number of claims that are initially being allowed, and the very conservative and ambiguous concept of "significant scientific agreement" that is required for any future health claim.

The wording of certain "model" claims is laborious, which can be a disincentive in the marketing of beneficial products. The net result will be to discourage the use of such health claims in the labeling and promotion of many products. An excellent example of the problem is the model claim for calcium and osteoporosis, which reads:

Regular exercise and a healthy diet with enough calcium helps teens and young adult white and Asian women maintain good bone health and may reduce their high risk of osteoporosis later in life.

C. Exclusions

The threshold of nutrient content provision prohibits the use of health claims unless the product contributes at least 10% of the Recommended Daily Intake for at least one of the specified nutrients without fortification. This so-called jelly bean rule is intended to prevent claims on products that FDA considers trivial. A related provision disallows health claims for foods like butter, that are normally eaten in small amounts. While it can be argued that these exclusions have merit, they never the less preclude the opportunity to provide useful nutrition information.

If a food exceeds the following levels of disqualifying substances, it is also prohibited from making a health claim. These levels are 13 g of fat; 4 g of saturated fat; 50 mg of cholesterol; or 480 mg of sodium per reference amount customarily consumed per label serving size, and per 50 g for foods with reference amounts customarily consumed of 30 g (1 tablespoon) or less.

D. Differing Jurisdictions

The inconsistencies between the FDA labeling requirements and those of USDA are beginning to sort themselves out. This issue is further complicated by USDA's newly issued "Safe Handling Policy" for ready-to-eat meat and poultry products. The latter is designed to protect consumers from possible pathogens that may have contaminated such products by recommending thorough cooking. It adds another space burden to what is already a very crowded label.

A comparison of FDA and USDA Nutrition labeling requirements represents the state of the two regulations as of mid-May, 1994 and is presented in Chapter 3 by John McCutcheon.

E. Supplement Labeling

The strict requirement of "significant scientific agreement" on health claims for a variety of dietary supplements presents a significant problem for that industry. As currently interpreted, it could very well exclude many supplements from making such claims. But in the words of one immortal baseball player—"This is deja vu all over again."

The champions of dietary supplements wish to continue their traditional way promoting these products, many of which have been on the market for years. They perceive NLEA as over-regulating and are calling on Congress to help.

In a draft bill currently in committee, the responsibility for deciding the appropriate level of scientific support necessary to justify a health claim would be shifted to the dietary supplement manufacturer instead of FDA. The standard would be markedly more lenient than that for health claims on a food product.

In the early 1970s, a similar situation arose when Senator Proxmire led a successful campaign to minimize FDA's authority over the dietary supplement industry. Now, NLEA has shifted the responsibility once again, providing FDA with both labeling and safety enforcement authority.

Other issues involving supplements include the listing of nutrients for which there is no RDI on the Nutrition Facts panel; labeling for high potency supplements; the status of powders added to milk or water; and, of course, toxicity or safety requirements. Much still needs to be done in both the legislative and regulatory arena to resolve these differences in implementing NLEA. Meantime, FDA continues to issue warning letters to the health food and the dietary supplement industries for products that are inappropriately labeled, or for which there is a safety concern.

V. EDUCATION

It cannot be emphasized enough that the true success of NLEA will not be realized until we have in place a comprehensive, ongoing and well-funded program of communication and education. All the tools of the trade need to be brought to bear in a consistent and reinforcing manner. No sporadic nor haphazard approach can be expected to have the needed impact. To borrow an old Latin phrase, "repetitio est mater studiorum."

The challenge is formidable. First is the need to identify and then to validate the nutrition messages that are important to communicate. Next is the process of segmenting the population to identify those groups with special needs (e.g., children, the elderly, those whose immune system has been compromised, etc.). Such groups will need a program specifically tailored for them, in addition to the broad-based thrust for the general population. The choice of vehicles of communication is extremely important. We need to reach the professional community such as nutritionists, dieticians, physicians, and the like; the media, both print and electronic, which will be a vital bridge to the consuming public; and finally, directly to consumers themselves. Each of these target audiences will have a unique role to play in helping the consumer to understand the new nutrition information and then to be motivated to put it to good use in making informed food choices consistent with individual preferences and lifestyles.

One very important step toward this education goal is the recently formed National Exchange for Food Labeling Education (NEFLE), sponsored initially by FDA and USDA's Food Safety and Inspection Service (FSIS). In cooperation with other public as well as private sector groups, NEFLE is producing educational materials, conducting research and engaging in a large number of media activities in an effort to bring the consumer up to date with these major changes in food labeling.

A comprehensive listing of these programs is available from the National Agricultural Library. Needless to say, the food industry is an active participant and is helping to devise ways to promote programs that can effectively educate and instruct the consuming public on the principles of good nutrition—balance, variety, and moderation—and how best to use the label to achieve these goals.

VI. NLEA TOMORROW

It is important that the experience gained in implementing today's nutrition labeling regulations be continually monitored and evaluated. Issues have already been identified and new ones will emerge. There must be an ongoing dialogue among the FDA, the food industry, and the consumer to determine if NLEA, as being implemented, is effectively serving the needs of the public.

Industry believes that it has the right to communicate the healthful dimensions of its products. It is committed to facilitating that process. If changes in the regulations or certain amendments to NLEA are needed to make nutrition labeling and education more effective, there will be strong support for such change.

Industry is also interested in embarking on programs of research and product development that will lead to the availability of new products that promote health. Investment in such programs can best be stimulated if the ground rules for the acceptance of appropriate claims are understood well in advance of such research.

This becomes increasingly important as biotechnology research plays a larger role in the development of new products. There is no argument for the need to publish such studies in peer-reviewed journals. But the requirement for significant scientific agreement, as currently interpreted, may not be realistic for new breakthroughs resulting form this new technology. There is another aspect to incentives for undertaking research on new products that promote health. That issue involves ways to provide a company with sufficient lead time to recoup the cost of such research.

Key to all of these issues is the question of reasonability and clarity. The standard that must be met in order to make a new health claim; the protocols to be followed in gathering and evaluating scientific evidence in support of health claims; the incentives needed to encourage research on new products; the rules for communicating the healthful dimensions of food products, whether on the label or elsewhere; and the educational programs that are needed to assist the consumer in effectively utilizing this nutrition information must all be debated and agreed upon.

A major step in that direction is currently underway with a policy dialogue program on food, nutrition and health. Representatives from government, academia, health professional groups, consumer interest organizations, and industry are engaged in a three-year dialogue under the leadership of the Keystone Center. Issues of research and of communication and education are under intense discussion with policy recommendations as the ultimate goal.

VII. SUMMARY

The regulations implementing NLEA are extremely complex and at times quite limiting. With over 800 pages in the *Federal Register* to consider, the food industry had to embark on a crash program of evaluation, and often redesign not only of the package and its label, but often the product itself. All of this took place in a short 18 months.

This chapter has reviewed the major elements of the NLEA nutrition labeling regulations and the issues they have raised from an industry perspective. Dietary supplement labeling has been addressed briefly, including efforts to enlist the aid of Congress in simplifying some of the current provisions. A strong educational effort was stressed, with progress being shown with the work of the National Exchange for Food Labeling Education.

Future research and development leading to new and improved products that promote health can be encouraged by clear direction as to the standards that need to be met for future health claims, and the research protocols needed to meet these requirements. These and other incentives were discussed. A major step in that direction has been the formation of a policy dialogue on food, nutrition, and health under the auspices of the Keystone Center. Participants include representatives from government, academia, health professionals, consumer interest groups, and industry.

7

A Comprehensive Review of the Nutrition Labeling and Education Act Regulations

Ralph Shapiro

Nutrition International, Dayton, New Jersey

I. INTRODUCTION

On October 26, 1990, the U.S. Congress passed the Nutrition Labeling and Education Act (NLEA) of 1990 (PL-101-535) and on November 8 of the same year, President Bush signed it into law. On January 6, 1993 the Food and Drug Administration (FDA) published final regulations implementing the law. Technical amendments were published on August 18, 1993.

The regulations are extensive (over 4,000 typed pages) (hundreds of pages in the *Federal Register*), detailed and complex. The purpose of this chapter is to highlight the main features of the 1990 NLEA and FDA's final rules. It has been written to facilitate reader understanding of the regulations by minimizing cumbersome legal terminology as it appears in the *Code of Federal Regulations*, Title 21 (21 CFR) and the *Federal Register* Vol. 58, No. (3), January 6, 1993 (FR)[1] without losing content or meaning. However, it is not meant to be all encompassing and the reader is encouraged to refer to the specific regulations whenever detailed information is required.

The chapter is divided into 13 sections. Sections I–III provide an overview of the regulations. Sections IV–XIII are considerably more detailed and represent a condensation of the rules. Each section is accompanied by appropriate CFR and *Federal Register* citations for easy reference. The sections are divided as follows:

- Sections I–III Overview
- Section IV Rules for mandatory nutrient declaration
- Section V Rules for voluntary nutrient declarations
- Section VI Label formats
- Section VII Reference Daily Intake (RDI) and Daily Reference Value (DRV)
- Section VIII Serving size
- Section IX Nutrient content claims
- Section X Standardized foods
- Section XI Requirements for health claims
- Section XII Ingredient declarations
- Section XIII Percent juice declarations
- Section XIV Nutrient content claims for butter
- Section XV Comparison of FDA and USDA nutrition labeling regulations

It should be noted that although not mandated by NLEA, the Food Safety and Inspection Service (FSIS) of the United States Department of Agriculture (USDA) simultaneously published separate nutrition labeling regulations on January 16, 1993, applicable to meat and poultry products under their purview. The FDA and USDA regulations are virtually the same, although there are some differences. While this chapter focuses on FDA rules, a section comparing the regulations of each agency is included. Chapter 3 by John McCutcheon focuses on USDA regulations.

Both FDA and USDA set compliance dates for products governed by the regulations. Product labels already complying with the rules published on January 6, 1993 (and prior to the August 13, 1993, Technical Amendments) may be used until the supply is exhausted. Thereafter, all newly printed labels must be updated. All products labeled (i.e., the date that the label is placed on the product or its container) after May 8, 1994, must comply. All USDA regulated

Table 7.1 Compliance Dates

November 8, 1994: Voluntary program for raw agricultural commodities and raw fish.
May 8, 1993:
  Ingredient labeling of standardized foods and certified colors
  Health claims
  Percent juice labeling
May 8, 1994[a]: Majority of FDA regulations common or usual name for juice
July 6, 1994: USDA regulated products

[a] Amended to August 8, 1994, by court order.

products must comply after July 6, 1994. A synopsis of the compliance dates is presented in Table 7.1.

II. MANDATORY NUTRITION LABELING

The main features of the 1990 NLEA are outlined in Table 7.2. The new regulations require nutrition labeling on virtually all foods which are intended for human consumption and for sale unless they are specifically exempt. Under these regulations, nutrition information must appear on packaged food labels or it must be prominently displayed at the point of purchase for unpackaged products (e.g., placards, pamphlets, tags, full display panels). It is estimated that approximately 90% of all processed food will carry nutrition information.

Exemptions are specified in the regulations and are listed in Tables 7.3 and 7.4. A product loses its exemption and full nutrition labeling becomes required if its label, display, or menu advertising contains any nutrition information, a health claim, or a nutrient content claim (descriptor) expressed explicitly or implicitly.

Under FDA and USDA voluntary point of purchase programs (November 8, 1991) nutrition labeling is presently voluntary for: the 20 most frequently eaten raw fruits, vegetables and fish and the 45 major cuts of raw, single ingredient meat and poultry products. While these programs are voluntary, both agencies have guidelines stating that if there is not "significant participation" (e.g., at least 60% of the stores evaluated must voluntarily label at least 90% of these products on the package or by display at the point of purchase, namely, by brochures, pamphlets, placards, monitors, etc.) they will consider initiating a mandatory program[2].

Table 7.2 Main Features of the 1990 NLEA and the FDA Final Rule

Requires nutrition labeling for virtually all processed food products unless specifically exempt.
Provides a revised uniform list of required nutrients and conditions for label declaration.
Establishes Recommended Daily Intake values (RDI) and Daily Values (DV).
Defines reference amounts and establishes serving size specifications.
Defines nutrient descriptors (nutrient content claims).
Defines health claims.
Standardizes and provides guidelines for nutrition label formats.
Defines ingredient labeling requirements.
Establishes Federal pre-exemption rules.
Provides for state and national enforcement.

Table 7.3 Exemptions from Mandatory Nutrition Labeling

Unconditional exemptions[a]
- Infant formulas
- Foods for infants and children less than 4 years of age other than infant formulas.
- Vitamin and mineral supplements.
- Medical foods
- Bulk foods intended for further processing or repackaging
- Products sold from bulk containers
- Shell eggs
- Game meats
- Packages of single ingredient products that consist of fish or game meat
- Unit containers in a multiunit retail package with overwrap labeling

Conditional exemptions[b]
- Small business[c]
- Food for immediate consumption (restaurants)
- Ready to eat foods not for immediate consumption (carry out establishments)
- Foods that contain insignificant amounts of all required nutrients.
- Foods in small packages (packages with a total surface area less than 12 square in.).
- Raw fruits, vegetables and fish subject to voluntary nutrition labeling.

[a] Unconditional exemptions: The exemption will remain in effect whether or not the label bears a nutrient content claim, health claim or other nutrition information.
[b] Conditional Claim: The exemption will be lost if the label bears a nutrition claim or provides other nutrition information.
[c] See Table 7.4 for details on small business exemptions.

Table 7.4 Requirements for Small Business Nutrition Labeling Exemption[a]

Maximum number of employees[b]	Number of manufactured units[c]	Exemption deadline
< 300	< 600,000	May 8, 1995
< 300	< 400,000	May 8, 1996
< 200	< 200,000	May 8, 1997
< 100	< 100,000	Until further notice

[a] To qualify for exemption companies must file with the FDA prior to the exemption deadline.
[b] Full-time equivalent employees.
[c] Package or form in which product is sold.

III. NUTRITION PANEL

A revised list of required nutrients and food components and the conditions for declaring them on labels are stipulated in the new regulations. These revisions were made to address the relationship of better nutrition to current health concerns.

Under the new regulations, certain nutrients must be declared on the label and others may be declared voluntarily. Only those nutrients or food components listed in Tables 7.5 and 7.6 as either mandatory or voluntary, respectively, may be on or included on the nutrition label. They must appear in a uniform, defined sequence together with the designated specified unit of expression. The regulations define quantitative increments of expression as shown in Table 7.7 for the mandatory nutrients and Table 7.8 for the voluntary nutrients.

Voluntary nutrients must be declared if they are added to a food or if a claim is made about the nutrient or related nutrient or food component. Table 7.9 outlines the conditions under which voluntary nutrient declarations become mandatory.

In some cases, because a "mandatory" nutrient or food component is present in a product in very low or "insignificant" amounts, a label declaration may not be required or an alternative statement may be permitted. The new regulations define an insignificant amount as the amount that may be rounded to zero

Table 7.5 Mandatory Nutrients

Calories[a]	Dietary fiber (grams)
Calories from fat	Sugars (grams)[e]
Total fat (grams)[b]	Protein (grams)
Saturated fat (grams)[c]	Vitamin A (% of DV)
Cholesterol (milligrams)	Vitamin C (% of DV)
Sodium (milligrams)	Calcium (% of DV)
Total carbohydrate (grams)[d]	Iron (% of DV)

[a] Energy content per serving may also be expressed in kilojoule units added in parenthesis immediately following the statement of the calorie content.
[b] Total fat: Defined as total lipid fatty acids expressed as triglycerides.
[c] Defined as the sum of all fatty acids containing no double bonds.
[d] Calculated by subtracting the sum of crude protein, total fat, moisture, and ash from the total weight of the food.
[e] Defined as the sum of all free mono- and disaccharaides.

Table 7.6 Voluntary Nutrients

Calories from saturated fat	Soluble fiber (grams)
Polyunsaturated fat (grams)[a]	Insoluble fiber (grams)
Monounsaturated fat (grams)[b]	Sugar alcohol (grams)[c]
Potassium (milligrams)	Other carbohydrate (grams)[d]
Vitamins and minerals for which there is an established RDI (% of daily value)	

[a] Polyunsaturated fats defined as *cis-cis* methylene-interrupted polyunsaturated fatty acids.
[b] Monounsaturated fats defined as *cis*-monounsaturated fatty acids
[c] Sugar alcohol defined as the sum of saccharide derivatives in which a hydroxyl group replaces a ketone or aldehyde group and whose use in a food is listed by the FDA (e.g., mannitol) or is generally recognized as safe (GRAS) (e.g., xylitol, sorbitol).
[d] "Other carbohydrate" defined as the difference between total carbohydrate and the sum of dietary fiber, sugars and sugar alcohol. If sugar alcohol is not declared (even if present), it is defined as the difference between total carbohydrate and the sum of dietary fiber and sugars.

Table 7.7 Guide to Numerical Expression of Mandatory Nutrients per Serving[a]

Nutrient	Incremental numerical expression	When "zero" is permitted
Calories	To nearest 5 calories if ≤ 50 calories; to nearest 10 calories if > 50 calories	< 5 calories
Calories from total fat	To nearest 5 calories if ≤ 50 calories; to nearest 10 calories if > 50 calories	< 5 calories
Total fat	To nearest 0.5 g if ≤ 5 g; to nearest 1g if > 5 g	< 0.5 g
Saturated fat	To nearest 0.5 g if ≤ 5 g; to nearest 1g if > 5 g	< 0.5 g
Cholesterol	To nearest 5 mg	< 2 mg
Total carbohydrate	To nearest 1 g	< 0.5 g
Sugars	To nearest 1 g	< 0.5 g
Dietary fiber	To nearest 1 g	< 0.5 g
Protein	To nearest 1 g	< 0.5 g
Sodium	To nearest 5 mg if ≤ 140 mg; to nearest 10 mg if > 140 mg	< 5 mg
Vitamins A, C; calcium; iron	To nearest 2% of RDI if ≤ 10%; to nearest 5% of RDI if > 10% and ≤ 50%; to nearest 10% of RDI if > 50%	< 2% RDI

[a] See Table 7.5 for definitions of total fat, saturated fat, total carbohydrates and sugars.

Table 7.8 Guide to Numeral Expression of Voluntary Nutrients per Serving[a]

Nutrient	Incremental numerical expression	When "zero" is permitted
Calories from saturated fat	To nearest 5 calories if ≤ 50 calories; to nearest 10 calories if > 50 calories	< 5 calories
Polyunsaturated fat	To nearest 0.5 g if ≤ 5 g; to nearest 1 g if > 5 g	< 0.5 g
Monounsaturated fat	To nearest 0.5 g if ≤ 5 g; to nearest 1 g if > 5 g	< 0.5 g
Potassium	To nearest 5 mg if ≤ 140 mg; to nearest 10 mg if > 140 mg	< 5 mg
Soluble fiber	To nearest 1 g	< 0.5 g
Insoluble fiber	To nearest 1 g	< 0.5 g
Other vitamins and minerals with RDI's	To nearest 2% of RDI if ≤ 10%; to nearest 5% of RDI if > 10%; to nearest 10% if > 50%	< 2% RDI
Sugar alcohol	To nearest 1 g	< 0.5 g
"Other" carbohydrate	To nearest 1 g	< 0.5 g

[a] See Table 7.6 for definition of polyunsaturated fat, monounsaturated fat, sugar alcohol and "other carbohydrate.

Table 7.9 When Does a Voluntary Nutrient Declaration Become Mandatory?

All voluntary nutrients	If added to a food
	If claim is made about the nutrient
Polyunsaturated fat	If monounsaturated fat is declared on label
	If fatty acid or cholesterol claims are made
Monounsaturated fat	If polyunsaturated fat is declared on label
	If fatty acid or cholesterol claims are made
Sugar alcohols	If claim is made about sugars when sugar alcohol is present in the food

Table 7.10 Alternative Label Declarations for Very Low Amounts of Nutrients/Food Components

Nutrient/food component	Amount per serving	Statement
Cholesterol	2–5 mg	"Contains less than 5 mg cholesterol per serving"
Total carbohydrate	< 1 g	"Contains less than 1 g carbohydrate per serving"
Dietary fiber	< 1 g	"Contains less than 1 g fiber per serving"
Sugars	< 1 g	"Contains less than 1 g sugar per serving"

(see Tables 7.7 and 7.8). In such cases, the statement "contains less than," "less than" or "not a significant source of" may be used for the appropriate nutrient or food component. In other situations where the nutrient or food component is present in a very low amount but not sufficiently low to be rounded to zero (e.g., not significantly low to qualify as "insignificant"), the alternative label declarations shown in Table 7.10 may be used.

IV. ALLOWABLE STATEMENTS FOR MANDATORY NUTRIENTS

The allowable statements for mandatory and voluntary nutrients are described below and in Section V, respectively. These declarations must appear in the "Nutrition Facts" section of the product label. As previously noted, Tables 7.7, 7.8, and 7.10 provide easy to use reference guides to their numerical expressions.

A. Calories [21 CFR 101.9 para (c)(1); FR p. 2175]

Caloric content per serving must be expressed to the nearest 5-calorie increment up to and including 50 calories, and to the nearest 10-calorie increment above 50 calories. Amounts less than 5 calories may be expressed as zero. Energy content per serving may also be expressed in kilojoule units, added in parentheses immediately following the statement of the calorie content.

Caloric content may be calculated by[3]:

- Using specific Atwater factors.
- Using the general factors of 4, 4, and 9 calories per gram for protein, total carbohydrate, and total fat, respectively.

- Using the general factors of 4, 4, and 9 calories per gram for protein, total carbohydrate (less the amount of insoluble dietary fiber) and total fat, respectively.
- Using data for specific food factors for particular foods or ingredients approved by FDA.
- Using bomb calorimetry data and subtracting 1.25 calories per gram protein to correct for incomplete digestibility.

B. Calories from Fat [21 CFR 101.9 para (c)(1)(ii); FR p. 2175]

The caloric content derived from total fat in a serving must be expressed to the nearest 5-calorie increment, up to and including 50 calories, and to the nearest 10-calorie increment above 50 calories. This declaration is not required on products that contain less than 0.5 g of fat in a serving and amounts less than 5 calories may be expressed as zero. If "calories from fat" is not required and, as a result, not declared, the statement "Not a significant source of calories from fat" must be placed at the bottom of the table of nutrient values in the same type size.

C. Total Fat [21 CFR 101.9 para (c)(2); FR p. 2176]

The amount of total fat in a serving (defined as total lipid fatty acids expressed as triglycerides) must be expressed to the nearest 0.5 g increment below 5 g and to the nearest gram increment above 5 g. If the serving contains less than 0.5 g, the content is expressed as zero.

D. Saturated Fat [21 CFR 101.9 para (c)(2) (i); FR p. 2176]

Saturated fat is defined as the sum of all fatty acids containing no double bonds. A declaration of saturated fat is not required for products that contain less than 0.5 g of total fat per serving if no claims are made about fat or cholesterol content, and if "calories from saturated fat" is not declared. If a statement of the saturated fat content is not required and, as a result, not declared, the statement "Not a significant source of saturated fat" must be placed at the bottom of the table of nutrient values in the same type size. Saturated fat content must be indented immediately under the total fat declaration and expressed as grams per serving to the nearest 0.5 g increment below 5 g and to the nearest gram increment above 5 g. If the serving contains less than 0.5 g, the content is expressed as zero.

E. Cholesterol [21 CFR 101.9 para (c)(3); FR p. 2176]

Cholesterol content in a serving must be expressed in milligrams to the nearest 5-milligram increment. This declaration is not required for products that contain less than 2 mg cholesterol per serving and make no claim about fat, fatty acids, or cholesterol content. Such products may state the cholesterol content as zero. If cholesterol content is not required and, as a result, not declared, the statement "Not a significant source of cholesterol" must be placed at the bottom of the table of nutrient values in the same type size. If the food contains 2 to 5 mg of cholesterol per serving, the content may be stated as "less than 5 milligrams."

F. Sodium [21 CFR 101.9 para (c)(4); FR p. 2176]

The amount of sodium in a specific serving of food must be expressed as zero when the serving contains less than 5 mg of sodium, to the nearest 5-milligram increment when the serving contains 5 to 140 mg of sodium, and to the nearest 10-milligram increment when the serving contains greater than 140 mg.

G. Total Carbohydrate [21 CFR 101.9 para (c)(6); FR p. 2176]

The carbohydrate content in a serving must be expressed to the nearest gram. If a serving contains less than 1 g, the statement "Contains less than 1 gram" or "less than 1 gram" may be used as an alternative. If the serving contains less than 0.5 g, the content may be expressed as zero. Total carbohydrate content is calculated by subtraction of the sum of the crude protein, total fat, moisture, and ash from the total weight of the food.

H. Dietary Fiber [21 CFR 101.9 para (c)(6) (i); FR p. 2176]

The amount of total dietary fiber in a serving, must be indented (under Total carbohydrate) and expressed to the nearest gram. If a serving contains less than 1 g, declaration of dietary fiber is not required or, alternatively, the statement "Contains less than 1 gram" or "less than 1 gram" may be used. If the serving contains less than 0.5 g, the content may be expressed as zero. If dietary fiber content is not required and, as a result, not declared, the statement "Not a significant source of dietary fiber" must be placed at the bottom of the table of nutrient values in the same type size.

I. Sugars [21 CFR 101.9 para (c)(6)(ii); FR pp. 2176–2177]

Sugars are defined as the sum of all free mono- and disaccharides (e.g., glucose, fructose, lactose and sucrose). Sugar(s) content must be indented (under Total carbohydrate and beneath Dietary fiber) and expressed to the nearest gram. If a serving contains less than 1 g, the statement "contains less than 1 gram" or "less than 1 gram" may be used as an alternative. If the serving contains less than 0.5 g, the content may be expressed as zero.

If a serving contains less than 1 g and no claims are made about sweeteners, sugar(s) or sugar alcohol content, declaration of sugar content is not required. If a statement of the sugar(s) content is not required and, as a result not declared, the statement "not a significant source of sugars" must be placed at the bottom of the table of nutrient values in the same type size.

J. Protein [21 CFR 101.9 para (c)(7); FR pp. 2177–2178]

The quantity of protein in a serving must be expressed to the nearest gram. If a serving contains less than 1 g, the statement "Contains less than 1 gram" or "less than 1 gram" may be used as an alternative. If the serving contains less than 0.5 g, the content may be expressed as zero.

1. Special Requirements

When the protein in foods for adults and children 4 or more years of age has a protein quality value (a protein digestibility-corrected amino acid score [PDCAAS]) less than 20% or when the protein in a food for children older than 1 but less than 4 years of age has a protein quality value (PDCAAS) less than 40%, either of the following must be placed adjacent to the declaration of protein content by weight:

- "Not a significant source of protein"
- The corrected amount of protein per serving calculated as a percentage of the Daily Reference Value (DRV) or Reference Daily Intake (RDI). This corrected value must appear in the Percent of Daily Value column (% DV).

When the protein quality in a food for infants as measured by the Protein Efficiency Ratio (PER) is less than 40% of the reference standard (casein), the statement "not a significant source of protein" must appear adjacent to the declaration of protein content.

A statement of the corrected amount of protein per serving, calculated as a percentage of the RDI or DRV for protein, and expressed as Percent of Daily Value, must be placed on the label, if a protein claim is made for the product, or if the product is intended for use by infants or children under 4 years of age. It may voluntarily be placed on the label in the absence of such claims.

When such a declaration is provided, it must be placed on the label adjacent to the statement of grams of protein and aligned under the column headed "Percent Daily Value," expressed to the nearest whole percent. However, the percentage of the RDI for protein shall not be declared if the food for infants has a protein quality value less than 40% of the reference standard.

2. Calculation of "Corrected" Protein Value

The "corrected amount of protein (gram) per serving" for foods for adults and children one or more years of age is equal to the actual amount of protein (gram) per serving multiplied by the PDCAAS. If the PDCAAS is greater than 1.00, it shall be set at 1.00. Acceptable protein digestibility values for a variety of foods are listed in Appendix 1, Table A1.2.

For infant foods, the "corrected amount of protein (gram) per serving" is equal to the actual amount of protein (grams) per serving multiplied by the relative protein quality value. The relative protein quality value is determined by dividing the PER value for the food by the PER value for casein. If the relative protein value is above 1.00, it shall be set at 1.00.

K. Vitamins and Minerals [21 CFR 101.9 para (c)(8); FR pp. 2177–2178]

The label must bear a statement of the amount per serving of the vitamins and minerals calculated as a percent of the RDI and expressed as % of Daily Value.

For purposes of labeling, the percent Daily Value (RDI) for infants, children less than 4 years of age, and pregnant or lactating women are shown in Table 7.11. For foods represented to be for both infants and children under 4 years of age, the percent of Daily Value must be presented by separate declarations based on the RDI values for infants from birth to 12 months of age and for children under 4 years of age. Similarly, the percent of Daily Value based on both the RDI values for pregnant and lactating women must be declared separately on foods for use by both. When such a dual declaration is used, it must be included in all labeling, and equal prominence must be given to both values. All other foods must use the RDI for adults and children 4 or more years of age (Table 7.12).

The declaration of vitamins and minerals as a percent of the RDI must include vitamin A, vitamin C, calcium, and iron, in that order, and must include any of the other vitamins and minerals when they are added as a nutrient supplement or when a claim is made about them. The declaration may also include any of the other voluntary vitamins and minerals listed in Table 7.12 when they are naturally occurring in the food. The additional vitamins and minerals must be listed

Table 7.11 Label Reference Values (RDIs)[a]

Vitamins and minerals		Infants	Children under 4 years of age	Pregnant or lactating women
Mandatory				
Vitamin A	IU	1500	2500	8000
Vitamin C	mg	35	40	60
Iron	mg	15	10	18
Calcium	g	0.6	0.8	1.3
Protein	g	14	16	60[b] (65)[c]
Voluntary				
Vitamin D	IU	400	400	400
Vitamin E	IU	5	10	30
Folic acid	mg	1	0.2	0.8
Thiamine	mg	0.5	0.7	1.7
Riboflavin	mg	0.6	0.8	2.0
Niacin	mg	8	9	20
Vitamin B_6	mg	0.4	0.7	2.5
Biotin	mg	0.05	0.15	0.30
Pantothenic acid	mg	3	5	10
Vitamin B_{12}	μg	2	3	8
Phosphorus	g	0.5	0.8	1.3
Iodine	μg	45	70	150
Magnesium	mg	70	200	450
Copper	mg	0.6	1.0	2.0
Zinc	mg	5	8	15

[a] These values may be used as guidance in the interim for labeling products to be used by infants, children under 4 years of age, or prgnant or lactating women. The FDA will make a final decision on these values following the provisions of the Diet Supplement (DS). This table is derived from the *Federal Register*, Vol. 58, No. 3, January 6, 1993, p. 2213.

[b] RDI for pregnant women.

[c] RDI for lactating women.

Table 7.12 Reference Daily Intakes (RDIs)[a]

Vitamins, minerals, and protein	Amount
Vitamin A (M)[b]	5000 IU
Vitamin C (M)[b]	60 mg
Thiamine	1.5 mg
Riboflavin	1.7 mg
Niacin	20 mg
Vitamin D	400 IU
Vitamin E	30 IU
Vitamin B_6	2.0 mg
Folic acid	0.4 mg
Vitamin B_{12}	6 μg
Biotin	0.3 mg
Pantothenic acid	10 mg
Calcium (M)[b]	1 g
Iron (M)[b]	18 mg
Phosphorus	1.0 g
Iodine	150 μg
Magnesium	400 mg
Zinc	15 mg
Copper	2 mg
Protein	50 g

[a] *Federal Register* Vol. 58, No. 3, January 6, 1993, p. 2227.
[b] "M" denotes mandatory declaration. All other vitamins and minerals may be declared voluntarily.

after the "iron" declaration in the following order: vitamin D, vitamin E, thiamine, riboflavin, niacin, vitamin B_6, folate, vitamin B_{12}, biotin, pantothenic acid, phosphorus, iodine, magnesium, zinc and copper. Only those voluntary nutrients that the manufacturer wants to declare need appear. However, only those vitamins and minerals designated above may appear in the Nutrition Facts panel.

Quantitative information on nutrients not stipulated above (i.e., omega 3 fatty acids or boron) may be declared on the label provided the information is truthful and not misleading. However, it must appear outside the Nutrition Facts panel and must not characterize the relative level of the nutrient (i.e., must not make a nutrient content claim).

The percentages for vitamins and minerals must be expressed to the nearest 2% increment up to and including the 10% level; the nearest 5% increment above 10% and up to and including the 50% level; and the nearest 10% increment above the 50% level. Amounts of vitamins and minerals present at less than 2% of the RDI are not required to be declared but may be declared by a zero or by the use of an asterisk (or other symbol) that refers to another asterisk (or symbol) that is placed at the bottom of the table and that is followed by the statement "Contains less than 2% of the Daily Value of this (these) nutrient(s)" or "Not a significant source of *name of vitamin(s) or mineral(s)*" in the same type size as nutrients that are indented. The percent of vitamin A that is present as beta-carotene may be declared to the nearest 10% increment immediately adjacent to or beneath the nutrient name [e.g., "Vitamin A (90% as beta-carotene)"].

The following synonyms may be added in parentheses immediately following the name of the nutrient or daily component:

Vitamin C: Ascorbic Acid	Thiamine: Vitamin B_1
Riboflavin: Vitamin B_2	Folate: Folacin

Vitamins and minerals that are required or permitted in a standardized food (e.g., thiamine, riboflavin, and niacin in enriched flour) and where that standardized food is included as an ingredient (e.g., component) in another food or included in a food solely for technological purposes and declared only in the ingredient statement need not be declared if neither the nutrient nor the component is otherwise referred to on the label.

V. ALLOWABLE STATEMENTS FOR VOLUNTARY NUTRIENTS

The allowable statements for voluntary nutrients are described below. As previously noted, a reference guide to their numerical expression is presented in Table 7.8. These declarations must appear in the "Nutrition Facts" section of the product label. Conditions which make voluntary nutrient declarations mandatory are described in Table 7.9.

A. Calories from Saturated Fat [21 CFR 101.9 para (c)(1)(iii); FR pp. 2175–2176]

If a manufacturer wishes to make a statement about the caloric content derived from saturated fat in a product, it must be expressed to the nearest 5-calorie increment up to and including 50 calories, and the nearest 10-calorie increment above 50 calories per serving. Amounts less than 5 calories may be expressed as zero. The statement must be indented under the statement of calories from fat.

B. Polyunsaturated Fat [21 CFR 101.9 para (c)(2)(ii); FR p. 2176]

Polyunsaturated fat is defined as *cis-cis* methylene interrupted polyunsaturated fatty acids. It may be declared voluntarily except it is required when monounsaturated fat is declared or when a label claim is made about fatty acids or cholesterol. Polyunsaturated fat content must be indented immediately under unsaturated fat declaration and expressed in grams per serving to the nearest 0.5 g increment below 5 g and to the nearest gram increment above 5 g. If the serving contains less than 0.5 g, the content shall be expressed as zero.

C. Monounsaturated Fat [21 CFR 101.9 para (c)(2)(iii); FR p. 2176]

Monounsaturated fat is defined as *cis*-monounsaturated fatty acids; it may be declared voluntarily except it is required when polyunsaturated fat is declared or when a label claim is made about fatty acids or cholesterol unless the product qualifies for a "fat free" claim.

Monounsaturated fat content must be indented immediately under the polyunsaturated fat declaration and expressed as grams per serving to the nearest 0.5 g increment below 5 g and to the nearest gram increment above 5 g. If the serving contains less than 0.5 g, the content shall be expressed as zero.

D. Potassium [21 CFR 101.9 para (c)(5); FR p. 2176]

The quantity of potassium in a specified serving of food may be declared voluntarily except it is required when a label claim is made about potassium content.

Potassium content must be expressed as zero when the serving contains less than 5 mg of potassium, to the nearest 5-mg increment when the serving contains less than or equal to 140 mg of potassium and to the nearest 10-mg increment when the serving contains more than 140 mg. The listing of potassium should appear in bold face directly under sodium. Since it is not a sub-group of sodium, it is not indented.

E. Soluble Fiber [21 CFR 101.9 para (c)(6)(i)(A); FR p. 2176]

A statement of the number of grams of soluble dietary fiber in a serving may be declared voluntarily except it is required when a label claim is made about soluble fiber.

Soluble fiber content must be indented under dietary fiber and expressed to the nearest gram. If a serving contains less than 1 g, the statement "less than 1 gram" may be used as an alternative. If the serving contains less than 0.5 g, the content may be expressed as zero.

F. Insoluble Fiber [21 CFR 101.9 para (c)(6)(i)(B); FR p. 2176]

A statement of the number of grams of insoluble dietary fiber in a serving may be declared voluntarily except it is required when a label claim is made about insoluble fiber.

Insoluble fiber content must be indented under dietary fiber and expressed to the nearest gram. If a serving contains less than 1 g, the statement "contains less than 1 gram" may be used as an alternative. If the serving contains less than 0.5 g, the content may be expressed as zero.

G. Sugar Alcohol [21 CFR 101.9 para (c)(6)(iii); FR p. 2177]

A statement of the number of grams of sugar alcohol in a serving may be declared voluntarily on the label except it is required when a label claim is made about sugar alcohol or sugars when sugar alcohols are present in the food.

For labeling purposes, sugar alcohols are defined as the sum of saccharide derivatives in which a hydroxyl group replaces a ketone or aldehyde group and whose use in the food is listed by FDA or is generally recognized as safe (GRAS). In lieu of the term "sugar alcohol," the name of the specific sugar alcohol (e.g., xylitol) present in the food may be used provided that only one sugar alcohol is present.

Sugar alcohol content must be indented under the sugar declaration and expressed to the nearest gram. If a serving contains less than 1 g, the statement "contains less than 1 gram" or "less than 1 gram" may be used as an alternative. If the serving contains less than 0.5 g, the content may be expressed as zero.

H. Other Carbohydrate [21 CFR 101.9 para (c)(6)(iv); FR p. 2177]

A statement of the number of grams of other carbohydrates may be declared voluntarily. "Other carbohydrates" is defined as the difference between total carbohydrate and the sum of dietary fiber, sugars and sugar alcohol. If sugar alcohol is not declared (even if present) it is defined as the difference between total carbohydrate and the sum of dietary fiber and sugars. "Other carbohydrate content" must be indented and expressed to the nearest gram. If a serving contains less than 1 g, the statement "contains less than 1 gram" or " less than 1 gram" may be used as an alternative. If the serving contains less than 0.5 g, the content may be expressed as zero.

VI. LABEL FORMATS

A. Nutrition Information and Label Format [21 CFR 101.9 para (d)(1–12); FR pp. 2178–2205]

The January 6 regulations contain detailed label format, graphics and printing requirements. They were developed from consumer studies conducted by FDA, USDA, and other groups to achieve uniform representation of nutrition information on all product labels and to help consumers make informed choices and understand how each food product fits into the total diet.

Most products will comply with the nutrition information panel format shown in Figure 7.1. However, there are a number of alternative formats which may be used, depending on the nutrient content and package size of the product[4].

The main features of the new label format are shown in Figure 7.1. The title *Nutrition Facts* at the top of the nutrition information panel is intended as a signal to the consumer that a new label is in place. The top of the nutrition panel must also include the serving size of the food expressed in household and metric units and the number of servings in the container. Serving size is still the basis for expressing nutrient content and is based on reference amounts commonly consumed (RACC Values) to achieve uniformity across a given product line and to reflect typical eating patterns. As previously noted, the mandatory and voluntary nutrients have been redefined and must be listed on the label in the order described in Figure 7.1 and page 138. The mandatory nutrients and food components must appear in the bold type on the middle left side of the panel. Allowable sub-components must be in lighter type and indented immediately below the major component. The amount must be expressed on a single serving basis, in metric units and appear in light print adjacent to each respective nutrient or food component (e.g., grams for total fat, saturated fat, carbohydrate, dietary fiber, sugars and protein; milligrams for cholesterol and sodium). In addition, the right column of the panel must list the Percent Daily Value (in bold type) corresponding to each respective nutrient.

The bottom section of most labels will have a standard footnote stating that the "Percent Daily Values are based on a 2,000 calorie diet" and that individual ". . . daily values may be higher or lower depending on . . . individual needs." The footnote must list the daily values for total fat, saturated fat, cholesterol, sodium, total carbohydrate, and dietary fiber based on 2,000 and 2,500 calories reference diets, respectively. A second footnote providing information for converting grams of fat, protein, and carbohydrate into calories is optional.

B. Printing and Graphics

To enhance legibility, the regulations stipulate many graphic requirements including easy to read type face, upper and lower case letters, bold and light

Nutrition Facts

Serving Size 1/2 cup (114g)
Servings Per Container 4

Amount Per Serving

Calories 260 Calories from Fat 120

	% Daily Value*
Total Fat 13g	**20%**
Saturated Fat 5g	**25%**
Cholesterol 30mg	**10%**
Sodium 660mg	**28%**
Total Carbohydrate 31g	**11%**
Dietary Fiber 0g	**0%**
Sugars 5g	
Protein 5g	

Vitamin A 4% •	Vitamin C 2%
Calcium 15% •	Iron 4%

* Percent Daily Values are based on a 2,000 calorie diet. Your daily values may be higher or lower depending on your calorie needs:

	Calories:	2,000	2,500
Total Fat	Less than	65g	80g
Sat Fat	Less than	20g	25g
Cholesterol	Less than	300mg	300mg
Sodium	Less than	2,400mg	2,400mg
Total Carbohydrate		300g	375g
Dietary Fiber		25g	30g

Calories per gram:
Fat 9 • Carbohydrate 4 • Protein 4

Figure 7.1 Standard label format. *Source: Federal Register* Vol. 58, No. 3, January 6, 1993, p. 2182.

imprinting, large relative type size, lines between nutrient listings, spacing between letters and lines. Various layouts and graphic enhancements are illustrated in Chapter 10 by Roy Parcels and Aaron Brody.

The following is a synopsis of the specifications:

1. Nutrient information must be set off in a box by use of hairlines and must be black or one color type, printed on a contrasting background.
2. All information within the nutrition label must use:

- A single easy to read type style.
- Upper and lower case letters.
- At least one point leading (i.e., space between two lines of text) except four point leading must be used under certain conditions as specified in the regulation.
- Type that is kerned to provide sufficient space between letters so they do not touch[5].
- Typesize for information in the Nutrition Facts Panel must be at least 6 point type except for "serving size," "servings per container," "calories" and all nutrient and food component declarations and the "not a significant source of protein" all of which must appear in a minimum of 8 point type.
- The "Nutrition Facts" heading must appear in larger type size than any other information on the panel at a minimum, larger than 8 point[6].

or

- Specific headings must be highlighted by bold or extra bold type or other highlighting that prominently distinguishes them from other information.
- Specific headings requiring bold or extra bold type:

Nutrition facts	Calories
Sodium	Amount Per Serving
Total Fat	Total Carbohydrate
% Daily Value*	Cholesterol
Protein	Percentage Amounts

- A hairline rule centered between the lines of text must separate "Amount Per Serving" from the required calorie statements and must separate each nutrient and its corresponding percent Daily Value from the nutrient and percent Daily Value above and below it.

C. Nutrition Facts

The information must be presented under the heading *Nutrition Facts* which must be in a type size larger than all other print size in the nutrition label and must be set the full width of the information panel unless otherwise specified.

D. Serving Size

- Information on serving size must immediately follow the heading and must include:
  "serving size": *(statement of serving size)*
  "servings per container": (number of servings per container) (not required for single serving containers as defined by the regulations).
- A subheading *Amount Per Serving* must be separated from serving size information by a bar.

E. Calories

- Information on calories must follow the heading *Amount Per Serving* and must be declared in one line, leaving sufficient space between the declaration of *Calories* and *Calories from Fat* to allow clear differentiation.
- If calories from saturated fat is declared, *Total Calories* must be at the top, followed by *Calories from Fat* (indented) and *Calories* from saturated fat (indented).

F. Percent Daily Value

- The column heading *% Daily Value** must be separated from information on calories by a bar. The position of this heading must allow for a list of nutrient names and amounts to the left and below this heading, respectively. Acceptable alternative headings for this column are: *Percent Daily Value**, *Percent DV** or *% DV**.
- The asterisk (*) must correspond to a footnote at the bottom of the nutrition information panel stating:

> Percent Daily Values are based on a 2,000 calorie diet. Your daily value may be higher or lower depending on your calorie needs.

G. Nutrient Declaration

Except for vitamins and minerals, the required nutrients must be declared as follows:

- The name of each nutrient, immediately followed by the quantitative amount by weight in "g" (for grams) or "mg" for milligrams
- The percent of the DV (right column of label) corresponding to each respective nutrient, expressed to the nearest whole percent in bold type. This numerical value must be followed by the symbol for percent (%).

H. Vitamins and Minerals

Vitamin and mineral information must be separated from other nutrient information by a bar and presented horizontally. When more than the four mandatory vitamins and minerals are declared, a vertical presentation with percentages listed under "% Daily Value" is permitted.

If beta carotene is declared, it must be indented and listed under vitamin A with its corresponding % Daily Value expressed in the same applicable increments as for the other vitamins and minerals.

If potassium is declared, it must be listed under sodium with its corresponding % Daily Value.

If the percent of Daily Value is presented for protein, it must be listed under dietary fiber.

If the space beneath the declaration of iron is not adequate for listing any remaining micronutrients, they may be listed to the right and set off by a line to distinguish them. The caloric conversion information may be presented beneath either side or along the full length of the panel.

I. Caloric Conversion Statement

Although FDA encourages its use whenever the general caloric conversion factors are used, the caloric conversion footnote is optional. If used it must appear at the very bottom of the nutrition information panel and must be separated from all other information by a hairline. This information may be presented horizontally or vertically in columns (e.g., Calories per gram: Fat 9 • Carbohydrate 4 • Protein 4).

If the space beneath the information on vitamins and minerals is not adequate to accommodate the required footnotes, the footnotes may be moved to the right of the nutrient information column and separated by a line setting it apart from the percent daily value information.

J. Dual Labeling

Dual labeling may be used to present nutrition information. Typical situations where dual labeling is applicable include:

- different forms of the same food (e.g., "as purchased" and "as prepared")
- common food combinations (i.e., dry ready-to-eat cereal with milk)
- different units (slices of bread or per 100 g)
- two or more groups for which RDI's are established (e.g., infants, children less than 4 years of age)

When dual labeling is used, equal prominence must be given to both sets of values and all information must be presented in a format consistent with the specifications previously cited in this section with the following modifications:

There must be two or more column headings accurately describing the:

- forms of the same food (e.g., "Mix and Baked")
- the combinations of food (e.g., "Cereal and Milk")
- the units (e.g., slices and grams)
- the RDI groups being declared (e.g., infants, children less than 4 years of age).

The column providing the packaged serving size information must remain to the left of the numeric columns.

When dual labeling is used, all indentations as previously described with quantitative amounts must be declared in columns aligned under the appropriate column headings.

Quantitative information by weight must be specified for the form of the product as packaged in accordance with the rules for serving size.

Quantitative information by weight may be included for other forms of the product represented by the additional column(s) either immediately adjacent to the required quantitative information or as a footnote. If provided immediately adjacent to the required quantitative information, it must be declared for all nutrients listed and differentiated by a comma; it must not be put in a separate column.

If additional quantitative information is given in a footnote, it must be declared in the same order as the nutrients are listed in the nutrition label. The additional quantitative information may state the total nutrient content of the product identified in the second column or the nutrient amounts added to the packaged product as packaged for only those nutrients present in different amounts than the amounts declared in the required quantitative information. The footnote must clearly identify which amounts are declared. Any declared subcomponents must be listed parenthetically after principal components (e.g., 1/2 cup skim milk contributes an additional 40 calories, 65 mg sodium, 6 g total carbohydrate [6 g sugars], and 4 g protein).

Total fat and its amount by weight must be followed by an asterisk (or other symbol) (e.g., "Total fat (2 g)*") referring to another asterisk (or symbol) at the bottom of the nutrition label identifying the form(s) of the product for which quantitative information is presented.

The percent of the daily value for each appropriate nutrient must appear under the subheading % Daily Value and directly under the applicable column.

K. Alternative Formats

Several alternate nutrition label formats may be used in lieu of the "typical" format shown in Figure 7.1. These alternatives and the conditions for their use are described in the regulations. In such cases, the nutrient content and the package size of the food product dictate the type of format that may be used.

Alternatives include a simplified format, a shortened format, and a tabular format. Strict rules govern the use of these options. In each case, five "core" nutrients (e.g., calories, total fat, sodium, total carbohydrate and protein) must be listed[4].

1. Simplified Format

Under FDA regulations a simplified format may be used when it contains insignificant amounts of seven or more of the mandatory nutrients:

Calories	Total Carbohydrate	Vitamin A
Total Fat	Dietary Fiber	Vitamin C
Saturated Fat	Sugars	Calcium
Cholesterol	Protein	Iron
Sodium		

For FDA regulated foods intended for children less than 2 years of age, a simplified label may be used when the food contains insignificant amounts of six or more of the following:

Calories	Dietary Fiber	Vitamin A
Total Fat	Iron	Vitamin C
Sodium	Sugars	Calcium
Total carbohydrate	Protein	

For USDA regulated products a simplified label may be used when the product contains an insignificant amount of any *one* of the following:

Calories from Fat	Vitamin C	Dietary Fiber
Cholesterol	Iron	Vitamin A
Sugars	Saturated Fat	Calcium

An "insignificant amount" is defined as the amount that allows a "zero" declaration or "less than 1 gram" according to the rounding rules described in Tables 7.7 and 7.8.

The simplified format must include:

- Serving size
- Number of servings per container
- Total calories, total fat, total carbohydrate, protein, sodium
- Any other mandatory nutrient present in more than insignificant amounts.
- Added vitamins and minerals

- Naturally occurring voluntary nutrients may be declared if present in more than insignificant amounts.

The simplified nutrition information panel must be presented in the same manner as the general format except that the "Percent Daily Value" and calorie conversion footnote statements are not required. When these are omitted, an asterisk must be placed at the bottom of the label with a statement "Percent Daily Values are based on a 2,000 calorie diet."

According to the August 18, 1993, Technical Amendments, the phrase "not a significant source" of the following *name of nutrient(s)* is required only when:

- Vitamins or mineral addition is necessary for a product with a standard of identity.
- Vitamins or minerals are voluntarily added to a formula.
- Voluntary nutrients that occur in more than significant amounts are declared.
- Nutrient content claims are made.

2. Shortened Format

This format may be used for FDA regulated products containing insignificant amounts of certain "mandatory nutrients or food components." In this format, the "mandatory component or nutrient" may be omitted from the column listing and replaced with a "not a significant source of. . . ." footnote. The Percent Daily Values or caloric conversion footnotes are not required.

3. Tabular Format

This format is permitted when dictated by package size constraints (e.g., when label size is less than 40 square inches . . . and or if there is insufficient continuous vertical space (i.e., approximately 3 inches) to accommodate the required nutrient components up to and including the declaration of iron. Authorized abbreviations are allowed and the Percent Daily Values footnote must be replaced with the statement "Percent Daily Values are based on a 2,000 calorie diet." The caloric conversion footnote is not required.

FDA-regulated products that have labels smaller than 12 square inches are exempt from the specified label requirements provided they do not make any nutrition claims. However, the manufacturer must have a "call statement" listing the address or telephone number where the required nutrition information can be obtained (e.g., "for nutrition information, write or call . . .")

4. Aggregate Display

Nutrition labels on the outer label of packages of products that contain two or more separately packaged foods that are intended to be eaten individually (e.g., variety packs of cereals or snack foods) or packages that are used interchangeably for the same type of food (e.g., round ice cream containers) may use an aggregate display. Aggregate displays must comply with the general format requirements described previously except that the identity of each food must be specified immediately under the *Nutrition Facts* title, and both the quantitative amount by weight and the Percent Daily Value for each nutrient must be listed in separate columns under the name of each food.

5. Bilingual Nutrition Label

When nutrition labeling appears in a second language, the nutrition information may be presented in a separate nutrition label for each language or on one label with the information in the second language following that in English. Numeric characters that are identical in both languages need not be repeated (e.g., "protein/Protein as 2 g"). All required information must be included in both languages.

L. Separate Packaging

Products with separately packaged ingredients or foods, with assortments of food, or to which other ingredients are added by the user may be labeled as follows:

- If a product consists of two or more separately packaged ingredients enclosed in an outer container or of assortments of the same type of food (e.g., assorted nuts or candy mixtures) in the same retail package, nutrition labeling must be located on the outer container or retail package to provide information for the consumer at the point of purchase.
- If two or more food products are simply combined together in such a manner that no outer container is used, or no outer label is available, each product must have its own nutrition information (e.g., two boxes taped together or two cans combined in a clear plastic overwrap).
- When separately packaged ingredients or assortments of the same type of food are intended to be eaten at the same time, the nutrition information may be specified per serving for each component or as a composite value.
- If a product consists of two or more separately packaged foods that are intended to be eaten individually and that are enclosed in an outer container (e.g., variety packs of cereals or snack foods), the nutrition information must be specified per serving for each food in a location that is clearly visible to the consumer at the point of purchase.
- If a package contains a variety of foods, or an assortment of foods, and is in a form intended to be used as a gift, the nutrition labeling must meet the

requirements and be in the formats previously described in this section but may be modified as follows:

- Nutrition information may be presented on the label of the outer package or in labeling within or attached to the outer package.
- The number of servings per container may be stated as "varied."
- Nutrition information may be provided per serving for individual foods in the package, or, alternatively, as a composite per serving for reasonable categories of foods in the package having similar dietary uses and similar significant nutritional characteristics. Reasonable categories of foods may be used only if accepted by FDA.
- Nutrition information is not necessary if:
  - The food is of small size within a gift package.
  - The food is not specifically listed in a promotional catalog as being present in the gift package.
  - The food is used in small quantities primarily to enhance the appearance of the gift.
  - The food is included in the gift package as a free promotional item.

M. Prepared Foods

If a food is typically combined with other ingredients or is cooked or otherwise prepared before eating, and directions for such combination or preparations are provided, another column of figures may be used to declare nutrition information on the basis of the food as consumed in the format required for dual labeling [e.g., a dry ready-to-eat cereal may be described with one set of Percent Daily Values for the cereal as sold—for examle, per ounce—and another set for the cereal and milk as suggested in the label—for example, per ounce of cereal and 1/2 cup of vitamin D fortified skim milk); and a cake mix may be labeled with one set of Percent Daily Values for the dry mix (per serving) and another set for the serving of the final cake when prepared]: provided that the type and quantity of the other ingredients to be added to the product by the user and the specific method of cooking and other preparation is prominently described on the label.

VII. REFERENCE DAILY INTAKES AND DAILY REFERENCE VALUES [21 CFR 101.9 para (c)(7)(8)(iv)(9); FR p. 2227]

The new regulations establish two sets of label reference values—Reference Daily Intakes (RDIs) and Daily Reference Values (DRVs)—for use in declaring the nutrient content of a food. These two sets of values do not appear on the label but instead are expressed on the label as Daily Value (DV) in order to simplify the label and enhance consumer understanding. The Daily Value is meant to help consumers use the food label to plan healthier daily diets and to facilitate comparison of the nutritional values of different foods. It is the basis for defining nutrient content claims (descriptors, e.g., high, low etc.). The percent Daily Value (% DV) is calculated from the DV and allows consumers to compare foods relative to a total daily diet intake of 2,000 calories and also to compare specific nutrient recommendations for good nutrition. As previously noted in Section VI, this declaration is required on all except simplified or shortened food labels.

The DV is derived from the RDI and DRV. The term "RDI" replaces the term "U.S. Recommended Daily Allowances" (US RDA) which was established in 1973 for vitamins, minerals, and protein. Except for protein, the values for US RDA and RDI remain the same; only the terminology has changed. Table 7.12 shows the RDIs for 12 vitamins, and 7 minerals and protein for adults and children over 4 years of age. Table 7.11 shows RDI values for children less than 4, infants, and pregnant and lactating women. It should be noted that under the provisions of the Dietary Supplements Act of 1992, the government can revise the RDIs after November 1993. This could result in new DVs for these nutrients.

The DRVs are reference values established by scientific concensus for those nutrients and food components required on the label but for which no set standards previously existed. They are based on a 2,000 calorie reference diet and are shown in Table 7.13. For purposes of labeling, the *Percent Daily Value* is calculated from the standard values or from RDI values as appropriate.

It should be noted that inconsistencies could occur between reported nutrient/food component values based on analytical data and the determination of

Table 7.13 Daily Reference Values (DRVs)[a]

Food component	DRV
Fat	65 g
Saturated fatty acids	20 g
Cholesterol	300 mg
Total carbohydrate	300 g
Fiber	25 g
Sodium	2400 mg
Potassium	3500 mg
Protein	50 g

[a] Based on a 2,000 calorie diet.

Table 7.14 Standard Label Footnote Declaring Percent Daily Values[a]

Percent Daily Values are based on a 2,000 calorie diet. Your daily value may be higher or lower depending on your calorie needs.

	Calories	2,000	2,500
Total fat	Less than	65 g	80 g
Saturated fat	Less than	20 g	25 g
Cholesterol	Less than	300 mg	300 mg
Sodium	Less than	2400 mg	2400 mg
Total carbohydrate		300 g	375 g
Dietary fiber		25 g	30 g

[a] Not required for simplified or shortened label formats.

%DV in accordance with regulatory guidelines. The August 18, 1993, Technical Amendments address this issue by allowing for the use of either rounded or unrounded values to calculate %DV, depending on which one results in the most consistent label.

As previously noted, in the section on formats, the footnote shown in Table 7.14 must appear on all except simplified or shortened labels.

VIII. SERVING SIZE [21 CFR 101.8, 101.9, 101.12 ; FR pp. 2291–2300]

A. Background

The 1990 NLEA required FDA to promulgate regulations standardizing serving sizes. The purpose was to achieve uniformity across product lines and to standardize nutrient claims, thereby enabling consumers to compare the nutritional value of products based on equivalent serving sizes. The main features of the serving size regulations are summarized in Table 7.15.

B. Definition of Serving Size [21 CFR 101.9 para (b)(1); FR p. 2291]

The term "serving" or "serving size" is defined as an amount of food customarily consumed per eating occasion by persons 4 years of age or older expressed in a common household measure appropriate to the food.

Table 7.15 Main Features of Serving Size Regulations

The new regulations:
• Define serving size on the basis of the amount of food customarily consumed per eating occasion.
• Establish reference amounts customarily consumed (RACC) per eating occasion for 139 food product categories.[a]
• Provide criteria for determining label serving sizes from the reference amount.
• Require the use of both common household and metric measures to declare serving sizes.
• Define a "single serving container".
• Require that nutrient content claims (descriptors) such as "low sodium" be based on the reference amount.
• Permit the declaration of serving size in U.S. measures (ounces, fluid ounces)
• Permit the optional declaration of nutrient content per 100 g or 100 ml.

[a] Appendix 1, Table A1.1, lists the RACC for 139 FDA regulated food product categories.

When the food is specially formulated for infants or toddlers, a serving is the amount of food customarily consumed per eating occasion by infants up to 12 months of age or children 1 through 3 years of age, respectively.

The term "eating occasion" refers to meals and snacks.

C. Reference Amounts Customarily Consumed (RACC) [21 CFR 101.12 para (a); FR p. 2293]

Working collaboratively with USDA and using data from national food consumption surveys, FDA established criteria for the declaration of serving sizes in nutrition labeling. Reference amounts were established for 139 different non-meat or poultry categories, 23 meat categories and 22 poultry categories. Appendix 1, Table A1.1, provides tables of the Reference Amount Commonly Consumed (RACC) for 139 FDA regulated food product categories.

Strict rules for converting reference amounts into serving sizes for labeling are defined in the regulations. The objective is to ensure that foods having similar dietary use, product characteristics, and customarily consumed amounts have a uniform reference amount and serving size.

1. General

A reference amount is the average amount of a given food customarily consumed at a single eating. It is based on the edible portion of the food and its major intended use. The reference amount for products consumed as an ingredient of

Table 7.16 Reference Amounts for Determining Serving Sizes

Reference amounts are based on historical food consumption data. They serve to standardize the amount of a given food consumed at one eating occasion. These amounts are published in 21 CFR in the *Federal Register*[a] and are transcribed in Appendix 1, Table A1.1, of this book.

↓

Serving size is derived from the reference amount. In most instances, serving size and reference amounts are comprable to one another. However, in some instances, special rules apply that cause them to be different. There are strict rules for converting reference amount to serving size.[b]

[a] *Source*: *Federal Register*, January 6, 1993 (Book 1), Vol. 58, No. 3, pp. 2294–2298.
[b] The rules governing reference amounts and serving sizes are described in this section.

other foods, which may also be consumed as purchased (e.g., butter), is based on use in the form purchased. While the term reference amount does not appear on the label, it is the basis for determining serving size (Table 7.16). It is also the basis for determining whether nutrient content claims and health claims satisfy requirements (See Section IX and XI of this chapter).

2. Products as Packaged or Purchased

The reference amount and consequently the serving size and information for the nutrition facts panel of the label are based on the food product as packaged or purchased with the exception of raw fish or game meat (for which there are special provisions) and foods packaged in liquids that are not customarily consumed (e.g., brine, oil, vinegar; sardines, maraschino cherries, pickled fruits, pickled vegetables). For these products, reference amounts, serving size and nutrient and food component declarations are based on the drained solids.

3. Unprepared Products [21 CFR 101.12 para (c)(f); FR p. 2299]

The reference amount for a product requiring preparation (e.g., batches, dry mixes, uncooked foods requiring cooking, the addition of water or other ingredients) is the amount of unprepared product required to make one reference amount of the finished product. For products representing two or more foods packaged together and intended to be consumed together (e.g., peanut butter and jelly, pancake and syrup), the reference amount is the sum of the reference amounts for each individual food in the package unless FDA has established and listed a reference amount for the combination. For example, macaroni and cheese, and spaghetti and tomato sauce are combinations measurable in household units that have listed reference amounts. Therefore, the reference amount for the combination rather than the sum of the reference amounts for each food applies for these products.

4. Multiple Use Products [21 CFR 101.12 para (a)(7) ; FR p. 2293]

If there is more than one use for a product, the manufacturer must use the reference amount appropriate to its major usage to establish serving size. For example, a cake mix with preparation directions for a two layer cake and cupcakes should use the reference amount for the two layer cake, if that is the major intended use.

5. Imitation or Substitute Foods [21 CFR 101.12 para (d); FR p. 2299]

The reference amount for an imitation or substitute food is the same as for the food for which it substitutes.

6. Conversion of Reference Amount to Serving Size

As previously noted, although reference amounts form the basis of serving sizes, only the term "serving size" may appear on the label. Conversion of reference amount to serving size depends on the nature of the food product. For purposes of converting reference amounts into serving size, FDA established three categories according to the shapes and characteristics of products and the way they are usually served. The three categories are:

- Discrete products (e.g., muffins, bagels, sliced bread, apples, burritos, individually packaged products within a multiserving package)
- Large discrete products (e.g., products usually divided for consumption: cake, pie, pizza, grapefruit, melon)
- Non-discrete bulk products (e.g., cereals, flour, sugar, dry mixes, concentrates)

Table 7.17 provides guidelines for determining serving size for these designated categories as well as other product and package types.

The commonly used acronym for reference amounts is "RACC" or "RACC Number"; this refers to the "Reference Amount Customarily Consumed." The RACC numbers for 139 FDA regulated food categories with suggested label statements are listed in Appendix 1, Table A1.1. The suggested statements in the last column on each page are intended to provide guidance on how serving

Table 7.17 Guidelines for Determining Serving Size

Product/package type	Serving size determination
Discrete unit product	The number of whole units closest to the RACC[a]
Large discrete unit products	The fractional slice of the food closest to the RACC
Non-discrete bulk products	The amount in household measure closest to the RACC
Meals/main dishes	
Single serving container	Entire content of package
Multi-serving container	Closest to RACC for product, or sum of RACCs of individual foods in the package
Variety pack	
Single serving container	According to single serving rule
Multi-serving container	Closest to RACC for product, or sum of RACCs of individual foods

[a] RACC: Reference Amount Customarily Consumed.

sizes for specific products should appear on the label. When possible, the household measure and equivalent metric measure per serving for each product are listed. The list is intended to help manufacturers identify the category in which their specific product(s) fit. Those who are unsure about which product category to use should contact FDA or USDA. New reference amounts may be established and existing ones may be amended by submitting a petition to FDA or USDA. The petition requirements are detailed in the regulations. The content of the RAAC tables is outlined in Table 7.18.

a. Discrete Units
[21 CFR 101.9 para (b)(2)(i); FR p. 2291]

Except for products that naturally vary in size (e.g., maraschino cherries, pickles, shellfish, whole fish, fish fillets) serving size is the number of whole units

Table 7.18 Content of RAAC Table

- Based on the edible portion of a product as packaged and as intended to be consumed.
- Lists 139 product categories.
- Provides reference amounts in metric units and household units.
- Includes a label statement column which suggests units of measure that can be used (i.e., pieces; non-discrete items [cups, tablespoons]); large discrete items (fraction); meals/main dishes (metric unit or household measure).

closest to the reference amount for the product category (e.g., sliced products, cookies, donuts, bagels, nuggets, apple muffins, individually packaged products within a multi-serving package).

Special additional rules are applicable to discrete products based on their unit weights:

- If a unit weighs 67% or more, but less than 200% of the reference amount, the serving size is one unit.
- If a unit weighs more than 50%, but less than 67% of the reference amount, the serving size may be one unit.
- A unit weighing 50% of the reference amount is by definition one-half of a serving.
- If a unit weighs 200% or more of the reference amount, the manufacturer may declare the whole unit as one serving if it can reasonably be consumed at a single eating occasion.
- If a product naturally varies in size (i.e., pickles, shellfish, whole fish, maraschino cherries, fish fillet), it must be expressed in ounces closest to the reference amount of the product category (except for maraschino cherries; see below) followed by a gram-equivalent weight and the appropriate number of pieces for small pieces (e.g., shrimp, scallops) or the dimension of a large piece (e.g., pickles, fish fillet). For example, a serving size for cooked shrimp may be declared as "3 oz (84 g/about 5 shrimp)," or for cooked fish "3 oz (84 g about 1 fillet)" or for large pickles "1 oz (28 g/about 1/2 pickle)." This provision does not apply to processed products where the manufacturer controls piece size (e.g., fish sticks).

The rules make special provisions for maraschino cherries. The serving size for maraschino cherries must be expressed as one cherry with a parenthetical metric measure equal to the average weight of a medium-sized cherry.

In those cases where product size is controlled by the manufacturer but where there is some variation from unit to unit within a package and/or between batches, the serving size must be based on the average gram weight equivalent of the unit or the fraction in the household measure.

b. Large Discrete Units
[21 CFR 101.9 para (2)(ii); FR p. 2292]

The serving size is the fractional slice of the food that is closest to the reference amount for the product category (e.g., products usually divided for consumption such as cake, pie, melon, pizza, grapefruit). Allowable fractions are: 1/2, 1/3, 1/4, 1/5, 1/6 or smaller fractions that can be derived by dividing by 2, 3 or 5 (e.g., 1/12 cake, 1/8 pie, 1/4 cabbage).

c. Non-Discrete Bulk Products
[21 CFR 101.9 para (b)(2)(iii); FR p. 2292]

The serving size is the amount in household measures that most closely approximates the reference amount for the product category (e.g., cereals, dry mixes, concentrates, flour, sugar). For example, if the reference amount is 50 g, the labeled serving size is the household measure of that food that comes closest to weighing 50 g (e.g., possibly 3/4 cup or 1 cup).

d. Meals and Main Dishes
[21 CFR 101.9 para (b)(3); FR p. 2292]

The definition of meals and main dishes are presented in Tables 7.19 and 7.20, respectively.

If the product comes in a single serving container (see item [e] below for definition of single serving container) the entire "edible portion" is the serving size.

The serving size of meal or main dish products that are sold in multi-serving containers are based on the reference amount applicable to the product, if listed.

Table 7.19 Definition of Meals

Definition

- Must make a major contribution to the total diet
- Must weigh at least 10 oz per serving
- Must contain at least three 40 g portions of food or combinations of food from 2 or more of the following food groups:
  1. Bread, rice, pasta, cereal
  2. Fruits and vegetables
  3. Milk, yogurt, cheese
  4. Meat, poultry, fish, dry beans, eggs, nuts
- Must be represented as a breakfast, lunch, dinner, or meal

Example

Food	Group	Amount
Pasta	1	100 g
Mixed vegetables	2	60 g
Cheese	3	40 g
Poultry	4	120 g
Total		320 g

Table 7.20 Definition of Main Dishes

Definition

- Must make a major contribution to a meal
- Must weigh at least 6 oz per serving
- Must contain at least 40 g of food from at least 2 of the following food groups:
  1. Bread, rice, pasta, cereal
  2. Fruits and vegetables
  3. Milk, yogurt, cheese
  4. Meat, poultry, fish, dry beans, eggs, nuts
- Must be represented as a main dish

Example

Food	Group	Amount
Rice	4	120 g
Fish vegetables	1	60 g
Total		180 g

If not listed, the serving size is based on the sum of reference amounts for each food item in the package.

For purposes of establishing serving size, frozen entrees packaged in separate pouches containing more than one distinct food in each package (e.g., spaghetti with meat sauce; rice with vegetables) are considered mixed dishes unless they satisfy the definition of a meal or main dish. Although the components are packaged separately, they are considered as one product and the appropriate rules for serving size apply.

e. Single Serving Containers
[21 CFR 101.9 para (b)(6); FR p. 2292]

A single serving container is one that is packaged and sold individually and contains less than 200% of the reference amount for the product category (e.g., canned or bottled soft drinks, ice cream bars, sandwiches). However, products that have reference amounts of 100 g (100 ml) or larger may declare the number of servings as either one or two when the package contains more than 150% but less than 200% of the reference amount (e.g., canned fruits and vegetables, soups and entrees).

Packages that contain 200% or more of the applicable reference amount may be labeled as a single serving if the entire content of the package can reasonably be consumed at a single eating occasion.

Regardless of package size, products that are clearly intended to be consumed at one "eating" must be labeled as one serving (e.g., muffins, ice cream bars, sandwiches and products with labels suggesting a single serving such as "single," "the perfect size for one," etc.).

There is no lower limit for a single serving container. Thus, a product that is packaged and sold individually and that contains less than the upper limit of the single serving container must be labeled with the entire contents of the package being one serving. For example, if a muffin weighing 45% of the reference amount is packaged and sold individually, it is a single serving product and the nutrition information is to be provided for the entire content of the container (e.g., one muffin). All nutrition information and claims, however, must be based on the listed reference amount.

f. Variety Packs
[21 CFR 101.9 para (b)(4); FR p. 2292]

A package containing several varieties of single serving units, and a product having two or more compartments each containing a different food must provide nutrition information for each variety of food per serving size. The serving size for each must be derived from the reference amount applicable for each variety following the procedures for converting the RACC to serving size.

g. Weight Control Products
[21 CFR 101.9 para (b)(2); FR p. 2291]

For products intended for weight control and available only through weight control programs, serving size may be determined by the manufacturer provided it is consistent with the meal plan of the program. Such products must declare "for sale only through the *name* program." However, all nutrient content or health claims will be evaluated based on the listed reference amount for the product.

h. Aerated Products
[21 CFR 101.12 para (e); FR p. 2299]

For aerated products where the density is lowered by more than 25%, the reference amount may be calculated by adjusting for the difference in the density of aerated food relative to the density of the appropriate reference food. In such cases, the RACC for the aerated product must be rounded to the nearest 5 g increment and the product must bear a descriptive term indicating that extra air has been incorporated (e.g., whipped aerated). When calculating the difference in density between aerated and regular food:

- The regular and aerated product must be the same shape, size, and volume.
- Sample selections for density measurements must be performed in accordance with specified provisions (21 CFR 101.9 [e]).

- Density measurements of both foods must be made by the same trained operator, using the same equipment, procedures and techniques under the same conditions.
- Density measurements must be replicated a sufficient number of times to ensure valid representation of both products.

D. RULES FOR DECLARING COMMON HOUSEHOLD MEASURES [21 CFR 101.9 para (b)(5); FR p. 2292]

The regulations require that serving size be expressed in "common household measures" (Table 7.21) followed parenthetically by the equivalent metric quantity (fluids in ml; solids in g). The acceptable household declarations are described in Table 7.22 except for single serving containers. For single serving containers, the parenthetical metric quantity (which is part of the net weight statement on the principal display panel) is not required except where nutrition information is required on a drained weight basis.

Wherever possible, cups, tablespoons or teaspoons should be used (except for beverages which may be expressed in fluid ounces). Cups must be expressed in 1/4 or 1/3 cup increments; tablespoons in whole number of tablespoons for quantities less than 1/4 cup but greater than or equal to 1 tablespoon; in 1/3, 1/2, and 2/3 tablespoons between 1 and 2 tablespoons; and teaspoons in whole number of teaspoons, for quantities less than 1 tablespoon but greater than or equal to 1 teaspoon and 1/4 teaspoon increments for quantities less than 1 teaspoon.

If these measures are not applicable, units such as piece, slice, tray, jar, and fractions should be used. If neither of these options is appropriate, "ounces" may be used with a "visual" unit of measure such as a dimension (e.g., one ounce [28 g/about 1/2 pickle]). Ounce measurements must be expressed in 0.5 ounce increments most closely approximating the reference amount with rounding indicated by the use of the term "about" (e.g., about 2.5 oz). Directions for using household measures are summarized in Table 7.23.

Table 7.21 Labeling of Serving Size in Household Measures

Household measures and metric units:

- The primary unit for serving size must be the common household measure most appropriate to the specific product (e.g., cups, tbsp., oz).
- Metric equivalents of the household measures must also be declared (e.g., 1 cup [55 g])
- Because many foreign countries use 100 g or 100 ml as the basis for their nutrition labeling, a complementary listing of values per 100 g or per 100 ml is permitted.

Table 7.22 Acceptable Serving Size Declarations

Household measures	
Cup	Fraction[a]
Tablespoon	Ounce
Teaspoon	Fluid ounce
Piece	Common house packaging (i.e., jar, tray)
Slice	
Metric units	
Fluids:	Milliliter
All other foods:	Grams
Optional	
Declaration of nutrient content per 100 g, 100 ml, 1 oz, or 1 fl oz	

[a] e.g., 1/4 pizza

When a serving size determined from the RAAC falls exactly halfway between two serving sizes expressed in household units (e.g., 2.5 tbsp), the serving size must be rounded up to the next incremental size.

The metric (g) equivalent to the household measure should be rounded to the nearest number except for quantities that are less than 5 g. Amounts between 2 and 5 g should be rounded to the nearest 0.5 g and if less than 2 g expressed in 0.1 g increments. In addition, the serving size may be declared in oz and fl oz in parenthesis, following the metric measure separated by a slash where other common household measures are used as the primary unit for serving size [e.g., 1 slice (28 g/1 oz) for sliced bread]. The oz quantity equivalent to the metric

Table 7.23 Directions for Using Household Measures

1. Whenever possible use cups, tbsp., tsp.
2. Express cups in 1/4 or 1/3 increments.
3. Express tbsp. in whole numbers for quantities < 1/4 cup and ≥ 1 tbsp.
4. Express tsp. in whole numbers for quantities < 1 tbsp. and > tsp.; express in 1/4 tsp. increments for quantities < 1 tsp.
5. If foregoing not applicable use:
   Slice, Piece, Tray
   Jar, Fraction, Oz
6. Oz should be accompanied with appropriate visual unit of measure (i.e., dimension of a piece).
7. Express oz in 0.5 increment closest to RACC; use term "about" to indicate rounding (i.e., about 2.5 oz.).

Table 7.24 Acceptable Abbreviations, Metric Equivalents, Rounding Rules

Declaration	Acceptable abbreviations	Metric equivalent
Tablespoon	tbsp	15 ml
Teaspoon	tsp	5 ml
Gram	g	g
Milliliter	ml	ml
Ounce	oz	28 g
Fluid ounce	fl oz	30 ml

Rounding rules

Serving size	Rounded value
< 2 g	nearest 0.1 g
≥ 2 to ≤ 5 g	nearest 0.5 g
> 5 g	nearest 1.0 g

quantity should be expressed in 0.1 oz increments. The acceptable household abbreviations, metric equivalents and applicable "rounding rules" are shown in Table 7.24.

To promote uniformity in label serving size in household measures, FDA developed guidelines for determining the gram weight of household measures. These guidelines can be obtained from Division of Nutrition (HFF 260), Center for Food Safety and Applied Nutrition, Food and Drug Administration, 200 C St., S.W., Washington, D.C. 20204.

1. Labeling of Food with Number of Servings [21 CFR 101.8 para (a); FR p. 2291]

The label of any food package declaring the number of servings it contains must state the net quantity (in terms of weight, measure or numerical count) of each such serving in the same size type. This statement may be expressed in terms that differ from the terms used in the required statement of net quantity of contents (e.g., cups, tablespoons) when the differing term is common to cookery and describes a constant quantity.

When nutrition labeling is required, the statement of the net quantity of each serving must be consistent with the requirements for serving size set forth in the regulations (e.g., ten 1-cup [240 ml] servings). A statement of the number of units in a package is not in itself considered a statement of the number of servings.

2. Rules for Determining Servings per Container [21 CFR 101.9 para (b)(8); FR p. 2292]

The regulations require declaration of the exact serving size but allow for declaration of an approximate number of servings per container. Detailed procedures are provided for determining the number of servings per container, including the increments in which the serving should be declared and procedures for rounding. Determination of the number of servings per container is based on the serving size of the product as follows.

- The number of servings must be rounded to the nearest whole number except for the number of servings between 2 and 5 servings and random weight products.
- The number of servings between 2 and 5 servings must be rounded to the nearest 0.5 serving.
- Rounding must be indicated by the use of the term "about" (e.g., about 2 servings; about 3.5 servings).
- When the serving size is required to be expressed on a drained solids basis and the number of servings vary because of a natural variation in unit size (e.g., maraschino cherries, pickles), the manufacturer may state the typical number of servings per container (e.g., usually 5 servings).
- For random weight products, a manufacturer may declare "varied" (e.g., usually 4 to 6 servings) for the number of servings per container provided the nutrition information is based on the reference amount expressed in ounces. The manufacturer may provide the typical number of servings in parenthesis following the "varied" statement. The FDA encourages manufacturers to declare the typical number of servings whenever possible.

3. Petition Process [21 CFR 101.12 para (h)(1–9); FR pp. 2299–2300]

The regulations set forth specific requirements for filing a petition to establish or amend a reference amount. These include:

- Description of the product and form as sold.
- Intended major use and population group and if used as an ingredient.
- Names of closely related products.
- Sample of proposed label.
- Suggested RAAC.

Details of the petition process are described in the references cited above. The regulations also provide detailed requirements for petitioning a new subcategory of food with its own RACC.

IX. NUTRIENT CONTENT CLAIMS [21 CFR 101.13; 101.54; 101.56; 101.60–101.62; 101.65; 101.69; FR pp. 2410–2426]

A nutrient content claim, sometimes referred to as a "descriptor" is any word or phrase used to characterize the nutrient content of a food. It is a claim about the level of a nutrient in a food in relation to its Reference Daily Intake (RDI) or Daily Reference Value (DRV). It may be expressed, implied, "absolute" or relative. To achieve uniformity across product lines, eliminate misleading claims, and reduce consumer confusion, strict regulations governing nutrient content claims were established by the FDA and USDA/FSIS (Table 7.25). Such claims may be made if:

- The food is labeled in accordance with the requirements for nutrition labeling.
- The claim complies with the requirements for nutrient content claims.
- The claim uses one of the terms defined in the regulations.

Table 7.25 Main Features of the Nutrient Content Claim Regulations

- Nutrient content claims trigger nutrition labeling
- The regulations establish definitions for specific nutrient content claims and conditions for their use.
- Nutrient content claims *must be stated only as defined by the regulations*
- The regulations establish definitions and specifications for:
  - Absolute claims[a]: "Free, low/very low, high, good source, lean, extra lean"
  - Relative claims[a]: "Reduced, less, more, light"
  - Calorie content
  - Sodium content
  - Fat, fatty acid, cholesterol content
  - Sugar content
  - Fiber content
  - Implied claims and related statements
- The regulations provide for the use of synonyms.
- Define and provide for the use of the term "fresh"

[a] Acceptable synonyms are shown in Table 7.26.

Table 7.26 Nutrient Content Claims: Acceptable Synonyms

Free	Free of, no, zero, without, trivial source of, negligible source of, dietarily insignificant source of
Low	Few, contains small amount of, low source of
Reduced	Reduced in, fewer, lower, lower in, less
High	Rich in, excellent source of
Good source	Contains, provides
More	Fortified, enriched, added

Summaries of definitions and conditions for nutrient content claims as well as allowable synonyms are shown in Tables 7.30 through 7.39 and Table 7.26, respectively.

A. General Principles [21 CFR 101.13; FR pp. 2410–2413]

A claim that expressly or implicitly characterizes the level of a nutrient (nutrient content claim) may not be made on a food label unless the term has been defined by the regulations or accepted by petition.

1. Expressed Claims

An expressed nutrient content claim is any direct statement about the level (or range) of a nutrient in the food (e.g., "low fat" or "contains 100 calories").

2. Implied Claims

An implied claim is any claim that:

- Describes the food or an ingredient therein in a manner that suggests that the nutrient is absent or present in a certain amount (e.g., "high in oat bran")
- Suggests that the food because of its nutrient content may be useful in maintaining healthy dietary practices and is made in close association with an explicit claim or statement about a nutrient (e.g., "healthy, contains 3 g of fat").

3. Nutrient Content Claims on Foods for Infants and Children Less Than 2 Years of Age

Except for claims regarding vitamins and minerals, no nutrient content claims may be made on food intended specifically for infants and children less than 2 years of age unless the claim is specifically provided for in the regulations.

4. Nutrition Panel Information

Information that appears as part of the nutrition panel is not a nutrient content claim and is not subject to the rules applicable to nutrient content claims. However, if such information is declared elsewhere on the label, it becomes a nutrient content claim and is regulated as such.

5. "Free" or "Low" Special Processing Requirement

Only if a food has been specially processed, altered, formulated, or reformulated to lower the amount of a nutrient may it bear a "free" or "low" claim.

Any claim for the absence of a nutrient in a food or that the food is low in a nutrient when the food has *not* been specially processed, altered, formulated, or reformulated to qualify for that claim must indicate that the food *inherently* meets the criteria and must refer to all foods of that type and not merely to the particular brand making the claim (e.g., "corn oil, a sodium-free food").

6. Referral Statement

A nutrient content claim must be followed by a referral statement. "See *appropriate panel* (with location) for nutrition information." This statement must appear prominently and in immediate proximity to the claim with no intervening material. The type size and style of the nutrient content claim must be no larger than twice that of the statement of identity. The regulation provides very detailed information regarding the placement and prominence of the referral statement (21 CFR 101.13 (g)(i)(2) or *Federal Register 58*(3), Jan 6, 1993, p. 2411).

7. Multiple Nutrient Content Claims

If a food panel contains multiple nutrient content claims or a single claim repeated several times, a single referral statement may be made. The statement must be adjacent to the claim that is printed in the largest type on that panel.

8. Disclosure Levels

When making a nutrient content claim the referral statement must also disclose the presence of the nutrients and food components specified in Table 7.27 if they exceed the levels shown. If a nutrient/food component is present in an amount that exceeds the disclosure level, the label must disclose as part of the referral statement that the nutrient exceeding the specified level is present in the food as follows:

> See *(appropriate panel)* for information about *(nutrient requiring disclosure)* and other nutrients (e.g., "See side panel for information about total fat and other nutrients.")

Table 7.27 Disclosure Levels. Nutrient Content Claims Trigger a Disclosures Statement if the Presence of the Following Nutrients Exceed Set Levels

	Disclosure level		
Nutrient	Per RA and per serving[a]	Meal[10]/ serving	Main dish[10]/ serving
Fat	13 g	26 g	19.5 g
Saturated fat	4 g	8 g	6 g
Cholesterol	60 mg	120 mg	90 mg
Sodium	480 mg	960 mg	720 mg

[a] Per 50 g for small reference amount (i.e., ≤ 30 g *or* ≤ 2 tbsp.)

9. Implicit Characterization of Nutrient Content

An implicit statement about the amount or percentage of a nutrient in a food is permitted if it satisfies the definition for a nutrient content claim as set forth in the regulations. If the implied statement does not satisfy the definition it must bear a disclaimer such as "the food is not low in [name of nutrient] or a good source of [name of nutrient] the nutrient" [e.g., "only 200 mg sodium per serving (the implied statement), not a low sodium food" (the disclaimer)]. Details of the location and print size of the disclaimer are provided in the regulations 21 CFR 101.13 (i)(2); *Federal Register 58*(3), Jan 6, 1993, p. 2412.

10. Relative Claims

A relative claim is one which compares the level of a nutrient in a food with the level in an appropriate reference food. "Light," "reduced," "less," "fewer," and "more" are relative claims. To bear a relative claim of "less," "fewer," or "more," the reference food may be a *dissimilar* food within a product category that can generally be substituted for one another (e.g., potato chips as a reference food for pretzels or vice versa).

For "light," "reduced," "added," "fortified," and "enriched" claims, the reference food must be a *similar* food (e.g., a potato chip reference for a potato chip). The reference food for "light claims" must be representative of the type of product that bears the claim. Table 7.28 lists the sources of representative reference foods that may be used for product comparisons.

In all cases, foods bearing relative claims must state the identity of the reference food and the percentage (or fraction) of the amount of the nutrient in the reference food by which the nutrient has been modified (e.g., "50% less fat

Table 7.28 Relative Nutrient Content Claims: Source of Reference Foods

Relative claims compare the nutrient level in a food to the level in a reference food. Representative reference foods are:

Manufacturers regular product
Competitors product
Value from valid database
Value from top 3 national brands
Value from "market leader"

Reference food for "*fortified, enriched, light, reduced, added*" must be similar food type.

Reference food for "*less, more*" may be similar or dissimilar food type.

than reference food" or "1/3 fewer calories than reference food." In addition, labeling for a relative claim must compare the amount of the subject nutrient in the product to the amount in the reference food.

Under no circumstances may a relative claim for decreased levels of a nutrient be made if the nutrient content of the reference food satisfies the definition of a "low" claim for that nutrient. Table 7.29 summarizes the requirements for making relative nutrient claims.

11. Relative Claims and Statement of Product Identity

The term "modified" may be used in the statement of identity of a food that bears a relative claim that complies with the regulations. In such cases, the term "modified" must be followed immediately by the name of the nutrient whose content has been altered (e.g., "modified fat cheesecake"). This statement must be immediately followed by the comparative statements noted above and in Table 7.26.

Table 7.29 Required Label Declarations for Relative Claims

- Identity of the reference food
- Percentage (or fraction) of the amount in the reference food by which the nutrient has been modified
- Comparison of the amount of the subject nutrient in the product to the amount in the reference food
- Relative claims to describe a nutrient reduction may *not* be used with foods that already qualify as "low" or "free" in that nutrient.

12. Nutrition Labeling Requirement

Nutrition labeling is required if a food label bears a nutrient content claim even if the product would otherwise be exempt.

13. Compliance

Compliance with requirements for nutrient content claims and nutrition labeling will be determined by the sampling and analytical methodology prescribed in the regulations or by use of an FDA approved database that has been computed using FDA guidelines[7,8].

14. Reference Amounts

Unless otherwise specified, the reference amount (RACC) must be used in determining whether a product meets the criteria for a nutrient content claim. If the serving size differs from the RACC and the amount of nutrient for which a nutrient content claim is made exceeds the maximum or minimum criterion for the descriptor, the claim must be followed by the criteria for the claim (e.g., "very low sodium, 35 mg or less per 240 ml [8 fl oz]").

15. Exemptions to Nutrient Content Claims Requirements

- Claims contained in a brand name in use before October 25, 1989, that have not been defined provided the brand name is not false or misleading. The exempt brand name must still comply with the requirements pertaining to the prominence of a claim, and the referral and disclosure statements.
- An implied claim may be part of a brand name if the claim has been authorized by FDA. Petitions requesting approval of such a claim may be submitted to the agency.
- A soft drink that used the term "diet" as part of its brand name before October 25, 1989[9].
- Soft drinks marketed after October 25, 1989, may use the term "diet." The label must not be false or mislead and it must comply with the requirements pertaining to prominence of a claim and the referral and disclosure statements.
- Claims that were part of the common or usual names of foods with a product name with a standard of identity on November 8, 1990.
- A statement describing the percentage of a vitamin or mineral in a food in relation to an RDI level unless the claim is specifically prohibited.
- Infant formulas
- The terms "fluoridated," "fluoride added," or "with added fluoride" to describe bottled water that contains added fluoride.
- Medical foods defined by Section 5(b) of the Orphan Drug Act

- Restaurant menus.

16. Restaurant Foods

A nutrient content claim used on food that is served in restaurants (except on menus) or other establishments in which food is served for immediate consumption or which is sold for sale or use in such establishments must comply with the requirements for nutrient content claims, but it is exempt from the requirements for referral and disclosure statements. Unique compliance requirements are described in the regulations.

A term or symbol may be used if it does not characterize the level of a nutrient and a statement clearly explaining the basis for its use is prominently displayed (e.g., "lite fare" followed by an asterisk referring to a note that makes it clear that in "this" restaurant "lite fare" means smaller portion sizes than usual).

A food item that states that it meets the criteria for dietary guidance established by a recognized dietary authority would not be considered a nutrient content claim.

B. Nutrient Content Claims for "High," "Good Source," and "More" [21 CFR 105.54 (a–e); FR p. 2414]

The requirements for "high," "good source," and "more" claims are summarized in Table 7.30.

1. "High" Claims [21 CFR 101.54 para (b); FR p. 2414]

The basic principles governing nutrient content claims for "high," "good source," and "more" are outlined in Table 7.30.

a. Individual Foods

The term "high" (or "rich in" or "excellent source of") may be used if the nutrient is present in the individual food at 20% or more of the RDI or DRV per RACC.

b. Meal and Main Dish Products[10]

A product qualifies for this claim if:

- It contains a food that provides at least 20% of the RDI or DRV per RACC.
- The label identifies the food component that is the subject of the claim (e.g., "the serving of broccoli in this product is high in vitamin C").

Table 7.30 Requirements for Nutrient Content Claims for "Good Source," "High," and "More"

	Individual foods	Meals/main dishes
High	Present at ≥ 20% of RDI or DRV per RACC	Present at ≥ 20% of RDI or DRV per RACC and label identifies subject nutrient
Good source	Present at 10–19% of RDI or DRV per RACC	Present at 10–19% of RDI or DRV per RACC and label identifies subject nutrient
More	Present at ≥ 10% of RDI or DRV per RACC versus reference food	Nutrient present at ≥ 10% of RDI or DRV of reference food per 100 g and label identifies subject nutrient

2. "Good Source" Claims [21 CFR 101.54 para (c); FR p. 2414]

a. Individual Foods

The terms "good source," "contains," or "provides" may be used if the nutrient is present in the individual food between 10 to 19% of the RDI or DRV per RACC.

b. Meal and Main Dish Products[10]

A product qualifies for this claim if:

- It provides 10 to 19% of the RDI or DRV per RACC, and
- The label identifies the food component that is the subject of the claim (e.g., "the serving of sweet potatoes in this product is a good source of fiber").

3. Fiber Claims [21 CFR 101.54 para (d); FR p. 2414]

A claim about dietary fiber is permitted if its level in a product (individual food, meal or main dish) qualifies for the claim (e.g., "high" or "good source," as defined above) provided the food is also low in total fat as defined in the regulations. If the subject food is not low in total fat, the label must disclose the level of total fat per labeled serving (e.g., "contains *amount of fat* of total fat per serving. See *name and location of panel* for nutrition information." The regulation stipulates the placement and the prominence requirements for this statement.

4. "More" Claims [21 CFR 101.54 para (e); FR p. 2414]

a. Individual Foods

A relative claim using the term "more" ("fortified," "enriched" or "added") may be used to describe the level of protein, vitamins, minerals, dietary fiber or potassium in an individual food if the subject nutrient is present in the food at 10% or more of its RDI or DRV per RACC compared to an appropriate reference food. The explanatory information required for relative claims (see Table 7.29) must accompany the claim. When the claim is based on a nutrient that has been added to the food, FDA policy on fortification must be followed (21 CFR 104.20).

b. Meal and Main Dishes Products[10]

A product qualifies for this claim if it contains at least 10% more of the RDI or DRV for the subject nutrient (e.g., protein, vitamins, minerals, dietary fiber, or potassium) per 100 g than an appropriate reference food. The explanatory requirements and fortification rules noted for individual foods also apply to meal or main dish products.

C. Nutrient Content Claims for "Light" or "Lite" [21 CFR 105.56 para (a–g); FR pp. 2414–2415]

The requirements for light calorie or light fat claims are summarized in Table 7.31.

1. Calories and Fat

a. Individual Foods

The term "light" or "lite" may be used to describe an individual food under the following conditions:

- If the food derives 50% or more of its calories from fat, its fat content must be reduced by 50% or more per RACC compared to an appropriate reference food.
- If the food derives less than 50% of its calories from fat, the number of calories must be reduced by at least one-third per RACC compared to an appropriate reference food, *or* its fat content must be reduced by 50% or more per RAAC compared to the reference.

b. Explanatory Information

As with all relative claims, appropriate explanatory information (Table 7.29) must accompany the claims and must appear in immediate proximity to the most prominent claim. For example, there must be a quantitative comparison of the level of calories and fat in the product serving size with that of the

Table 7.31 Requirements for Light Calorie or Fat Claims[a]

Claim	Individual foods	Meals/main dishes
Light	For foods with ≥ 50% of calories from fat, the *fat content* must be reduced by ≥ 50% of RACC versus reference food. For foods with < 50% of calories from fat, *caloric content* must be reduced by ≥ 1/3 per RACC vs. reference food *or* fat content must be reduced by 50% per RACC vs. reference food.	Must meet requirements for: *Low fat*: ≤ 5 g total fat per 100 g and ≤ 30% of calories from fat or *Low calorie*: ≤ 120 calories per 100 g Must identify the nature of the claim: e.g., whether light measures low in calories, low in fat, or low in both

[a] *Stipulations*: 1. Relative claims statement. (See Table 7.29 for relative claims information that must appear on the label.) 2. Not allowed if reference food is low in fat and calories. 3. If food contains < 40 calories or < 5 g fat per RACC, the percentage reduction need not be declared.

reference food (e.g., "lite cheese cake contains one third fewer calories and 50 percent less fat than our regular cheese cake; light cheese cake contains 200 calories, 4 g fat; regular cheese cake contains 300 calories, 8 g fat per serving"). This must be declared adjacent to the most prominent claim on the information panel.

If the food contains less than 40 calories or less than 5 g of fat per RACC (i.e., qualifies as a low calorie or low fat product), the percentage reduction for that nutrient need not be declared as part of the explanatory statement.

c. Meal and Main Dish Products[10]

The term "light" ("lite") may be used on the label of a meal or main dish product provided that:

- The food meets the definition of "low calorie" or "low fat."
- The principal display panel explains whether "light" is used to mean "low fat," "low calories," or both (e.g., "*name of product*, low fat meal").
- The print size, placement, and prominence requirements specified in the regulations are satisfied.

d. Non-Permissible Claim

A light claim may not be made if the reference food meets the definition of "low fat" *and* "low calorie."

2. Light Claims and Sodium Content

The requirements for "light" claims and sodium content are summarized in Table 7.32.

a. Individual Foods

A product for which the reference food contains 40 calories or less and 3 g fat or less per RACC may use the term "light" without qualification if the sodium content is reduced by 50% or more per RACC compared to the reference food. The label must disclose the actual nature of the nutrient modification made to the food (e.g., "light in sodium," "light in fat," or "light in calories").

Explanatory information (Table 7.29) must accompany the claim (e.g., "Brand X Light Soy Sauce," contains "50% less sodium than our regular soy sauce; lite soy sauce: 500 mg sodium per serving; regular soy sauce: 1,000 mg sodium per serving." This information must appear in immediate proximity to the most prominent claim. The placement and prominence requirements for these statements must be satisfied.

If the reference food contains more than 40 calories or more than 5 g of fat per RAAC, the term "light in sodium" may be used to designate a reduction of 50% or more. In such cases, the term "light" must be in immediate proximity with "in sodium" (e.g., light in sodium) and the entire claim must be in uniform

Table 7.32 Requirements for Light Sodium Claims

Claim	Individual foods
	Definition
Light	A product for which the reference food contains ≤ 40 calories and ≤ 5 g fat per RACC *and* ≥ 50% reduction in sodium per RACC compared to reference food.
Light in sodium	A product for which the reference food contains > 40 calories *or* > 5 g fat per RACC if ≥ 50% reduction in sodium per RACC compared to reference food.
Lightly salted	Product containing < 50% added salt compared to the reference food.
	Stipulations
Light	Relative claims statement.[a]
Light in sodium	Not allowed if reference food is low in sodium.
Lightly salted	If not low sodium, requires statement "not a low sodium food."

[a] See Table 7.29 for relative claims information that must appear on the label.

type, size, color, and prominence. The typical explanatory information required for relative claims must be provided (Table 7.29).

b. Meals and Main Dishes[10]

The term "light" ("lite") in sodium may be used if the food is "low" in sodium as defined in the regulations and the printing, placement, and prominence requirements are satisfied.

c. Non-Permissible Claim

A "light in sodium" claim is not permitted when the reference food satisfies the "low sodium" definition.

3. Light as a Non-Nutrient Content Claim

The regulations permit certain non-nutrient content claim uses for the term "light" ("lite"), for example,

- To describe a physical or organoleptic attribute of the food such as texture or color (e.g., "light in color" or "light in texture").
- To describe a term that has been commonly used with a particular food to reflect a particular physical or organoleptic attribute (e.g., light brown sugar, light corn syrup, light molasses) to the point where it has become part of the statement of identity.
- To describe a product as "lightly salted" when half as much (50%) sodium is added than is normally added to the reference food. However, if the product is not "low in sodium," it must state "not a low sodium food."

D. Nutrient Content Claims for the Calorie Content of Foods[9] [21 CFR 101.60 para (a)(b); FR pp. 2415–2416]

The requirements for calorie content claims are summarized in Table 7.33.

1. Calorie Free Claims

The terms "calorie free," "free of calories," "no calories," "zero calories," "without calories," "negligible source of calories," or "dietarily insignificant source of calories" may be used if:

- The food contains less than 5 calories per RACC and per label serving size.
- The food meets this condition as the result of special processing, alteration, formulation, or reformulation. Otherwise, the label must disclose that calories are not usually present in the food (e.g., "cider vinegar, a calorie free food").

Table 7.33 Requirements for Calorie Content Claims

Claim	Individual foods	Meals/main dishes
	Definition	
Calorie free	< 5 calories per RACC and per serving	Not defined
Low calorie	≤ 40 calories per RACC[a]	≤ 120 calories per 100 g
Reduced calorie	≥ 25% calorie reduction per RACC compared to reference food	≥ 25% fewer calories per 100 g than reference food
	Stipulations	
Calorie free, low calorie	Qualifying statement if inherently low or free	
Reduced calorie	Relative claims statement.[b] Not allowed if reference food is low calorie.	

[a] < 40 calories per 50 g for small serving size foods (e.g., where RACC is ≤ 30 g or ≤ 2 tablespoons).
[b] See Table 7.29 for relative claims information that must appear on the label.

2. Low Calorie Claims

a. Individual Foods

The terms "low calorie," "few calories," "contains a small amount of calories," "low source of calories," or "low in calories" may be used on the label of individual foods if:

- The individual food contains no more than 40 calories per RACC and per 50 g (except for sugar substitutes) if it is a small serving size food (e.g., the RACC is 30 g [or 2 tablespoons] or less).[11]
- The food qualifies for the claim as the result of special processing, alteration, formulation or reformulation. Otherwise, it must be accompanied by an appropriate qualifying statement disclosing that the product is inherently low in calories.

b. Meal and Main Dish Products[10]

A product qualifies for this claim if:

- It contains 120 calories or less per 100 g

- The food qualifies for the claim as the result of special processing, alteration, formulation, or reformulation. Otherwise, it must be accompanied by an appropriate qualifying statement disclosing that the product is inherently low in calories.

3. Reduced Calorie Claims

a. Individual Foods

The terms "reduced calorie," "reduced in calories," "calorie reduced," "fewer calories," "lower calories," or "low in calories" may be used on the label of individual foods provided:

- The food contains at least 25% fewer calories per RACC than appropriate reference food.
- The claim is accompanied by explanatory information required for relative claims (Table 7.29) (e.g., Reduced calorie cupcakes: "one third fewer calories than regular cupcakes; calorie content has been reduced from 150 to 100 calories per serving.")

b. Meal and Main Dish Products[10]

A product qualifies for this claim if:

- It contains at least 25% fewer calories per 100 g than an appropriate reference food.
- The claim is accompanied by explanatory information required for relative claims (Table 7.29) (e.g., "Brand X" Reduced Calorie Lasagna, "25% fewer calories per 3 oz than our regular lasagna; calorie content has been reduced from 108 calories per 3 oz to 83 calories per 3 oz.")

c. Non-Permissible Claim

A reduced calorie claim is not permitted for any food if the reference food meets the definition of "low calorie."

E. Nutrient Content Claims for the Sugar Content of Foods[9] [21 CFR 101.60 para (c); FR pp. 2416–1417]

The requirements for sugar content claims are summarized in Table 7.34. "Sugar free," "free of sugar," "no sugar," "zero sugar," "without sugar," "sugarless," "trivial source of sugar," "negligible source of sugar," or "dietarily insignificant source of sugar" may be used if:

- The food contains less than 0.5 g of sugar per RACC (and per labeled serving size) or per serving for meals or main dish products.

Table 7.34 Requirements for Sugar Content Claims

Claim	Individual foods	Meals/main dishes
	Definition	
Sugar free	< 0.5 grams of sugar per RACC and per serving	< 0.5 grams of sugar per serving
No added sugar	No sugar or sugar containing ingredient is added	No sugar or sugar containing ingredient is added
Reduced sugar	≤ 25% less sugar per RACC compared to reference food	≤ 25% less sugar per 100 g versus reference food
	Stipulations	
Sugar free, no sugar added	The food may not contain any ingredient that is a sugar or generally understood to contain sugars (unless accompanied by statement "adds a trivial amount of sugar") The food must be "low calorie" or "reduced calorie" otherwise must declare "not a low calorie food"	
No sugar added	The food that it resembles normally contains added sugars	
Reduced sugar	Relative claims statement[a]	

[a] See Table 7.29 for relative claims information that must appear on the label.

- The food does not contain any ingredient that is a sugar or is generally understood by consumers to contain sugars unless the ingredient as declared on the ingredient statement is accompanied by an asterisk that refers to the statement "adds a trivial amount of sugar" or a similar specified statement.
- The food is labeled "low calorie" or "reduced calorie" or bears a relative claim of special dietary usefulness. Otherwise the term "sugar free" must be immediately accompanied by the statement "not a reduced calorie food," "not a low calorie food," or "not for weight control."

1. No Sugar Added Claims

The terms "no added sugar," "without added sugar," or "no sugar added" may be used if:

- No amount of sugars or any other ingredient that contains sugars that functionally substitute for added sugars is added during processing or packaging.
- The product does not contain an ingredient containing added sugars such as jam, jelly, or concentrated fruit juice.
- The sugar content has not been increased by use of enzymes or similar means, except where the intended functional effect of the process is not to increase

the sugars content of the food and a functionally insignificant increase in sugars results.
- The food that it resembles and for which it substitutes normally contains added sugars.

In addition, if the food does not qualify as "low calorie" or "reduced calorie," it must declare "not low calorie" or "not calorie reduced" followed by a referral statement directing attention to the nutrition panel for further information on sugar and calorie content.

The requirements for "sugar free" claims do *not* apply to a factual statement that a food is unsweetened, or contains no added sweeteners in the case of foods inherently containing a substantial amount of sugar.

2. Reduced Sugar Claims

The term "reduced sugar" (or "reduced in sugar," "sugar reduced," "less sugar," "lower sugar," or "lower in sugar) may be used if the sugar content in the individual food is reduced by 25% per RACC as compared to an appropriate reference food. The claim must be accompanied by the explanatory information required for relative claims. For meal and main dish products, the 25% reduction in sugars is measured per 100 g as compared to an appropriate reference food. The requirements for an individual food are otherwise applicable.

F. Nutrient Content Claims: Sodium/Salt [21 CFR 101.61 para (b); FR pp. 2417–2418]

The requirements for sodium content claims are summarized in Table 7.35.

1. Sodium Content Claims

The terms "sodium free," "free of sodium," "no sodium," "zero sodium," "without sodium," "trivial source of sodium," "negligible source of sodium," or "dietarily insignificant source of sodium" may be used if:

- The food contains less than 5 mg of sodium per RACC and per labeled serving size or in the case of a meal or a main dish product, less than 5 mg of sodium per labeled serving.
- The food contains no ingredient that is sodium chloride or is generally understood by consumers to contain sodium, unless the listing of the ingredient in the ingredient statement is followed by an asterisk that refers to the statement: "Adds a trivial amount of sodium," "adds a negligible amount of sodium," or "adds a dietarily insignificant amount of sodium."

Table 7.35 Requirements for Sodium Content Claims

Claim	Individual foods	Meals/main dishes
	Definition	
Sodium free	< 5 mg sodium per RACC and per serving	< 5 mg sodium per serving
Very low sodium	≤ 35 mg sodium per RACC *or* per 50 g for small serving size foods[a]	≤ 35 mg sodium per 100 g
Low sodium	≤ 140 mg per RACC *or* per 50 g for small serving size foods	≤ 140 mg sodium per 100 g
Reduced sodium	≤ 25% sodium per RACC compared to reference food	≤ 25% sodium per 100 g compared to reference food
	Stipulations	
Sodium free	Cannot contain any added ingredient that is sodium chloride or generally understood to contain sodium unless accompanied by statement "adds a trivial amount of sodium"	
Sodium free, very low sodium, low sodium	Qualifying statement if inherently low or free	
Reduced sodium	Relative claims statement.[b] Not allowed if reference food is low sodium.	

[a] Small serving size food: a food with a RACC of ≤ 30 g or ≤ 2 tablespoon.
[b] See Table 7.29 for relative claims information that must appear on the label.

- The food meets these conditions as the result of special processing, alteration, formulation, or reformulation. Otherwise the label must disclose that sodium is not usually present in the food (e.g., "leaf lettuce, a sodium free food").

2. Very Low Sodium Claims

a. Individual Foods

The terms "very low sodium" or "very low in sodium" may be used provided an individual food contains 35 mg or less per RACC (and per 50 g if it is a small serving size food).[11] If the food qualifies for this claim without the benefit of special processing, alteration, formulation, or reformulation to vary the sodium content, an appropriate qualifying statement must accompany the claim (e.g., "potatoes, a very low sodium food").

b. Meal and Main Dish Products[10]

For a meal or main dish product, the sodium content is measured per 100 g. The requirements for an individual food are otherwise applicable.

3. Low Sodium Claims

a. Individual Foods

The term "low sodium" (or "low in sodium," "little sodium," "contains a small amount of sodium," or "low source of sodium") may be used provided the individual food contains 140 mg or less sodium per RACC (and per 50 g if it is a small serving size food).[11]

b. Meal and Main Dish Products[10]

For meal and main dish products, the food qualifies for a claim if it contains 140 mg or less sodium per 100 g. The requirements for an individual food are otherwise applicable.

4. Reduced Sodium Claims

a. Individual Foods

The terms "reduced sodium," "reduced in sodium," "sodium reduced," "less sodium," "lower sodium," or "lower in sodium" may be used if:

- The individual food is reduced at least 25% in sodium compared to an appropriate reference food.
- The claim is accompanied by explanatory information required for relative claims (Table 7.29) (e.g., reduced sodium; *name of food*, 50% less sodium than regular *name of food*: "sodium content has been lowered from 300 to 150 mg per serving").

b. Meal and Main Dish Products[10]

For meal and main dish products, the 25% reduction in sodium is measured per 100 g as compared to an appropriate reference food. The requirements for an individual food are otherwise applicable.

c. Non-Permissible Claim

No food may bear a "reduced sodium" claim if the reference food qualifies for a "low sodium" claim.

5. Salt Claims

"Salt" is not considered synonymous with "sodium." Salt refers to sodium chloride. References to salt content, such as "unsalted," "no salt," and "no salt added" are considered potentially misleading and are subject to regulation. The term "salt free" may only be used if the food is "sodium free." The terms "unsalted," "without added salt," and "no salt added" may be used provided:

- No salt is added during processing.
- The food that it resembles and for which it substitutes is normally processed with salt.
- It bears a the statement "not a sodium free food" or "not for control of sodium in the diet" if it is not "sodium free."

G. Nutrient Content Claims for Fat, Fatty Acid, and Cholesterol Content [21 CFR 101.62 para (b); FR pp. 2418–2423]

1. Fat Free Claims [21 CFR 101.62 para (b)(i–iii); FR p. 2418]

The requirements for fat content claims are summarized in Table 7.36.

The terms "fat free," "free of fat," "no fat," "zero fat," "without fat," "non-fat," "trivial source of fat," "negligible source of fat," or "dietarily insignificant source of fat" may be used if the individual food contains less than 0.5 g of fat per RACC or per serving for meal or main dish products. In addition, the food may not contain any added ingredient that is a fat or is generally understood to contain fat unless the ingredient, as declared in the ingredient statement, is accompanied by an asterisk that refers to the statement "adds a trivial amount of fat," or similar specified statement (e.g., "negligible source of fat" or "dietary insignificant source of fat"). If the food meets these conditions without the benefit of special processing, alteration, formulation, or reformulation, it must disclose that fat is not usually present in the food (e.g., "broccoli, a fat free food").

Table 7.36 Requirements for Fat Content Claims

Claim	Individual foods	Meals/main dishes
Fat free	< 0.5 g fat per RACC per serving	< 0.5 g fat per serving
Low fat	≤ 5 g per RACC *or* per 50 g for small serving foods[a]	≤ 5 g fat per 100 g *and* ≤ 30% of calories from fat
Reduced fat	≥ 25% fat per RACC compared to reference food	≤ 25% less fat per 100 g compared to reference food
	Stipulations	
Fat free	Cannot contain any added ingredient that is fat or generally understood to contain fat unless accompanied by statement such as "adds a trivial amount of fat."	
Fat free, low fat	Qualifying statement if inherently low or free.	
Reduced fat	Relative claims satement[b]. Not allowed if reference food is low in fat.	

[a] Small serving size food: a food with a RACC of ≤ 3 g or ≤ 2 tbsp.
[b] See Table 7.29 for disclosure levels for fat.

2. Low Fat Claims [21 CFR 101.62 para (b)(2)(i,ii) (3)(i,ii); FR p. 2418]

a. Individual Foods

The terms "low fat," "low in fat," "contains a small amount of fat," "low source of fat," or "little fat" may be used if the individual food contains 5 g or less of fat per RACC (and per 50 g if it is a small serving size food).[11] If the food qualifies for the claim without the benefit of special processing, it must disclose that all foods of its type and not merely the brand are low in fat (e.g., frozen perch, a low fat food).

b. Meal and Main Dish Products[10]

A meal or main dish product must meet the 5 g or less criterion per 100 g and not derive more than 30% of calories from fat. The requirements for an individual food are otherwise applicable.

3. Reduced Fat Claims [21 CFR 101.62 para (b)(4); FR pp. 2418–2419]

a. Individual Foods

The terms "reduced fat," "reduced in fat," "fat reduced," "less fat," "lower fat," or "lower in fat" may be used if the individual food is reduced at least 25% in fat compared to the reference food. The claim must be accompanied by the explanatory information required for relative claims (Table 7.29) (e.g., "reduced fat," *name of product* "contains 50% less fat than our regular brownies;" "fat content has been reduced from 8 g to 4 g per serving").

b. Meal and Main Dish Products[10]

For meal or main dish products, the 25% reduction in the fat is measured per 100 g as compared to an appropriate reference food. The requirements for an individual food are otherwise applicable.

c. Non-Permissible Claim

No food may bear a "reduced fat" claim if the reference food qualifies for a "low fat" claim.

4. Percent Fat-Free Claim [21 CFR 101.62 para (b)(6); FR p. 2419]

A percent fat-free claim is restricted to products that qualify as "low fat." The percent reduction and the words "fat free" must be in uniform type size. A 100% "fat-free" claim may only be made on foods that meet the criteria for "fat-free," contain less than 0.5 g of fat per 100 g, and contain no added fat.

5. Saturated Fatty Acid Content Claims [21 CFR 101.62 para (c); FR p. 2419]

The requirements for saturated fat content claims are summarized in Table 7.37.

Labels that bear claims about the level of saturated fat must disclose the level of total fat and cholesterol in the food in immediate proximity to such claim each time the claim is made.

Declaration of cholesterol content may be omitted when an individual food contains less than 2 mg of cholesterol per RAAC or per serving for meal or main dish products.

Declaration of total fat may be omitted in connection with a saturated fat free claim when the food satisfies the requirements for a fat free claim (e.g., contains 0.5 g or less of total fat per RACC for an individual food or per serving for a meal or main dish product).

Declaration of total fat may also be omitted when using the terms "low in saturated fat" or "reduced saturated fat" if the food satisfies the requirements for a low fat or reduced fat claim, respectively.

In addition to these general requirements, the regulation establishes specific requirements for the saturated fatty acid and cholesterol claims cited below.

6. Saturated Fat Free Claims [21 CFR 101.62 para (c)(1); FR p. 2419]

The terms "saturated fat free," "free of saturated fat," "no saturated fat," "zero saturated fat," "without saturated fat," "trivial source of saturated fat," "negli-

Table 7.37 Requirements for Saturated Fat Content Claims

Claim	Individual foods	Meals/main dishes
	Definition	
Saturated fat free	< 0.5 g saturated fat per RACC and per serving; *trans* fatty acids ≤ 1% of total fat	< 0.5 g saturated fat per serving; *trans* fatty acids ≤ 1% of total fat
Low in saturated fat	≤ 1g SFA per RACC and ≤ 15% of calories from SFA[a]	≤ 1 g SFA per 100g *and* < 10% of calories from saturated fat
Reduced saturated fat	≤ 25% saturated fat per RACC compared to reference food	≤ 25% saturated fat per 100 g compared to reference food
	Stipulations	
Saturated fat free	May not contain any ingredient that is a SFA or is generally understood to contain SF[a] unless accompanied by statement "adds a trivial amount of SF"	
Saturated fat free, low saturated fat	Qualifying statement if inherently low or free	
Reduced saturated fat	Relative claims statement. Not allowed if reference food is "low saturated fat"	

[a] SFA: Saturated fatty acids; SF: Saturated fat.

gible source of saturated fat," or "dietarily insignificant source of saturated fat' may be used provided that:

- The food contains less than 0.5 g of saturated fat per RACC and the level of trans fatty acids does not exceed 1% of the total fat, or per serving for meal or main dish products.
- The food contains no ingredient generally understood to contain saturated fat unless the listing of the ingredient in the ingredient statement is followed by an asterisk that refers to a statement below the list of ingredients which states, "adds a trivial amount of saturated fat," "adds a negligible amount of saturated fat," or "adds a dietarily insignificant amount of saturated fat."
- The food meets these conditions as the result of special processing, alteration, formulation, or reformulation. Otherwise the label must disclose that saturated fat is not usually present in the food.

7. Low Saturated Fat Claims

a. Individual Foods

The terms "low in saturated fat," "low saturated fat," "contains a small amount of saturated fat," "low source of saturated fat," or "a little saturated fat" may be used labeling individual foods provided that:

- The individual food contains 1 g or less of saturated fatty acids per RACC and not more than 15% of calories from saturated fatty acids.
- The food meets these conditions as the result of special processing, alteration, formulation, or reformulation. Otherwise the label must disclose that saturated, fat is not usually present in the food (e.g., "raspberries, a low saturated fat food").

b. Meal and Main Dish Products[10]

The term "low saturated fat" may be used on the label of meal or main dish products provided:

- The product contains 1 g or less of saturated fatty acids per 100 g and less than 10% calories from saturated fat.
- The food meets these conditions as the result of special processing, alteration, formulation, or reformulation. Otherwise the label must disclose that saturated fat is not usually present in the food.

8. Reduced Saturated Fat Claim [21 CFR 101.62 para (c)(4); FR p. 2419]

a. Individual Foods

The terms "reduced saturated fat," "reduced in saturated fat," "saturated fat reduced," "less saturated fat," "lower saturated fat," or "lower in saturated fat" may be used if:

- The individual food contains 25% less saturated fat per RACC than an appropriate reference food.
- The explanatory information required for relative claims must accompany the claim (e.g., "reduced saturated fat, *name of product*. Contains 50% less saturated fat than the national average for non-dairy creamers; saturated fat reduced from 3 g to 1.5 g per serving).

b. Meal and Main Dish Products[10]

For a meal or main dish product, the food must contain 25% less saturated fat per 100 g than an appropriate reference food. The requirements for an individual food are otherwise applicable.

c. Non-Permissible Claim

No food may bear a "reduced saturated fat" claim if the reference food qualifies for a "low saturated fat" claim.

9. Cholesterol Free Claim [21 CFR 101.62 para (d)(i); FR p. 2420]

The requirements for cholesterol content claims are summarized in Tables 7.38 and 7.39.

Table 7.38 Requirements for Cholesterol Content Claims *not* Exceeding Disclosure Level for Fat[a]

Claim	Definition	Special requirements
Cholesterol free	< 2 mg cholesterol *and* ≤ 2 g SFA[b] per RACC[c] and per serving	May not contain any ingredient generally understood to contain cholesterol; otherwise must state "adds a trivial amount of cholesterol"
Low cholesterol	≤ 20 mg cholesterol *and* ≤ 2 g SFA per RACC[d] (and per 50 g for small serving size)	
Reduced cholesterol	≤ 25% than reference food and ≤ 2 g SFA per RACC[d] versus reference food	Relative claims statement. Reference food must not be low cholesterol
Cholesterol free, low cholesterol, reduced cholesterol		Reference food must have significant market share (e.g., ≥ 5% of market). Qualifying statement if inherently low or free

[a] Disclosure levels for fat are different for individual foods, meals, and main dishes. See Table 7.27.

[b] SFA: Saturated fatty acid.

[c] Per serving for meal and main dish products.

[d] Per 100 g for meal and main dish products.

Table 7.39 Requirements for Cholesterol *Exceeding* Disclosure Level for Fat in Individual Foods[a]

Claim	Definition	Special requirements
Cholesterol free	Must disclose fat level < 2 mg cholesterol *and* ≤ 2 g SFA[b] per RACC[c] and per serving	May not contain any ingredient generally understood to contain cholesterol; otherwise must state "adds a trivial amount of cholesterol"
Low cholesterol	Must disclose fat level < 20 mg cholesterol and < 2 g SFA per RACC[d] (and per 50 g for small serving size)	
Reduced cholesterol	Must disclose fat level ≤ 25% less than reference food and contain ≤ 2 g SFA per RACC[d]	Relative claims statement. Reference food must not be low cholesterol. Reference food must have significant market share (e.g., ≥ 5% of market)
Cholesterol free, low cholesterol, reduced cholesterol		Qualifying statement if inherently low or free

[a] Disclosure levels for fat are different for individual foods, meals, and main dishes. See Table 7.27.
[b] SFA: Saturated fatty acids.
[c] Per serving for meal and main dish products.
[d] Per 100 g for meal and main dish products.

The requirements for use of the term "cholesterol free," "free of cholesterol," "zero cholesterol," "without cholesterol," "no cholesterol," "trivial source of cholesterol," "negligible source of cholesterol," or "dietarily insignificant source of cholesterol" are governed by whether the fat content of the food exceeds the applicable disclosure level (Table 7.27). If total fat content is *below* the disclosure level, the claim can be made if:

- The food contains less than 2 mg of cholesterol per RACC or per serving for a meal or main dish product.
- The food contains no ingredient generally understood to contain cholesterol, unless the ingredient, as declared in the ingredient statement, is accompanied by an asterisk which refers to a statement below the list of ingredients "adds

a trivial amount of cholesterol," " adds a negligible amount of cholesterol," or "adds a dietarily insignificant amount of cholesterol."
- The food contains 2 g or less of saturated fatty acids per RACC or for a meal or main dish product or per labeled serving.
- The food meets these conditions as the result of special processing, alteration, formulation, or reformulation. Otherwise, the label must disclose that saturated fat is not usually present in the food (e.g., "applesauce, a cholesterol free food").

If a food *exceeds* the disclosure level for total fat, the label must also disclose the level of fat in a serving in addition to satisfying the foregoing requirements. Placement and prominence instructions are specified in the regulations.

If the food qualifies for the claim without the benefit of processing alteration, formulation or reformulation, it must include an appropriate qualifying statement, e.g., "corn oil, a cholesterol free food contains 14 g of fat per serving."

10. Low In Cholesterol Claims [21 CFR 101.62 para (d)(2); FR p. 2421–2422]

a. Individual Foods

The terms "low in cholesterol," "low cholesterol," "contains a small amount of cholesterol," "low source of cholesterol," or "little cholesterol" may be used to describe an individual food that does not exceed the applicable disclosure level for total fat if:

- The food contains 20 mg or less of cholesterol per RACC (and per 50 g if it is a small serving size food).
- The food contains 2 g or less of saturated fatty acids per RACC (and per 50 g if it is a small serving size food.)[11]
- The food meets these conditions as the result of special processing, alteration, formulation, or reformulation. Otherwise, the label must disclose that saturated fat is not usually present in the food (e.g., "low fat cottage cheese, a low cholesterol food").

If a food exceeds the disclosure for total fat, the label must disclose the level of total fat in a serving, in addition to the other requirements described above. Placement and prominence requirements for the statement of total fat are specified in the regulations.

b. Meal and Main Dish Products[10]

A meal or main dish product may bear a "low cholesterol" claim if it meets the requirements set forth above for individual foods. However, the disclosure levels for total fat differ (Table 7.27). In addition, the 20 mg cholesterol and 2 g saturated fat criteria are measured per 100 g, rather than per RACC.

11. Reduced Cholesterol Claims [21 CFR 101.62 para (d)(4); FR pp. 2422–2423]

a. Individual Foods

The terms "reduced cholesterol," "reduced in cholesterol," "cholesterol reduced," "less cholesterol," "lower cholesterol," or "lower in cholesterol" may be used for an individual food that does not exceed the disclosure level for total fat provided:

- The food has been specially formulated, altered, or processed to reduce its cholesterol content by at least 25% from the reference food it resembles and for which it substitutes.
- The reference food has a significant market share (e.g., 5% or more of a national or regional market)
- It contains 2 g or less of saturated fatty acids per RACC.
- It is accompanied by the explanatory information required for relative claims (e.g., "25% less cholesterol than *reference food*; cholesterol lowered from 55 mg to 30 mg per serving. Contains 13 g fat per serving.")

Reduced cholesterol claims may be made for foods that exceed the disclosure level for total fat so long as the level of total fat in a serving is disclosed and the requirements specified above are satisfied. The placement and prominence requirements for the fat content statement are specified in the regulations.

b. Meal or Main Dish Product[10]

A meal or main dish product that does not exceed the disclosure level for total fat may bear a "reduced cholesterol" claim if it meets the requirements set forth for individual foods. However, the disclosure levels for total fat differ (Table 7.27). In addition, the 2 g saturated fatty acid stipulation is based on per 100 g rather than per RACC.

For meals or main dishes where the total fat exceeds the disclosure level the level of fat in a serving must be declared on the label.

c. Non-Permissible Claim

A "reduced cholesterol" claim is not permitted for any food if the reference food qualifies for a "low cholesterol" claim.

12. Lean and Extra Lean Claims [21 CFR 101.62 para (e); FR p. 2423]

a. Lean Claims

The term lean may be used for seafood and/or game meat provided the individual food as packaged contains:

- Less than 10 g total fat
- 4.5 g or less saturated fat
- Less than 95 mg cholesterol

per RACC and per 100 g.

A meal or main dish product qualifies for this claim if it satisfies the same specifications per 100 g and per labeled serving.

b. Extra Lean Claims

FDA will permit the use of the term "extra lean" for individual seafood or game products provided the food as packaged contains

- Less than 5 g total fat
- Less than 2 g saturated fat
- Less than 95 mg cholesterol

per RACC and per 100 g.

A meal or main dish product qualifies for this claim if it satisfies the same specifications per 100 g and per labeled serving. The requirements for lean and extra lean claims are identical to those of USDA for the meat and poultry products under its jurisdiction.

H. Petitions for Nutrient Content Claims [21 CFR 101.69 para (a–o); FR pp. 2423–2426]

The regulations set forth detailed specifications and formats for those seeking to establish a nutrient content claim, a synonymous term or a use of an implied claim in a brand name not already defined in the regulations. They also stipulate time frames for FDA to act on the petition.

The petition should be submitted in quadruplicate to:

Regulations and Industry Activities Branch (HFF-312)
Food and Drug Administration
Department of Health and Human Services
Washington, D.C. 20204

1. Petitions for New Nutrient Content Claims

The following information must be provided:

- A statement identifying the new terms and the level of the pertinent nutrient.
- An explanation of why the term will not be misleading
- Examples of how the claim will be used on labels and the applicable food types.
- The level at which the nutrient must be present and other conditions that must be satisfied for appropriate use of the new term.
- Factors that would make use of the term inappropriate.

- A detailed explanation, supported by data of why the pertinent nutrient is of importance to human nutrition at the levels described by the claim.
- The nutritional benefit(s) to be derived from use of the claim and why such benefit(s) is not available through use of existing terms.
- If the claim is intended for a specific population group, the specific nutritional needs of the group should be addressed with supporting scientific data.
- Analytical data showing the amount of the subject nutrient in the types of foods for which the claim is intended.
- A detailed analysis of the potential effect of the use of the proposed claim on food consumption and any corresponding changes in nutrient intake (specifically the intake of nutrients that have positive or negative effects).

2. Petitions for a Synonymous Term

The petition must provide a statement identifying the synonymous descriptive term and the existing term with which the new term is considered to be consistent. The statement should:

- Address why the proposed synonymous term is consistent with the already defined term and why its use will not be misleading.
- Provide examples of how the claim will be used on labels and the applicable types of foods.
- Specify whether any limitations not applicable to the use of the defined term apply to the use of the synonymous term.
- Provide a detailed explanation, supported by data, of why the proposed term is requested, including an explanation of whether the existing defined term is inadequate for the purpose of effectively characterizing the level of a nutrient.
- Describe the nutritional benefit to be derived from use of the proposed claim and why such benefit is not available through use of existing term.
- If the claim is intended for a specific group within the population, the analysis should specifically address the nutritional needs of such group, and include supporting scientific data.

3. Petitions for the Use of an Implied Nutrient Content Claim in a Brand Name

The petition must provide a statement identifying the implied nutrient content claim, the pertinent nutrient, the corresponding term for characterizing the level of such nutrient as defined by regulation and the brand name. The statement should:

- Address why the use of the brand name as proposed will not be misleading.
- Address what information is required to accompany the claim or other ways in which the claim meets the requirements of Sections 201(n) and 403(a) of the Food, Drug and Cosmetic Act.

- Provide examples of the types of foods on which the brand name will appear.
- Include data showing that the actual level of the nutrient in the food qualifies the food to bear the corresponding term defined by regulation.
- Provide a detailed explanation, supported by data, of why the proposed brand name is requested. This should include the nutritional benefit to be derived from use of the proposed brand name.
- If the branded product is intended for a specific group within the population, the petition should specifically address the nutritional needs of such group with supporting scientific data.

I. Implied Nutrient Content Claims and Related Label Statements [21 CFR 101.65 para (a–d); FR p. 2423]

1. General Requirements

An implied nutrient content claim can be made on the label only if it complies with the general requirements for nutrient content claims and the food is labeled in accordance with the nutrition labeling requirements.

2. Label Statements That Are Implied Claims [21 CFR 101.65 para (b); FR p. 2423]

The regulations set forth label statements that will be regulated as implied nutrient content claims. These include:

- Claims about a food or an ingredient in a food that suggest that a nutrient or an ingredient is absent or present in a certain amount (e.g., "high in oat bran").
- A product name (e.g., a statement of identity) that identifies a characterizing ingredient associated with a nutritional benefit (e.g., "Corn Oil Margarine," "Oat Bran Muffins," "Whole Wheat Bagels") and implies that a nutrient is present or absent in a certain amount (e.g., "Free," "low," "good source," or "high").
- The phrase "contains the same amount of *name of nutrient* as a *name of food* and as much *nutrient* as a *food*". The amount of the nutrient in the reference food qualifies it as a "good source" of that nutrient. The implication is that the labeled food, on a per serving basis, is also an equivalent, good source of that nutrient (e.g., "as much fiber as an apple," "contains the same amount of vitamin C as an eight ounce glass of orange juice").
- A claim that a food contains or is made with an ingredient that is known to provide a particular nutrient or that is prepared in a way that effects the content of a particular nutrient in the food, if the finished food is labeled either "low in", or a "good source of" that nutrient.

3. Label Statements That Are Not Implied Claims [21 CFR 101.65 para (b); FR p. 2423]

The regulations identify label statements about a product that are not considered to be nutrient content claims. The following types of label statements will *not* be regulated as implied claims:

- A claim that a specific ingredient or food component is absent from a product, provided the purpose of such claim is to facilitate avoidance of the ingredient because of food allergies, food tolerance, religious beliefs, or dietary practices such as vegetarian or other non-nutrient related reasons (e.g., "100% milk free").
- A claim about a substance that is non-nutritive or that does not have a nutritive function (e.g., "contains no preservatives," "no artificial colors").
- A claim about the presence of an ingredient that is perceived to add value to the product (e.g., "made with real butter," "made with whole fruit," "contains honey").
- A statement of identity where an ingredient constitutes essentially 100% of a food (e.g., "corn oil," "oat bran").
- A statement of identity that names as a characterizing ingredient an ingredient associated with a nutrient benefit (e.g., "corn oil," "margarine," "oat bran muffins," or "whole wheat bagels"), unless the claim is made in a context which suggests that a nutrient is present or absent in a certain amount.
- A label statement that notes that a food has a special dietary usefulness relative to a physical, physiological, pathological, or other condition where the claim identifies the special diet of which the food is intended to be a part.

4. General Nutritional Claims [21 CFR 101.65 para (d); FR p. 2423]

A claim that suggests that a food, because of its nutrient content may be useful in maintaining healthy, dietary practices and is made in association with an explicit claim or statement about a nutrient (e.g., "healthy, contains three grams of fat") is considered an implied nutrient content claim and accordingly must comply with the regulations for implied claims.

J. Specific Requirements for Descriptive Claims That Are Neither Nutrient Content Claims nor Health Claims [21 CFR 101.95 para (a–c); FR p. 2426]

1. Fresh, Freshly Frozen, Fresh Frozen, and Frozen Fresh Claims

While not mandated by NLEA, FDA established a new regulation for use of the term "fresh," "freshly frozen," "fresh frozen," and "frozen fresh."

Uses of the term "fresh" are not subject to the requirements specified if the term does not suggest or imply that a food is unprocessed or unpreserved. For example, "fresh" used to describe pasteurized whole milk is not covered by the requirements because the term does not imply that the food is unprocessed (e.g., consumers commonly understand that milk is nearly always pasteurized). However, the term "fresh" to describe pasta sauce that has been pasteurized or contains pasteurized ingredients is covered by the regulation because the term implies that the food is not processed or preserved.

a. Fresh

The term "fresh," when used on a food label in a manner that suggests or implies that the food is unprocessed, means that the food is in its raw state and has not been frozen or subjected to any form of thermal processing or any other form of preservation, except as described below:

- The addition of approved waxes or coatings.
- The post-harvest use of approved pesticides.
- The application of a mild chlorine wash or mild acid wash on produce.
- The treatment of raw foods with ionizing radiation not to exceed the maximum dose of one kilo Gray in accordance with regulations (21 CFR para 179.26).

b. Fresh Frozen

The terms "fresh frozen" and "frozen fresh" mean that the food was quickly frozen while still fresh (e.g., the food has been recently harvested when frozen). Blanching of the food before freezing does not preclude use of the term "fresh frozen" to describe the food. "Quickly frozen" refers to a freezing system such as blast-freezing (sub-zero Fahrenheit temperature with fast moving air directed at the food) that assures the food is frozen, even to the center of the food, quickly and that virtually no deterioration has taken place.

The regulations further provide that a refrigerated food meeting the definition of "fresh" described above is not precluded from using this term.

K. Label Statements on Foods for Special Dietary Use [21 CFR 105.66 para (a–f); FR p. 2430]

Under NLEA, certain provisions concerning label statements on foods for special dietary use in reducing or maintaining caloric intake or body weight are defined as nutrient content claims. Therefore, FDA amended the regulations to address this issue. These changes are described below.

1. Label Statements Relating to Usefulness in Reducing or Maintaining Body Weight[9]

Any food for special dietary use in reducing or maintaining body weight must bear nutrition labeling (unless specifically exempt) and a conspicuous statement of the basis upon which the food claims to be of special usefulness.

a. Non-Nutritive Ingredients
[21 CFR 105.66 para (b); FR, p. 2430]

If the food achieves its special dietary usefulness by use of a non-nutritive ingredient (e.g., one not utilized in normal metabolism), the label must contain a statement that discloses the non-nutritive ingredient and percentage by weight, except if such an ingredient is a non-nutritive sweetener, in which case the percentage by weight is not required. If a nutritive sweetener(s) as well as non-nutritive sweetener(s) is added, the statement must indicate the presence of both types (e.g., "sweetened with nutritive sweetener(s) and non-nutritive sweetener(s)").

b. Low Calorie and Reduced Calorie Foods
[21 CFR 105.66 para (c,d); FR, p. 2430]

A food claiming to be "low calorie," "reduced calorie," or otherwise containing fewer calories than a reference food must comply with the nutrient content claim regulations (Section IX-D of this chapter).

c. Label Terms Suggesting Usefulness as Low Calorie or Reduced Calorie Foods
[21 CFR 105.66 para (e); FR, p. 2430]

Except as noted below, a food may be labeled with terms such as "diet," "dietetic," "artificially sweetened," or "sweetened with a non-nutritive sweetener" if the claim is not false and misleading and the food is labeled "low calorie" or "reduced calorie" or bears another nutrient content claim in accordance with the nutrient content claim regulations.

Exemptions to this include:

- The use of a term specifically authorized by regulations governing a particular food or unless otherwise restricted by regulation, to any use of the term "diet" that clearly shows that the food is offered solely for a dietary use other than regulating body weight (e.g., for low sodium diets).
- The use of such terms on a formulated meal replacement or other food that is represented to be of special dietary use as a whole meal.
- The use of such terms on soft drinks as authorized in the regulations for nutrient content claims.

d. Sugar Free
[21 CFR 105.66 para (f); FR, p. 2430]

The use of the terms "sugar free" and "no sugar added" must adhere to the criteria established for nutrient content claims for sugar as previously described in Section IX-E of this chapter).[9]

X. FOOD STANDARDS: REQUIREMENTS FOR FOODS NAMED BY USE OF A NUTRIENT CONTENT CLAIM AND A STANDARDIZED TERM[12] [21 CFR 130; FR, pp. 2431–2447]

A. Allowable Name Modifications

The General Provisions for food standards were amended to allow a food with a standard of identity to include in its name a defined nutrient content claim in conjunction with the standardized product name. Modified versions that bear descriptive names of standardized foods are also permitted.

The regulations also apply to a food that substitutes for a standardized food and uses the name of that food in the statement of identity but does not comply with the standard of identity because of a deviation that is described by an expressed nutrient content claim (e.g., "low fat").

B. Requirements [21 CFR 130.10 para (a–f); FR, pp. 2446–2447]

These name modifications may be used provided:

- The nutrient content claim satisfies regulatory definition.
- The product complies with the relevant standard in all other respects except:
  - Nutrients may be added to achieve nutritional equivalency to the standardized food.
  - Deviations from non-ingredient provisions (e.g., moisture content, food solids content, processing conditions) are permitted to ensure that the substitute has the same or similar performance characteristics as the standardized food.
- All deviations must be the minimum necessary to qualify for the nutrient content claim while maintaining performance characteristics similar to the standardized food.

- The performance characteristics (e.g., physical properties, flavor characteristics, functional properties, shelf life) of the food must be similar to the standardized food. If there is a substantial difference in performance characteristics that materially limits the uses of the food compared to uses of the standardized food, the label must include a statement informing the consumer of such difference (e.g., "not recommended for cooking"). However, the modified product must perform at least one of the principal functions at least as well as the standardized product.

C. Ingredients

Ingredients used in the product must be those provided for by the standard except that safe and suitable ingredients may be used to improve texture, add flavor, prevent syneresis, extend shelf life, improve appearance, or add sweetness so that the product is not inferior in performance characteristics to the standardized food.

An ingredient or component of an ingredient that is specifically required by the standard must not be replaced or exchanged with a similar ingredient from another (e.g., oil must not replace milkfat in light sour cream).

An ingredient or component of an ingredient that is specifically prohibited by the standard must not be added to a substitute food.

An ingredient that is specifically required must be present in the product in a significant amount. (A significant amount of an ingredient or component of an ingredient is at least that amount that is required to achieve the technical effect of that ingredient in the food.)

Water and fat analogs may be added to replace fat and calories in accordance with regulations.

Each of the ingredients used in the food must be declared on the label as required by regulation.

Ingredients not provided for and ingredients used in excess of the levels provided for by the standard, must be identified as such with an asterisk in the ingredient statement. However, ingredients added to restore nutrients to the product do not require an asterisk.

XI. GENERAL REQUIREMENTS FOR HEALTH CLAIMS [21 CFR 101.14 para (a–g); FR, pp. 2533–2534]

A. Background

During the last twenty years, consumers have become increasingly aware of the importance of nutrition in disease prevention and in maintaining good health. Nevertheless, it was not until 1987 that FDA proposed formal rules permitting health claims on food labels. Until that time, FDA considered foods with health messages to be drugs and, consequently, subject to drug regulations.

In 1984, FDA relaxed its position and in 1987 it passed rules which permitted health messages provided they were:

- Supported by reliable scientific evidence.
- Consistent with generally accepted medical and nutritional principles.
- Accompanied by full nutrition labeling.

As scientific support for diet-disease related relationships and public interest in nutrition and health continued to grow, new claims appeared on foods. Some were considered misleading and others only partly satisfied FDA's criteria. In response to these factors, the Congress, through the 1990 NLEA, directed the agency to determine whether the following health claims were appropriate for food labeling:

- Calcium and osteoporosis
- Sodium and hypertension
- Lipids (fats and fat-like substances) and cardiovascular (heart and blood vessel) disease
- Lipids and cancer
- Dietary fiber and cardiovascular disease
- Dietary fiber and cancer
- Folic acid and neural tube defects
- Antioxidant vitamins and cancer (antioxidants inhibit or prevent oxidation—a chemical reaction whose effect in the body is not well understood)
- Zinc and immune function in the elderly
- Omega-3 fatty acids and heart disease.

After receiving over 6,000 letters and reviewing 1,400 scientific studies and authoritative reports, a final rule was established. The rule authorizes eight health (disease prevention) claims (Table 7.40). One or more claims may be made for a food if it qualifies. The final regulations do not apply to dietary supplements of vitamins, minerals, herbs, or other similar nutritional substances.

B. Health Claims: General Requirements [21 CFR 101.14; FR, p. 2553]

1. Definitions [21 CFR 101.14 para (a); FR, p. 2533]

A health claim is any claim made on the label of a food, or dietary supplement, that expressly or by implication, including "third party" references, written statements (e.g., a brand name including a term such as "heart"), symbols (e.g.,

Table 7.40 Health Claims

Definition: A statement endorsement or symbol that characterizes the relationship of a substance to a disease or health related condition.

Rule: Health claims are permitted on foods only where claims are specifically provided for by regulations and where no other nutrient is present in an amount that increases the risk of disease. The claim must be supported by publicly available scientific evidence and there must be significant scientific agreement among qualified scientists.

Permitted claims

- Calcium and osteoporosis
- Sodium and hypertension
- Saturated fat and cholesterol and heart disease
- Fat and cancer
- Fiber containing grain products, fruits, vegetables, and cancer
- Fiber containing grain products, fruits, vegetables, and coronary heart disease
- Fruits, vegetables, and cancer

a heart symbol), or vignettes, characterizes the relationship of any substance to a disease or health-related condition. Implied health claims include those statements, symbols, vignettes, or other forms of communication that suggest, within the context in which they are presented, that a relationship exists between the presence or level of a substance in the food and a disease or health-related condition.

Claims about general health or food classes are not considered health claims (e.g., the Food Guide Pyramid logo [a pyramid-shaped depiction of the Dietary Guidelines for Americans], valentine candy in a heart shaped box, and "Eat five servings of fruits and vegetables a day for good health"). However, FDA would consider these examples health claims if a specific nutrient and disease were introduced (e.g., the term "low fat" and a heart logo).

2. Eligibility
[21 CFR 101.14 para (b); FR, p. 2533]

To be eligible for a health claim, a food or food component must be associated with a disease or health-related condition for which the general U.S. population or an identified population sub-group (e.g., elderly) is at risk.

If the substance will be consumed as a component of a conventional food at decreased dietary levels, it must be a nutrient specified by NLEA or a nutrient that FDA requires be included in labeling. On the other hand, if the substance is to be consumed at other than decreased dietary levels, it is eligible for a health claim if it contributes taste, aroma, nutritive value, or any technical effect provided in the food additive regulations, and it retains that attribute when consumed at levels necessary to justify the claim. The substance must be a food, food ingredient, or a component of a food ingredient used at levels necessary to justify a claim which has been demonstrated to be safe and lawful.

FDA will authorize a health claim only when it determines, based on the totality of publicly available scientific evidence, that there is significant scientific agreement among qualified experts that the claim is supported by that evidence. If FDA determines that a health claim is valid it will propose a regulation to authorize the use of that claim.

A health claim may be made provided:

- It is consistent with the conclusions in the regulation.
- It is limited to describing the value that ingestion (or reduced ingestion) of the substance, as part of the total dietary pattern, may have on a particular disease or health-related condition.
- The claim is complete, truthful, and not misleading.
- All required information appears in one place without other intervening material (except that the principal display panel may a refer to the location on the label concerning the health claim and information about the relationship between *the nutrient* and *disease*). In addition, if graphic material is used (e.g., a heart symbol) constituting an explicit or implied health claim, the reference statement or the complete claim must appear in immediate proximity to the graphic material.
- The claim enables the public to comprehend the information provided and understand the relative significance of that information in the context of the total daily diet.
- If the claim is about the effects of consuming the substance at decreased dietary levels, the level of the substance in the food is sufficiently low to justify the claim or, if the claim is about the effects of consuming the substance at other than decreased dietary levels, the level of the substance is sufficiently high and in an appropriate form to justify the claim.

If the food bearing the claim satisfies the low or high nutrient level described above based on the applicable RACC, but the labeled serving size differs from that amount, a statement explaining that the claim is based on the reference amount, not the serving size, must follow (e.g., "Diets low in salt and sodium may help lower blood pressure in many people. A serving of *amount* ounces of this product conforms to such a diet").

If the food bearing a health claim is sold in a restaurant or in other establishments where food ready for consumption is sold, the firm must have a valid reason to believe that the food satisfies the claim requirements and must be able to present that information upon request. Claims made on menus are not subject to the rule. Any food about which a health claim is made must bear nutrition labeling according to the general nutrition labeling rules or, for restaurant foods, the applicable nutrition labeling regulations.

If a food meets or exceeds a disqualifying (disclosure level) level established for specified nutrients, it may not bear a health claim, unless an applicable exemption exists. Disqualifying nutrient or food component levels (disclosure levels) are shown in Table 7.27.

3. Prohibited Claims [21 CFR 101.14 para (e); FR, p. 2534]

No expressed or implied health claim may be made unless:

- The claim is specifically provided for by regulation.
- The claim conforms to the general provisions of the nutrition labeling regulations and the requirements of the specific rule adopted for that claim.
- None of the disqualifying levels are exceeded, unless alternative levels have been specifically established.
- No substance is present at an inappropriate level.
- The label does not represent or purport to be food for infants or toddlers less than two years of age.
- For conventional food, the food contains 10% or more of the reference daily intake or daily reference value for vitamin A, vitamin C, iron, calcium, protein, or fiber per RACC prior to any nutrient addition.

Health claims do not apply to:

- Exempt infant formulas
- Foods intended for children under 2 years
- Medical foods
- Foods regulated as drugs

4. Petitions for Health Claims [21 CFR 101.70 para (a–j); FR, pp. 2534-2536]

Any interested person may petition FDA to issue a regulation regarding a health claim. The petition should be submitted in quadruplicate to:

Food and Drug Administration
Regulatory Affairs Staff (HFF-204)
Office of Nutrition and Food Sciences
200 C St., SW
Washington, D.C. 20204

The regulations set forth detailed specifications and formats to petition for new health claims. All petitions are available for public review. The information required in the petition is outlined below.

a. Eligibility

An explanation of how the food substance conforms to the eligibility requirements for health claims, to the definition of the term "substance" in the regulations [21 CFR 101.14 (a)(2)] and to the "safe and lawful" requirements of the Food, Drug, and Cosmetic Act.

b. Scientific Data

A summary of scientific data providing the basis upon which authorizing a health claim can be justified. The summary must establish that, based on the totality of publicly available scientific evidence, there is significant scientific agreement among experts qualified by scientific training and experience to evaluate such claims and that the claim is supported by such evidence.

c. Public Health Benefit

The summary must state what public health benefit will derive from use of the claim as proposed. If the claim is intended for a specific group within the population, the summary must specifically address nutritional needs of such group and must include scientific data showing how the claim is likely to assist in meeting such needs.

Issues addressed in the summary must include answers to such questions as:

- Is there an optimum level of the particular substance to be consumed beyond which no benefit would be expected?
- Is there any level at which an adverse effect from the substance or from foods containing the substance occurs for any segment of the population?
- Are there certain populations that must receive special consideration?
- What other nutritional or health factors (both positive and negative) are important to consider when consuming the substance?

d. Effect on Food Consumption

In addition, the summary of scientific data must include a detailed analysis of the potential effect of the use of the proposed claim on food consumption, specifically any change due to significant alterations in eating habits and corresponding changes in nutrient intake resulting from such changes in food consumption. The latter item must specifically address the effect on the intake of nutrients that have beneficial and negative consequences in the total diet.

e. Subpopulation

If the claim is intended for a significant subpopulation, within the general U.S. population, the analysis must specifically address the dietary practices of such group, and must include data sufficient to demonstrate that the dietary analysis is representative of such group (e.g., adolescents or the elderly).

If appropriate, the petition must explain the prevalence of the disease or health related condition in the U.S. population and the relevance of the claim in the context of the total daily diet.

f. Model Health Claims

The petition must include one or more model health claims that represent label statements that may be used on a food label to characterize the relationship between the substance in a food to a disease or health-related condition that is justified by the summary of scientific data provided in Section C of the petition. The model health claim must include:

- A brief statement of the relevant conclusions of the summary.
- A statement of how this substance helps the consumer to attain a total dietary pattern or goal associated with the health benefit that is provided.

The petition must include the following attachments:

- Copies of any computer literature searches done by the petitioner (e.g., Medline).
- Copies of articles cited in the literature searches and other information as follows:
  - All information relied upon for the support of the health claim, including copies of publications or other information cited in review articles and used to perform meta-analyses.
  - All information concerning adverse consequences to any segment of the population (e.g., sensitivity to the substance).
  - All information pertaining to the U.S. population

g. Environmental Impact

The petitioner is also required to submit an environmental assessment. The petition must include a statement signed by the person responsible for the petition that, to the best of his/her knowledge, it is a representative and balanced submission that includes unfavorable information as well as favorable information, known to him/her to be pertinent to the evaluation of the proposed health claim.

5. Health Claims: Fiber Containing Grain Products, Fruits, and Vegetables, and Cancer [21 CFR 101.76 para (a–e); FR, p. 2548]

a. Summary

The regulations do not authorize the use of health claims relating to an association between dietary fiber and cancer. However, health claims relating diets low in fat and high in fiber-containing grain products, fruits, and vegetables to a reduced risk of cancer are authorized.

b. Specific Requirements

A health claim associating diets low in fat and high in fiber-containing grain products, fruits, and vegetables with reduced risk of cancer may be made provided that:

- The claim states that diets low in fat and high in fiber-containing grain products, fruits, and vegetables "may" or "might" reduce the risk of some cancers.
- In specifying the disease, the claim uses the following terms: "some types of cancer" or "some cancers."
- The claim is limited to grain products, fruits, and vegetables that contain dietary fiber.
- The claim indicates that development of cancer depends on many factors.
- The claim does not attribute any degree of cancer risk reduction to diets low in fat and high in fiber-containing grain products, fruits, and vegetables.
- In specifying the dietary fiber component of the labeled food, the claim uses the term "fiber," "dietary fiber" or "total dietary fiber."
- The claim does not specify types of dietary fiber that may be related to risk of cancer.
- The food must be or must contain a grain product, fruit or vegetable.
- The food must meet the nutrient content claims definition for a "low fat" food.
- The food must meet, without fortification, the nutrient content requirements for a "good source" of dietary fiber.

The following model health claims may be used on the food label to characterize the relationship between diets low in fat and high in fiber-containing grain products, fruits, and vegetables and cancer risk.

- "Low fat diets rich in fiber-containing grain products, fruits and vegetables may reduce the risk of some types of cancer, a disease associated with many factors."
- "Development of cancer depends on many factors. Eating a diet low in fat and high in grain products, fruits and vegetables that contain dietary fiber may reduce your risk of some cancers."

6. Health Claims: Fruits, Vegetables and Grain Products (Containing Fiber, Particularly Soluble Fiber), and the Risk of Coronary Heart Disease [21 CFR 101.77 para (a–e); FR, pp. 2578-2580]

a. Summary

The regulations do not authorize the use of health claims relating to an association between dietary fiber and cardiovascular disease (CVD). However, health claims relating diets low in saturated fat and cholesterol and high in fruits, vegetables, and grain products that contain dietary fiber (particularly soluble fiber), to reduced risk of coronary heart disease (CHD) are authorized.

b. Specific Requirements

A health claim associating diets low in saturated fat and cholesterol and high in fruits, vegetables, and grain products that contain fiber, particularly soluble fiber, with reduced risk of heart disease may be made on the label provided:

- The claim states that diets low in saturated fat and cholesterol and high in fruits, vegetables, and grain products that contain fiber "may" or "might" reduce the risk of heart disease.
- In specifying the disease, the claim uses the following terms: "heart disease" or "coronary heart disease."
- The claim is limited to those fruits, vegetables, and grains that contain fiber.
- In specifying the dietary fiber, the claim uses the term "fiber," "dietary fiber," "some types of dietary fiber," "some dietary fibers," or "some fibers." The term "soluble fiber" may be used in addition to these terms.
- In specifying the fat component, the claim uses the terms "saturated fat" and "cholesterol."
- The claim indicates that development of heart disease depends on many factors.
- The claim does not attribute any degree of risk reduction for coronary heart disease to diets low in saturated fat and cholesterol and high in fruits, vegetables, and grain products that contain fiber.
- The food must be or must contain a fruit, vegetable, or grain product.
- The food must meet the nutrient content claims requirements of for a "low saturated fat," "low cholesterol," and "low fat" food.
- The food must contain, without fortification, at least 0.6 g of soluble fiber per RACC.
- The content of soluble fiber must be declared in the nutrition information panel, consistent with nutrition labeling regulations.

c. Model Health Claims

The following model health claims may be used to characterize the relationship between diets low in saturated fat and cholesterol and high in fruits, vegetables, and grain products that contain soluble fiber:

- "Diets low in saturated fat and cholesterol and rich in fruits, vegetables, and grain products that contain some types of dietary fiber, particularly soluble fiber, may reduce the risk of heart disease, a disease associated with many factors."
- "Development of heart disease depends on many factors. Eating a diet low in saturated fat and cholesterol and high in fruits, vegetables, and grain

products that contain fiber may lower blood cholesterol levels and reduce your risk of heart disease."

7. Health Claims: Fruits and Vegetables and Cancer [21 CFR 101.78 para (a–e); FR, pp. 2639–2640]

a. Summary

The regulations do not authorize the use of health claims relating to an association between antioxidant vitamins and cancer. However, they do authorize health claims relating substances in diets low in fat and high in fruits and vegetables (foods that are low in fat and may contain dietary fiber, vitamin A, and vitamin C) to a reduced risk of cancer.

b. Specific Requirements

A health claim associating substances in diets low in fat and high in fruits and vegetables with reduced risk of cancer may be made on the label or labeling of a food provided that:

- The claim states that diets low in fat and high in fruits and vegetables "may" or "might" reduce the risk of some cancers.
- In specifying the disease, the claim uses the following terms: "some types of cancer," or "some cancers."
- The claim characterizes fruits and vegetables as foods that are low in fat and may contain vitamin A, vitamin C, and dietary fiber.
- The claim characterizes the food bearing the claim as containing one or more of the following, for which the food is a good source of dietary fiber, vitamin A, or vitamin C.
- The claim does not attribute any degree of cancer risk reduction to diets low in fat and high in fruits and vegetables.
- In specifying the fat component of the labeled food, the claim uses the term "total fat" or "fat."
- The claim does not specify types of fats or fatty acids that may be related to risk of cancer.
- In specifying the dietary fiber components of the labeled foods, the claim uses the term "fiber," "dietary fiber," or "total dietary fiber."
- The claim does not specify types of dietary fiber that may be related to risk of cancer.
- The claim indicates that development of cancer depends on many factors.
- The food must be or must contain a fruit or vegetable.
- The food shall meet the nutrient content claim requirements for a "low fat" food.
- The food must meet, without fortification, the nutrient content requirements for a "good source" of at least one of the following: vitamin A, vitamin C or dietary fiber.

The following model health claims may be used in food labeling to characterize the relationship between substances in diets low in fat and high in fruits and vegetables, and cancer.

- "Low fat diets rich in fruits and vegetables (foods that are low in fat and may contain dietary fiber, vitamin A, and vitamin C) may reduce the risk of some types of cancer, a disease associated with many factors. Broccoli is high in vitamins A and C, and it is a good source of dietary fiber."
- "Development of cancer depends on many factors. Eating a diet low in fat and high in fruits and vegetables, foods that are low in fat and may contain vitamin A, vitamin C, and dietary fiber, may reduce your risk of some cancers. Oranges, a food low in fat, are a good source of fiber and vitamin C."

8. Health Claims: Calcium and Osteoporosis [21 CFR 101.72 para (a–e); FR, pp. 2676–2677]

a. Summary

The regulations authorize health claims relating to an association between adequate calcium intake and osteoporosis.

b. Specific Requirements

A health claim associating calcium with a reduced risk of osteoporosis may be made on the label or labeling of a food provided that:

- The claim must make clear that adequate calcium intake throughout life is not the only recognized risk factor in this disease by listing specific factors, including sex, race, and age that place persons at risk and stating that an adequate level of exercise and a healthful diet are also needed.
- The claim does not state or imply that the risk of osteoporosis is equally applicable to the general U.S. population. The claim must identify the populations at particular risk.
- The claim states that adequate calcium intake throughout life is linked to reduced risk of osteoporosis through the mechanism of optimizing peak bone mass during adolescence and early adulthood.
- The claim does not attribute any degree of reduction in risk of osteoporosis to maintaining an adequate calcium intake throughout life.
- The claim states that a total dietary intake greater than 200% of the recommended daily intake (2,000 mg of calcium) has no further known benefit to bone health. This requirement does not apply to foods that contain less than 40% of the recommended daily intake of 1,000 mg of calcium per day or 400 mg of calcium per RACC or per total daily recommended supplement intake.
- The food must meet or exceed the requirements for a "high" level of calcium.

- The calcium content of the product must be assimilable.
- Dietary supplements must meet the United States Pharmacopeia (USP) standards for disintegration and dissolution applicable to their component calcium salts. Dietary supplements for which no USP standards exist must exhibit appropriate assimilability under the conditions of use stated on the product label.
- A food or total daily recommended supplement intake shall not contain more phosphorus than calcium on a weight per weight basis.

The following model health claims may be used in food labeling to describe the relationship between calcium and osteoporosis:

- *Appropriate for Most Conventional Foods.* "Regular exercise and a healthy diet with enough calcium helps teen and young adult white and Asian women maintain good bone health and may reduce their high risk of osteoporosis later in life."
- *Appropriate for Foods Exceptionally High in Calcium.* "Regular exercise and a healthy diet with enough calcium helps teen and young adult white and Asian women maintain good bone health and may reduce their high risk of osteoporosis later in life. Adequate calcium intake is important, but daily intakes above about 2,000 mg are not likely to provide any additional benefit."

9. Health Claims: Dietary Saturated Fat and Cholesterol and Risk of Coronary Heart Disease [21 CFR 101.75 para (a–e); FR, pp. 2757–2758]

a. Summary

The regulations authorize health claims relating to an association between dietary lipids (specifically, saturated fat, and cholesterol) and cardiovascular disease (specifically, coronary heart disease [CHD]).

b. Specific Requirements

A health claim associating diets low in saturated fat and cholesterol with reduced risk of coronary heart disease may be made on the label or labeling of a food provided:

- The claim states that diets low in saturated fat and cholesterol "may" or "might" reduce the risk of heart disease.
- In specifying the disease, the claim uses the terms "heart disease" or "coronary heart disease."
- In specifying the nutrient, the claim uses the terms "saturated fat" and "cholesterol" and lists both.
- The claim does not attribute any degree of risk reduction for coronary heart disease to diets low in dietary saturated fat and cholesterol

- The claim states that coronary heart disease risk depends on many factors.

The food meets all of the nutrient content requirements of a "low saturated fat," "low cholesterol," and "low fat" food; except that fish and game meats (e.g., deer, bison, rabbit, quail, wild turkey, geese, and ostrich) may meet the requirements for "extra lean."

The following are model health claims that may be used in food labeling to describe the relationship between dietary saturated fat, cholesterol, and risk of heart disease.

- "While many factors affect heart disease, diets low in saturated fat and cholesterol may reduce the risk of this disease."
- "Development of heart disease depends upon many factors, but its risk may be reduced by diets low in saturated fat and cholesterol and healthy lifestyles."
- "Development of heart disease depends upon many factors, including a family history of the disease, high blood LDL-cholesterol, diabetes, high blood pressure, being overweight, cigarette smoking, lack of exercise, and the type of dietary pattern. A healthful diet low in saturated fat, total fat, and cholesterol, as part of a healthy lifestyle, may lower blood cholesterol levels and may reduce the risk of heart disease."
- "Many factors, such as a family history of the disease, increased blood and LDL-cholesterol levels, high blood pressure, cigarette smoking, diabetes, and being overweight, contribute to developing heart disease. A diet low in saturated fat, cholesterol and total fat may help reduce the risk of heart disease."
- "Diets low in saturated fat, cholesterol and total fat may reduce the risk of heart disease. Heart disease is dependent upon many factors, including diet, a family history of the disease, elevated blood LDL-cholesterol levels, and physical inactivity."

10. Health Claims: Dietary Fat and Cancer [21 CFR 101.73 para (a–e); FR, pp. 2801–2819]

a. Summary

FDA authorizes health claims relating to an association between dietary fat and cancer.

b. Specific Requirements

A health claim associating diets low in fat with reduced risk of cancer may be made on the label or labeling of a food provided that:

- The claim states that diets low in fat "may" or "might" reduce the risk of some cancers.

- In specifying the disease, the claim uses the following terms: "some types of cancer" or "some cancers."
- In specifying the nutrient, the claim uses the term "total fat" or "fat."
- The claim does not specify types of fat or fatty acid that may be related to the risk of cancer.
- The claim does not attribute any degree of cancer risk reduction to diets low in fat.
- The claim indicates that the development of cancer depends on many factors.

The food shall meet all the nutrient content claim requirements for a "low fat" food; except that fish and game meats may meet the requirements for "extra lean."

The following model health claims may be used in food labeling to describe the relationship between dietary fat and cancer:

- "Development of cancer depends on many factors. A diet low in total fat may reduce the risk of some cancers."
- "Eating a healthful diet low in fat may help reduce the risk of some types of cancers. Development of cancer is associated with many factors, including a family history of the disease, cigarette smoking, and what you eat."

11. Health Claims: Sodium and Hypertension [21 CFR 101.74 para (a–e); FR, pp. 2836–2837]

a. Summary

The regulations authorize health claims relating to an association between dietary sodium and high blood pressure.

b. Specific Requirements

A health claim associating diets low in sodium with reduced risk of high blood pressure may be made on the label or labeling of a food provided that:

- The claim states that diets low in sodium "may" or "might" reduce the risk of high blood pressure.
- In specifying the disease, the claim uses the term "high blood pressure."
- In specifying the nutrient, the claim uses the term "sodium."
- The claim does not attribute any degree of reduction in risk of high blood pressure to diets low in sodium.
- The claim indicates that development of high blood pressure depends on many factors.

The food must meet all of the nutrient content claim requirements for a "low sodium" food.

The following are model health claims that may be used in food labeling to describe the relationship between dietary sodium and high blood pressure.

- "Diets low in sodium may reduce the risk of high blood pressure, a disease associated with many factors."
- "Development of hypertension or high blood pressure depends on many factors. *This product* can be part of a low sodium, low salt diet that might reduce the risk of hypertension or high blood pressure."

12. Healthy

The term healthy may be used on processed foods that contain:

- ≤ 1 g fat per serving
- ≤ 60 mg cholesterol per serving
- ≤ 480 mg sodium per serving[22]

For raw meal and poultry, the term may be used if the product satisfies the definition of "extra lean" (i.e., < 5 g fat; < 2 g saturated fat; < 95 mg cholesterol per 100 g.

May 8, 1994 was the compliance date for FDA regulated products labeled with a "healthy" claim for the first time. For pre-existing products, FDA compliance date is January 1, 1996. The USDA compliance date is November, 1995, for all meat and poultry products making the claim.

XII. DECLARATION OF INGREDIENTS

A. General Requirements

Responding to the 1990 NLEA mandate and to make ingredient labeling more useful to consumers, FDA amended the regulations to require label declaration of mandatory as well as optional ingredients. Many changes were made in the ingredient labeling regulations and to individual standards of identity to reflect this requirement. The final rules:

- Require listing of the common or usual names of all ingredients in standardized foods.
- Require listing all color additives required to be certified by FDA.
- Require declaration of all protein hydrolysates including the identification of the food source (e.g., "hydrolyzed soy protein").
- Require identification of caseinate as a milk derivative when used in foods that claim to be non-dairy foods.
- Require declaration of sulfiting agents that have a functional effect in a standardized food or that are present in a standardized food at a detectable level (10 ppm or more).
- Permit inclusion of the food source in the names of sweeteners defined by the food standards (e.g., "corn sugar anhydrous," "corn sugar monohydrate").

- Provide a uniform format for voluntary declaration of percentage ingredient information.
- Provide certain exemptions for listing the specific common or usual names of preservative coatings on fresh fruits and vegetables.

B. Food: Designation of Ingredients [21 CFR 101.4 para (a–e); FR, p. 2875]

1. Common Name

Ingredients required to be declared on the label of a food, including foods that comply with standards of identity, (except those ingredients exempted by 21 CFR 101.10), must be listed by common or usual name in descending order of predominance by weight on either the principal display panel or the information panel.

2. Sugar

For purposes of ingredient labeling, the term "sugar" refers to sucrose, which is obtained from sugar cane or sugar beets.

3. Coatings

Wax and resin ingredients on fresh produce when stored must be declared collectively as "coated with food-grade animal-based wax, to maintain freshness" or the phrase "coated with food-grade vegetable-, petroleum-, beeswax-, and/or shellac-based wax or resin, to maintain freshness" as appropriate. The terms "food-grade" and "to maintain freshness" are optional. The term "lac-resin" may be substituted for the term "shellac."

4. Non-Dairy Products

When foods characterized on the label as "nondairy" contain a caseinate ingredient, the caseinate declaration must be followed by a parenthetical statement identifying its source. For example, if the manufacturer uses the term "nondairy" on a creamer that contains sodium caseinate, it must include a parenthetical term such as "a milk derivative" after the listing of sodium caseinate in the ingredient list.

5. Percent Ingredient Declaration

If the percentage of an ingredient is included in the statement of ingredients, it must be in parentheses following the name of the ingredient and expressed in terms of percent by weight. Percentage declarations must be expressed to the nearest 1%, except that when ingredients are present at levels of 2% or less, they may be grouped together and expressed accordingly.

C. Foods: Labeling of Spices, Flavorings, Colorings, and Chemical Preservatives [21 CFR 101.22 para (h)(k); FR, pp. 2875–2876]

1. Protein Hydrolysates

A protein hydrolysate must be declared by specific common or usual name that identifies the food source from which it was derived (e.g., "hydrolyzed wheat gluten," "hydrolyzed soy protein," "autolyzed yeast extract").

Unacceptable names include "hydrolyzed milk protein" (e.g., hydrolysates can be prepared from different milk proteins) and hydrolyzed vegetable protein (e.g., this does not identify the food source). No protein hydrolysate used in food for its effect on flavor may be declared simply as "flavor," "natural flavor," or "flavoring."

2. Colors

The label of a food to which any color (or the lack of such color additive), has been added must declare the coloring in the statement of ingredients. It is not necessary to include the prefix "FD&C" or the term "No." in the declaration, but the term "Lake" must be included, if applicable (e.g., Blue 1 Lake). An appropriate alternative name for the certified color additive may be listed parenthetically (e.g., Yellow 5 [tartrazine]).

Noncertified colors may be declared as "artificial color," "artificial color added," "color added," or "colored with *name of color*." Colors added to butter, cheese, or ice cream do not have to be declared in the ingredient statement unless such declaration is required by a color additive regulation to ensure safe conditions of use.

D. Sulfites in Standardized Food [21 CFR 130.9 para (a); FR, p. 2876]

Sulfiting agents must be identified by common or usual name(s) when added to standardized foods in accordance with the standard of identity. If indirectly added, they must be declared in the ingredient statement if they are present at 10 parts per million or more.

E. Label Designations of Ingredients for Standardized Foods [21 CFR 130.11; FR, p. 2876]

Some definitions and standards of identity for foods require that designated optional ingredients (e.g., spices, flavorings, colorings, emulsifiers, flavor enhancers, stabilizers, preservatives, sweeteners) be conspicuously declared in a specified manner on the label wherever the name of the standardized food appears. This requirement applies only when it is necessary to distinguish between two or more foods the optional forms or varieties, the packing medium, or the significant characterizing ingredients present in standardized foods that comply with the same standard.

The rules provide an extensive list of revisions and amendments for ingredients in standardized foods. An abbreviated list of ingredients is described below. In addition, the final rules establish and amend the ingredient labeling requirements for milk, dairy, maple syrup, and canned tuna products with standards of identity; declaration of sweeteners in soft drinks[9]; and use of the term "contains glutamate as part of the common or usual name of yeast extracts and certain hydrolyzed proteins."

- Cereal flours
- Macaroni and noodle products
- Canned fruits
- Canned fruit juices
- Fruit butters, jellies
- Preserves and related products
- Fruit pies
- Canned vegetables
- Frozen vegetables
- Eggs and egg products
- Fish and shellfish
- Cocoa products
- Treenut and peanut products
- Margarine
- Sweeteners and table syrups
- Food dressings and flavorings

For milk, dairy products, and maple syrup, FDA amended the standards of identity to require the listing of the common or usual names of all ingredients in the standardized foods. Products that are affected are listed in 21 CFR 131, 133, 135, and 168.

XIII. PERCENTAGE JUICE DECLARATION FOR FOODS PURPORTING TO BE BEVERAGES THAT CONTAIN FRUIT OR VEGETABLE JUICE [21 CFR 101.30 (a–l); FR, pp. 2925–2926]

FDA amended the food labeling regulations to establish requirements for the declaration of the percentage juice in beverages that "purport" to contain fruit or vegetable juice. A beverage covered by this regulation may be carbonated or noncarbonated, concentrated, full-strength, diluted, or contain no juice. Model disclosure statements and prominent placement requirements for the information and prominent display panels are described in the regulations.

A. Percent Juice Declaration [21 CFR 101.30 para (b); FR, p. 2925]

- If the beverage contains fruit or vegetable juice, the percentage must be declared by the words "Contains _____ percent (or %) *name of juice* or _____ percent (or %) *juice*," or a similar phrase, with the blanks filled in with the percentage expressed as a whole number not greater than the actual percentage of the juice vegetable (e.g., "Contains 50% apple juice" or "50% juice").
- If the beverage contains less than 1% juice, the total percentage juice must be declared as "less than 1% juice" or "less than 1% *name of juice*."
- If the beverage contains 100% juice and also contains non-juice ingredients that do not reduce the juice soluble solids or, in the case of expressed juice, change the volume, and a 100% juice declaration appears on the label without an accompanying ingredient statement, a disclosure must be made concerning the addition of such ingredients [e.g., "with added *name of ingredient(s)*"]. However, if the presence of the non-juice ingredient is declared as part of the statement of product identity, this qualifying statement is not required.

B. Beverages Containing Minor Amounts of Juice for Flavoring [21 CFR 101.30 para (c); FR, p. 2925]

- If a beverage contains minor amounts of juice for flavoring and is labeled with a flavor description using terms such as "flavor," "flavored," or "flavoring" with a fruit or vegetable name and does not mention the term "juice" on the label other than in the ingredient statement, or carry an explicit vignette depicting the fruit or vegetable from which the flavor derives or bear a specific physical resemblance to juice or distinctive juice characteristic (such as pulp) then the total percentage juice declaration is not required.

C. Implied Juice Claims [21 CFR 101.30 para (d); FR, p. 2925]

- If the beverage does not meet the foregoing criteria (item B above) for exemption from total juice percentage declaration and contains no fruit or vegetable juice, but the label or color and flavor of the beverage represents, suggests or implies that fruit or vegetable juice may be present, the label must carry a statement indicating that no fruit or vegetable juice is contained in the product (e.g., "contains zero percent (or %) juice"). Alternatively, the label may declare "Containing (or contains) no *name of fruit or vegetable* juice," or "juice does not contain *name of fruit or vegetable* juice." If there is a general suggestion that the product contains fruit or vegetable juice, such as the presence of fruit pulp, the word *fruit* or *vegetable* as applicable must appear (e.g., "contains no fruit juice," or "does not contain fruit juice.")

D. Label Display [21 CFR 101.30 para (e); FR, p. 2925]

- If the beverage is sold in a package that does not have with an information panel, the percent juice declaration must be prominently placed on the principal display panel in a type size not less than the "net quantity" statement and near the name of the food with no other printed information appearing above the statement except the brand name, logo, or universal product code.
- In the presence of an information panel, the percentage juice declaration may also be placed on the principal display panel, provided the declaration is consistent with that on the information panel and complies with the graphics requirements described in the regulations.

E. Enforcement [21 CFR 101.30 para (h); FR, pp. 2925–2926]

- For enforcement purposes, the labeled percentage of juice will be evaluated from concentrate found in a juice or juice beverages using the minimum Brix levels listed in the regulations. If a Brix level is not specified, the percentage juice will be calculated on the basis of the soluble solids content of the single strength (unconcentrated) juice used to produce the concentrated juice. Table 7.41 lists minimum Brix values for various single strength (100%) juices.
- Juices directly expressed from a fruit or vegetable (e.g., not concentrated and reconstituted) are considered to be 100% juice and must be declared "100% juice."
- Calculations of the percentage of juice in a juice blend or a diluted juice product made directly from expressed juice (e.g., not from concentrate) must be based on the percentage of the expressed juice in the product computed on a volume/volume basis.

F. Modified Juices [21 CFR 101.30 para (k); FR, p. 2926]

- If a product is a beverage that contains a juice whose color, taste, or other organoleptic properties have been modified to the extent that the original juice is no longer recognizable or if its nutrient profile has been diminished

Table 7.41 Minimum Brix Values for Single Strength Juices

Juice	100% Juice[a]	Juice	100% Juice[a]
Acerola	6.0	Guanabana (soursop)	16.0
Apple	11.5	Guava	7.7
Apricot	11.7	Honeydew melon	9.6
Banana	22.0	Kiwi	15.4
Blackberry	10.0	Lemon	4.5[b]
Blueberry	10.0	Lime	4.5[b]
Boysenberry	10.0	Loganberry	10.5
Cantaloupe melon	9.6	Mango	13.0
Carambola	7.8	Nectarine	11.8
Carrot	8.0	Orange	11.8
Casaba melon	7.5	Papaya	11.5
Cashew (Caju)	12.0	Passion fruit	14.0
Celery	3.1	Peach	10.5
Cherry, dark, sweet	20.0	Pear	12.0
Cherry, red, sour	14.0	Pineapple	12.8
Crabapple	15.4	Plum	14.3
Cranberry	7.5	Pomegranate	16.0
Currant (black)	11.0	Prune	18.5
Currant (red)	10.5	Quince	13.3
Date	18.5	Raspberry (black)	11.1
Dewberry	10.0	Raspberry (red)	9.2
Elderberry	11.0	Rhubarb	5.7
Fig	18.2	Strawberry	8.0
Gooseberry	8.3	Tangerine	11.8
Grape	16.0	Tomato	5.0
Grapefruit	10.0	Watermelon	7.8
		Youngberry	10.0

[a] Indicates Brix value unless other value specified.
[b] Indicates anhydrous citrus acid percent by weight.

to a level below the normal nutrient range for the juice, then that juice to which such a major modification has been made must not be included in the total percentage juice declaration.

G. Related and Unrelated Juice Declarations [21 CFR 101.30 para (l); FR p. 2926]

- A beverage required to bear a percentage juice declaration on its label must not bear any other percentage declaration that describes the juice content of the beverage on its label (e.g., "100% natural" or "100% pure"). However, the label may declare percentages clearly unrelated to juice content (e.g., "provides 100% of U.S. RDA of vitamin C").

H. Beverages that Contain Fruit or Vegetable Juice [21 CFR 102.33 para (a–g); FR, p. 2926]

1. General
[21 CFR 102.23 para (a); FR, p. 2926]

For a carbonated or non-carbonated beverage that contains less than 100% and more than 0% fruit or vegetable juice, the common or usual name must satisfy the requirements for common or usual names. If the common or usual name uses the word "juice," it must include a qualifying term to advise the consumer that the product is less than 100% juice (e.g., "diluted grape juice beverage" or "grape juice drink").

2. Diluted Multiple Juice Beverages and Blends of Single Strength Juices
[21 CFR 101.33 para (b); FR, p. 2926]

If the product is a diluted multiple-juice beverage or blend of single-strength juices and names (other than in the ingredient statement), more than one juice, the names of those juices, must be in descending order of predominance by volume (except in the ingredient statement) unless the name specifically shows that the juice with the represented flavor is used as a flavor (e.g., raspberry-flavored apple and pear juice drink). The presence of added natural flavors is not required to be declared in the name of the beverage unless the declared juices alone do not characterize the product before the addition of the added flavors.

3. Multiple Juices or Blends: Implied Claims
[21 CFR 102.23 para (c); FR, pp. 2926]

If a multiple-juice beverage or blend of single-strength juices contains a juice that is named or implied other than in the ingredient statement ("represented juice"), and also contains a juice other than the named or implied juice ("non-represented juice"), the common or usual name for the product must indicate that the represented juice is not the only juice present (e.g., "Apple blend; apple juice in a blend of two other fruit juices").

4. Multiple Juices or Blends: Common or Usual Names [21 CFR 102.33 para (d); FR, p. 2926]

In a multiple juice beverage or blend of single-strength juices where one or more, but not all, of the juices are named on the label other than in the ingredient statement, and where the named juice is not the predominant juice, the common or usual name for the product must:

- Indicate that the named juice is present as a flavor or flavoring (e.g., "Raspcranberry"; raspberry and cranberry flavored juice drink); or
- Declare the amount of the named juice in a 5% range (e.g., Raspcranberry; raspberry and cranberry juice beverage, 10- to 15% cranberry juice and 3- to 8% raspberry juice).

5. Modified Juice Beverages [21 CFR 102.33 para (e)(f); FR, p. 2926]

The common or usual name of a juice that has been modified must include a description of the exact nature of the modification (e.g., "acid-reduced cranberry juice," "deflavored, decolored grape juice").

If the juice has been modified to the extent that the original juice is no longer recognizable at the time processing is complete (e.g., color, taste, or other organoleptic properties), or if its nutrient profile has been diminished to a level below the normal nutrient range for the juice, then the source fruits or vegetables from which the modified juice was derived may not be depicted on the label by vignette or other pictorial representation.

6. Beverages Made from Concentrate [21 CFR 102.33 para (g); FR, p. 2926]

If one or more juices in a juice beverage is made from concentrate, the name of the juice must include a term indicating that fact (e.g., "from concentrate," or "reconstituted"). Such terms must be included in the name of each individual juice or it may be stated once, adjacent to the product name so that it applies to all juices (e.g., "cherry juice [from concentrate] in a blend of two other juices" or "cherry juice in a blend of two other juices [from concentrate]").

XIV. USE OF NUTRIENT CONTENT CLAIMS FOR BUTTER [21 CFR 101.67 para (a–c); FR, p. 2455]

The new regulations allow products to deviate from the statutory standard for butter if they satisfy the conditions cited below and if the product label bears appropriate qualifying terminology.

- The claim must comply with nutrient content claim requirements (see Section IX of this chapter). In determining whether a claim is appropriate, the calculation of the percent fat reduction must be based on the statutory 80% milkfat standard for butter.
- The product must contain cream or milk, including milk constituents (including, but not limited, to whey, casein, modified whey, and salts of casein) or both, with or without safe and suitable other ingredients such as colorings, salt, bacterial cultures to improve texture, prevent syneresis, add flavor, extend shelf life, improve appearance, and improve sweetness. The product may contain water to replace milk fat as long as the amount of water does not exceed the amount of milk, cream, or milk constituents.
- The product must not be nutritionally inferior to butter.
- If the product would violate the statutory standard for butter but for the nutrient content claim that characterizes the level of nutrients, that claim must be an explicit claim that is included as part of the common or usual name of the product.
- Deviations from the statutory standard for butter must be the minimum necessary to achieve the performance standard or the food will be considered adulterated. These performance characteristics include, but are not limited to, physical properties, organoleptic characteristics, functional properties, shelf life. If there is a significant difference in performance characteristics, the label must disclose the difference (e.g., "not recommended for baking."
- Each ingredient of the product must be declared. If water or other ingredients are added, the ingredient(s) must be identified with an asterisk following a disclosure statement (e.g., "not in regular butter").

XV. COMPARISON OF FDA AND USDA REGULATIONS

The FDA regulates the labeling of foods other than meat and poultry. The U.S. Department of Agriculture Food Safety and Inspection Service (FSIS) regulates the labeling of meat and poultry products. While the 1990 NLEA mandated the FDA to issue revised nutrition labeling regulations, the Congress did not require USDA/FSIS to issue regulations for meat and poultry products. Nevertheless, on its own initiative and with the purpose of achieving uniformity in the labeling of all foods, FSIS also issued final nutrition labeling regulations on January 6, 1993. Most of these regulations are the same as those promulgated by FDA; however, there are a number of differences. A comparison of FDA and FSIS Nutrition Labeling Requirements is shown in Appendix 1, Table A1.3[13].

XVI. SUMMARY

The new nutrition labeling regulations represent an opportunity to the food industry to provide important dietary information to the general public and enhance their understanding of nutrition and the importance of proper dietary choices to health and disease prevention. The effectiveness with which this is accomplished will be a paramount factor in determining the success of a product in a marketplace that has been redefined by uniformity in nutrition labeling (i.e., standardization of serving sizes; definition of nutrient content and health claims; format specifications). The knowledgeable and responsive food manufacturer, through the vehicles of improving the nutritional quality of a product, education and accurate, informative promotion, will enhance sales and gain a competitive edge in the new atmosphere of consumer awareness created by the 1990 NLEA.

This chapter provides an overview of the NLEA regulations promulgated by FDA. It highlights the main features of the regulations and should serve as an easy to read guideline for compliance by the food packager, distributor and retailer. It is not a legal document nor a legal interpretation of the final rules; rather, it is intended for educational purposes and to facilitate reader understanding and implementation of the cumbersome *Code of Federal Regulations*.

The chapter is divided into sections each providing a condensed version of an important and practical aspect of the regulations. Each section contains specific references to the CFR and *Federal Register* from which the condensed version is derived. For further clarification, the reader should contact the Food and Nutrition Information Center, National Agricultural Library, 10301 Baltimore Boulevard, Beltsville, Maryland 20705-2351 (Telephone: 301-504-5719 or Fax: 301-504-6409) or individual contact persons at FDA and FSIS. A list of these individuals with areas of specialization are provided in Appendix 1, Table A1.4.

XVII. ENDNOTES

1 Copies of the *Federal Register* and the *Code of Federal Regulations* may be obtained from the United States Government Printing Office, Superintendent of Documents, Washington, D.C. 20402 (202-783-3238).

[2] Details of the Regulations for Nutrition Labeling of Raw Products and Fish are presented by Jean Pennington, with Virginia L. Wilkening, Mary M. Bender, and Jung Keun Lee in Chapter 8.

[3] The technical amendments issued on August 18, 1993, require the caloric conversion to be applied to the *actual* (not rounded) amounts of protein, fat and carbohydrate.

[4] Details on package design, graphics, type sizes and acceptable abbreviations with illustrations are presented in Chapter 10 by Roy Parcels and Aaron Brody.

[5] A "kerned value" specifies the amount of space that must appear between adjacent letters in a word. The August 18, 1993, Technical Amendments revised the original requirement for a kerned value no tighter than -4 to a more general requirement for better separation to accommodate type styles permitted by the regulations.

[6] These type size specifications represent revisions from the January 6, 1994, rules and appear in the August 18, 1993, amendments.

[7] Details of the regulations governing compliance are presented in 21 CFR 101.9 para (g) and the *Federal Register* V. 58, No. 3, January 6, 1993, p. 2183.

[8] Chapter 23 by Jim Heimbach and Kathryn Egan describes details and criteria of procedures for compliance. Chapter 24 by Ralph Shapiro describes applicable analytical methodology and use of databases.

[9] Additional information on the labeling of soft drinks, calories, sugar, and calorie control products is presented in Chapter 16 by Lyn O'Brien Nabors.

[10] See Table 7.19 and 7.20 for definition of meal and main dish products, respectively.

[11] For dehydrated foods that are typically consumed when rehydrated with water, or a dilutent that has an insignificant amount of all nutrients (defined as an amount that may be declared as zero per RACC) the per 50 g refers to the as prepared form.

[12] Additional information on the impact of nutrition labeling on standardized foods and details on the labeling of dairy products and amendments to their standards of identity are presented in Chapter 17 by Tom Balmer and Kimberlee J. Carson.

[13] An overview of Nutrition Labeling of Meat and Poultry Products is presented in Chapter 3 by John McCutcheon.

8

Regulations for Nutrition Labeling of Raw Produce and Fish

Jean A.T. Pennington, Virginia L. Wilkening, Mary M. Bender, and Jung Keun Lee

Food and Drug Administration, Washington, D.C.

I. NUTRITION LABELING OF RAW PRODUCE AND FISH

A. Retailer Perspective

Nutrition labeling has not traditionally been available for raw produce and fish sold in retail stores. Nutrition labeling of these foods is not required under the regulations which were issued on January 6, 1993, (1) or under the previous nutrition labeling regulations. Under current regulations, raw produce and fish are exempt from mandatory nutrition labeling. Most foods for which nutrition labeling is mandatory are processed and packaged in a manufacturing plant, where the company has the opportunity to control the production of the product and to initiate chemical analyses for nutrient content. Raw produce and fish are not processed or have only minimal processing, and they are generally brought to retail stores directly from growers, producers, or distributors. Perhaps more important with regard to nutrition labeling are the facts that raw produce and fish tend to be wet, not always clean, and usually without packaging. Thus, raw produce and fish require special consideration with regard to nutrition labeling because responsibility for providing the nutrition information is not clear and attachment of the nutrition labeling information to the food poses some challenges.

B. Consumer/Educational Perspective

The Dietary Guidelines for Americans (2) encourages a varied diet and stresses the importance of "plenty" of fruits and vegetables. Consumers need nutrition labeling information for raw produce and fish to recognize their nutritional attributes. Consumers also need information on the nutrient content of these foods if they are to compare these products with their processed counterparts and with other raw products (e.g., to compare the nutrient content of shrimp with rainbow trout) to make wise nutritional choices in the marketplace.

II. THE NLEA CHALLENGE

Specific requirements of the Nutrition Labeling and Education Act (NLEA) (3) provided a solution to the needs of consumers for nutrition labeling information for raw produce and fish and the designation of a group responsible for providing the information. The NLEA directed the Food and Drug Administration (FDA) to develop a voluntary nutrition labeling program for raw fruit, vegetables, and fish. Specifically, the NLEA required that the FDA identify the 20 most frequently consumed raw fruits, vegetables, and fish in the United States; develop guidelines for food retailers for the voluntary nutrition labeling of these foods; and define substantial compliance with respect to adherence by food retailers to the guidelines.

The NLEA provided for a period of voluntary compliance by retailers with the guidelines established by the FDA. At the end of that period, the FDA was required to determine the level of compliance with the guidelines. If compliance was substantial, the voluntary program would continue and be reevaluated every two years. If the agency did not find substantial compliance, the NLEA required that the agency make nutrition labeling of raw produce and fish mandatory. The first assessment of substantial compliance was to be reported by May 8, 1993.

III. THE 1991 PROPOSAL AND FINAL RULE

The FDA published the proposal concerning nutrition labeling of raw fruit, vegetables, and fish in the *Federal Register* on July 2, 1991 (4). The comment period for this proposal ended on August 1, 1991. Forty responses were received from trade and retail associations, government agencies (of the United States and other countries), retailers, consumer groups, state organizations (e.g., Departments of Public Health and Agriculture), private organizations (e.g., nutrition consultants and a manufacturing company), professional societies, and one university. In general, the responses were favorable to the FDA proposal. Some comments discussed issues relating to label content and format, time frame for implementation, program cost, data sources, serving sizes, identification of fish species, statistical treatment of data, and definition of substantial compliance. Issues raised in the comments were summarized, discussed, and resolved in the final regulations, which were published in the November 27, 1991, *Federal Register* (5).

The following defines the FDA's interpretation of the NLEA requirements and summarizes specific aspects of the guidelines presented in the November 27, 1991, final regulations.

A. Definitions

The FDA interpreted "raw fruit and vegetables" to be those that receive minimal or no processing and no heat treatment, but may be packaged, waxed, or both. Fruit and vegetables that are dried, canned, frozen, or otherwise processed are subject to mandatory nutrition labeling. The FDA interpreted "raw fish" to include raw or raw and frozen whole or filleted fish that are packaged by the retailer or unpackaged; fish that are alive in the retail store (e.g., lobster and crab); shrimp that have been shelled and deveined; and lobster, crab, and shrimp that have been thermally processed or shelled but not otherwise processed or prepared. Nutrition labeling is mandatory for fish that are canned or smoked; that undergo processing such as breading, flaking, or pressing; or that are packaged by a manufacturer.

FDA interpreted "most frequently consumed" to mean "consumed in the largest quantities" and used national retail sales data for raw produce and fish plus information from national food consumption surveys to identify the top 20 items in each of the three categories (6–13). The 20 most frequently consumed raw fruit, vegetables, and fish identified at this point are listed in Table 8.1.

B. Presentation of Information

In accordance with the NLEA, the regulations specified that retailers shall be permitted to make the required information available in "brochures, notebooks, leaflets, or posters, and may supplement the information by videos, live demonstrations, or other media." The FDA encouraged experimentation with various presentations of nutrition labeling for raw fruit, vegetables, and fish. Retailers, trade associations, and other groups were encouraged to conduct surveys to ascertain which types of presentations are most useful to consumers.

Table 8.1 The 20 Most Frequently Consumed Raw Fruit, Vegetables, and Fish (1991)

Fruit	Vegetables	Fish
Banana	Potato	Shrimp
Apple	Iceberg lettuce	Cod
Watermelon	Tomato	Pollock
Orange	Onion	Catfish
Cantaloupe	Carrot	Scallops
Grapes	Celery	Salmon, Atlantic/Coho
Grapefruit	Sweet corn	Flounder
Strawberries	Broccoli	Sole
Peach	Green cabbage	Oysters
Pear	Cucumber	Orange roughy
Nectarine	Bell pepper	Mackerel, Atlantic/Pacific & Jack
Honeydew melon	Cauliflower	Ocean perch
Plum	Leaf lettuce	Rockfish
Avocado	Sweet potato	Whiting
Lemon	Mushrooms	Clams
Pineapple	Green onion	Haddock
Tangerine	Green (snap) beans	Blue crab
Sweet cherries	Radishes	Rainbow trout
Kiwifruit	Summer squash	Halibut
Lime	Asparagus	Lobster

C. Label Content and Format

Nine nutrients were to be included on the label [calories, fat, protein, carbohydrate, sodium, and percent of the U.S. Recommended Daily Allowance (RDA) for vitamin A, vitamin C, calcium, and iron]. Dietary fiber was voluntary for fruits and vegetables; saturated fat and cholesterol were voluntary for fish. One major change from pre-NLEA regulations was that thiamin, riboflavin, and niacin were voluntary and not required for compliance with the guidelines. The information was allowed in a chart format for posters and brochures; however, if individual foods were labeled, the format should follow the pre-NLEA nutrition labeling format for processed foods. One item not required in the nutrition labeling format was "number of servings per package."

D. Serving Sizes

The serving portions were as consistent as possible with the proposal for reference amounts issued on November 27, 1991 (14). Basically, the amounts (in edible portions) were 4 to 5 oz for raw fruit, 2 oz for raw fruit used as ingredients, 10 oz for watermelon, 3 to 5.5 oz for raw vegetables, 1 oz for raw vegetables used as ingredients, and 3 oz for cooked fish. The fish data reflect cooking without added fat or seasoning, and this fact is to be noted on the labeling information.

The FDA believes that values for cooked fish are more useful for consumers and that presenting values for cooked fish is safer because including values for raw fish might mislead consumers into believing that raw fish should be consumed. In addition, the use of values for cooked fish was consistent with Seafood Nutri-Facts (15), an in-store nutrition labeling program. The NLEA (3) specified that the FDA consider current in-store programs in developing the nutrition labeling guidelines for raw fruit, vegetables, and fish.

IV. DEVELOPMENT OF THE DATABASE

The FDA provided nutrition labeling values for the 20 most frequently consumed raw fruit, vegetables, and fish in the November 27, 1991, final regulations (5). These values were available for use by retailers from the November 27, 1991, publication date and are to be used until the first revision of the database is completed. The FDA provided retailers with data for 10 nutrients for raw fruit and vegetables and 11 nutrients for fish to use in nutrition labeling. Sources of the data included the Produce Marketing Association (PMA) (16), Agriculture Handbooks 8-9, 8-11, and 8-15 (17-19), and several other databases (20-22).

The nutrition labeling values were provided for serving portions that were consistent with the FDA's proposal on serving sizes (14), which provided the most current information on serving sizes at that time. Nutrition labeling data were provided on a raw basis for fruits and vegetables and on a cooked basis for fish.

It must be noted that the nutrition labeling values were provided prior to the publication of the January 6, 1993, regulations (1) dealing with label content, label format, and serving sizes. Therefore, the regulations for the voluntary program were devised to be as consistent as possible with the pre-NLEA labeling regulations, but also to be forward-looking toward the new ones.

V. REQUIREMENTS FOR COMPLIANCE

To comply with the voluntary nutrition labeling guidelines, retailers must provide consumers with information on nine nutrients (energy, protein, fat, carbohydrate, sodium, and percent U.S. Recommended Daily Allowance for vitamin A, vitamin C, calcium, and iron). Listing cholesterol and saturated fat for fish and dietary fiber for fruit and vegetables was voluntary. The FDA provided these additional values because of their particular interest to consumers. As indicated above, information for fruit and vegetables should be provided on a raw basis and values for fish on a cooked basis.

Individual stores selected for evaluation were considered to be in compliance if the store provided nutrition labeling for at least 90% of the 20 most frequently consumed raw fruit and vegetables routinely sold by the store or 90% of the 20 most frequently consumed raw fish routinely sold by the store; if the information was provided in the produce and fresh fish sections of the retail store; and if the nutrition labeling complied with the guidelines. Compliance for fruit and vegetables was assessed separately from that for fish. Compliance assessments were based only on the identified foods that retailers routinely sold. Retailers were not required to provide values for items among the 20 most frequently consumed foods that they did not sell. FDA defined substantial compliance of retailers to mean that at least 60% of all stores that are evaluated are in compliance with the guidelines.

VI. FIRST ASSESSMENT OF RETAIL COMPLIANCE

In the *Federal Register* of May 18, 1993, the FDA announced the availability of its "Report on Voluntary Compliance of Food Retailers in Providing Nutrition Labeling Information for Raw Fruit and Vegetables and for Raw Fish" (23). This report was based on a survey of a scientifically drawn nationally representative sample of approximately 2,000 retail stores, conducted in November and December of 1992. The survey was done under an FDA contract. The survey design represented U.S. stores on the basis of sales volume, store type, state, and county size, criteria mandated by the NLEA. Two levels representing average annual sales (i.e., $2 million or more in sales vs. sales falling between $500,000 and $2 million) were established in the sample design to account for the sales volume of food retailers, a factor that is indicative of both the size of the store and the portion of the market served by the stores. Store type was based on the definition of each food store as a chain (i.e., four or more stores under common ownership) or an independently operated store. All states in the continental United States plus the District of Columbia were included in the sample design. The inclusion of county size allowed for representation of food retailers from highly urbanized, urbanized, rural, and very rural counties.

The results of the survey indicate that compliance of retailers with the nutrition labeling guidelines was substantial for both fruit and vegetables (75.7% compliance) and for fish (73.2% compliance). Annual commodity volume

(ACV) compliance estimates (reflecting compliance on the basis of sales volume) were similar to compliance based on the number of stores surveyed. For raw fruit and vegetables, stores in compliance accounted for 76.9% of annual sales of all food stores. For raw fish, the estimate was 74.3%.

ACV data indicate that approximately three-fourths of U.S. consumers are exposed to nutrition labeling information for raw fruit and vegetables and raw fish. These percentages may represent a minimum estimate because some consumers shop in several different retail stores, which increases their chance of exposure to nutrition labeling of raw produce and fish. In accordance with the NLEA, the nutrition labeling of raw fruit, vegetables, and fish will remain voluntary for another two years (i.e., until May 8, 1995).

Results from the survey indicate that nutrition information for raw produce and fish was provided primarily on posters (68% for produce and 61% for fish) and in brochures (11% for produce and 16% for fish). There was a substantial increase in the availability of nutrition information for raw produce and raw fish since the 1991 baseline estimates of 0.5% for produce and 0.0% for fish. The initial estimates indicated that little information was available to consumers for these raw products. The increase in the availability of nutrition information for these foods is believed to be due primarily to the voluntary nutrition labeling program provided by the NLEA.

VII. JULY 1994 PROPOSAL

On July 18, 1994, the FDA issued a proposal in the *Federal Register* (24) to:

- Revise the guidelines for the voluntary nutrition labeling of raw fruit, vegetables, and fish to make them as consistent as possible with the January 6, 1993, nutrition labeling requirements for processed, packaged foods (especially as regards label content and format and serving size);
- Revise the listing of the 20 most frequently consumed raw fruit, vegetables, and fish;
- Update and revise the nutrition labeling values for the 20 most frequently consumed raw fruit, vegetables, and fish; and
- Revise the definition for compliance with the guidelines for the voluntary nutrition labeling of raw fruit, vegetables, and fish.

The proposal had a 60-day comment period which ended on September 16, 1994. The comments will be considered and incorporated or otherwise addressed in the final regulation which is expected to issue in 1995.

The following summarizes several of the changes in the proposed regulations:

The FDA proposed several changes to the listing of the 20 most frequently consumed fish—namely, to combine the species flounder and sole, to add the subspecies chum/pink and sockeye for salmon, and to add swordfish. Flounder and sole are both classified as "flatfish," and the data for these two subspecies are aggregated together in the National Nutrient Databank (25) of the U.S. Department of Agriculture (USDA), which was the source of the data for these fish. The subspecies of salmon are listed separately because they have different nutrient contents, especially different fat levels. The three groups of salmon on the revised list (Atlantic/coho, chum/pink, sockeye) are those most commonly available in the raw state in the U.S. retail market. Swordfish was added to keep the list at 20 fish as required by the NLEA. Swordfish was the next fish to be added to the list on the basis of consumption data provided to the FDA from the National Fisheries Institute (10).

The FDA proposed that compliance with the guidelines requires that retailers use the data provided by the FDA for the 20 most frequently consumed raw fruit, vegetables, and fish. Currently, retailers may use the data provided by the FDA or other data that they believe accurately reflect the nutrient content of these foods. the FDA feels that consistent nutrition labeling values for these foods in the marketplace will be of benefit to consumers.

In the July 18, 1994 proposal, the FDA again requested that organizations submit data to the agency on the nutrient content of the 20 most frequently consumed raw fruit, vegetables, and fish. The FDA also encouraged the development of nutrition labeling data for raw fruit, vegetables, and fish not included among the 20 most frequently consumed items and for raw fruit, vegetables, and fish having unique nutritional characteristics (e.g., farmed vs. wild catfish; valencia vs. navel oranges) that might be overlooked in composite data.

The FDA intends to provide one set of data for each of the 20 most frequently consumed raw fruit, vegetables, and fish to provide uniformity and consistency of values among stores. The FDA will review and evaluate submitted data on the basis of variable identification, analytical methods, sampling design, and statistical treatment of data. The FDA has prepared a manual for the development of databases and estimation of labeling values (26) and a guide to the FDA database review system (27). Sampling and compositing methods should reflect identified variables, i.e., season, geographic region, and different variety, species, or cultivar. Nutrient data to be used for nutrition labeling of raw fruit, vegetables, and fish should be subjected to the FDA compliance calculations (26). These are the same procedures used for the nutrition labeling of processed foods. Label values based on these FDA procedures give 95% assurance that the nutrients contained in the foods are at least equal to 80% of the label values for protein, vitamins, and most minerals and not more than 120% of the label values for energy, cholesterol, fat, and sodium.

VIII. FIRST REVISION OF THE DATABASE

The content of the nutrition label for mandatory nutrition labeling was discussed in the proposal which was issued on November 27, 1991, (28) and the final regulations which were issued on January 6, 1993 (1). The nutrition labels for raw fruit, vegetables, and fish should conform as closely as possible to the nutrient content requirements of this final rule so that labeling for all foods in the marketplace will be as consistent as possible. The first revision of the interim data attempts to make the labeling information for raw fruit, vegetables, and fish consistent with the final rules for nutrition labeling of processed foods with regard to label content, serving size, and daily intake reference values.

In addition, at least every two years, the FDA will publish revisions of the nutrition labeling data for the 20 most frequently consumed raw fruit, vegetables, and fish or a notice indicating that the values have not changed. The revisions or notices are to be published in the *Federal Register* with an opportunity for public comment. The FDA may use the current values or replace them with values that have been determined to be more accurate for labeling purposes.

In the 1991 final regulations for raw fruit, vegetables, and fish, the FDA invited the submission of food composition data for the 20 most frequently consumed raw fruit, vegetables, and fish so that the agency could effectively revise the data during the next two years.

The nutrient values for the first revision of the database have been updated by:

- Adding nutrients or components required for processed foods under the January 6, 1993, nutrition labeling regulations (1) (i.e., calories from fat, saturated fat, cholesterol, sugars, and dietary fiber);
- Using the rounding rules specified by the January 6, 1993, nutrition labeling regulations (1) for processed foods;
- Adding newer data for tangerines provided by the PMA (16);
- Replacing the data for American grapes with data for European grapes because the European varieties are most commonly consumed raw while the American types are further processed before consumption; and
- Subjecting all data from the USDA (where possible) to the FDA compliance calculations.

Sources of data for the revised values (shown in Tables 8.2 and 8.3) include the PMA (16), the USDA National Nutrient Databank (25), Agriculture Handbooks No. 8-9, 8-11, and 8-15 (17-19), supplements to these Handbooks (29), and Home Economics Research Report No. 48 (30).

A background document (31) provides the documentation for the proposed values. It identifies the source for each value and includes the 1991 interim

Table 8.2 Proposed Nutrition Labeling Values for Raw Fruits and Vegetables[1,2]

Raw, edible portion	Calories	Calories from fat	Total fat (g)	Total fat (%)	Saturated fat (g)	Saturated fat (%)	Cholesterol (mg)	Cholesterol (%)	Sodium (mg)	Sodium (%)
Banana, 1 medium (126 g/4.5 oz)	110	0	0.5	1	0.0	0	0	0	0	0
Apple, 1 medium (154 g/5.5 oz)	80	10	1.0	2	0.0	0	0	0	0	0
Watermelon, 1/18 medium melon; 2 cups diced pieces (280 g/10.0 oz)	90	0	0.0	0	0.0	0	0	0	10	0
Orange, 1 medium (154 g/5.5 oz)	80	0	0.0	0	0.0	0	0	0	0	0
Cantaloupe, 1/4 medium melon (134 g/4.8 oz)	50	0	0.0	0	0.0	0	0	0	35	1
Grapes, 1 1/2 cups (138 g/4.9 oz)	90	10	1.0	2	0.0	0	0	0	0	0
Grapefruit, 1/2 medium (154 g/5.5 oz)	70	0	0.5	1	0.0	0	0	0	0	0
Strawberries, 8 medium (147 g/5.3 oz)	70	0	0.5	1	0.0	0	0	0	0	0
Peach, 1 medium (98 g/3.5 oz)	40	0	0.0	0	0.0	0	0	0	0	0
Pear, 1 medium (166 g/5.9 oz)	100	10	1.0	2	0.0	0	0	0	0	0
Nectarine, 1 medium (140 g/5.0 oz)	70	0	0.5	1	0.0	0	0	0	0	0
Honeydew melon, 1/10 medium melon (134 g/4.8 oz)	50	0	0.0	0	0.0	0	0	0	45	2
Plums, 2 medium (132 g/4.7 oz)	80	10	1.0	2	0.0	0	0	0	0	0
Avocado, California, 1/5 medium (30 g/1.1 oz)	60	50	6	9	1.0	5	0	0	0	0
Lemon, 1 medium (58 g/2.1 oz)	20	0	0.0	0	0.0	0	0	0	10	0
Pineapple, 2 slices, 3" diameter, 3/4" thick (112 g/4 oz)	70	0	0.0	0	0.0	0	0	0	10	0
Tangerine, 1 medium (109 g/3.9 oz)	80	10	1.0	2	0.0	0	0	0	5	0
Sweet cherries, 21 cherries; 1 cup (140 g/5.0 oz)	90	10	1.0	2	0.0	0	0	0	0	0
Kiwifruit, 2 medium (148 g/5.3 oz)	100	15	1.5	2	0.0	0	0	0	0	0
Lime, 1 medium (67 g/2.4 oz)	20	0	0.0	0	0.0	0	0	0	0	0
Potato, 1 medium (148 g/5.3 oz)	120	0	0.0	0	0.0	0	0	0	5	0
Lettuce, 1/6 medium head (89 g/3.2 oz)	20	0	0.0	0	0.0	0	0	0	10	0

Potassium		Total carbo-hydrate		Dietary fiber		Sugars	Protein	Vitamin A	Vitamin C	Calcium	Iron
(mg)	(%)	(g)	(%)	(g)	(%)	(g)	(g)	(%)	(%)	(%)	(%)
390	11	29	10	1	4	21	1	0	0	0	0
160	5	24	8	4	16	20	0	0	6	0	0
230	7	26	9	1	4	25	1	10	25	0	2
250	7	21	7	5	20	14	1	0	120	4	0
210	6	13	4	1	4	11	1	80	80	2	2
270	8	24	8	1	4	23	1	2	25	2	2
210	6	18	6	5	20	10	1	10	80	4	0
220	6	17	6	3	12	8	1	0	130	2	2
190	5	10	3	2	8	9	1	2	10	0	0
210	6	25	8	4	16	17	1	0	10	2	0
300	9	16	5	2	8	12	1	4	15	0	2
290	8	14	5	1	4	12	1	0	40	0	2
220	6	19	6	2	8	10	1	6	20	0	0
105	3	2	1	1	4	0	0	0	2	0	0
65	2	6	2	1	4	1	0	0	35	2	0
100	3	17	6	1	4	13	0	0	25	0	0
120	3	20	7	3	12	12	0	0	35	2	0
260	7	23	8	3	12	19	1	0	8	2	0
450	13	25	8	4	16	16	2	2	200	6	4
75	2	7	2	2	8	0	0	0	35	0	0
680	19	27	9	2	8	3	3	0	40	0	6
85	2	3	1	1	4	2	1	2	4	0	0

Table 8.2 (Continued)

Raw, edible portion	Calories	Calories from fat	Total fat (g)	Total fat (%)	Saturated fat (g)	Saturated fat (%)	Chol-esterol (mg)	Chol-esterol (%)	Sodium (mg)	Sodium (%)
Tomato, 1 medium (148 g/5.3 oz)	35	10	1.0	2	0.0	0	0	0	5	0
Onion, 1 medium (148 g/5.3 oz)	60	0	0.0	0	0.0	0	0	0	5	0
Carrot, 7" long, 1 1/4" diameter (78 g/2.8 oz)	40	0	0.0	0	0.0	0	0	0	50	2
Celery, 2 medium stalks (110 g/3.9 oz)	25	0	0.0	0	0.0	0	0	0	125	5
Sweet corn, kernels from 1 medium ear (90 g/3.2 oz)	80	10	1.0	2	0.0	0	0	0	0	0
Broccoli, 1 medium stalk (148 g/5.3 oz)	50	0	0.5	1	0.0	0	0	0	70	3
Green cabbage, 1/12 medium head (84 g/3.0 oz)	25	0	0.0	0	0.0	0	0	0	25	1
Cucumber, 1/3 medium (99 g/3.5 oz)	15	0	0.0	0	0.0	0	0	0	0	0
Bell pepper, 1 medium (148 g/5.3 oz)	30	0	0.0	0	0.0	0	0	0	0	0
Cauliflower, 1/6 medium head (99 g/3.5 oz)	25	0	0.0	0	0.0	0	0	0	40	2
Leaf lettuce, 1 1/2 cups shredded (85 g/3.0 oz)	15	0	0.0	0	0.0	0	0	0	40	2
Sweet potato, medium, 5" long, 2" diameter (130 g/4.6 oz)	130	0	0	0	0.0	0	0	0	45	2
Mushrooms, 5 medium (84 g/3.0 oz)	20	0	0.0	0	0.0	0	0	0	0	0
Green onion, 1/4 cup chopped (25 g/0.9 oz)	10	0	0.0	0	0.0	0	0	0	5	0
Green (snap) beans, 3/4 cup cut (83 g/3.0 oz)	25	0	0.0	0	0.0	0	0	0	0	0
Radishes, 7 radishes (85 g/3.0 oz)	20	0	0.0	0	0.0	0	0	0	30	1
Summer squash, 1/2 medium (98 g/3.5 oz)	20	0	0.0	0	0.0	0	0	0	0	0
Asparagus, 5 spears (93 g/3.3 oz)	20	0	0.0	0	0.0	0	0	0	0	0

[1] Sources of data for revised values: Produce Marketing Association (PMA); the USDA National Nutrient Data Bank; Agricultural Handbooks No. 8-9 and 8-11; Supplements to these handbooks; and Home Economics Research Report No. 48.

Potassium		Total carbo-hydrate		Dietary fiber		Sugars	Protein	Vitamin A	Vitamin C	Calcium	Iron
(mg)	(%)	(g)	(%)	(g)	(%)	(g)	(g)	(%)	(%)	(%)	(%)
300	9	7	2	1	4	4	1	15	35	0	2
200	6	16	5	3	12	9	1	0	15	4	0
220	6	9	3	2	8	5	1	220	8	2	0
300	9	5	2	2	8	0	1	2	10	4	2
240	7	18	6	3	12	5	3	2	10	0	2
480	14	9	3	4	16	3	4	10	200	6	4
170	5	6	2	2	8	3	1	0	60	4	0
160	5	3	1	0	0	2	1	4	8	2	2
240	7	7	2	2	8	4	1	6	150	0	0
250	7	5	2	2	8	2	2	0	100	2	2
210	6	3	1	1	4	2	1	30	4	2	0
350	10	33	11	4	16	7	2	440	30	2	2
280	8	3	1	1	4	0	2	0	0	0	0
70	2	2	1	1	4	1	0	2	8	0	0
190	5	5	2	3	12	2	1	2	8	4	0
180	5	4	1	0	0	2	0	0	30	0	0
240	7	4	1	1	4	2	1	4	25	2	2
210	6	5	2	2	8	2	2	10	10	0	0

[2] (g)-grams; (mg)-milligrams; (oz)-ounces; (%)-percent Daily Value. Percent Daily Values are based on a 2,000 calorie diet.

Table 8.3 Proposed Nutrition Labeling Values for Fish[1,2]

Edible portion, cooked[3] (84 g/3.0 oz)	Calories	Calories from fat	Total fat (g)	Total fat (%)	Saturated fat (g)	Saturated fat (%)	Cholesterol (mg)	Cholesterol (%)
Shrimp	80	10	1.0	2	0.0	0	165	55
Cod	90	0	0.5	1	0.0	0	45	15
Pollock	90	10	1.0	2	0.0	0	80	27
Catfish	170	80	9.0	14	1.5	8	55	18
Scallops (about 6 large or 14 small)	120	10	1.0	2	0.0	0	55	18
Salmon								
Atlantic/Coho	160	60	7.0	11	1.0	5	50	17
Chum/Pink	130	35	4.0	6	1.0	5	70	23
Sockeye	180	80	9.0	14	1.5	8	75	25
Flounder/Sole	100	14	1.5	2	0.5	3	60	20
Oysters (about 12 medium)	100	35	3.5	5	1.0	5	115	38
Orange roughy	80	10	1.0	2	0.0	0	20	7
Mackerel, Atlantic/Pacific	210	120	13.0	20	6.0	30	60	20
Ocean perch	110	25	3.5	5	0.0	0	75	25
Rockfish	100	20	2.0	3	0.0	0	40	13
Whiting	110	25	3.0	5	0.5	3	70	23
Clams (about 12 small)	100	15	1.5	2	0.0	0	55	18
Haddock	100	10	1.0	2	0.0	0	80	27
Blue crab	100	10	1.0	2	0.0	0	90	30
Rainbow trout	140	50	6.0	9	2.0	10	60	20
Halibut	110	20	2.0	3	0.0	0	35	12
Lobster	80	5	0.5	1	0.0	0	60	20
Swordfish	130	35	4.5	7	1.0	5	40	13

[1] Sources of data for revised values: the USDA National Nutrient Data Bank; Agricultural Handbook No. 8-15; Supplements to this handbook; and Home Economics Research Report No. 48.

Sodium		Potassium		Total carbo-hydrate		Dietary fiber		Sugars	Protein	Vitamin A	Vitamin C	Calcium	Iron
(mg)	(%)	(mg)	(%)	(g)	(%)	(g)	(%)	(g)	(g)	(%)	(%)	(%)	(%)
190	8	140	4	0	0	0	0	0	18	0	0	2	15
60	3	450	13	0	0	0	0	0	20	0	0	2	2
110	5	360	10	0	0	0	0	0	20	0	0	0	2
40	2	350	10	0	0	0	0	0	21	0	0	0	0
260	11	280	8	2	1	0	0	0	22	0	0	2	2
50	2	490	14	0	0	0	0	0	22	0	0	0	4
65	3	410	12	0	0	0	0	0	22	2	0	0	2
55	2	320	9	0	0	0	0	0	23	4	0	0	2
90	4	290	8	0	0	0	0	0	21	0	0	2	2
190	8	390	11	4	1	0	0	0	10	0	0	6	45
70	3	330	9	0	0	0	0	0	16	0	0	0	0
100	4	400	11	0	0	0	0	0	21	0	0	0	5
200	8	330	9	0	0	0	0	0	20	0	4	4	2
70	3	430	12	0	0	0	0	0	21	4	0	0	2
95	4	320	9	0	0	0	0	0	19	2	0	6	0
95	4	530	15	0	0	0	0	0	22	10	0	6	60
85	4	340	10	0	0	0	0	0	21	0	0	2	6
320	13	360	10	0	0	0	0	0	20	0	0	8	4
35	1	370	11	0	0	0	0	0	21	4	4	6	2
60	3	490	14	0	0	0	0	0	23	2	0	4	4
320	13	300	9	1	0	0	0	0	17	0	0	4	2
100	4	310	9	0	0	0	0	0	22	2	2	0	4

[2] (g)-grams; (mg)-milligrams; (oz)-ounces; (%)-percent Daily Value. Percent Daily Values are based on a 2,000 calorie diet.

[3] Cooked without fat, breading, or seasoning.

nutrition labeling values for comparison. It also explains the priority for selecting the data from the various sources.

IX. RETAILER PRESENTATION OF INFORMATION

Several trade associations, including the Food Marketing Institute (FMI), the PMA, and the United Fresh Fruit and Vegetable Association, and private organizations incorporated the 1991 interim nutrition labeling values provided by the FDA into posters and leaflets and made them available for retail use. For example, the FMI used the 1991 interim values in its revision of the Nutri-Facts program (32). These program materials are available free to FMI members and at a nominal charge to non-members. The FMI plans to revise the Nutri-Facts program when the FDA's revised values are published in the final regulations.

Retailers may also be in compliance with the FDA guidelines by developing their own charts and/or brochures which display the FDA nutrition labeling values for raw fruit, vegetables, and fish. These charts may be mounted overhead, free-standing in the store aisles, attached to the walls, or placed above the food bins. Brochures with the information could be placed in holders near the food bins or on walls adjacent to the foods. Notebooks containing the nutrition labeling values could be chained to produce and fish sections of the stores.

There are also supplemental ways that retailers may make the nutrition labeling information available to consumers. These methods do not by themselves constitute compliance. They are additional methods that may be used along with charts, brochures, or notebooks. They include providing the nutrition labeling values on the small plastic bags available for consumers to wrap produce and fish and/or on the larger plastic or paper grocery bags at the checkout counter, showing in-store videos that include the nutrition information, and placing the charts and brochures in other areas of the store, such as the checkout counter.

X. IMPLICATIONS OF THE REGULATIONS

The voluntary nutrition labeling program for raw produce and fish will increase the availability of nutrition information for these foods at the retail level and will provide information that is consistent from store to store across the United States. Consumers will have needed information to do comparative shopping and make informed food selections. They will be able to use in-store nutrition labeling information to compare the nutrient values of raw produce and fish with those of other foods. Nutrition labeling information will serve as an educational tool for individuals on restricted diets. Dietitians and nutritionists are expected to encourage their patients and clients to use the nutrition information to adhere to specific diet therapies.

The voluntary nutrition labeling program provides an opportunity for trade associations to assist retailers by making the FDA nutrition labeling values available in charts, brochures, or other forms for retail use. Trade associations may also gather the pertinent nutrient information or initiate nutrient analyses for the products they represent and ensure that the information reaches the FDA. Trade associations benefit by emphasizing the advantages of their products and may be encouraged to conduct more research to develop more reliable nutritional profiles.

By making charts and brochures readily available to retailers, trade associations help bring stores into compliance. The retailer can be assured that the use of trade association-provided materials will ensure compliance. Trade associations can also assist retailers by reminding them to display the charts and brochures, especially during holidays and other busy seasons.

XI. ACKNOWLEDGMENTS

Jean A.T. Pennington, Ph.D., R.D. was the primary author for the 1991 proposal and final regulations and the 1993 proposal and developed the 1991 interim database and the 1993 database; Virginia L. Wilkening, M.S., R.D. reviewed and provided comments on drafts of the proposals and regulations; Mary M. Bender, Ph.D. monitored the contract and authored the report concerning assessment of retailer compliance; and Jung Keun Lee, Ph.D. completed the compliance calculations on data from the USDA National Nutrient Databank. The authors acknowledge Thomas O'Brien for typing the tables for the 1993 nutrition labeling values and Carole Adler, M.S., R.D. for reviewing the revised database and background document for this database.

XII. REFERENCES

1. Food Labeling; Mandatory Status of Nutrition Labeling and Nutrient Content Revision, Format for Nutrition Label. *Federal Register* 58:2079–2205, January 6, 1993. Technical Amendments to these regulations. *Federal Register* 58:44063–44090, August 18, 1993. Food Labeling; Serving Sizes. *Federal Register* 58:2229–2300, January 6, 1993. Technical Amendments to these regulations. *Federal Register* 58:44039–44059, August 18, 1993.

2. U.S. Department of Agriculture, U.S. Department of Health and Human Services, *Nutrition and Your Health. Dietary Guidelines for Americans*, third edition, Home and Garden Bulletin No. 232, 1990.
3. Nutrition Labeling and Education Act (the 1990 amendments), Pub L No. 101-535.
4. Food Labeling; Nutrition Labeling of Raw Fruit, Vegetables, and Fish; Proposed Rule, *Federal Register* 56:30468–30482, July 2, 1991.
5. Food labeling; Nutrition Labeling; Guidelines for Voluntary Nutrition Labeling; and Identification of the 20 Most Frequently Consumed Raw Fruit, Vegetables, and Fish; Definition of Substantial Compliance; Final Rule, *Federal Register* 56:60880–60891, November 27, 1991; amended 57:8174–8178, March 6, 1992.
6. Produce Marketing Association, Newark, DE, Data on retail sales of fruits and vegetables provided to FDA, Washington, D.C., January 4, 1991.
7. United Fresh Fruit and Vegetable Association, Supply Guide, Monthly Availability of Fresh Fruit and Vegetables, 13th ed., Alexandria, VA, 1989.
8. Average per Capita Consumption, 1987–89, Economic Research Service, USDA, Washington, D.C.
9. 43rd Consumer Expenditure Study. Supermarket Business, September 1990.
10. National Fisheries Institute, Data on fish production provided to FDA, Washington, D.C., January 4, 1991.
11. Food Marketing Institute Speaks, Table 144B, Washington, D.C., 1990.
12. Food Marketing Institute, phone survey data provided to FDA, Washington, D.C., January 4, 1991.
13. Human Nutrition Information Service, USDA, "Nationwide Food Consumption Survey (1987–88), Individual Intakes, 3 Days," computer tape PB 90-504044, National Technical Information Service, Springfield, VA, 1990.
14. Food Labeling; Serving Sizes; Proposed Rule. *Federal Register* 56:60394–60421, November 27, 1991; amended *Federal Register* 57:8179, March 6, 1992.
15. Food Marketing Institute, National Fisheries Institute, *Seafood Nutri-Facts*, National Fisheries Education and Research Foundation, Washington, D.C., 1988.
16. Produce Marketing Association, 1500 Casho Mill Road, Newark, DE 19714-6036.
17. S.E. Gebhardt, R. Cutrufelli, and R.H. Matthews, *Composition of Foods; Fruits and Fruit Juices, Raw, Processed, Prepared*, Agriculture Handbook No. 8–9, U.S. Government Printing Office, Washington, D.C., 1982.
18. D.B. Haytowitz and R.H. Matthews, *Composition of Foods: Vegetables and Vegetable Products, Raw, Processed, Prepared*, Agriculture Handbook No. 9–11, U.S. Government Printing Office, Washington, D.C., 1984.
19. J. Exler, *Composition of Foods: Finfish and Shellfish Products, Raw, Processed, Prepared*, Agriculture Handbook No. 8–15, U.S. Government Printing Office, Washington, D.C., 1987.
20. A.A. Paul and D.A.T. Southgate, *McCance and Widdowson's The Composition of Foods*, 4th revised edition, Elsevier/North-Holland Biomedical Press, New York, NY, 1978.
21. J. Davies and J. Dickerson, *Nutrient Content of Food Portions*, The Royal Society of Chemistry, Cambridge, England, 1991.

22. V.D. Sidwell, *Chemical and Nutritional Composition of Finfishes, Whales, Crustaceans, Mollusks, and Their Products*, U.S. Department of Commerce, National Oceanic and Atmospheric Administration Technical Memorandum NMFS F/SEC-11, Washington, D.C., 1981.
23. Report on Voluntary Compliance of Food Retailers in Providing Nutrition Labeling Information for Raw Fruit and Vegetables and for Raw Fish, Food and Drug Administration, Center for Food Safety and Applied Nutrition, Washington, D.C., May 8, 1993.
24. Food labeling; Nutrition Labeling of Raw Fruit, Vegetables, and Fish; Proposed Rule, *Federal Register* 59:36379–36394, July 18, 1994; amended 59:37190, July 21, 1994.
25. National Nutrient Databank, Maintained at the Human Nutrition Information Service, United States Department of Agriculture, 6505 Belcrest Road, Hyattsville, MD 20782, 301 436-8491.
26. F.D. McClure and J.K. Lee, *FDA Nutrition Labeling Manual: A Guide for Developing and Using Databases,*, Division of Mathematics, Office of Toxicological Sciences, Center for Food Safety and Applied Nutrition, Food and Drug Administration, 1993.
27. J.A.T. Pennington, J.I. Rader, and F.D. McClure. *Guide to FDA's Database Review System*, Food and Drug Administration, Washington, D.C., June 1994.
28. Food Labeling; Reference Daily Intakes and Daily Reference Values; Mandatory Status of Nutrition Labeling and Nutrient Content Revision, *Federal Register* 56:60366–60394, November 27, 1991.
29. L.E. Dickey. *Composition of Foods—Raw, Processed, Prepared*, 1989 Supplement (1990), 1990 Supplement (1991), 1991 Supplement (1992). US Department of Agriculture, Washington, D.C.
30. R.H. Matthews, P.R. Pehrsson, and M. Farhat-Sabet, Sugar Content of Selected Foods: Individual and Total Sugars, U.S. Department of Agriculture Home Economics Research Report No. 48, September 1987.
31. J.A.T. Pennington. *FDA Voluntary Nutrition Labeling Program for Raw Fruit, Vegetables, and Fish: Documentation for the Proposed 1993 Nutrition Labeling Values*. Office of Food Labeling, Center for Food Safety and Applied Nutrition, Food and Drug Administration, Washington, D.C., October 1993.
32. *Nutri-Facts—1992 Update. Produce, Seafood, Meat & Poultry*, Food Marketing Institute, Washington, D.C., 1992.

9

Strategic Planning and Management of Changing Packages to Comply with Nutrition Regulations

Aaron L. Brody

Rubbright•Brody, Inc., Devon, Pennsylvania

I. INTRODUCTION

Implementation of the 1990 nutrition labeling law and its regulations will demand major investments in food company managerial and supplier resources. Changing all of an organization's packages is always difficult. Changing all packages virtually simultaneously to comply with the nutrition labeling law and its implementing regulations appears to be a monumental task. Such tasks may be ordered from management with little appreciation for the immensity of the operation. Achievement of the objective requires significant investment of resources coupled with a redeployment of these resources.

These actions must be preceded by a plan derived from a strategy—a somewhat mystical concept which, like packaging design and development itself, is perceived to be elementary to many managers who are non-practitioners.

Much confusion obviously reigns about the concept, but the application of strategy and tactics to comply with nutrition labeling regulations is sound, meaningful, and beneficial to members of organizations required to perform the task. Whether it is called planning, strategy, strategic planning, tactics, forward thinking, management mechanics, or anything else, the notion of making decisions today on the basis of expectations is a key tool in guiding professional, economic and business action in packaging—and in particular, in a massive, all-product move.

In the relatively superficial worlds of the nation's business and packaging periodicals, strategy deals with total corporate efforts—the large corporations such as Kraft General Foods and General Mills, Inc. that recognized the need for strategy and pioneered its use in business. If observers have difficulty translating from the multibillion dollar Grand Met to companies like Little Papoose Pea Canners of Lost Lake, Maine, how can a rational business person even consider applying strategy or planning to packaging and nutrition labeling, especially under the time constraints involved? Strategic planning is advisable for the entire food packaging community with its thousands of diverse companies.

One reaction to this proposal to plan before acting might be, "How could strategic planning be applied to my two-, ten-, or twenty-person laboratory, commissary, plant, supermarket, or import agency?" This reaction is virtually the same as is encountered within large corporate business units. Another common response is "Since I am already planning, why introduce a formalized procedure? Since I am already developing packaging, why do I need a plan?" Another response is that computer graphics can generate the requisite plans, nutrition data, package graphic presentation, etc.—so why bother? Still another is that the packaging manager is only taking direction from marketing, R&D, legal, or production, and so is only a cog in a much larger plan.

In addition to the universal human fears of change and organizational obstacles to altering the structural fabric, a long list of rationalizations on why strategy or planning is not for the package's nutrition labeling implementation may be conjured up more easily than an explanation of how such strategic planning benefits the individual and the organization, and facilitates the implementation of nutrition labeling.

Had the concepts of strategic planning been implemented a generation or more ago, the counterproductive abuses of the cost-is-everything tunnel vision would never have been, and the present residual consequences of fully amortized inflexible packaging equipment would not be limiting profitable progress. Perhaps strategic alliances and contract purchasing might never have come into vogue—and, possibly even, there would have been no need for nutrition labeling.

The world of packaging in the "good old days" was clearly "put the product into a bottle and put the bottle, pouch into a box, etc." Choices were limited to determining structural design, closure, and suppliers.

In those times, suppliers were very price competitive. Then, nearly 40 years ago, Coca-Cola came in only one size of glass and in a single multiple, and was distributed to one class of retail outlet. Today, the 6-1/2 oz., returnable, wasp-waisted Coca-Cola green bottle has been replaced by containers that are 10 oz., 12 oz., 16 oz., 32 oz., 0.5 liter, 1, 2 and 3 liters in size; of glass, metal, plastic, and aluminum in 6, 8, 12, and 24 packs; returnable and non-returnable; and so forth. Labels were either whatever the manufacturer wanted (pre-1938) or circumscribed by minimum regulation (1938–1964) or more rigidly proscribed (1964 to early 1980s) or even tighter when the optional information label appeared to near chaos with the words healthy, light, natural, preservative-free, fat-free, and their brethren hurled about with little apparent basis in fact—and much basis in implied consumer perception—which might in part account for confusion with government, consumers, retailers, health care, etc.

Retailer groups dictate the type and count, delivery time, shelving, and labeling, but now have been mandated on in-store nutrition information signage. If carbonated beverage packages do not fit the image of complexity for nutrition labeling, consider breakfast cereals, with carton sizes, sugar-sweetened, non-sugar sweetened, and their incessant promotions; coextrusions; warehouse club store distribution; convenience stores; salad bars; delicatessens; and so forth.

Examining packaging from the perspective of the food packager shows internal company influences from marketing, production, engineering, finance, government via the legal department, and so forth, even if the company has no specific person assigned, to the extent that absolutely no agreement exists as to where the packaging department belongs within the organization, or whose in whose jurisdiction nutrition labeling responsibility lies—marketing, legal, consumer affairs, or as we have so often heard in retail establishments—let our suppliers of meat and fresh produce do it.

Packaging and its development is at least as complicated as any other function within a food product manufacturing/marketing firm. Packaging is probably more difficult than other functions because its outputs particularly now are so visible. When the packaging department or professional does not function well or economically, or is under regulatory oversight, the message is communicated instantly. Unlike the accounting or legal departments, there is no way to hide the mistakes of either omission or commission.

If the food marketer or retailer or department requiring the nutrition labeling resists and therefore does not have the time to do the job right the first time, they certainly do not have the time to correct it again and again. And think of the numerous occasions in which errors have to be rectified again and again! In the great American game of "Criticize Your Own Company," it should be remembered that the organization today is the result of whatever strategy or planning was or was not conducted in the past.

The foregoing preamble, intended to justify the use of strategic and tactical planning for the implementation of nutrition labeling, has until this juncture purposely omitted the prerequisite of all strategic planning: its definition.

II. DEFINITION OF STRATEGIC PLANNING

Strategic planning is the process that establishes the objectives, defines the policies, generates the philosophies, develops the implementation details, enumerates the resources required and the benefits to be derived, and links today's issues to actions to be taken next week, next month, next year, and even into the next decade as certainly nutrition labeling requires. Strategic planning is the "futurity of current decisions," making certain that the future, whether tomorrow morning or next year, is realistically included in packaging decisions and actions. If strategic planning were simplistic, it would have been made an integral component of every business long ago. Strategic planning is not simplistic, but rather simple: the concept is easy to comprehend and put into practice, but the process of reaching this point requires thought and effort. Tactical implementation is the how to do it aspect—once having decided the objective and the route by which this objective is to be reached, the actual mechanics are the tactics. Merely complying with the letter of the nutrition labeling regulation is a strategy. The selection of a graphic designer to position the front label qualifier is tactical. Color selection is a tactical detail worked out among marketing, graphic designer and converter—with some regulatory oversight, of course, to ensure contrast and position. Subsidiary tactics are printing plate preparation and ink acquisition.

A. Responsibilities for the Future

An immediate reaction of strategic planning is that the future is upper management's responsibility and does not belong to operating staff such as packaging managers. Indeed, at all the various levels within the organization, the scope of the future is the ultimate responsibility of upper management. All executives should worry about very long-term implications for the entire organization; his or her decisions can affect millions of dollars, thousands of people, and the very existence of the organization.

Marketing management should be concerned with the organization's products, their mix and their definition, positioning, pricing, selling, shelving, use, and, of course, packaging—all in the future which can be six months or two years ahead. Marketing sets the objectives based on distribution and consumer input and, for many, the restrictions. Marketing management then expects every manager of production, design, engineering, sales, and packaging, of course, to perform in his or her specialty. Each functional organization such as packaging is assigned the duty today to produce something tomorrow or next week. And that assignment usually includes "packaging the product."

Packaging does not happen by itself; packaging is not conjured up by magic; and it is certainly not a function that can be performed with any credibility by anyone hired in from the street, although of course, some organizations have been accused of using such people. In general, with few exceptions, large food manufacturing firms have packaging departments and/or packaging department managers. A manager, by title and responsibility, manages a function that might or might not include people, factories, machines, etc. Thus the packaging manager manages packaging—its structure, design, use, synergy with distribution, equipment, retailers and consumers, cost, safety, environmental impact, in this context, the transition into nutrition regulatory compliance. If the packaging manager does not manage these elements of packaging, then who will? And if these elements are not managed, then the result is chaos, distortion, nonfunctionality or, in our context, a paucity of regulatory compliance.

III. WHAT IS MANAGEMENT?

The next obvious question is "what is management?"—especially in the context of nutrition labeling. A manager's functions include:

- *Establish goals and objectives.* What do the FDA/USDA, consumer, the retailer, and marketing and the company's own management expect the package and its new nutrition information to do?
- *Plan the strategy for achieving the goals and objectives.* The packaging manager should decide whether to depend on material suppliers, do-it-yourself, hire outside graphic designers, use the advertising agency artists, imitate the competition if there is enough time, self-manufacture, use a package that is already in the line, adhere an overlabel, use an inside or outside laboratory, or a data base, etc.
- *Generate a philosophy of operation, a determination of how strategy is to be implemented.* Decide whether to emphasize compliance with the letter or the spirit of the regulation, design, appearance, cost, or engineering, quantitative or qualitative considerations, a scientific or an instinctive approach, etc. Is nutrition labeling an opportunity to change all of the designs—or even structures?
- *Set the policy through action plans.* Detail the objective, the strategy and the nature of the approach, and the specifics of who does what and when, in writing, with schedules and contingencies to comprehend and implement the nutrition labeling regulations. What will be the source of nutrition information to be incorporated into the labels? How many different packages are involved? Are any face panels involved, and, if so, are these to be changed? What is the schedule for design, new plates, printing, scrapping of obsolete designs?—contingencies in case of serious problems or obstacles.
- *Plan the structure of the organization that must execute the policy,* even if this means coordination of resources outside of the control of the packaging

manager. If the packaging development manager has a complete staff of designers, engineers, chemists, etc., plus the required laboratories and pilot plants, then these resources should be deployed for maximum efficiency. If not, then the required internal and external resources must be tapped and integrated structurally to achieve the objectives within the prescribed time frame. Most food companies, however, do not have internal technical resources to analyze every product for nutritional values—and the comprehension of data bases is not an easy task. Nor do they have access to designers to change designs. Further, of course, plate makers and converters cannot produce everything instantly.

- *Provide and actuate the staff.* After all the managerial activity is done, staff people or external resources must do the actual work required. The people assigned or requested, suppliers, consultants, etc., to perform the task must be motivated to produce effectively and in a timely fashion.
- *Provide the physical resources.* The designs, the nutritional analyses, the printing plate preparation or at least their sources, facilities, press time, consumer focus groups, etc., must be made available when and where they are required; the staff people, designers and consultants must have the necessary tools.
- *Establish procedures and schedules.* When does the staff start, especially since time is short? Who is tracking? What are the contingencies? Where is information recorded? Who prepares the reports and to whom are reports sent? What mechanisms are used to communicate with suppliers and consultants involved?
- *Budget.* Within organization financial guidelines, the budget must be constructed, approved, and followed. To achieve the objective, this project has an associated cost. What happens if the budget runs out and the nutrition labeling is incomplete?
- *Set standards of performance.* By word and deed, by dictate or by example, demonstrate the type of output that is tolerated, expected, or appreciated. Whoever is actually executing the label changes must have standards of performance.
- *Control.* No one in a manager's span of control should have greater knowledge of the overall scope of programs and progress than the manager. Others might and probably should have deeper knowledge on the specific aspect such as the legal interpretation or the meaning of the nutritional data, or the mechanics of making or mounting printing plates. The manager is wholly responsible for knowing the progress of each element and of using that information to improve, accelerate, retard, or modify the action to be taken. Feedback on progress must be comprehensive, timely and based on the facts, especially in this tight time frame. Remember that no change occurs without problems, and nutrition labeling requires at the very least graphic changes

on virtually every package. The more packages, the more potential and real problems. So be prepared to literally manage change.

These key operating functions do not include public relations, completing expense reports, reporting upward to the boss, smoothing the perpetual irritants among the staff members, and a thousand other things we all encounter each year. For each of the multiple nutrition labeling tasks, the same functions apply.

Management cannot be by the tried and true past, intuitive, reactive, or even opportunistic approach—especially in the mechanics of implementing nutrition labeling. A packaging manager cannot muddle through on the strength of his or her own wits and perform. A packaging manager's responsibility to protect his/her company's products is not an adaptation of some instinctive response to a stimulus. Shooting from the hip and personal preferences have no place on responsible management of nutrition label development for a food company's packaging.

Nutrition labeling or providing consumers with nutrition information is the result of systematic strategic planning: the process that sets objectives, defines the means with which to attain those objectives, and details the steps required to follow those means, linking today's decisions to the future result and optimizing the resources and outcomes. Once understood, the meaning of strategic planning is essentially common sense uncommonly applied. Strategic planning does not include *making* future decisions. Rather, it provides the framework within which today's and tomorrow's required decisions are and will be made by the persons responsible.

Strategic planning has a higher probability of delivering the requisite nutrition labeling on the food company's package surfaces than the alternatives because the process forces the packaging manager to ask and answer critical questions of time, design, placement, color, etc., relevant to the issue at hand. This is not a Machiavellian scheme for personal gain—or a distasteful task that can be handed off to some gnome, or a job that can be ignored. Our government, whether you agree or not, has established a time limit and a framework that is tighter than any in the past for any packaging task, and the information on nutritional content of each food must be delivered to the packaging manager in an accurate and timely fashion.

IV. STRATEGIC PLANNING BENEFITS

By compelling questions and answers, strategic planning for packaging and implementing nutrition labeling introduces forces to the decision-making process that are indispensable to the process.

- Strategic planning simulates the future in an objective manner. By establishing a scenario, a model that is the basis for packaging development is generated. An action plan with costs, trigger points and contingencies is almost indispensable in this program.
- It applies a systematic approach—a uniform, detached methodology for information accumulation and analysis such as for nutritional data, for projecting requirements for printing press time, for stocking the warehouse or scheduling just-in-time delivery, etc. This is an organized mechanism for appraisal and for structuring an approach to achieving an objective. System implies a beginning, an end and an orderly path between, coherence and logic to all the components—a visible pathway from today's packages with unacceptable information and those which will comply.
- It compels the establishment of an objective. Having a precisely defined objective is a key beginning for strategic planning. So simple is this element that it is often overlooked in favor of the simplistic "comply with the regulations" or "keep the costs low." The objective is clear here.
- Strategic planning identifies and aids comprehension of opportunities and threats. By forcing the manager to examine the future in a pragmatic rather than a visionary fashion, strategic planning allows realistic scenarios of what will probably occur to be prepared. In general, of course, the future is a series of incremental steps from the present. By having the regulatory schedules for both FDA and USDA, the planner is able to compress or expand the required actions—and to predict the probability of a new means to measure nutrition information or to be concerned about a converter's ability to schedule a press run.
- Strategic planning offers a basis for other decisions. A foundation of sound information, analyzed in an orderly fashion, provides a solid rationale for decisions at the core of and peripheral to packaging itself. Knowing exactly when and how the regulations are to be implemented, decisions on marketing the benefits may be made. A forward looking marketing management can communicate the benefits of the new information on the package: "look at the nutritional value of this food on the back, side, etc., of this package."
- It gives a measure of performance. By establishing objectives, the packaging manager knows at the start what he/she must do, in what time frame, and for how much. Thus he/she has established the mileposts—what week will the designs, mechanicals, plates, etc., be ready, by which the project, his staff, and measure—no more post facto analysis, Monday morning quarterbacking, finger pointing in this situation—except, of course, from those who have pre-asserted that it cannot be done within the government dictated time frame.

It should be clear by this point that the implementation of the nutrition labeling regulations is amenable to and should be performed with strategic planning. The assembly of nutrition information on freshly made foods and foods after (x) days or months of shelf life should be underway now—with sufficient resources allocated so that all will be ready by the dictated date. Graphic designers, typesetters, printing press time, etc., should be penciled in at the converter's schedule. "Label" inventories and run-out times should be known by now. There are numerous combinations and permutations possible but the time to evaluate all is not available; therefore planning is required to eliminate the need for every single graphic and position variable to be specifically evaluated in comparison with every other variable.

Clearly stated objectives and problem definitions are necessary to communicate the problem to the many and varied specialists, schedules, designers, analysts, distribution managers, etc., who are involved in this transition.

A key feature of strategic planning for packaging managers that differs from that usually performed by and for upper and/or marketing management is that the packaging manager formulates, implements, and is responsible for the planning in packaging. A corporate business planner or marketing strategist who holds this position for more than a one year cycle is an exception. It is entirely possible that the brand manager present now will not be in position when the nutrition information must be on packages and/or in retail display.

V. PROBLEMS OF STRATEGIC PLANNING

Particularly in the 1990s, nothing is perfect. Strategic planning has its limitations, the most important of which is that it provides no guarantee of a positive result, but rather the notion that the probability of positive results is significantly greater with a properly formulated and implemented strategy and plan than without. Among the limitations to strategic planning are the following:

- The social, political, economic, market, or regulatory environment might prove to be different than was planned. In the past, both USDA and FDA granted extensions. Throughout the period since the regulations were first issued, changes have been made. Such variables should be incorporated into contingency plans—what if they change the font, position, size, etc.?
- Planning is initially expensive, requiring the attention and effort of several persons, including operating managers and a full-time coordinator or planner, purchasing agent, inventory control, R & D, nutritionists, etc. Beyond the planning process are inevitably the changes dictated by implementation of plans once actions are underway, or if the lawyers add a twist—or if the

government changes a regulation from that proposed. Planning might be viewed as a beginning cost or as an investment. Either way, at the outset, monetary outlays are required—sort of analogous to a capital investment to install equipment on a packaging line.

- Internal resistance: in no human endeavor has there ever been an absence of resistance to change, to more efficient operations, or to the perceived threat to job or security. Obstacles to planning can range from subtle failure to understand the simplest assignments, to refusal to cooperate, to overt actions intended to undermine the process. Most disconcerting of all, however, are those who go through the planning process and concur with the paperwork, but then ignore the plan and proceed along their normal paths. In this context, the most disturbing might be persons who ignore the reality of the regulations.
- Day-to-day operations: breathes there a planner who has never encountered managers who do not have time for such fads as planning? Every manager seems to be more concerned with solving today's immediate problems than with tomorrow on the grounds that if he/she does not survive today, there will be no tomorrow. The less managerially inclined the manager is, the greater the incentive will be to use today's crises as a rationale to defer or ultimately avoid planning. One might view nutrition labeling as a today crises—but it is so long and complex that it cannot be treated as if it will be resolved on the food processing plant floor packaging line by "slapping on" pressure sensitive labels. What about those who do not know the nutrient values of their products—or the resources to readily and economically acquire them? What is their alternative to generating nutrition label data? What about those who own two years' supply of printed packaging materials—or those who have never had nutrition information on their packages and have "no space" available on the package surface—or those whose product brand or name will be banned by the new regulation? Their planning with alternatives—and costs per alternative—begins immediately.
- Difficulty: Strategic planning is not an easy task. Strategic planning requires combinations of pragmatism and vision, reasoning and imagination, courage and patience, qualitative and quantitative skills, and communications both up and down that are really the domain of the experienced and disciplined. Strategic planning is neither simple nor simplistic; it is a tough, systematic process of analyzing and projecting how the labeling regulations function today and how to implement them.

Strategic planning offers no guarantee that rigid adherence to the rules or to the plans will provide a trouble-free ride. No manager can become a bureaucratic drone (with his inviolable sacred rule book) with his plan for this action and be assured that the objectives will be achieved. The world is not yet a sufficiently rigid environment for a given input in one spot to always provide predicted outputs elsewhere. The use of strategic planning will, however, significantly increase the probability that the decisions made and actions taken at the outset will be correct in the context of the function, the organization's objectives and the regulations.

Strategic planning forces the manager to introduce organized thinking, to anticipate what might happen, to assign probabilities to future events, to establish systems through which the needed functions may be performed, to forecast and be ready for contingencies, and, above all, to plot the courses of action to be taken.

VI. MODELS

How does a packaging manager who has the responsibility of implementing nutrition labeling and who invariably does not have the luxury of a planning assistant perform strategic planning? Planning is analogous to the universe according to Einstein—frightening from a distance, but so orderly when understood that it is almost elementary.

The four basic strategic planning models are the premises, formulation, implementation, and information flow. Establish what you must do and how to do it, do it, and make certain that you know you have done it or know how to correct what went wrong.

Expanding the basic conceptual model into a typical strategic planning model that might be used by a packaging manager reveals sequential but overlapping steps:

- Formulate the tasks that must be accomplished over the long term for the organization. An accurate information panel must be printed on all packages by May 1994—dictating design, input data, printing data, printing plates, printing, distribution, etc. Determine the sources of design, printing, etc., the information and analyses.
- Obtain inputs on the past and present environments—e.g., what the marketing and purchasing departments and especially package material purchasing and production want. What resources, i.e., consultants, designers, etc. are available, and which are informative, and which require reinforcement?
- Enumerate and evaluate alternatives for use in the strategic plans—for example, add to the staff, use more supplies, hire contract designers, go offshore for printing or plates, transfer some tasks to other departments, discard all leftover packaging in May 1994, and take a write-off.
- Define the principal objectives: number of packages to be labeled, specifications to be prepared, tasks to be performed, materials to be printed, suppliers to be prodded, etc. Remember, distribution packaging does not need to have nutrition information, but perhaps the SKU number will need to be changed.

- Formulate the major strategies, such as schedules, methodologies, number of staff members, budgets, etc.
- Decide precisely what is needed immediately to begin implementing the plans—such as, for example, data on nutrient values of foods.
- Monitor progress and performance in accordance with pre-established standards and schedules—how close are you to meeting the May 1994 schedule deadline and who must be prodded?
- Recycle the process regularly—and after May 1994, reviewing and improving the process and implementing it more in line with the strategic plan—because it is inevitable that not everything will proceed perfectly—and besides, you will have made mistakes.

This process of developing a plan, whether formal or informal, imposes a discipline on both the process and its implementation that improves chances that the functions will benefit other departments and the organization, and that the proper labels will be on all packages by the deadline date.

VII. PREPARING THE PLAN

Although the sequence described is typical, it is neither the ideal nor the only one. There is neither a perfect strategic plan not perfect means to produce a perfect plan. Just as the strategic plan is a means to achieve obtain an end, producing a good strategic plan requires planning. Budget preparation is one part of planning that is an arithmetic product of the planning process. If the food company has a designated planner, the packaging manager could abrogate his responsibility in this key area to that person, who of course probably has almost no understanding of the packaging function, the packaging process, or nutrition labeling and its implications. The packaging manager can delegate the planning to one of his/her staff members. If planning is to be performed, it should be performed well, or else the program is counterproductive—and in this context, potentially a major problem to the organization.

A plan prepared in a vacuum can be wasteful. Although in many organizations the planner and the executives are charged with making sure that the plans of the packaging department coincide and coordinate with those of the company and of other departments, the packaging manager cannot depend on others for this skillful interlocking.

The packaging manager should expect that designated planners will be looking at broader "long range" scenarios and financial issues, and thus may brush over functional plans such as for nutrition labeling that might be regarded as mundane. Thus, the packaging manager should undertake the task and responsibility of making sure that his/her nutrition labeling plans indeed coincide with those of marketing, R & D, engineering, and production managements, at the very least. He/she should also encourage the participation of his own staff in the planning processes, since the staff is to be a major tool in the execution of the plan. With these actions, the packaging manager at least alerts those who are part of the plan to what his/her ideas are. He/she will want to know how many new products, packaging changes, promotions, etc., marketing plans, over and above the current list to be labeled, to introduce during the plan cycle. He/she will want to know whether engineering is designing a new line that will have a new filler and thus require a change in a package specification, which would necessitate another panel for nutrition information. He/she will want to know if "human resources" has recommended a reduction in the packaging development department staff because the proposition that its financial benefit or need in this regulatory context is essential to the organization has not been adequately presented.

In searching for support for the plan, the packaging manager will, of course, encounter opposition and an inevitable absence of support. The process of strategic planning, by its very existence, alters relationships between departments, changes the information flow, brings new ways of decision making, and highlights differences. The process also, of course, heightens the perceived risk and fear of failure and security makes new demands upon people and in some persons, generates greater, rather than less, uncertainty than the process is designed to do—especially in this realm of nutrition labeling which may be regarded as an intense short term action.

When the packaging manager engages in dialogue with all the people involved, a greater understanding can be achieved and the threats can be alleviated, if not eliminated.

It is important to point out that, in general, excellence in planning is the basis for excellence in implementation, and the best planning takes place in organization with the best management.

VIII. THE DEVELOPMENT OF A STRATEGIC PLAN FOR IMPLEMENTING THE NUTRITION LABELING REGULATIONS

To this point, strategic planning has been defined, justified, circumscribed, qualified and planned. The process itself must now be developed.

A. Strategic Planning

Strategic planning usually begins with a situation analysis or audit. By whatever name, the exercise involves an in-depth examination of the present and recent past. How many primary packages must be labeled? Are any multiple or secondary packages involved? How many specifications must be prepared? How much must your company spend for the newly labeled packaging? How many products already have nutrition information and how many will require new labeling? What do the latest regulations say? What consumer inputs dictate label information? Which retailer suggestions or demand on the topic have been revealed by surveys or field observations? Which suppliers are available? A situation analysis is a representation of a data base.

B. Current Resources

An inventory should be made of who, what, and how much is available, both internally and externally. How many people and of what type? What kind of equipment? What kinds of consultants? Can talents, computers, equipment, production, time, etc., be borrowed from other departments, from suppliers, etc.? What resources are the trade or technical associations offering or suggesting?

C. Competitive Environment

What kinds of nutrition labeling will be used by the direct competitors or can you obtain such information? What packaging development resources do the competitors have? How much do they pay for packaging and its changes? What type of marketing development do they have underway, according to your intelligence?

An audit of the packaging used by direct competitors on an organized periodic basis, and especially during this period of package change, is invaluable.

The sporadic, reactive examination of one or two competitive packages does not provide the meaningful perspective that comprehensive study does. Regular audits should also be performed on competitive items to identify trends that will eventually have an impact.

D. Supplier Audit

There should be an updated appraisal of base material and converter supplier performance, plants, activities, and trends. Each supplier should be queried on the outlook for his material—the costs, availability, and any plans that will directly affect packaging. For example, one might ask: How will the supplier handle the flood of printing changes anticipated all scheduled for a single date?

Almost all good suppliers are pleased to present their situation and plans. It is vital for the packaging manager to be an active participant in these presentations, and to solicit them if by chance they are not offered gratuitously or if they are delivered only to the package purchasing departments.

E. Analysis

Acronyms of various types have been generated for the self-analysis process, such as SOWT (strengths, opportunities, weaknesses, threats) and WOTS (weaknesses, opportunities, threats, strengths). Examine the packaging department for its strong and weak points. Examine the nutrition labeling problems presented and projected for the threats and opportunities. Each of the key issues, whether internal department or packaging matters, should be analyzed in terms of the four key areas: strengths, opportunities, weaknesses, and threats.

F. Develop Objectives

"The Past is Prologue." A data base has been established through a situation analysis. The environment has been described. Everything has been analyzed in the harsh light of what can or cannot be done.

Now set that all aside and decide where you want to go or what you want to do. If you are fortunate, the marketing or production departments will have already provided input on this topic. As every packaging manager knows, with an infinite amount of resources, money and suppliers, these imposed objectives could be achieved in a finite time. In the perspective provided by as complete a knowledge as is available on the subjects, a meaningful mission statement must be formulated. Is the packaging department a leader, a structural design group, a technical service group, a testing agency, a specification writer, an identifier of true trends, or a vestige of past organizations? Regardless of what it is, what do you intend it to be in the long- and medium-term future?

The indispensable long-range objective of 100% nutrition labeling by May 1994 is measurable, feasible, understandable, flexible, and, above all, limited to just that—do not simultaneously add in a major design or name change. A problem defined is a problem half solved. An objective clearly formulated defines the roles of the packaging department and its associated organizational units.

IX. ALTERNATIVE STRATEGIES

Various packaging strategies have been proposed that, in one or another situation, work well: matrix, life-cycle, niche, adaptation, limitation, intuition, synergy, and serendipity or chance.

A. Serendipity

Since luck favors those who are prepared for it, serendipity is the only acceptable form of chance. Serendipity is, however, really an extension of planning that short-circuits the excruciating process of development. If it happens, great; if it does not, then the plan is still in place to be followed. But you cannot plan for serendipity. Gamble that a format will be unchanged, call in a designer and a printer—and get your order converted—right now!

B. Imitation

Imitation of competitive packaging is an acceptable strategy if it fits the marketing strategy. Imitation is, of course, following the leader, which reduces the risk and is often successful, but this time, everyone starts at the same time, and so imitation is difficult, if not impossible for the moment. Just copy what the first competitor appears to be doing. If the competitor indicates it will place a label on the left rear face in black on white in a specific size with no front face claim on one of the printed health claims, an imitative strategy would be to do the same. If competitive intelligence indicates the use of a data base, imitation suggests using an accepted data base.

C. Adaptation

Adaptation of existing or competitive packaging is a common packaging department strategy which can yield imitation and, at the same time take advantage of the competitor's food and bad points, but can he/she afford to wait? Since development is usually incremental, anyway, adaptation in its broadest sense can be a good strategy.

D. Leadership

A leadership strategy is not at all unusual. Innovation requires time, money, and people and carries with it high risk, but everyone is a leader in nutrition labeling.

Leadership implies stepping forward and, for example, emphasizing the presence of the information panel early, and taking advantage of every permitted claim. Leadership implies early conversion of all packages, pre-empting competitors into design studios and mechanical art, onto presses and onto packaging machines. Leadership links the new package design to other forms of marketing communication such as promotional brochures.

E. Subsidiary Strategies

Strategy depends on the objectives, resources and the environment, and every organization has its own matrix of these given variables. Developing a coherent strategy depends on knowing and analyzing these parameters and deciding which strategy fits best at this time of package change. Corollary to the basic strategy are subsidiary strategies and tactics. Although a stand-pat, follow-the-leader approach might be used for one product line, an innovative strategy, might be used for new products. Perhaps some products will be dropped rather than invest in converting their labeling to comply with the regulations.

No strategy in packaging can be static. All packaging strategies must be dynamic, ready to alter direction when the situation warrants, but not abruptly or reactively.

The basic strategy must be thoroughly evaluated, as the choice among the alternatives suggested above might to some degree be dictated by the marketing or production departments. Regardless, those functional departments or staff directly or indirectly affected by the selected packaging strategy should be informed so that their inputs are included. A marketing department might object to a strategy of just replacing the existing label, whereas a production department always resists the change that would be necessitated by an innovative strategy and the purchasing department wants to go to zero inventory.

Within the food packaging company, the strategy rotates about such issues as use of suppliers, other departments, technical orientation, and testing protocols for generating nutrition data. The basic objective will have considerable influence on the internal strategy. Often, however, strategy is based on the philosophy, training, or experience of the packaging manager which may be good or bad. Other packaging strategies are an integral part of the company or marketing strategy. Thus, one packaging manager might be heavily oriented toward a sound, in-depth consumer research effort prior to any field or market testing, as for example, on a face violator. Another might opt for very rapid development, drawing heavily on suppliers, with the bulk of the development taking place as rectification after the first printing. Some programs divide the activity, with the packaging department performing only the technical product-protection testing while marketing handles graphic design and traffic oversees the inventory and costs. In many companies, the division of effort is in terms of appearance, marketing, economics, purchasing, materials, packaging, etc. The problem with such a strategy is evident: no one is responsible for the total, and, in packaging, the whole is much greater than the sum of the parts.

F. Medium-Range Plan Formulation

This step in the strategic planning process might be described as functional planning or tactics. Having established the objective and strategy, it is necessary to develop directions for the various pieces that must come together to create a cohesive label program. If the packaging department has several groups, plans should be made for each of the groups for the entire time period. If other departments are involved, then their roles relative to nutrition labeling should be developed.

Each known action should be listed and scheduled, with the various predictable elements incorporated. In essence, this step means enumerating actions initiated by packaging, developmental marketing, production, packaging, etc., and incorporating all into a schedule and allocating resources on this basis.

G. Translation into Current Decisions

Besides planning, what must be done today to initiate the plan? Describe in systematic detail who is going to do what, when, where, etc. Which suppliers, consultants, contractors, staff members, etc., are expected to perform which designers, at which times? Who is going to follow up? From this scheduling, the budget can be prepared. Linkage to other programs and departments structurally and in terms of the budget must be ensured.

H. Contingency Plans

What happens if everything does not proceed according to plan as is inevitable? List all the possible events, both positive and negative, that might occur that are not already in the plan. What would be the scenario if these events were not to take place? At what point would the counteraction be triggered by an unplanned event? For example, what action would be taken if no ink were available, or if the nutrition data were not present on time. Several items have been removed precipitously. In contingency plans, only those events that have a finite probability of occurring should be included. Contingency plans are critical because they are planned responses and not reactions to unexpected events.

The steps described are neither definitive nor all-inclusive. Rather, they are intended as a structure that could be used to prepare a comprehensive strategic plan. They could be somewhat truncated or greatly expanded. These steps or their equivalent serve to impose a discipline on the process that is vital for obtaining a meaningful result.

X. THE PLANNING PROCESS

The steps may be analyzed and charted qualitatively, quantitatively, or even, in many more formalized organization, by using computer modeling techniques. Since all results are measured in some manner, quantitative methods usually have more meaning to some managers and to the executive staff. In essence, quantification means reducing the substance of the plan to mathematical data and incorporating these data into chart form. Time and financial data are the references most commonly employed, although, of course, percentages, value added, productivity, etc., can be used if these data express the concept more definitively.

Thus, a plan for package nutrition labels might incorporate such data as the number of packages to be converted, the cost per conversion, staffing requirements, the return per unit of investment expressed in the present value system, etc. By reducing the operation to relatively simplified mathematics, the sensitivities of varying inputs can be readily computed. Quantitative techniques might be traditional, involving an income statement, a budget, or cash flow analysis, or they might employ sophisticated modeling and forecasting tools.

In contemporary business, since profit, return on investment and cash flow are simultaneously driving forces and principal measures, the packaging manager or planner is often compelled to convert strategic planning into fiscal terms. In this manner, executives can readily comprehend the returns that will be derived from the investment in or cost of nutrition labeling.

Even if the packaging department is treated as a cost center, its benefit to the organization may ultimately be converted into quantitative terms. To make certain that the cost of nutrition labeling is not subverted to other alternative investments, the packaging manager is forced to portray the ultimate benefit to be derived from this activity. Estimates of the value or merit of nutrition labeling represents an area in which quantitative techniques might be readily employed.

In preparing the strategic plan, it is instructive to grasp the importance of the information used and communicated in the preparation. The basic plan should be realistic, with some stretch or optimistic objectives, because invariably, someone upstream will find a rationale to reduce the budget. Information is never perfect. The packaging manager has one information bank on a subject which may be different than the facts; he/she may request further information which will invariably be different from both his/her own, and from what the informant will perceive. Just as information is imperfect, so also is communication. If the information base can be incomplete and distorted, it will be generally to a greater extent than the packaging manager realizes. It is therefore vital that the information used in the strategic plan be as accurate and complete as possible—and, if possible, slightly inflated in the expectation that the plan that is submitted will be contracted.

A strategic plan has been likened to an instruction book. Open the book and follow the instructions and instant success.

A strategic plan is basically a road map or a flight plan. It gives a lot of direction, but it does not tell you how to fly or how to compensate for the minor deviations that are inevitable.

Thus, when a strategic plan is followed, it acts as an excellent guide but not as a rigid rule book. The manager's talent, skill, and action are indispensable for implementation.

Incorporated into this process of implementation must be the mechanisms to measure the progress and results and to control the process through feedback. The manager can never lose sight of his/her staff members and suppliers, all of whom are human and each of whom is different. The key to implementing any strategic plan is to have people who can perform and who respond to the effective decisions and actions of their manager.

The dangers in strategic planning are multifold and can never be minimized as a threat. Strategic planning for packaging is ultimately the responsibility of the manager and as such basically cannot be delegated. Many, if not most, of today's problems with nutrition labeling are the result of yesterday's paucity of effective planning.

Although the universe is ultimately very orderly, it is now often difficult to fathom the order because our span of specific knowledge and ability to comprehend in a finite time is limited. In management, we have resolved this dilemma by delegating to experts each of the individual steps and have the responsibility of interpreting the pieces into a coherent whole.

So it is with the packaging manager. A generation ago, the packaging professional, if he or she existed, could be the marketing and technical manager and handle the packaging job on alternate Tuesdays during coffee break. In those allegedly simpler times, the packaging manager called in or asked the purchasing agent to call in the supplier and was offered a graphic comprehensive of a package into which the product is thrown. Someone else worried about the machines.

Some food companies still operate in a similar manner: give the product to the packaging manager or a purchasing agent and have him or her slap on a label. Those who have no appreciation of marketing, costs and efficiencies, receive the results that this method delivers.

The professional packaging manager, inundated with regulatory information on one side, with new demands from the market on another side, and with pressures relative to economics, performance, environmentalism, compliance, etc., knows very well that his/her task is demanding and taxing. We have passed the era of simple materials and single converters. In the event that one system does not perform well, a viable alternative source exists in a totally different material system. There are alternatives to applying the new nutrition label information to a package surface. We did not invent packaging complexity as a scheme to thwart the consumer or the government. Just the reverse—complexity is a product of an expanding knowledge of what to do, how to do it better, and what each of us as individual consumers wants.

Today's business and packaging development environment clearly demands managerial talent, which in turn needs tools that enable us to not just cope, but rather to effectively use the knowledge available to us.

Strategic planning has been demonstrated to be a powerful tool in extracting the greatest amount from the vast body of information and requirements that face the packaging manager. Strategic planning is not a panacea nor does it compensate for a lack of skill, knowledge, experience, and diligence. Neither is it an academic exercise to soothe a turbulent operation. Packaging managers know that no single design can satisfy all product, regulatory, market or environmentalist requirements and that the perfect design is elusive. Similarly, strategic planning, when treated as an effective tool, given proper attention and used to think through the future consequences of today's decisions, will return rich rewards to the packaging manager who really uses it.

Every packaging manager knows the benefits of using every available alternative. Strategic planning is a managerial tool that enables the manager to take full advantage of all available resources.

Tomorrow's effective nutrition label is the result of the strategic planning begun today.

10

Designing Packages for a Healthier America and the Nutrition Labeling and Education Act

J. Roy Parcels

Dixon & Parcels Associates, Inc., New York, New York

Aaron L. Brody

Rubbright•Brody, Inc., Devon, Pennsylvania

Almost since the first food was put into packages for easier transportation to the marketplace and for the convenience of the end user, the producer has used the outside of the package to identify and inform. In the nineteenth century, the generic name of the product, the source of the product and the brand name were the most obvious and desirable pieces of information. As time passed to the late twentieth century, the information appearing on the surfaces of packages expanded to reflect enhancements made in the product, to encourage a range of uses, to increase product appeal, and to accommodate to self-service marketing.

A historical review of the milestones in U.S. food labeling legislation shows that beginning with the Federal Food and Drug Act and the Federal Meat Inspection Act in 1906, a series of acts have been promulgated by the various levels of government to require ingredients labeling and the labeling of meats and poultry (a brief history of FDA regulation relating to the Nutrient Content of Food is presented in Chapter 1 by Peter Barton Hutt.). Much of this legislation had already been encouraged by the food industry, particularly the canning industry. The lobbying efforts of the industry were constant because they felt that if the consuming public knew the ingredients of the products they were buying, they would be more confident of their purchases.

In 1938, the Federal Food, Drug, and Cosmetic Act replaced the 1906 Food and Drug Act. Among other things, it required the label of every processed, packaged food to contain the name of the food, its net weight and the name and address of the manufacturer or distributor. A list of ingredients was also

Figure 10.1 Example of "Retaining Brand Equity." An example of a product with very strong equity. Package on the left is a well known and readily identified Jolly Time Light Pop Corn. The package on the right, following the NLEA regulations, required a reduction in the height of "Light" and the inclusion of the referral and reference copy. This was done while maintaining strong resemblence to the original package.

required on most products. The law also prohibited statements in food labeling that are false or misleading. Then, in 1966, the Fair Packaging and Labeling Act required all consumer products in interstate commerce to contain accurate information to facilitate value comparisons. In 1973, the FDA issued regulations requiring nutrition labeling in a specific format on food containing one or more added nutrients or whose label or advertising included claims about the food's nutritional properties or its usefulness in the daily diet. Nutrition labeling was voluntary for almost all other foods. These regulations were an outgrowth of the 1969 White House Conference on Food, Nutrition, and Health.

Confusion often developed in the minds of the consumer when words used to describe one product were interpreted differently when used on different products. To address this and to standardize the presentation, the Nutrition Labeling and Education Act (NLEA) was passed in 1990 with regulations issued by the FDA and, separately, by the USDA in 1993 for implementation in 1994.

With each act, the manufacturer has been faced with making changes in the information on the packaging of its products. With issuance in 1993 of the regu-

Figure 10.2 Label in the background is prior to the NLEA and label in the foreground shows the manufacturer has chosen to eliminate the sell copy on the back panel and place the Nutrition Facts immediately to the right of the face panel and repeated the face panel.

lations of the NLEA, major challenges and opportunities impact the design of today's food packaging. Figures 10.1 and 10.2 show examples of package labels before and after NLEA.

I. THE PACKAGE

Incorporating the NLEA regulations into food package designs presents almost as many challenges as there are packages. To start with, there are many different forms of packaging and the definition of important panels of each that are affected by the new regulations.

Briefly, we have the folding carton—the package that has six definitive sides or faces. Then there are the labels on round, oval, and even rectangular bottles and cans as well as on pouches, bags, and wrappers that must be considered. The other panels are also important communication areas featuring product promotion, preparation and safe handling, desirable to the marketer and, of course, the nutrition facts, ingredients information, and the name and address of the manufacturer required by government regulation.

The shape, size, and constraints of the various packages present the package designer with the need to know how the new NLEA and USDA regulations interpret each panel of the package. The task is further complicated by the nature of the product and the determination of which government agency has jurisdiction over the content of the label. The Food and Drug Administration's broad focus is on nutrition, health claims, ingredients, product safety, additives, weight, volume, and source of processed foods and drugs. The United States Department of Agriculture oversees the safety inspection and labeling of meats and poultry. The Bureau of Alcohol, Tobacco & Firearms concerns itself with beer, wine, and distilled beverages which have not, to date, come under nutrition labeling regulations while seafood comes under the U.S. Department of Commerce at this time. There are areas of packaging copy, because of truth in advertising, where even the FTC has some jurisdiction. FTC has indicated its expectation to become involved in the marketing communications aspects of nutrition information flowing from the package information. (The effects of the Nutrition Labeling and Education Act on FTC Advertising Enforcement Policy is presented in Chapter 4 by former Commissioner Deborah K. Owen and Joyce E. Plyler.)

The NLEA legislation is broad in its impact and affects nutrient claims, reference and referral copy, nutrition panel formats, order of ingredients, type sizes and the relative prominence of statements and claims. Even the product names and trademarks on existing packaging now have to be reviewed for acceptability and compliance and must be an important and serious consideration for all new product packaging. Because the communication graphics and copy on every package must function as a whole, it is important to consider all the possibilities. With the six panel carton, it is easy to establish the Primary Display Panel, the right and left panels, the back panel, and the bottom and top panels. In some packages like sausage chubs, cans and irregularly-shaped jars, snack and confection wrappers, and pouches with and without side panels, the panels are less clearly defined. In earlier times, the canning industry settled on a general formula for label content by dividing the available usable space into percentages. 40% for the principal display area, 40% for serving suggestions and other promotional copy and 20% for all legal requirements. Today in the package design the breakdown of the space has to be evaluated in relation to the demands of the new labeling act.

II. THE NLEA AND THE PACKAGE

The goal of the NLEA is to take the guesswork out of good nutrition and to educate "if" the consumer chooses to use the information that will now be on 90% of food packages in the marketplace. The goal of the marketer and package designer is to create a package that will attract attention and will encourage the consumer to read the information on the package and to ultimately buy the product to enrich their lives. The package must also foster safe distribution of the product from the manufacturer to its point of use, sometimes to adequately protect unused portions for future use. By standardizing the presentation of ingredients and nutrition facts and by defining the parameters for using specified descriptors and disclaimers, the FDA believes that the public will be less confused by the information on the package and be in a better position to make informed food choices that will be beneficial to their health. However, in the process, these new regulations will impact the graphic design of all food packages.

III. PACKAGE DESIGNERS' CONTRIBUTIONS

Designers and design managers and decision makers must have an understanding of the public's eating trends; an appreciation for the history of containing, protecting, informing, identifying and selling packaged food; a clear idea of what motivates the consumer to buy; and, of course, knowledge of the regulations.

When designing the package, a clear definition of the target audience is vital. Age, income level and lifestyle of a particular audience helps the package designer to make informed decisions about what the overall design concept should be.

IV. PACKAGE DESIGN CHECKLIST

There is much to be considered when designing the package which will enter the marketplace in a timely and effective manner. For example, a minimum of 15 steps is required for the graphic design or redesign of food packages to meet food companies' objectives and be in compliance with government regulations. These are:

- Prepare a product analysis or consult a nutrition data base to arrive at proper nutrition information. Double check the ingredient copy to insure compliance with the NLEA. Validate data base information by selective chemical analysis.

- Develop copy consistent with marketing and communication goals to present and market the product, and to distinguish it from competitors in the marketplace.
- Follow the FDA and USDA regulations for exact copy requirements for nutrition claims.
- Design the package graphics exploring design and color alternatives.
- Package graphics and copy approval by the packager's various departments such as legal, marketing, sales, technical, etc.
- Conduct consumer evaluations of complete "dummy" packages to determine consumer attitude for the total design and to determine if the required information is understood.
- If required, refine the package design based on the results of the consumer evaluation.
- Prepare reproduction art for the retail package including reproduction photography and/or illustrations if they are part of the design. If claims are made, prepare the graphic design of the shipping cases and/or shipper displays in keeping with the individual package requirements.
- Government Approval. If the packaged product comes under the jurisdiction of the USDA a black and white print of the proposed package must be submitted to the USDA for "Sketch Approval." The FDA does not require their approval but will comment on the package design, if desired. For comments, fax number is (202) 260-0794, for questions, call (202) 205-5657.
- Revisions and resubmissions, if suggested.
- A critique of the color separation proofs by the package designer to assess the reproduction quality of all design and copy elements.
- Packager's final approval of copy and design elements.
- Supervise initial printing of packaging material to establish color and quality standards.
- Package the product for shipping and distribution to the marketplace.
- Review the packages in their selling environment, i.e., supermarkets, mass marketers, club stores and convenience stores to determine if the package design will require design or color refinements in the next printing of the packages.

V. TRADE DRESS AND THE PRINCIPAL DISPLAY PANEL

In redesigning a package for compliance, the designer has had to address adjustments in the presentation information, type sizes and interrelationships in relative size of statements specified by the act while attempting to preserve the integrity of the Trade Dress.

A growing trend is protection of the Trade Dress. The Trade Dress refers to the overall look of a label or package—its shape, color, size, texture and graphics. Most often the Principal Display Panel is the dominant feature in the company's Trade Dress. It is appropriate to say that Trade Dress is the package design. The United States Supreme Court ruled that there are three things companies must prove to protect their Trade Dress. The first is that the Trade Dress is non-functional to the package; that it is a design. Second, there has to be a likelihood of confusion between your package and a competitor's. Finally, your design must have achieved a "secondary meaning," that is, consumers have begun to associate your design with your company or brand.

The Trade Dress is the ultimate "salesperson." It is that part of the label most likely to be examined under customary conditions of display for retail sale. It is that part of the package that invites the consumer to stop for a second, reach out for the package and to make the acquisition decision.

VI. THE PRINCIPAL DISPLAY PANEL, THE NLEA, AND THE DESIGNER

Again, the most important panel from a marketing and consumer standpoint is the Principal Display Panel (PDP). While the NLEA emphasis in the regulations and in the media has been on the content, design and location of the Nutrition Facts Panel, there are also regulations that pertain to nutrient or health claims which dramatically affect the design of the PDP. These will have an effect on what the package designer must do with copy on all panels of the package. (Details of the NLEA Regulations are presented in Chapter 7 by Ralph Shapiro.)

VII. THE NUTRIENT CONTENT CLAIM

Whether it is cereal carton information the consumer scans at the breakfast table or a nutrient claim on a can hoping to trumpet "reduced this" or "more of that," the NLEA regulations require that all food processors, in each food category, use the same claim descriptors in the same way and within the same parameters.

Both the FDA and the USDA have specific regulations for using descriptors on labels and packages. Most of the criteria are similar, but it is important to check the appropriate regulations. Their main objective is to help guide consumers to foods that may be useful in meeting current dietary standards. They have been carefully selected to make sure the pictures that certain key food words evoke in consumers' minds are consistent with the standard. In addition,

Figure 10.3 An illustration of a package without side panels. The referral statement directs the consumer to the Nutrition Facts information, which appears immediately to the right of the Principal Display Panel.

these terms provide a basis for comparing the level of a nutrient in one food to its level in another food.

In addition, the nutrient claim must be supported by information on a Nutrition Facts Panel. A "referral statement" directing the consumer to information on this Nutrition Facts Panel must appear in the immediate proximity of the Nutrient Content Claim regardless of where this claim appears, except when it appears on the same panel as the Nutrition Facts Panel and there must be no intervening copy. Later in this chapter, there is a reference to the exact nature of the claims that may be made and the size and type requirements of both the Claim and the Referral Statement. An illustration of the use of a referral statement on a package without side panels is shown in Figure 10.3.

VIII. SIZE OF ALL NUTRIENT CLAIMS

When using the acceptable words, the nutrient claim may appear in type no larger than two times (2×) the size of the statement of identity (the generic name of the product) and shall not be unduly prominent in type style relative to the statement of identity.

It is further stated that disclaimer and quantitative statements such as "not a low sodium food," or "1/3 less fat than (referenced food)" must be at least the size of the net weight statement.

If the nutrient claim is less than twice the size of the net weight statement, the disclaimers and quantitative statements must be at least half the size of the claim but no less than 1/16 inch unless the package is very small in which case the referral statement may be in type of not less than 1/32 inch.

Only one referral statement is necessary per panel even if multiple claims are made on the panel.

If the claim is made on the panel containing the Nutrition Facts, no referral statement is necessary.

IX. THE DESCRIPTORS

The NLEA rules set forth definitions for specific *nutrient content claims* in order to achieve consistent label representation. These include definitions for "*explicit claims*," i.e., high, low, free, good source, light/lite, less/fewer and more, and "*relative claims*" that extol "more than" and "reduced." The regulations also provide qualifying and explanatory information including detailed placement and prominence requirements. Examples of the use of nutrient content claims on package labels are illustrated in Figures 10.4 through 10.8.

While most claims are straightforward, the claim of "light" or "lite" has several interpretations:

- A nutritionally altered product contains one-third fewer calories or half the fat of the reference food. If the food derives 50 percent or more of its calories from fat, the reduction must be 50 percent of the fat.
- The sodium content of a low-calorie, low-fat food has been reduced by 50 percent. The term "light in sodium" is allowed if the food has at least 50 percent less sodium than a reference food. If the food still does not meet the definition for "low sodium," the label must include the disclaimer "not a low-sodium food."

Figure 10.4 Examples of "LOW FAT," "REDUCED FAT," "99% FAT FREE," and "LESS FAT Directions." Less fat is used in connection with preparing the product. Elfin Delights Low Fat is an example of the referral statement coming after the claim. Reduced Fat Wheat Thins is an example of the claim and referral copy preceding the nutirtion claim "Reduced Fat." It is the reference statement and referral statement "Fat" in Wheat Thins. One marketer is using the claim "Less Fat" in the preparation directions.

Figure 10.5 Examples of three processed meat labels in the market prior to the NLEA regulations. These show three different uses of the Fat Free claim. Though there are many ways to express Fat Free in the meat section of the market, these three examples show distinctive ways of showing Fat Free as of May 1, 1994.

Figure 10.6 Examples of packages in the marketplace May 1, 1994, featuring Fat Free and No Fat claims.

Figure 10.7 Examples of packages in the market place May 1, 1994, featuring the attributes of No Cholesterol, No Salt Added, Low Fat, and 100% Natural.

Figure 10.8 Examples of packaging using Light and Lite as part of the trademark. Lightstyle and Lite Fruit Cup appear with a ® for registration of the name and Light as the descriptor for a drink mix on packages in the marketplace May 1, 1994.

"Light" will be allowed to describe color or texture, provided qualifying information is included. However, names that have a long history of use, such as "light brown sugar," can still be used without qualifying information.

Later in this section, there is a discussion of the exact nature of the claims that may be made and the size and type requirement of both the claim and the referral statement.

X. THE REFERRAL STATEMENT

The referral statement shall be immediately adjacent to the nutrient content claim and may have no intervening material other than, if applicable, other information in the statement of identity or any other information required with the claim, i.e., percent or fraction difference for a relative claim; total fat and/or cholesterol if required due to fat/cholesterol content for a fatty acid/cholesterol claim. The referral statement may precede the nutrient claim. An illustration of the use of a referral statement on a package without side panels is shown in Figure 10.3.

The referral statement shall be set in print or type and in a size no less than that required for the net quantity of contents statement, except where the size of the claim is less than two times the required size of the net quantity of contents statement, in which case the referral statement shall be no less than one-half the size of the claim but no smaller than one-sixteenth (1/16) inch. In a small package, the referral statement may be in type of not less than one thirty-second (1/32) inch.

The regulation also states that the referral statement "shall be in an easily legible, boldface print or type in distinct contrast to other printed or graphic matter."

When considering the very small packages, those with 12 square inches or less of space available for labeling, on which no nutrient claim is made and there is no space for the Nutrition Facts Panel, an address and telephone number that a consumer can use to obtain the required nutrition information must be provided. In all other cases, the "referral statement" must refer to the position of the Nutrition Facts Panel where the nutrition information is detailed.

XI. IMPLIED CLAIMS

Implied claims are explicitly covered by the regulations.

"Made with oat bran" and "no tropical oils" are examples of statements that may be implied nutrient content claims. Such claims are prohibited when they wrongfully imply that a food contains or does not contain a meaningful level of a nutrient. They are allowed if the food's nutrient content meets the definition for appropriate nutrient content descriptors that are implied by the claim.

For example, FDA considers statements about some types of oil as an ingredient, such as "made with canola oil" or "contains corn oil," to imply that the oil in the products is low in saturated fat. Therefore, to carry that claim, a food would have to meet the definition of "low saturated fat."

The statement "made only with vegetable oil" implies that because vegetable oil is used instead of animal fat, the oil component contributes no cholesterol and is low in saturated fat. In this case, the claim could be used only if the food meets the definition of "cholesterol free" and "low saturated fat."

The statement "contains no oil" implies that the product contains no fat and thus is fat free. Such a claim on a product that contained another source of fat, such as animal fat, would be misleading. Therefore, this statement would be allowed only if the food is truly "fat free."

Claims that imply a product contains a particular amount of fiber, such as "high in oat bran," can be made only if the food actually meets the definition for "high" fiber or "good source" of fiber, whichever is appropriate.

Statements that do not fall under the rules for nutrient content implied claims and therefore are still allowed are:

- Those that help consumers avoid certain foods because of religious beliefs or dietary practices—for example, a "milk-free" claim
- Those about non-nutritive ingredients, such as "no preservatives" or "no artificial colors"
- Those about ingredients that provide added value, such as "contains real fruit"
- Statements of identity, such as "Colombian coffee" and "100 percent corn oil"

A. Fresh

Although not mandated by the Nutrition Labeling and Education Act of 1990 as regulations for the other nutrient content claims are, the FDA has issued a regulation for the term "fresh." Under this regulation, "fresh" can be used only on a food that is raw, has never been frozen or vigorously heated, and contains no preservatives. "Fresh frozen," "frozen fresh," and "freshly frozen" can be used for foods that are quickly frozen while still fresh. Blanching (brief scalding before freezing to prevent nutrient breakdown) is allowed.

Other uses of the term "fresh," such as in "fresh milk" or "freshly baked bread," are not affected.

B. Healthy

The federal government defined the rule for the use of the term "healthy." In an unusual move the FDA and the USDA agreed on a single approach to a definition for "healthy" that had presented many problems.

The goal for the use of the word "healthy" was to define products low in fat, low in saturated fat, and restricted in sodium as a true and dependable guide for consumers.

Specifically, the qualifications for using "healthy" as a descriptor are that the product be low in fat, i.e. no more than 1 gram per serving; that cholesterol is restricted to 60 milligrams per serving and sodium to 480 milligrams per serving. In addition the product must contain at least 10% of the Daily Value of either vitamin A, vitamin C, iron, calcium, protein, or fiber.

Where meal-type products are concerned there is currently a 600 milligram limit on sodium. After November 1997 for the USDA and after January 1, 1998, for the FDA the limit on sodium is lowered to 360 milligrams for individual foods and 480 milligrams for meal-type foods.

Using "healthy" for raw meats and poultry the USDA requires meeting the "extra lean" definition or products less than 5 grams of fat, 2 grams of saturated fat and be under 95 milligrams of cholesterol for a serving of 100 grams of product. The FDA has similar requirements for seafood and game meats.

Compliance with the regulations for the proper use of "healthy" for all FDA regulated foods (everything but meat and poultry) is May 8, 1994, except for pre-existing foods where the compliance date is January 1, 1996. The USDA, which regulates meat and poultry, is allowing processors and packers until November 1995 to reformulate products and labels to meet the definition.

Lower sodium levels will be phased in requiring compliance with USDA regulations by November 1997 and with the FDA by January 1, 1998.

XII. HEALTH CLAIMS

Health claims are permitted only where claims are specifically provided for by the regulations and where no other nutrient is present in an amount that increases the risk of disease. The claim must be supported by publicly available scientific evidence and there must be significant scientific agreement among qualified scientists.

The FDA lists six Health Claims with certain associations that can be made. These are:

- *Calcium and Osteoporosis:* Linking calcium intake and osteoporosis is authorized for food containing at least 200 mg of calcium per serving.
- *Low Fat—High Fiber and Cancer:* While scientific evidence does not exist to justify linking dietary fiber to a reduced risk of cancer, the FDA allows the diets which are low in fat and high in fiber to be associated with a reduced risk of some types of cancer.
- *Low Fat—High Fiber and Heart Disease:* The FDA will not allow a health claim relating the intake of dietary soluble fiber to a decreased risk of heart disease. However, it will allow a health claim on foods low in saturated fat and cholesterol and high in fruits, vegetables and grain foods that contain fiber, particularly soluble fiber, to a reduced risk of heart disease.
- *Dietary Lipids and Cancer:* There is significant scientific evidence to support the claim that diets low in fat may reduce the risk of some cancers. In order to make a health claim such as "Development of cancer depends on many factors. A diet low in total fat may reduce the risk of some cancers," the fat percentage must be less than 30 percent of total calories.

- *Sodium and High Blood Pressure:* The FDA will permit health claims showing a relationship between the intake of dietary sodium and high blood pressure. The FDA is satisfied that there is enough evidence to suggest that diets low in sodium may help lower blood pressure in many people.
- *Dietary Saturated Fat and Cholesterol and Coronary Heart Disease (CHD):* There is sufficient evidence to support health claims that link diets high in saturated fat and cholesterol with increased risk of CHD. To make this health claim, the food would have to meet all the nutrient content requirements for "low saturated fat," "low cholesterol," and "low fat." Fish and game meats must be "extra lean."

In summary, the Principal Display Panel copy is dictated by a number of regulations which control elements which have never before had restrictive regulation. Within these regulations, however, there is ample room for the creative designer to complement the marketing goals of the food marketer. Being intimate with the regulations as they stand now and familiar with the changes that are bound to appear will make the challenge easier.

XIII. NUTRITION FACTS

One of the requirements of the new NLEA which has received a great deal of attention and publicity is the Nutrition Facts Panel. According to the regulations, *the Nutrition Facts Panel must appear on the next usable panel to the right of the Principal Display Panel (PDP).* In some cases, this will be a side panel while on those packages with very small or non-existent side panels, the information would appear on the back panel. For labels which wrap around cylindrical containers and "soft" packages the rule of "immediately adjacent to the right of the PDP" would present a challenge for the designer as far as positioning is concerned. In most cylindrical packages, this would mean that the Nutrition Facts Panel would immediately abut the PDP area and to the right. The top and the bottom of a cylindrical package are not recommended since they usually do not offer enough space and, in the case of the bottom, they are seldom looked at by the consumer.

Equally important to the positioning of the Nutrition Facts Panel are the format, content and type specifications.

The mandatory information on the Nutrition Facts Panel consists of information about the following:

- Serving size
- Servings per container

- Calories, which should be based on unrounded amounts of fat, carbohydrates and protein
- Calories from fat, expressed in grams and as a percent of Daily Values. Daily Values is a newly created term to describe the amount of a given nutrient the average person should consume on a daily reference 2,000 calorie diet.
- Saturated fat, expressed in grams and as a percent of Daily Values
- Cholesterol, expressed in grams and as a percent of Daily Values
- Sodium, expressed in grams and as a percent of Daily Values
- Total carbohydrate, expressed in grams and as a percent of Daily Values
- Dietary fiber, expressed in grams and as a percent of Daily Values
- Sugar, expressed in grams
- Protein, expressed in grams
- Vitamin A, expressed in percent of Daily Values
- Vitamin C, expressed in percent of Daily Values
- Calcium, expressed in percent of Daily Values
- Iron, expressed in percent of Daily Values

The regulations also require the declaration of all vitamins and minerals, that are otherwise optional, when they are added to the food as a nutrient supplement or when package copy bears content claims concerning their presence. While it is the responsibility of the manufacturer to provide the package designer with the actual figures for the nutrition content of each item listed, anyone responsible for package design should be aware of the nutrition facts requirement so that they may act as a check on the information when it is ultimately inserted in the final package design.

XIV. USDA REGULATIONS

When the NLEA was enacted in November, 1990, the law mandated nutrition labeling on most foods regulated by the FDA, but did not affect the labeling of meat and poultry products which fall under the jurisdiction of the USDA. In an effort to create consistency between all types of food labels, the USDA published its own nutrition labeling regulations.

One difference between the USDA's regulations and the FDA's is USDA's requirement for information about safe handling of packaged uncooked meat and poultry. For example, a package of fresh sausage will bear the the instructions as they appear in Figure 10.9.

According to the USDA regulation all products produced in federally inspected meat or poultry plants must bear nutrition labeling. There are exceptions. Among the exceptions are:

Figure 10.9 Example of Safe Handling instructions for fresh meat products.

- Single ingredient raw product provided the label for the product does not bear nutrient content claims or information
- Products intended for further processing, such as pepperoni used on pizzas produced at another establishment
- Institutional use products
- Products intended for export
- Sample products not for sale
- Custom slaughtered or prepared products
- Multi-ingredient products processed at a retail establishment, such as sausage

Even though these exemptions exist, cuts of single-ingredient raw meat and poultry products, including frozen products, may, on a voluntary basis, have nutrition information on the package. When this information is given, it may either be as "packaged" or as "consumed." In either case, all the requirements of the mandatory nutrition labeling program shall apply.

Because the USDA is concerned about uncooked meat and/or poultry and that these products are often sold unpackaged, the USDA has special rules about packaged uncooked meat and poultry. Among the rules are:

- If a nutrient content claim is displayed on a label, that label must bear nutrition labeling even if the label would otherwise be exempt from nutrition labeling
- Reference Amounts Customarily Consumed (RACCs) are used to determine if the criteria for a nutrition content claim are met
- The new regulations established specific print sizing requirements for nutrient content claims. These include: claims can be no larger than twice the size of the product names they modify; disclaimers and quantitative statements such as "not a low sodium food," must be at least the size of the net weight statement; and if the claim is less than twice the size of the net weight statement, the disclaimers and quantitative statements must be at least half the size, but no less than 1/16 inch in height. The FDA has basically the same rule

In some instances, the food processor will feel it is a marketing advantage to include information on the Nutrition Facts Panel which is not required by the regulations. Both the FDA and USDA regulations list the additional nutrient information which may be included and detail the conditions and the placement of this information. It should be noted that these voluntary declarations become mandatory when specific nutrient claims are made for them.

XV. FORMAT AND TYPE SIZES—THE NUTRITION FACTS PANEL

Format and type sizes for the information to be found in the Nutrition Facts Panel are offered in various layouts to accommodate different size packages and required nutrient declarations (see illustrations).

Since the introduction of the NLEA regulations in January 1993, significant amendments and clarifications to the Act have been implemented concerning the Nutrition Facts Panel. Among them are:

- Any legible type face may be used. This will allow for type style consistency with other copy on the package. See Figures 10.10 and 10.11.
- Reverse type can be used. In fact, there are no color restrictions on the type except that only one color may be used and words or phrases cannot be "called out" by using a second color. Type sizes, however, have been specified. See Figure 10.12.
- The background of the Nutrition Facts Panel does not have to be white. Whatever the color, the copy must be legible. This enables the Nutrition Facts Panel to become an integral part of the overall design of the package.
- The Nutrition Facts Panel can be any size as long as the type face used is the proper size and conforms to the regulations. The regulations state that

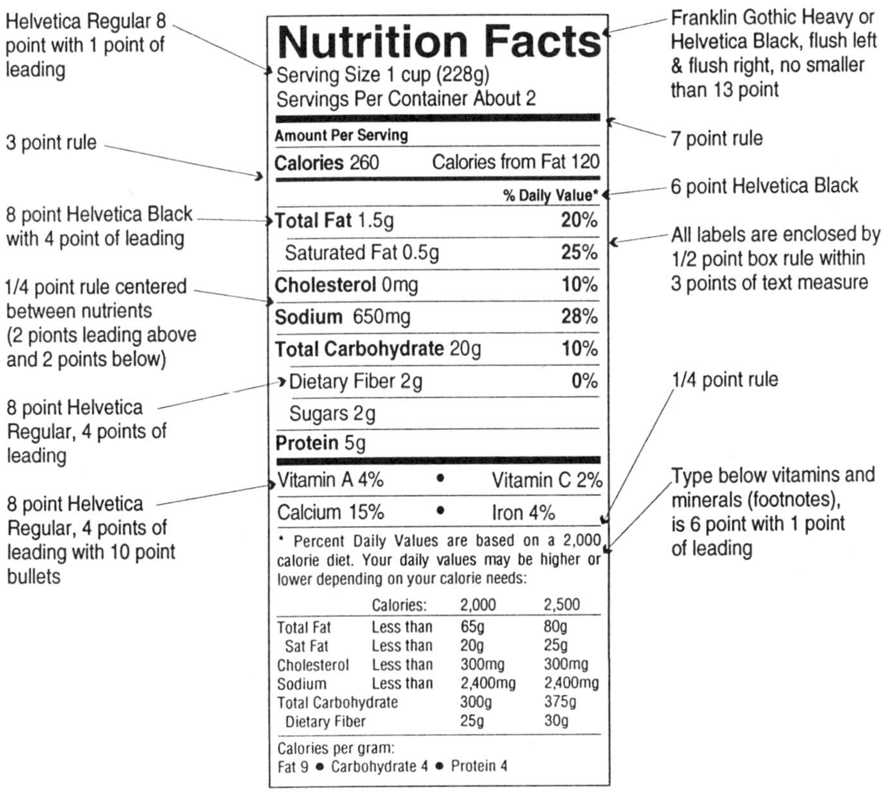

Figure 10.10 Example of typestyle for label in basic format.

Nutrition Facts

Serving Size 1 cup (228g)
Servings Per Container About 2

Amount Per Serving

Calories 260 Calories from Fat 120

	% Daily Value*
Total Fat 1.5g	**20%**
Saturated Fat 0.5g	**25%**
Cholesterol 0mg	**10%**
Sodium 650mg	**28%**
Total Carbohydrate 20g	**10%**
Dietary Fiber 2g	**0%**
Sugars 2g	
Protein 5g	

Vitamin A 4%	•	Vitamin C 0%
Calcium 10%	•	Iron 6%

* Percent Daily Values are based on a 2,000 calorie diet. Your daily values may be higher or lower depending on your calorie needs:

	Calories:	2,000	2,500
Total Fat	Less than	65g	80g
Sat Fat	Less than	20g	25g
Cholesterol	Less than	300mg	300mg
Sodium	Less than	2,400mg	2,400mg
Total Carbohydrate		300g	375g
Dietary Fiber		25g	30g

Calories per gram:
Fat 9 • Carbohydrate 4 • Protein 4

Figure 10.11 Example of serif typestyle for label in basic format.

Nutrition Facts

Serving Size 1 cup (228g)
Servings Per Container About 2

Amount Per Serving

Calories 260 Calories from Fat 120

	% Daily Value*
Total Fat 1.5g	**20%**
Saturated Fat 0.5g	**25%**
Cholesterol 0mg	**10%**
Sodium 650mg	**28%**
Total Carbohydrate 20g	**10%**
Dietary Fiber 2g	**0%**
Sugars 2g	
Protein 5g	

Vitamin A 4% • Vitamin C 2%
Calcium 15% • Iron 4%

* Percent Daily Values are based on a 2,000 calorie diet. Your daily values may be higher or lower depending on your calorie needs:

	Calories:	2,000	2,500
Total Fat	Less than	65g	80g
Sat Fat	Less than	20g	25g
Cholesterol	Less than	300mg	300mg
Sodium	Less than	2,400mg	2,400mg
Total Carbohydrate		300g	375g
Dietary Fiber		25g	30g

Calories per gram:
Fat 9 • Carbohydrate 4 • Protein 4

Figure 10.12 Example of reverse typestyle for label in basic format.

Tabular Full Size

Nutrition Facts
Serving Size 2 slices (56g)
Servings Per Container 10

Calories 140
Calories from Fat 10

Amount / Serving	% Daily Value*	Amount / Serving	% Daily Value*
Total Fat 1.5g	2%	**Total Carbohydrate** 26g	9%
Saturated Fat 0g	0%	Dietary Fiber 2g	8%
Cholesterol 0mg	0%	Sugars 1g	
Sodium 280mg	12%	**Protein** 4g	

Vitamin A 0% • Vitamin C 0% • Calcium 6% • Iron 6%
Thiamin 15% • Riboflavin 8% • Niacin 10%

* Percent Daily Values are based on a 2,000 calorie diet. Your daily values may be higher or lower depending on your calorie needs:

	Calories:	2,000	2,500
Total Fat	Less than	65g	80g
Sat Fat	Less than	20g	25g
Cholesterol	Less than	300mg	300mg
Sodium	Less than	2,400mg	2,400mg
Total Carbohydrate		300g	375g
Dietary Fiber		25g	30g

Calories per gram:
Fat 9 • Carbohydrate 4 • Protein 4

Tabular

Nutrition Facts
Ser.Size 1/3 cup (56g)
Servings about 3
Calories 80
Fat Cal. 10
*Percent Daily Values (DV) are based on a 2,000 calorie diet.

Amount/serving	% DV*	Amount/serving	% DV*
Total Fat 1g	**2%**	**Total Carb.** 0g	**0%**
Sat.Fat 0g	**0%**	Fiber 0g	**0%**
Cholest. 10mg	**3%**	Sugars 0g	
Sodium 200mg	**8%**	**Protein** 17g	

Vitamin A 0% • Vitamin C 0% • Calcium 0% • Iron 6%

Simplified Tabular (Pickle Slices)

Nutrition Facts
Serving Size 1oz.
(28g/about 12 slices)
Servings Per Container 16
Calories 20
*Percent Daily Values are based on a 2,000 calorie diet.

Amount / Serving	% Daily Value*
Total Fat 0g	0%
Sodium 190mg	8%
Total Carbohydrate 5g	2%
Sugars 5g	
Protein 0g	

Figure 10.13 Example of label in tabular display.

Nutrition Facts
Serving Size 1 cup (245g)
Servings Per Container 2

Amount Per Serving

Calories 55	Calories from Fat 20
	% Daily Value*
Total Fat 1g	**2%**
Sodium 800mg	**33%**
Total Carbohydrate 31g	**11%**
Dietary Fiber 4g	**16%**
Sugars 0 g	
Protein 2g	

Vitamin A 20% • Vitamin C 4% • Iron 2%

Not a significant soutce of saturated fat,cholestrol and calcium.

* Percent Daily Values are based on a 2,000 calorie diet. Your daily values may be higher or lower depending on your calorie needs:

	Calories:	2,000	2,500
Total Fat	Less than	65g	80g
Sat Fat	Less than	20g	25g
Cholesterol	Less than	300mg	300mg
Sodium	Less than	2,400mg	2,400mg
Total Carbohydrate		300g	375g
Dietary Fiber		25g	30g

Calories per gram:
Fat 9 • Carbohydrate 4 • Protein 4

Figure 10.14 Example of label in shortened format display.

the words "Nutrition Facts" must fit the width of the label flush left and flush right. The panel may be oriented perpendicular to the base of the package.

- The tabular display (horizontal) may be used on any package with 40 or less square inches of available label space and on larger packages that do not have sufficient continuous vertical space (over 3") to accommodate the vertical format. An unusual package shape may also qualify for the tabular display (i.e., a 2 lb. pork sausage chub wrapper). See Figure 10.13.
- The linear display may only be used for packages under 40 square inches when the tabular display does not fit in the vertical position along with other required information on the label. See Figure 10.14.
- To determine total surface area, unusable areas of the packaging may be excluded. However, packages that provide information on these areas should include these areas when calculating available label space.

XVI. REGULAR SIZE PACKAGES

Those packages with more than 40 square inches of space available for labeling are expected to use the primary full format indicated in Figures 10.10 and 10.11 (full panel illustration with dimensions and type sizes)

{Note 1: The 13 point type specification for Nutrition Facts is optional. The only requirement is that the type size be larger than all other type of the panel. However, it is to be set full width of the label.}

{Note 2: The thickness of the rules separating the groupings (i.e. 7 point and 3 point) is optional. A minimum size is not mentioned in the regulations.}

XVII. SMALL PACKAGES

For those packages that have less than 40 square inches of surface available for labeling and where the height of the standard vertical format cannot be accommodated, there are two options.

1. The first is identified as the divided Nutrition Facts Panel. This layout breaks up the information into two adjacent columns. An example of this format is shown in Figure 10.15.
2. The second is the Tabular Display layout. See Figure 10.13.
3. The third option is described as a Linear Display where the nutrition information is presented as a series of continuous lines. See Figure 10.14.
   {Note: The Linear Display can only be used if the tabular display cannot be accommodated because of the package's small size or shape.}

It should be noted that when using the Linear Display, the bolding requirements of the other formats is optional and, to further save space, there is the option of using the following abbreviations for elements of the Nutrition Facts Panel (Table 10.1). These abbreviations can also be used in the other layouts.

XVIII. SIMPLIFIED FORMAT

When the product contains insignificant amounts of seven or more of the required nutrients (listed under Nutrition Facts in this chapter), the regulations allow the use of a simplified format for the Nutrition Facts Panel. (Figures 10.16 and 10.17).

An "insignificant amount" has specified parameters which should be consulted before using this form.

Nutrition Facts

Serving Size 1 cup (228g)

Servings Per Container About 2

Amount Per Serving	
Calories 260	Calories from Fat 120
	% Daily Value*
Total Fat 1.5g	**20%**
Saturated Fat 0.5g	**25%**
Cholesterol 0mg	**10%**
Sodium 650mg	**28%**
Total Carbohydrate 20g	**10%**
Dietary Fiber 2g	**0%**
Sugars 2g	
Protein 5g	

Vitamin A 4%	•	Vitamin C 2%
Calcium 15%	•	Iron 4%

* Percent Daily Values arc based on a 2,000 calorie diet. Your daily values may be higher or lower depending on your calorie needs:

	Calories:	2,000	2,500
Total Fat	Less than	65g	80g
Sat Fat	Less than	20g	25g
Cholesterol	Less than	300mg	300mg
Sodium	Less than	2,400mg	2,400mg
Total Carbohydrate		300g	375g
Dietary Fiber		25g	30g

Calories per gram:
Fat 9 • Carbohydrate 4 • Protein 4

Figure 10.15 Example of label in divided format.

Table 10.1 Allowable Abbreviations

Serving size	=	Serv. size
Servings per container	=	Servings
Calories from saturated fat	=	Sat. fat cal.
Saturated fat	=	Sat. fat
Monounsaturated fat	=	Monounsat. fat
Polyunsaturated fat	=	Polyunsat. fat
Cholesterol	=	Cholest.
Total carbohydrate	=	Total carb.
Dietary fiber	=	Fiber
Soluble fiber	=	Sol. fiber
Sugar alcohol	=	Sugar alc.
Other carbohydrates	=	Other carb.

Nutrition Facts	
Serving Size 1 Tbsp (14g)	
Servings Per Container About 2	
Amount Per Serving	
Calories 260	Calories from Fat 120
	% Daily Value*
Total Fat 1.5g	**22%**
Saturated Fat 2g	**10%**
Polyunsaturated Fat 4g	
Monounsaturated Fat 8g	
Sodium 0mg	**10%**
Total Carbohydrate 20g	**0%**
Protein 0g	
Not a significant source of cholesterol, dietary fiber, sugars, vitamin A, vitamin C, calcium, and iron.	
* Percent Daily Values are based on a 2,000 calorie diet.	

Figure 10.16 Example of label in simplified format for a product that contains fat.

Nutrition Facts	
Serving Size 1 cup (228g)	
Servings Per Container About 2	
Amount Per Serving	**% Daily Value***
Calories 145	0%
	1%
Total Fat 0g	12%
Sodium 20mg	
Total Carbohydrate 36g	0%
Sugars 36g	
Protein 5g	
* Percent Daily Values are based on a 2,000 calorie diet.	

Figure 10.17 Example of label in simplified format for a soft drink.

XIX. INGREDIENTS AND SOURCE ADDRESS

The regulations insist that the ingredients and the name and address of the manufacturer, distributor, marketer or other identifiable source must be placed immediately below or adjacent to the Nutrition Facts label and be presented in legible type no smaller than 1/16 inch when measured against the height of a lower case "o".

XX. INGREDIENTS LABELING

- Listing of common or unusual names in descending order by weight
- Declaration and source of protein hydrolysates
- Identification of caseinate as milk derivative for non-dairy products
- Uniform format for voluntary percentage ingredient products listing
- Permits certain exemptions for fruit and vegetable preservative coatings
- Permits inclusion of food source of sweeteners

Vegetable and fruit juice beverages must describe the percent of each juice on the information panel.

XXI. FEDERAL PREEMPTION

State and local governments must adhere to FDA's NLEA rules. National Uniformity becomes effective upon enactment.

XXII. NLEA'S IMPACT ON TRADEMARKS

Does your trademark make a nutrition claim or contain descriptors for which the NLEA requires referral statements?

The regulations spell out eleven core terms that are considered descriptors requiring referral statements, and it is possible that many trademarks, present and considered, will have these descriptors as an integral part of the mark.

The eleven terms are: free, low, lean, extra lean, high, good-source, reduced, less, light, fewer and more.

Consider for a moment a hypothetical trademark for a line of cookies that last year would not have been questioned. Reduced Fat Cookies. To conform to the new regulations, there would have to be additions.

To recap, the federal government has enacted the most significant changes in food labeling since the passage of the Federal Food, Drug and Cosmetic Act of 1938. These sweeping changes were primarily designed to enable the consumer to make intelligent food choices in the marketplace. The ultimate goal is to provide the avenue for creating a healthier America, one where fewer medical problems develop from poor diets. Good nutrition is being publicized in many venues by the government. This new labeling law will complement education programs being implemented in the schools where nutritionists teach, where the Food Pyramid Chart is being distributed . . . where the leaders of tomorrow are being formed.

The NLEA has certainly helped manufacturers and food processors to tell the story of their nutritious foods and by standardizing the nomenclature and presentation of nutrition information, the consumer will become the "educated consumer" able to choose and serve nutritious foods in the home. With standardized language for claims and disclaimers, the manufacturer has the opportunity to show benefits when benefits truly exist.

This has resulted in a flow of new and reformulated food products into the marketplace. Many national, regional, and store brand packages are featuring nutrition information as a means of attracting new consumers to their respective brands. We believe this is only the beginning of a trend that will continue for the next several years that will contribute to FDA's goal—to help create a healthier America.

11

NLEA: Linking Education to Regulation

Naomi Kulakow

Food and Drug Administration, Washington, D.C.

Although the Nutrition Labeling and Education Act (NLEA) of 1990 mandates consumer education for all Americans, it will take more than legislation to convert the law into reality. The following highlights some of the compelling reasons for Congressional action, describes the government's multi-year educational campaign, and offers general guidance for teaching consumers how to use the new label.

I. WHY A NEW FOOD LABEL?

The Nutrition Labeling and Education Act, on which the new labeling regulations are based, has been justifiably called one of the most far-reaching pieces of food labeling legislation ever adopted. A number of salient reasons prompted Congress to pass this legislation.

Since the end of World War II, and in particular during the last two decades, nutrition science has made enormous advances. Dietary modifications, such as increasing the consumption of some foods (fruit, vegetables, and grains) and decreasing the consumption of others (foods high in fat, cholesterol and sodium), were found to reduce the risk factors for serious illnesses like coronary heart disease, stroke, some cancers, and osteoporosis. This vital new information

---

This chapter was written in December, 1993.

linking diet and health needed to be reflected in the food label if consumers were to act on the dietary advice experts were offering. With a few exceptions, however, nutrition labels had remained basically unchanged since the 1970s.

Another reason involved the confusion and abundance of unauthorized and exaggerated content and health claims found on food labels in the last decade. Congress sought to dispel this confusion and protect consumers from deception by giving FDA the authority to require scientifically rigorous standards that the agency could enforce. Although it will take time, an essential goal of the Food Labeling Education Campaign is to instill public trust and confidence in the label, especially regarding nutrient content and health claims.

A third reason for the new food label is also connected to the marketplace. By providing a level playing field, Congress has given manufacturers a new incentive to produce healthier food products and to compete by the quality as well as the price of their goods. The proliferation of reduced and low fat, low sodium products already showing up on retail shelves has resulted in a multi-billion dollar business, thus attesting to the foresight of this reasoning.

Finally, consumers told us and research verified that a new kind of food label was needed to motivate people to read, understand, and use the label to shop for more healthful foods and plan more nutritious meals. They told us that labels needed to be on all food products in a format that was easy to read and understand. And information on more nutrients (e.g., fiber) and how they relate to the total diet was also needed.

The new food label answers many of these needs. Now consumers can use the label to improve their diets because they will find the information they need on all packaged foods, even high fat, high calorie products; they can also believe the claims they see, like "lite" and "low fat," because they are based on standard references defined by the government. And thanks to the Nutrition Facts panel, the information on the new label is not only easier to read and understand, but also easier to apply. By consulting the percent Daily Value, consumers have immediate access to information that helps them evaluate their food choices and determine quickly and accurately which nutrients or food components they should increase or decrease in their diet. Promoting these benefits to consumers represents the basis for the government's Food Labeling Education Campaign. Hopefully, Americans will come to consider the food label a familiar, indispensable tool to improving their diets and their health.

II. THE TRANSITION PERIOD

The transition period between old and new labeling presents a challenge when educating the public about the new label. Although food labeling regulations went into effect in August 1994, new food labels did not appear on all packages at that time because foods labeled before the required date could still be held in stock and sold. During the transition period, the government is helping educators understand and accurately interpret the new regulations, and, as new labels appear on food products, it is also emphasizing label reading skills, including how to distinguish new from old labels by looking for the words "Nutrition Facts" on the back or side of product panels.

III. THE FOOD LABELING EDUCATION CAMPAIGN

Undertaking the massive revision of the food labeling regulations reflects only part of the responsibility Congress has placed on the Government. NLEA also calls for activities that educate consumers about the availability of nutrition information on the food label and the importance of using that information to maintain healthful dietary practices. To carry out this mandate, the Federal government, along with other public and private-sector groups, has initiated a major education campaign to help consumers use the new label to plan a healthy diet. This multi-year initiative, called "The New Food Label—Check It Out!", is jointly headed by the Department of Health and Human Services' (DHHS) Food and Drug Administration (FDA) and the Department of Agriculture's Food Safety and Inspection Service (FSIS).

A. The National Exchange on Food Labeling Education

A key goal of the education campaign is to reach all American consumers, yet no one organization or government agency can realistically do this alone. Thus, to maximize resources and achieve the greatest impact, the Government is building extensive partnerships and sharing information by holding national conferences, joining diverse nutrition education networks, and encouraging joint projects. FDA and FSIS call this effort the National Exchange on Food Labeling Education (NEFLE). By means of NEFLE, various public and private sector organizations (including other government agencies, the food industry, trade associations, consumer groups, health professionals, and educators) are encouraged to pool their ideas and, in some cases, their funds, to ensure that all consumers are educated about the new food label. NEFLE holds periodic national conferences to help participants learn effective ways of reaching diverse audiences and build what they learn into materials and programs. These meetings also seek to identify educational gaps, communicate research findings, and discuss new programs.

Another important goal of NEFLE is to deliver the new food labeling facts to special populations such as older Americans, children, and people on restricted diets or with limited reading skills. Thanks to an extensive network of government, community-based and professional organizations, many activities are underway to develop and evaluate educational information designed for a multicultural population.

B. Education Information Center

FDA and FSIS have jointly established a special Food Labeling Education Information Center, located in the Food and Nutrition Center at the National Agricultural Library, 10301 Baltimore Bouleard, Room 304, Beltsville, Maryland 20705-2351 (Tel.: 301-504-5755). The Center, created in 1992, encourages groups to exchange information about their food labeling education activities and aids them in initiating and finding partners for label education programs.

The cornerstone of the Information Center is a computer database of activities, materials, research, and publications specifically relevant to food labeling education. The staff collects information from diverse sources, ranging from books, brochures, and audiovisuals to exhibits, computer materials, research reports, and bibliographies. Listings in the database, which change regularly, provide detailed descriptions of each project, demographics of the target audience, and ordering information. Center staff can search the educational and research activities for projects that meet requestors' specific needs. Listings include:

- reproducible and non-reproducible print materials (books, fact sheets, articles, posters, brochures, newsletters)
- audiovisuals (slide sets, transparencies, tapes)
- children's materials (games, comic books, interactive video)
- media kit pieces (ad slicks, press releases, scripts)
- programs and program materials (exhibits, conferences, calendars, health fairs)
- computer materials (CD-ROMs, databases, software)
- research results (reports, studies, bibliographies)

It is possible to view the most current version of the complete database if one has a computer, modem, and communications software. An updated listing from the database, as well as many other labeling publications, is available from the National Agricultural Library's electronic bulletin board, the Agricultural Library Forum. Help and information on electronic access is available from the Labeling Education Information Center (301-504-5719).

The Center also helps organizations and educators track projects and identify gaps. To keep groups current on all these activities, new research findings, and materials developed, it publishes periodic updates and sends them to organizations on its mailing list.

C. Research

Another key component of the campaign is research which is used to guide the development of educational materials and determine that they are properly targeted, carry clear and understandable messages, and are widely disseminated. The labeling education campaign depends heavily on research to clarify the key variables associated with desirable behavior and to correct misunderstandings. Some research findings have been particularly significant, for example, we know that people may change their behavior in ways consistent with a healthier diet without necessarily improving the overall healthfulness of their diet. Consumers may, for instance, decrease their consumption of meat and eggs, but also increase their intake of salad oils, cheese, and ice cream, the net effect being no change in the percentage of calories from fat in their diet.

Currently FSIS, FDA, and the Office of the Assistant Secretary for Planning and Evaluation, DHHS, are collaborating on a study of consumers' use of food labels. A nationally representative sample of adults will be contacted by phone to identify the reasons why they use food labels; the parts of the label they read; whether they believe what they read on the label; their sources for health and diet information; and their self-reported health condition and health behaviors. The results of this study will provide a baseline for government officials to measure progress in achieving the education goals.

In addition, FDA and USDA's Human Nutrition Information Service (HNIS) is examining data from HNIS' Diet and Health Knowledge Survey (DHKS) and the Continuing Survey of Food Intakes by Individuals (CSFII) to learn how food knowledge and awareness of the Dietary Guidelines affect consumers' use of labels and their eating habits. Government agencies also review and evaluate research by other organizations, share research tools with them, and disseminate relevant research findings through the Food Labeling Education Information Center and national NEFLE conferences.

D. Educational Materials

Recognizing the need for accurate, attractive basic information that is easy to use and obtain, FDA and USDA have prepared numerous print and video materials designed for consumers and educators. Because many of these products are free, and not copyrighted, anyone can use or replicate them for their own educational purposes, thus saving interested parties both time and money and ensuring that accurate information and consistent messages are being disseminated (see Table 11.1 for the list of government publications and visuals and how to order them.)

Among the Government materials offered are pamphlets in English and Spanish (over 3 million were distributed in the first six months of the campaign); brochures in large print and materials designed for those with limited reading

Table 11.1 How to Order Government Materials

FDA/USDA Food Label Education Campaign Government Publications and Audio-visuals

Materials listed are now available (or will be soon) from the source listed at the top of each section and are FREE unless otherwise stated.

For single copies: Food and Drug Administration, Office of Public Affairs, 5600 Fishers Lane, HFE-88, Rockville, MD, 20857, 301-443-3220.

For multiple copies, slides, or to borrow videos/negatives: Food and Drug Administration, Office of Public Affairs, 5600 Fishers Lane, HFI-40, Rockville, MD, 20857, 301-443-2071.

FDA Consumer May 1993 Focus on Food Labeling for consumers and "multipliers." One free copy while supplies last. Then, $5 from Government Printing Office.

For multiple-copy needs, color negatives and black-and-white reproducibles also available.

FDA/FSIS Backgrounder "Read the Label, Set a Healthy Table: Public Education Campaign on the New Food Label"—gives an overview of campaign goals and activities.

1. Read the Label, Set a Healthy Table—brochures for consumers. Basic brochure opens into a poster of the Nutrition Facts panel with explanatory text. 2. How The New Food Label Can Help You Plan a Healthy Diet; DHHS Publication No. (FDA) 94–2273. Low-literacy brochure, written at the fifth-grade reading level, gives brief, simple introduction to the new food label. Large-print brochure offers basic information in a convenient format for the elderly and visually-impaired. 3. Using the New Food Label to Choose Healthier Foods; DHHS Publication No. (FDA) 94–2276. Color negatives and black-and-white reproducibles also available.

The New Food Label at a Glance—8 1/2" x 11" black-and-white reproducibles of sample front label and sample Nutrition Facts panel. Useful for publications.

Food Label Education Campaign logo—black-and-white reproducibles of the campaign logo and slogan, "Read the Label, Set a Healthy Table." Color reproductions may be produced as needed.

Ready-to-use articles—A camera-ready, two-part series on the new food label.

FDA video: Food Labels—6 1/2 minute, news-feature style introduction to the new food label. Borrow from National Agricultural Library (see other side), or Limited distribution available through FDA at number above.

How to Read the New Food Label—A brochure for consumers produced with the American Heart Association, FDA 93–2260.

USDA Home & Garden Bulletin Number 266.

The Food Label, The Pyramid, and You. Brochure for consumers.

USDA, Food Safety and Inspection Service Public Awareness Office Room 1180-S 14th and Independence Ave., SW, Washington, DC 20250 202-690-0351

Table 11.1 (Continued)

Food News for Consumers Spring/Summer 1993 The New Food Label and You, for consumers and "multipliers." Black-and-white reprints available.

FSIS Backgrounder. "Nutrition Labeling of Meat and Poultry Products"—summarizes regulations.

Nutrition Labeling: A Comparison of FDA and FSIS Nutrition Labeling Requirements

FDA/FSIS Backgrounder "Read the Label, Set a Healthy Table: Public Education Campaign on the New Food Label"—overview of campaign goals and activities.

FDA/USDA Food Labeling Education Information Center National Agricultural Library 10301 Baltimore Blvd., Room 304 Beltsville, MD 20705-2351 301-504-5719 /Fax 301-504-6409 /Internet gmcneal@nalusda.gov

Information Center Fact Sheet/Database form—Explains services; collects information for database.

NEFLE Update, Spring/Summer 1993—reports resources, research and activities of the Exchange.

The Communication Process: Research, Materials, Audience—summary of September 16, 1992 NEFLE conference.

FSIS Backgrounder "Nutrition Labeling of Meat and Poultry Products"—summarizes regulations.

FDA Backgrounder "The New Food Label"—summarizes regulations.

Nutrition Labeling: A Comparison of FDA and FSIS Nutrition Labeling Requirements

FDA/FSIS Backgrounder "Read the Label, Set a Healthy Table: Public Education Campaign on the New Food Label"—An overview of campaign goals and activities.

The New Food Label Summaries—FDA 30-page set. Technically-oriented summaries of each of the final rules and proposals.

FDA Summary of Consumer Research on Health and Diet Attitudes and Knowledge and Use of Food Labels—29 pages. Includes bibliography.

FDA speech—ready-to-use talking points on the new food label, for use with non-technical audiences.

Video of FDA/FSIS/USDA Extension Service Teleconference—two-hour videotape of the April 1, 1993 teleconference, Understanding the New Food Label from the Consumer's Perspective. Intended for educators. Available through interlibrary loan: Videocassette no. 1565.

FDA slide set: Nutrition Labeling—The technical, 90-slide set (with script) is most useful for educating professionals/industry. Available through inter-library loan: Slide no. 412.

FDA video: Food Labels—6 1/2 minute, news-feature style introduction to the new food label. Available through interlibrary loan: Videocassette no. 1491.

Guidelines/Messages—Suggested education activities and key messages for consumers, from FDA.

skills; a general pamphlet that opens into a poster; special issues of *FDA Consumer*; and a consumer brochure, explaining how to use the new label to meet the Dietary Guidelines for Americans; and videotapes, including a 6-minute tape that summarizes the new label. Exhibits, educational curricula for high schools, and a feature-length educational video have also been developed. In May, 1994, FDA launched a major muliti-media campaign to alert, excite, and inform the public about the new label and how it can help them make healthier food choices in and out of the home. It included TV, radio, and print public service announcements; placement of stories, press conferences and other events, and a media kit with video materials.

In addition to the considerable efforts of FDA and USDA to develop needed information, many public and private-sector groups have joined with the Government or have independently prepared materials, such as label-learning guides for professionals, education curricula, exhibits, newsletters, brochures for different target audiences, media kits, books, and video productions. Information about these projects is also available from the Food Labeling Education Information Center.

E. The Educational Challenges: Four Key Messages

One of the biggest challenges facing the Campaign is how to help educators motivate consumers to use the food label to make more nutritious food selections that fit into their total daily diet. Motivation represents a particularly difficult challenge to nutrition educators because the reasons why people choose or avoid the foods they eat often has little to do with good nutrition or even knowing what advice to follow; it depends, instead, on such things as taste, cost, accessibility to specific foods, religious beliefs and entertainment value. Consumers can also be discouraged from reading labels if they equate label reading with the need to have a sophisticated understanding of nutrition or with the need to keep track of cumulative amounts of nutrients consumed per day.

Given this fact, it is especially important that when teaching individuals how to use the new label, educators in both the public and private sector avoid overwhelming consumers with complex dietary guidance. Consumers must believe that understanding and accurately using the new label does not require an in-depth knowledge of nutrition, metric measures, percentages or the need to perform complicated mathematical conversions. With just a little guidance many people should be able to consult the label and make quick, healthier food selections to meet their personal dietary needs because the new label gives them an easy-to-use tool to determine the nutritional content of the foods they eat. It provides them with a relatively simple means for converting the information on labels into healthier dietary behavior. To convey these and other points, the public education campaign has developed information, including four key messages that underscore trust, usability, comprehension, and dietary guidance.

The first message deals with the trust factor.

1. You can believe the claims on the package.

Some consumers who may want to tailor their diets but have little time or interest examining the label in any depth may only refer to the claims on a food product and never look at any other part of the package for additional nutrition information. These individuals, as well as their diet-conscious counterparts, can still get a quick, reliable indication of the product's characteristics by paying attention to the claims on the principal display panel of the product. They can rely on the credibility of claims that say a food is "low fat" or which link a particular nutrient to a specific disease because government regulations now define and regulate the terms and claims. Claims such as "low," moreover, mean the same for all food groups, making them especially useful for fast comparisons between items.

The second message is equally simple to convey.

2. You can more easily compare products because serving sizes will be more comparable for similar food products.

Now, for the first time, the information on the label will reflect an average serving in amounts people customarily consume. Because the serving size for each product is defined in the regulations, product comparisons are easier to make and are more meaningful than they were previously. Someone looking at snacks, for instance, can compare not just one package of potato chips with another, but can also compare a serving of chips with a different but comparable snack like pretzels (see Figure 11.1 for sample label of potato chips and pretzels). This person can also compare the nutrients in a serving, for example, if fat is a problem, the consumer will see immediately that pretzels (like those in this example) are a better choice because they are fat free; however, if sodium is a concern, one can see at once that there is twice as much (sodium) in a serving of pretzels than in a serving of potato chips. Although the new serving size information will be useful to consumers in many ways, it is important that consumers know that it represents amounts people customarily eat, not what is recommended by experts, such as the servings found on the Food Guide Pyramid.

From an educator's perspective the percent Daily Value, which is the focus of the third message, represents one of the most important and useful innovations on the new label.

Nutrition Facts

Serving Size 1 oz (28g/about 10 pieces)
Servings Per Container 2

Amount Per Serving

Calories 150 Calories from Fat 90

	% **Daily Value***
Total Fat 10g	**15**%
Saturated Fat 1g	**5**%
Cholesterol 0mg	**0**%
Sodium 280mg	**12**%
Total Carbohydrate 15g	**5**%
Dietary Fiber 0g	**0**%
Sugars 0g	
Protein 2g	

Vitamin A	0%	•	Vitamin C	10%
Calcium	0%	•	Iron	2%

*Percent Daily Values are based on a 2,000 calorie diet.

(a)

Nutrition Facts

Serving Size 2 pieces (31g)
Servings Per Container 4

Amount Per Serving

Calories 100 Calories from Fat 0

	% **Daily Value***
Total Fat 0g	**0**%
Saturated Fat 0g	**0**%
Cholesterol 0mg	**0**%
Sodium 590mg	**25**%
Total Carbohydrate 23g	**8**%
Dietary Fiber 1g	**4**%
Sugars 3g	
Protein 3g	

Vitamin A	0%	•	Vitamin C	0%
Calcium	0%	•	Iron	2%

*Percent Daily Values are based on a 2,000 calorie diet.

(b)

Figure 11.1 Potato Chips (a) and Pretzels (b).

3. *By using the percent Daily Value, you can quickly determine if a product is high or low in a nutrient.*

A glance at this column allows consumers to quickly and easily compare the nutrients from one product to another and at the same time see if that amount is a lot (a high percentage) or a little (a low percentage) of what they should eat in a day (exemplified in Figure 11.1). The goal is for consumers to choose foods that together give them close to 100% of each nutrient for a day, or average about 100% over a few days. The percent Daily Value helps consumers to learn simple ways to follow dietary advice so that they can quickly and accurately make nutritious dietary trade-offs that relate to the whole day's worth of food consumed. Once this is understood, it can serve as a meaningful motivator as well as a powerful learning tool that may result in consumers being more willing and better able to alter their eating habits in healthful ways.

For many people, using the example of "budgeting" their daily intake of food represents an easy-to-understand way to communicate the critical, but for some, complex, concept of Daily Values. Consider, for example, the percent Daily Value as analogous to a travel budget where one has up to $100 a day to spend on food and incidental expenses. Most people would recognize that by budgeting their money, they do not have to deprive themselves of the pleasure of eating at an expensive restaurant that might cost as much as $60 or $70 a meal if they control how much they spend the remainder of that day or over the next few days. This concept is compelling when one is concerned about motivation, for it means consumers do not have to equate healthy eating with totally renouncing favorite foods, like ice cream or cake, which are high in fat, calories, or other nutrients one is told to limit or avoid. With the exception of those people who must closely adhere to a medically-prescribed diet, alert label readers can "have their cake and eat it too"; they can still maintain a generally healthy diet if they "budget" (i.e., control) the portion size of the favorite food and balance (reduce or eliminate) the amount of fat or other nutrients they eat the rest of that day or the next.

While no one expects individuals, even diet-conscious consumers, to monitor or track the exact day's worth of nutrients they wish to decrease or increase, one can nonetheless urge them to use the percent Daily Value as a simple and quick reference tool for determining which foods to eat more or less of on a daily basis.

Depending on one's needs, then, the percent Daily Value can serve as an interpreter or calculator, providing a tool that significantly simplifies the label learning process by tying knowledge to behavior:

- It evaluates nutrient content and ingredient claims so that consumers can easily determine if a claim of "reduced calories" is higher or lower than "lite" or "low calorie" without a person having to learn technical definitions.

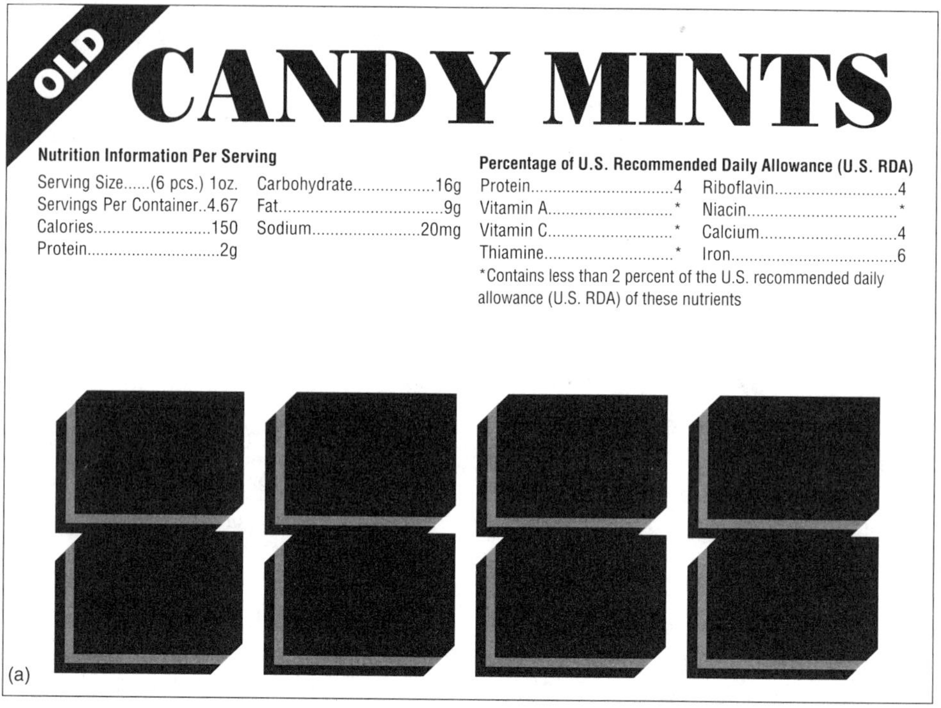

Figure 11.2 Candy Mints label: old (a) and new (b).

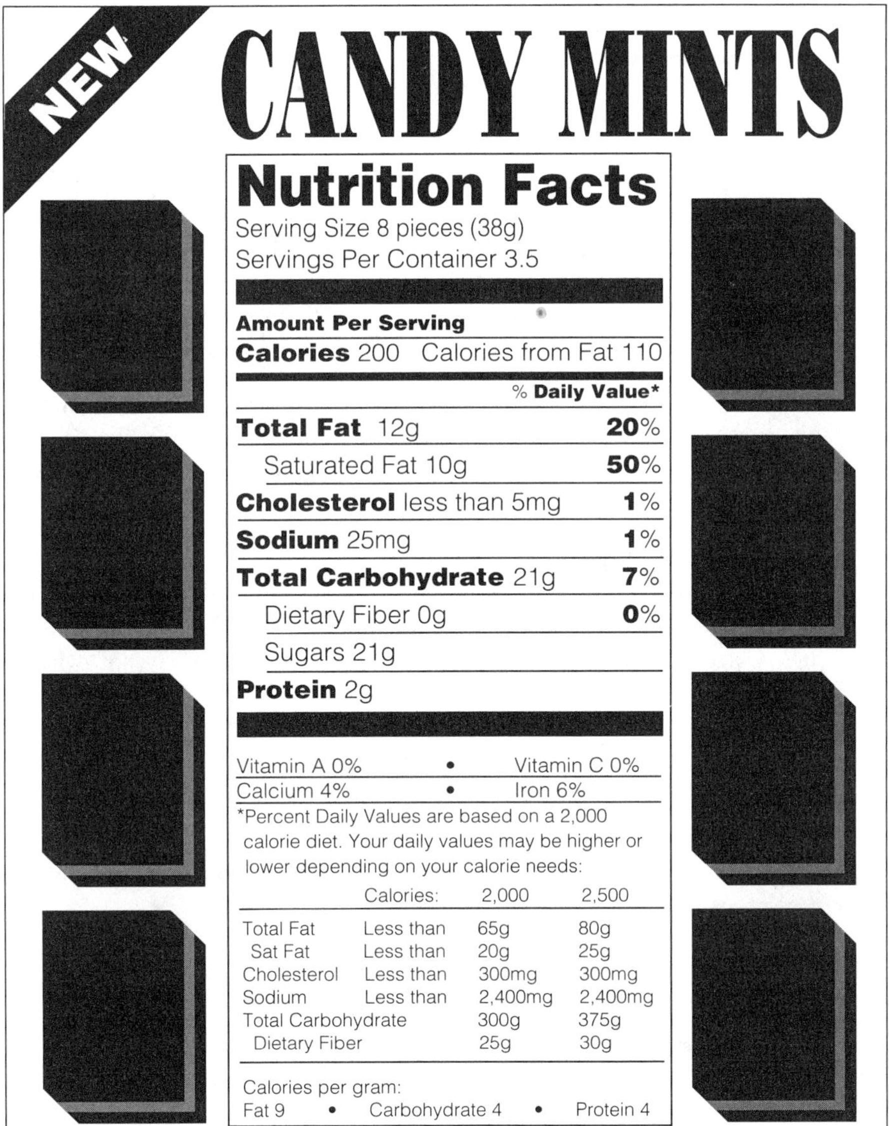
NEW
CANDY MINTS
Nutrition Facts
Serving Size 8 pieces (38g)
Servings Per Container 3.5
Amount Per Serving
Calories 200 Calories from Fat 110
% Daily Value*
Total Fat 12g 20%
Saturated Fat 10g 50%
Cholesterol less than 5mg 1%
Sodium 25mg 1%
Total Carbohydrate 21g 7%
Dietary Fiber 0g 0%
Sugars 21g
Protein 2g
Vitamin A 0% • Vitamin C 0%
Calcium 4% • Iron 6%
*Percent Daily Values are based on a 2,000 calorie diet. Your daily values may be higher or lower depending on your calorie needs:
Calories: 2,000 2,500
Total Fat Less than 65g 80g
Sat Fat Less than 20g 25g
Cholesterol Less than 300mg 300mg
Sodium Less than 2,400mg 2,400mg
Total Carbohydrate 300g 375g
Dietary Fiber 25g 30g
Calories per gram:
Fat 9 • Carbohydrate 4 • Protein 4

(b)

- It translates numbers into meaningful information. On old labels, for instance, if a mother went to buy a snack like candy mints for her active grade school child (see Figure 11.2 for sample labels), she would see that one serving (i.e., six pieces) had 9 grams of fat and might wonder if this amount is significant. A glance at the new label would reveal some unexpected findings. Not only does the label translate the 12 grams total fat (the serving size is slightly larger) into a context of 20% Daily Value, but it also discloses 10 grams of saturated fat. This translates into 50% of the Daily Value or half the limit of saturated fat a person eating 2000 calories daily should have for an entire day. If this parent compared the mints to an alternative chocolate roll snack, she would find that a serving of the latter had 2 grams of total fat (3% Daily Value) and no saturated fat—a much better choice.
- By expressing values as percentages it interprets metric numbers so that consumers can differentiate grams from milligrams in a meaningful health context. In some circumstances this is especially important—according to FDA food label research, consumers often attribute a more positive value to large numbers like those in milligrams, compared to the smaller ones in grams. For people on restricted diets, choosing foods based on the size of a number, without knowing how to interpret its meaning, could have harmful, if not dangerous, health consequences.
- It computes for the consumer, converting the amount of a key nutrient in a serving into a percentage that represents the amount one should consume in a day. Consumers who eat approximately 2000 calories a day and wish to limit their calories from fat to no more than 30%, do not need to divide the appropriate numbers to achieve this goal nor, for that matter, do they have to know the 30% figure. The label does the mathematical conversions for them. (The percent Daily Value calculates the percentages).

The fourth label education message focuses on the information on the lower part of the label which, unlike the upper part, never changes with the product. (See bottom of label, Figure 23-2.)

4. *By consulting the Daily Values, you can determine how much (or how little) of the major nutrients you should eat on a daily basis.*

The Daily Values, which appear only on larger packages, serve as a reference for dietary guidance, helping consumers understand how much of a nutrient they should eat at a minimum (e.g., for fiber) or at a maximum (for nutrients such as fat and sodium). The Daily Values, based on the Dietary Guidelines for Americans, are listed for people who eat approximately 2000 calories a day (many older adults, children, and sedentary women) or 2500 calories a day (active men, teenage boys and very active women). Unless one is on a very closely monitored diet, keeping track of calories, grams, or milligrams in a day's worth of food often leads to inaccurate results and extreme frustration for the average consumer. It is suggested, therefore, that the Daily Values listed on the lower portion of the label be used as a general reference, not as an inflexible set of guidelines, to help consumes gauge what they should eat more or less of that day.

These four messages and accompanying information represent only part of the information consumers need to become capable label users. Many factors, including the relatively low level of literacy prevalent in the United States, will certainly complicate the label learning process. Yet in our efforts to help Americans better understand and use label information to achieve healthier lives, we should at the same time cautiously avoid sacrificing accurate interpretations of nutrition information in the name of simplicity. Consumers can be easily misled by oversimplified dietary advice, making false assumptions about the health benefits of a food based on what they mistakenly think is accurate information. In their desire to reduce dietary fat and calories, consumers might assume, for example, that the claim "low cholesterol" means the total amount of fat in the serving is low, when in actuality only the saturated fat must be limited, or that the claim "low fat" can be automatically equated with a low-calorie food, when sweeteners that enhance taste may be added to compensate for the loss of fat. By regularly consulting the Nutrition Facts panel, consumers will find they can often avoid these and other misleading assumptions.

IV. SUMMARY

The massive reform of America's food labeling regulations will undoubtedly change the ways consumers use and understand the nutrition information on food labels. But the move to a new food label will not occur overnight for either the industry or for consumers. Even after all products are relabeled, it will still take an extraordinary effort to create national awareness of the label changes and educate people to make more informed food choices that will improve their diets, their health, and the quality of their lives. This is the ultimate goal of those working with the government. Yet, we in Government can only ensure the integrity and accuracy of the information disclosed on food labels. It will be up to consumers to decide whether and how they will use this information for their personal health benefit. If they do make more healthful food choices, not only they, but society as a whole, will be the beneficiary.

V. REFERENCES

1. U.S. Department of Health and Human Services, Public Health Services, *The Surgeon General's Report on Nutrition and Health*, Washington, D.C.: U.S. Government Printing Office, 1988, pp. 727.
2. National Research Council, *Diet and Health : Implications for Reducing Chronic Disease Risk*. Report of the Committee on Diet and Health, Food and Nutrition Board, Commission on Life Sciences, Washington, D.C.: National Academy Press, 1989, pp. 749.
3. Levy, A.S., *Key Findings for Nutrition Education: Quantitative Research at FDA and USDA*, FDA/USDA Food Labeling Education Information Center "UPDATE," Fall/Winter, 1993, pp. 2,4.
4. Division of Consumer Studies, Center for Food Safety and Applied Nutrition, Food and Drug Administratin, *Summary of Consumer Research on Health and Diet Attitudes and Knowledge and Use of Food Labels*. Washington, D.C., Food and Drug Administration, September 1992.
5. Kulakow, N., Durham, K., *Guidelines for Educating Consumers on the New Food Label*, FDA/USDA Food Labeling Education Information Center "Update," Fall/Winter, 1993, pp. 1,8.
6. Levy, A.S., Fein, S.B., Schucker, R.E., *Evaluation of Nutrition Label Formats FDA Study 2*, March 1992.

12

Meeting the NLEA Education Challenge: A Consumer Research Perspective

Brenda M. Derby and Sara Beck Fein

Food and Drug Administration, Washington, D.C.

I. BACKGROUND

A. Overview

When the U.S. Congress passed the Nutrition Labeling and Education Act of 1990 (NLEA), it culminated a process within and outside the Food and Drug Administration (FDA) to revise the food label dramatically. NLEA has been called "the most important piece of legislation about food and the way it's labeled since the Federal Food, Drug, and Cosmetic Act was passed in 1938" (1).

NLEA addressed both deficiencies in food labeling regulations, such as limitations in coverage and content, and controversial issues related to nutrient content claims and health claims.[1] The framework for the food label changes was provided by two landmark reports, the Surgeon-General's Report on Nutrition and Health (2) and the National Research Council's Diet and Health: Implications for Reducing Chronic Disease (3), and by the application of these findings in the revised Dietary Guidelines for Americans (4). These reports revealed the role nutrition labels can play in helping consumers to implement dietary practices based on their awareness of the links between dietary behavior and chronic diseases.

A unique feature of NLEA is that it included a mandate to educate consumers about:

- The availability of nutrition information in the label or labeling of food

- The importance of that information in maintaining healthful dietary practices[2]

By including this mandate in the NLEA, the Congress explicitly recognized that the benefits of the new food labels will be enhanced by education. Education is seen as necessary to ensure that consumers notice labeling changes and understand how to use new food labels to achieve healthier diets.

The identified deficiencies in food labels have been corrected by the new regulations. The challenge now is to help American consumers use the new food labels effectively, and the ultimate goal is healthier food choices. It is important to recognize that the presentation of relevant information by itself will not produce healthier diets. Changing dietary behavior is inherently difficult; dietary habits and preferences are complex, and less healthful diets are not simply a result of missing information or lack of nutrition knowledge.

B. Role of Research

The new food label presents fresh opportunities to educators, food companies, supermarkets, health care providers, and others who deal directly with consumers. The label is eye-catching and highly recognizable, so consumers should have little difficulty finding new labels once they become aware of them. A variety of educational programs and mass media events were planned in conjunction with the implementation date of the new labels and thereafter, providing opportunities to inform and motivate consumers. The success of these education and media efforts can be maximized by reviewing what is currently known about consumer concerns, knowledge, and behaviors. It is the role of research to describe these areas, and in the process to illuminate potential barriers and to suggest methods for overcoming them.

Levy (5) described how consumer research can help in the development and implementation of nutrition education campaigns. He identified three purposes of research and their importance for consumer education: (a) to describe the current situation, the market, and trends—a necessary starting point for education; (b) to clarify the important variables associated with healthful behaviors—necessary for developing messages and defining target audiences; and (c) to identify differences among target audiences—necessary to tailor messages and to allocate limited resources efficiently.

II. CONTEXT FOR LABEL USE

Consumer use of food labels represents a motivated behavior, guided by awareness of diet-disease links, nutrition knowledge and beliefs, and dietary management strategies, which form the context of label use. A number of consumer surveys and other studies that have been conducted by the Federal government, industry, and professional organizations describe this context.[3] Two of the major sources used in this chapter are the from the FDA and the United States Department of Agriculture (USDA). Since 1982, the FDA has conducted an approximately biennial Health and Diet Survey (HDS) that provides trend measures of awareness, beliefs, knowledge, and label use among American adults. Beginning in 1989, the USDA added a Diet and Health Knowledge Survey (DHKS) to their Continuing Survey of Food Intake by Individuals (CSFII), which permits analysis of detailed food consumption data in relationship to participants' attitudes, beliefs, concerns, and knowledge.[4]

A. Awareness of Diet-Disease Relationships

American consumers' awareness of relationships between their diets (what they eat or do not eat) and the prevention of serious chronic diseases such as heart disease and cancer is generally high, particularly for diet-disease relationships that have been targeted by major public health campaigns. For example, since 1978, at least three major campaigns have been conducted to inform the public about the relationships between diet and specific chronic diseases (sodium/hypertension, cholesterol/heart disease, fiber/cancer) in an effort to improve American dietary practices (6). These public health campaigns have been effective in raising knowledge and awareness levels across the population (6) and have reached less educated as well as more educated people. However, recent evidence indicates that some diet and health concerns may have peaked and some may even be declining (7–9).

Knowledge and awareness levels are related to food label use and to nutrient intake. People who are aware of diet-disease relationships are more likely to use food labels (10), and awareness of diet-disease relationships is an important predictor of specific nutrient intake (e.g., overall fat consumption) (11). The research results related to four diet-disease relationships are described below, along with the implications for food label education.

1. Sodium and Hypertension

The relationship between sodium consumption and hypertension was one of the first to be widely publicized in a government-endorsed public health campaign, and FDA surveys have tracked consumer awareness of this relationship. In 1978, few consumers (12 percent) were aware of a link between sodium consumption and hypertension. In 1981, the FDA and the National Heart, Lung and Blood Institute (NHLBI) jointly sponsored a sodium initiative to educate the public and to encourage manufacturers to display sodium content on food labels. The success of this campaign was apparent in the large increase in awareness by 1982 (34 percent). Awareness peaked in 1988, with 50 percent of Americans aware; by 1990, awareness of the sodium-hypertension relationship had dropped to 44 percent.[5] USDA research found similar results; in the 1989 DHKS, main meal planners or preparers were asked if they had heard of health problems related to salt or sodium intake. More than half identified hypertension as a health problem related to sodium intake (12).[6]

Some of the FDA results from the sodium initiative evaluation are of particular relevance to other nutrition education campaigns. Although education level is highly correlated with awareness (more of those with high education are aware), the campaign produced comparable gains across education groups (9). In 1978, 10 percent of consumers with less than a high school education were aware of a link between sodium consumption and hypertension, compared with 19 percent of consumers who had more than a high school education. By 1988, the percent aware had tripled or more than tripled in each education category (36 percent aware among the least educated consumers; 57 percent aware among the most educated). However, awareness levels fell by 1990 for all consumers (27 percent aware among those with less than high school education; 51 percent aware among those with more than high school education) (9).

2. Fiber and Cancer

Few Americans (9 percent in 1983) were aware of a relationship between fiber consumption and reduced risk of cancer prior to the National Cancer Prevention Awareness Program, initiated by the National Cancer Institute (NCI) in 1984. Awareness of the fiber-cancer link increased significantly from 1983 to 1986 (up to 32 percent) but decreased to 21 percent by 1990 (10). A pattern similar to that of sodium and hypertension was found for fiber and cancer awareness by education level (9). Fewer of the respondents with less than a high school education were aware of the relationship, but the increase in awareness was greater among this group than among the those with more than a high school education.

In a 1987 study, the NCI found that 28 percent of Americans were able to identify fiber and whole grains as something you should eat more of to prevent cancer (9). Consistent with the HDS findings, education level was strongly related to awareness; 38 percent of adults with more than a high school education and 13 percent of those with less than a high school education identified fiber and whole grains as something to eat more of to prevent cancer (9).

3. Fat and Heart Disease

During the 1980s, Americans showed a progressive and large increase in top-of-the-mind awareness of a link between fat consumption and the risk of heart disease. Only 29 percent of Americans were aware that dietary fats and fatty foods were a major cause of coronary heart disease in 1983, but 55 percent were aware by 1988 (6). Similarly, a growing number of Americans agreed that eating fewer high fat foods would have a "large effect" in preventing heart disease (73 percent in 1990, up 7 percent from 1984) (10).

The link between saturated fat and heart disease appears to be less well known. Main meal planners or preparers were more likely to identify heart disease as a health problem associated with total fat intake than with saturated fat intake (12).

4. Dietary Cholesterol and Heart Disease

NHLBI initiated a consumer education campaign in 1985 to educate the public about prevention of coronary heart disease by lowering blood cholesterol through diet. The HDS has tracked trends in cholesterol awareness since 1983, prior to the campaign. The percent of Americans who were aware of a link between dietary cholesterol and heart disease showed a large gain just after the campaign began and continued to increase: 26 percent aware in 1983, 40 percent aware in 1986, and 45 percent aware in 1988 (6).

In the 1989 DHKS, meal planners or preparers were asked if they had heard of health problems related to eight nutrients: sodium, fat, saturated fat, fiber, iron, calcium, and sugar. The link between dietary cholesterol and heart disease was the best known nutrient-disease relationship; about three in four identified this relationship (12).

Since 1983, the HDS has also asked consumers whether various dietary actions would help control high blood cholesterol. An increasing number believed that the following dietary choices would help (the percent in 1990 who believed the action would help is given in parentheses, followed by the increase since 1983 except as otherwise noted):

- Eating less fat (86 percent, up 17 percent) and eating less cholesterol (83 percent, up 14 percent);
- Reducing consumption of specific high-cholesterol and high-fat foods, such as sausage, bacon, and luncheon meats (81 percent, up 18 percent), eggs (75 percent, up 14 percent), and cheese (65 percent, up 15 percent since 1986);
- Eating more foods with lower fat and cholesterol such as skim milk (76 percent, up 16 percent), fish (74 percent, up 6 percent since 1986), and poultry (71 percent, up 8 percent) (13).

5. Concern about Specific Nutrients

The level of consumer awareness of nutrient-disease relationships is one indicator of the potential level of interest in product-specific nutrition information, such as that displayed on the nutrition label. Related questions from other surveys provide indicators of concerns about specific nutrients.

The 1993 Food Marketing Institute (FMI) Trends survey showed changes in dietary concerns over more than a decade (8), and these changes were consistent with the recent decrease in awareness of diet-disease relationships. Fewer shoppers said they were "very concerned" about the nutrition content of food (54 percent in 1993, a 10 percent drop from the five-year high in 1992). This drop in concern was more pronounced for less educated shoppers: 48 percent of those with a high school education or less were very concerned in 1993 (down 12 percent from 1992), while 59 percent of those with at least some college were very concerned in 1993 (down 9 percent).

Changes in concern about the specific nutrients of greatest interest to consumers are summarized below.

a. Fat

Concern about fat reached an all-time high in 1993; 54 percent of shoppers volunteered fat as their greatest concern and a similar proportion rated fat as a serious health hazard (57 percent), more than rated cholesterol, salt, or sodium as serious health hazards. Fat content was also the most frequently mentioned label information of interest (8).

b. Sodium

Concern about sodium peaked in 1990, when 30 percent of shoppers said sodium concerned them most. Sodium concern fell to about 20 percent for two years, but by 1993 it was the second most frequently cited concern (26 percent), following fat (8).

c. Cholesterol

Cholesterol was the nutrient of greatest concern to shoppers in 1989, and second to fat for the following three years. By 1993, cholesterol had fallen to third place in the ranking (23 percent) (8).

B. Knowledge and Beliefs

Awareness of diet-disease relationships may motivate consumers to pay attention to label information, but a working knowledge about the nutrients of concern is required for them to correctly use the information to guide food choices. Nutrition information and advice can seem contradictory and can be complex or too general for consumers to apply easily. The focus of nutrition guidance is often nutrients rather than foods, and the burden is on the consumer to know which foods are high or low sources of the nutrients. Research has documented that many consumers lack adequate knowledge to translate their awareness into healthier diets.

The new food label should enable most consumers to make healthful food choices without the level of nutrition understanding that was necessary in the past. However, even though the new label was designed to be more informative and easier to use, it still requires some effort on the part of the consumer. Realities of time and convenience mean that effective use of the new labels will be enhanced when the consumer is more knowledgeable. The food label provides product specific information, although basic dietary guidance is incorporated into the information provided by the Percent Daily Values, and dietary guidance appears in the footnote found on large packages. But it is up to the consumer to determine what nutrients are personally most relevant and what tradeoffs to make in selecting foods to eat.

Consumers understand the essence of dietary messages, such as to eat a variety of foods, eat more fruits and vegetables, and limit fat. But they often lack the specific knowledge necessary to translate this advice into effective actions, such as what is a serving and how many servings are needed per day, how many calories and what levels of other nutrients should be consumed daily, which foods are good sources of the desired nutrients, and which foods have moderate levels of the nutrients that should be limited.

1. Number of Servings and Serving Size

Awareness of the recommended number of servings of fruits and vegetables to eat each day was low in 1991; only 8 percent of adults believed they should eat at least five servings of fruits and vegetables a day (14), and 66 percent of adults thought one or two servings a day was sufficient (15). But awareness improved considerably after the NCI initiated a nationwide campaign in 1991 to encourage Americans to eat more fruits and vegetables (15). A follow-up survey in 1993 found that 29 percent were aware of the recommended number of servings (14).

Awareness of the amount in a serving was also low. In focus groups on fruit and vegetable consumption conducted for NCI, participants showed little consensus in their definitions of "one serving" of a fruit or vegetable. Interpretations ranged from one cup to a large spoonful to whatever is served to you (15).

2. Fats and Cholesterol

Research about consumer knowledge of dietary fats and cholesterol generally shows increasing levels of knowledge. Paradoxically, the same research suggests that misunderstandings may also be increasing.

In 1990, the HDS showed that fewer than one in ten consumers had not heard of high blood or serum cholesterol. In addition, results showed increases in the proportion who knew the desirable blood cholesterol level for adults and in the proportion who reported having had a blood cholesterol test. However, another survey found that some consumers were confused about the distinction between dietary cholesterol and blood cholesterol, especially older (over age 50) and less educated consumers (16).

The HDS measured knowledge about dietary fat and cholesterol at four points between 1983 and 1990 (10,17). Eleven multiple-choice questions asked about specific facts, such as in what kinds of foods cholesterol is found; whether polyunsaturated or saturated fats are more likely to be a liquid; and whether a product labeled as containing only vegetable oils would be low in saturated fat, high in saturated fat, or either low or high. Each question allowed the response, "not sure."

A trend analysis showed that for almost all questions, knowledge increased over the seven year period, sometimes substantially. By 1990, half or more of consumers knew that saturated fats are usually found in animal products, polyunsaturated fats are usually found in vegetable oils, saturated fats are more likely than polyunsaturated fats to raise people's blood cholesterol level, cholesterol is not the same thing as saturated fat or polyunsaturated fat, and a food that is cholesterol free could be either high or low in saturated fats. Fewer people, about a fourth, knew that saturated fats and polyunsaturated fats have the same number of calories and that if oil is hydrogenated, it becomes more saturated.

An interesting finding was that the proportion of consumers who answered "not sure" declined steadily and sharply for many questions. But over time for some of the questions, more respondents moved to an incorrect response than to the correct response. Either people were less willing to say they were not sure, so they guessed, or more people thought they knew about fats when they did not. Two questions important for dietary management on which an increasing percentage of people answered incorrectly involved cholesterol, specifically, which kind of fat is more likely to raise blood cholesterol and where cholesterol is found. Another question with an increased percentage incorrect involved labeling. If a label said the product contained only vegetable oils, many consumers mistakenly assumed it would also be low in saturated fat.

The pattern of results for fat and cholesterol suggests that one way consumers deal with the complexity of nutrition information is to rely on a simpler heuristic device to guide them, such as categorizing foods as "good" or "bad" based on the presence or absence of a key nutrient. Errors suggested that consumers generalize, so that a food high in a "bad" nutrient like cholesterol is assumed to be high in fat and, conversely, a food low in one "bad" nutrient is also assumed to be low in other "bad" nutrients (17).

Beliefs about good and bad nutrients may lead to similar categorizing of foods. In 1993, three-quarters of Americans agreed that "nutritionally speaking, I believe there are such things as 'good foods' and 'bad foods,'" unchanged from 1991 (18).

3. Sources of Fat and Fiber

Knowledge of the food sources of specific nutrients is critical for consumers to act on current dietary recommendations. Several surveys have examined the level of consumer knowledge about foods high or low in two important nutrients, fat and fiber. The results showed that knowledge of food sources of fat was greater than knowledge of food sources of fiber, but misunderstandings occurred for both nutrients.

Data from a 1987 study of over 20,000 American adults were used to examine progress toward a government goal that 70 percent of Americans will be able to identify foods that are low in fat and/or are good sources of fiber (19). Many respondents correctly identified high fat foods: fried chicken (81 percent), doughnuts (71 percent), luncheon meat (67 percent), and peanut butter (46 percent); and many identified low fat foods: broiled fish (94 percent), bananas (90 percent), soda (76 percent), and white bread (76 percent).

Errors such as classifying soda and white bread as high in fat (about one in five made this mistake), suggested that some consumers may have confused high-fat with fattening. Variation in knowledge appeared to arise from the foods considered rather than characteristics of the respondents, such as education level (19).

The 1993 FMI/Prevention survey (20) included a National Nutrition Quiz on fat and fiber. In this survey, many shoppers had identified fat as a concern, and many had recognized it as being linked to serious chronic diseases. Shoppers were asked how much fat was contained in 19 different foods ("a lot," "some," or "almost none"). They underestimated the fat content of some foods, including skinless chicken, lean ground beef, seafood, and eggs. They thought margarine contained less fat than butter and misunderstood the fat content of varieties of milk. Half the shoppers thought skim milk had some fat, while one in twelve thought 2% milk had no fat.

The possible effects on dietary changes of consumer confusion about sources of fat was demonstrated in an analysis of USDA dietary intake data. During a period when fat consumption was decreasing, female household heads made significant reductions in their percent of calories from fat, but their fat consumption remained well above the recommended level (21). Total fat intake by women did not differ across education levels. But better educated women, who were most aware of the health consequences of dietary fat, made systematic changes in the sources of their fat intake by significantly reducing their consumption of red meat and eggs. At the same time, they increased their fat intake from sources such as fats, oils, and dairy products.

These findings showed that even women who had better access to nutrition information failed to reduce their total fat consumption to recommended levels, although they had altered their sources of fat. One explanation is that the women may have converted nutrition information into a simpler heuristic to guide food choices, such as eliminating "bad" foods (17). This is easier for consumers than trying to keep a running total of their intake of a nutrient. New food labels will make it easier for concerned consumers to identify significant sources of nutrients so that one high source is not mistakenly substituted for another.

Two surveys that have asked about sources of dietary fiber have found substantial confusion. In the 1987 study of over 20,000 Americans, a majority correctly categorized low fiber foods: hamburgers (86 percent), white rice (69 percent), corn flakes (68 percent), and lettuce (52 percent); but fewer were able to correctly categorize foods that were high in fiber. A majority correctly classified bran flakes (82 percent) and apples (53 percent) as high-fiber, but fewer classified carrots (36 percent) or baked beans (19 percent) as high in fiber (19).

The 1993 FMI/Prevention Magazine survey, conducted six years later, also showed that consumers were confused about sources of fiber. A slim majority knew that fruits (56%), vegetables (56%), and beans (55%) contained "a lot of fiber," but fewer recognized nuts as having a lot of fiber (28%). "Some fiber" was incorrectly thought to be contained in beef (53%) and chicken (55%) (20). The FMI/Prevention Magazine survey asked about general food categories, such as vegetables, whereas the 1987 survey asked about specific foods, such as lettuce and carrots. Because of the substantially different question wording, trends are not possible to infer from these sets of results.

4. Recommended Nutrient Levels

Most consumers do not know the recommended intake levels for specific nutrients. In the recent FMI/Prevention Magazine survey (20), many shoppers underestimated the number of calories they needed each day; those likely to need the most calories (active men and very active women) were the most likely to underestimate their calorie needs. In contrast, sedentary women and seniors were as likely to overestimate as underestimate their calorie needs.

A 1993 American Dietetic Association (ADA) survey (18) found that barely half (49 percent) of Americans could correctly identify a recommended level of calories per day for themselves.

Most consumers did not know the recommendation that no more than 30 percent of calories in the diet come from fat (22). The FMI/Prevention Magazine survey found that more than one in three shoppers thought less than 20 percent of calories from fat was recommended, and 18 percent could not answer. Only 14 percent correctly answered that 26–30 percent of calories was appropriate. Another 1993 survey found that only nine percent of American adults could correctly identify the dietary recommendation for total fat. A similar proportion correctly identified the guideline for saturated fat (8 percent) (18).

Even if consumers learn the calories from fat guideline, they may misinterpret it. A majority of consumers mistakenly agreed with the statement "All foods should contain less than 30% of calories from fat" (16).

In the 1993 ADA survey, most consumers did not know the intake recommendations for any nutrients. From a list of eight nutrients, only the recommended guideline for calcium was correctly identified by more than half the participants (18). Although sodium content has been required on nutrition labels since 1986, and a majority of consumers were aware of sodium as a risk factor for hypertension, none of the participants could correctly identify the recommended guideline for sodium (18).

C. DIETARY MANAGEMENT STRATEGIES

Consistent with rising awareness and concern, consumers have made changes in what they eat or avoid eating, the ways they prepare foods, and the types of special diets they adopt to reduce their risk of particular health conditions. The changes generally have brought diets closer to the dietary guidelines, although some improvements peaked then declined, like awareness of some diet-disease relationships. These changes in dietary management strategies were revealed in dietary intake studies and have also been tracked by a number of surveys. In addition, survey results have shown which sources consumers commonly use for information about health and diet and have tracked changes in their level of interest in receiving more information on this topic. Interest in receiving more information has been steady or declining, congruent with other health and diet measures.

1. Trends in Consumption

In a recent report (9), USDA described 12-year dietary changes, based on the 1977–78 Nationwide Food Consumption Survey and the 1989–90 CSFII. Two major trends were documented:

- Declines in consumption of high-fat foods such as beef (–48%), pork (–40%), whole milk (–35%), eggs (–26%), and luncheon meats/sausage (–10%)
- Increases in consumption of lower fat foods such as low fat and skim milk (+122%), dark green vegetables (+38%), grains (+24%), and meat/poultry/fish mixtures (+20%)

Consistent with these food consumption changes, dietary intake surveys have also documented progress toward nutrient intake objectives, such as reduction of calories from total fat to 30 percent or less. Calories from fat dropped from 42 percent in the mid-1960s to 36 percent in 1988 (9). Data from the 1989–90 CSFII showed a further drop to 35 percent of calories from fat (12). Supporting this decreased trend, population estimates from the latest National Health and

Nutrition Examination Survey (1988-91) were that 34 percent of total calories came from fat (23). While fat consumption declined, overall consumption (total calories) did not, a result that is consistent with reports of an increasing number of overweight American adults (23). These dietary intake trends confirm that behaviors have changed in the directions expected from self-reported changes in eating habits.

Several surveys have asked consumers about changes in their eating habits and have found consistent results. Compared with 1992, shoppers in 1993 who said they changed their diets in the past five years were more likely to say that they now eat less fat (59 percent, up 15 percent) and eat more fruits and vegetables (25 percent, up 8 percent), but fewer said they had reduced cholesterol (16 percent, down 11 percent) or eat less salt/sodium (13 percent, down 19 percent) (20).

Similar changes were reported in the 1993 FMI Trends study (8), reflecting responses to the question, "what, if anything, are you eating more or less of to ensure that your diet is healthy?" The shoppers' most common responses were the following:

- More fruits/vegetables (62%)
- Less meat/red meat (30%)
- Less fats/oils (26%)
- Less sugar (26%)
- More chicken/turkey/white meat (15%)

In the 1993 NCI survey, two out of three adults said they were trying to eat more fruits and vegetables, but, paradoxically, the proportion who reported that they actually ate more has decreased since 1991 (one in eight in 1993, down from one in four in 1991) (14).

Studies also indicate that consumers are increasingly dissatisfied with the healthfulness of their diets. In 1993, fewer Americans said "I am satisfied with the way I currently regulate my diet" (39 percent, down 7 percent from 1991) (18). Major obstacles were the following:

- I don't want to give up the foods I like (39 percent).
- It takes too much time to keep track of my diet (22 percent).
- There are so many new or conflicting studies on nutrition that I don't know what's good for me anymore (22 percent) (18).

The 1993 FMI/Prevention Magazine survey found that a majority of shoppers believed their own diets could be at least somewhat healthier (74 percent), even though many stated that they had made major changes in their diets in recent years (60 percent) (20). Health concerns were the main reason for these dietary changes, in particular to lower cholesterol, to reduce the risk of heart problems, to control weight, to stay healthy, and to control hypertension (20).

The 1993 ADA survey showed that while many consumers said diet and nutrition were extremely important to them personally (54 percent, up 5 percent since 1991), fewer said they "carefully select what [they] eat to achieve balanced nutrition and a healthy diet" (39 percent, down 5 percent) (18).

The marketplace has also changed; it now offers an increasing number of reformulated and new products with reduced fat or calories. A marketing manager estimated an increase from 275 reduced fat products in 1988 to 1,600 in 1992 (24). One market research firm estimated a 39 percent increase in new low-fat/low-cholesterol food products in just one year, 1991 to 1992 (25).

These types of new products are increasingly accepted by consumers. The Calorie Control Council reported that, in 1993, more Americans said they ate reduced fat products (73%) and sugar-free, low-calorie products (58%) than in a 1991 survey (24). The top reason for using these products was "to stay in better health," and weight loss was identified as a reason by far fewer consumers.

Targeted information campaigns that identify specific types of food or individual food products can be highly effective at modifying purchase behavior. Consumers responded to a comprehensive health promotion marketing program initiated in 1984 by making major shifts toward the purchasing of high-fiber cereals and away from lower fiber cereals, particularly granolas. The program included media advertising that highlighted an NCI dietary fiber recommendation and identified All-Bran cereal as responsive to that recommendation. Sales of All-Bran and other high-fiber cereals increased substantially and prompted changes in promotion and packaging of competing high-fiber cereals (26).

2. Special Diets

One indicator of consumer attempts to affect their health through their diet is the prevalence of self-initiated diets to reduce the risk of chronic disease, such as low-sodium and cholesterol-reducing diets. The HDS has tracked prevalence of both self-initiated and doctor-initiated diets to reduce sodium intake and blood cholesterol levels.

The pattern that emerged for self-initiated efforts to reduce sodium and cholesterol paralleled the trends for concern about sodium and cholesterol described earlier. The prevalence of such diets increased or was sustained through the late 1980s, then dropped by 1990. About one in four Americans reported trying to reduce sodium consumption in 1982; by 1986, nearly one in three was doing so. This level was maintained through 1988. By 1990, however, the prevalence had returned to the 1982 level (9). The pattern for self-initiated cholesterol-lowering diets was similar. In 1983, about one in ten Americans was trying to reduce cholesterol on their own initiative. Twice as many said they were trying to lower their cholesterol by 1988 (22 percent). By 1990, only 15 percent were doing so (9).

In contrast to the self-initiated diets, the trend for doctor-initiated diets showed a steady increase. By 1990, 15 percent of consumers were on a doctor-prescribed limited sodium diet (up 3 percent from 1982), and 9 percent were on a doctor-prescribed cholesterol-lowering diet (up 6 percent from 1983) (9).

3. Sources of Information

The 1990 HDS asked a series of questions about how much diet and health information consumers acquire from various sources (10). The results showed that several sources are widely used. The most frequently reported sources of information were the following:

- News stories and programs (44 percent)
- Health organizations (41 percent)
- Food labels (35 percent)

The sources used less frequently were the following:

- Physicians and other health providers (30 percent)
- Food advertisements (23 percent)
- Grocery store handouts or displays (10 percent)

Fewer consumers with less than a high school education reported receiving information from news media, health organizations, and food labels. But these less educated consumers were as likely as others to use health care providers for diet and health information (10).

Most consumers are interested in information about health and diet, but trend results from two surveys suggests that, like concern, the level of interest is steady or has peaked. A plurality of consumers acknowledged that they needed to know "a little more" about nutrition, but the proportion saying they needed to know a little more has not increased over the past 15 years (27). The HDS results indicated a sharp decrease in the proportion who said they wanted *more* information on diet and health. In 1988, 83 percent of consumers said they wanted more diet and health information, but in 1990, only 56 percent said they wanted more information (10).

Levy summarized some possible reasons for the downturn in interest in diet and health:

- It could be the natural and inevitable consequence of ten years of rising interest.
- The American public may feel overloaded by the increasing number of diet-disease messages.
- The public may have become skeptical as a result of exaggerated and questionable claims.

- The public may have become discouraged by the difficulty and effort required to make effective dietary change (5).

III. CURRENT USE OF FOOD LABELS

Educational programs are more likely to be effective if they take into consideration patterns of consumer label use. Knowledge of these patterns can help in defining successful messages and targeting them to particular audiences. Patterns include the prevalence of label use, characteristics of people who use and do not use labels, consumer confidence in the information presented on labels, and the purposes for which consumers currently use labels.

A. Prevalence of Label Use

A substantial proportion of American adults use food labels at least some of the time, but several studies have shown that use is either steady or declining. When label use is measured by observation in shopping situations, prevalence estimates are substantially lower than when it is measured by self-reports in surveys. However, food labels may be used at times other than when the product is purchased, such as when it is cooked or consumed, so that use in supermarkets underestimates total use.

The HDS has tracked food label use for nearly a decade, and results show increasing use through 1990. Most consumers reported using the ingredient list and the nutrition label "always" or "sometimes." From 1982 to 1990, use of the ingredient list increased only slightly (78 percent, up 3 percent), while use of the nutrition label increased significantly (76 percent, up 8 percent) (28). Use was greater the first time a product was purchased than on subsequent occasions. In 1990, 34 percent of consumers said they "always" used the nutrition label the first time they purchased a product, compared with 22 percent who said they "always" read the label at other times (10).

Although surveys have found some variation in label use since 1990, no systematic increase seems to have occurred. According to the 1993 FMI Trends survey of grocery shoppers (8), reported label use has not increased. Most shoppers said they "always" or "sometimes" used the ingredient label (85 percent, down from 88 percent in 1990) and the nutrition label (78 percent, down from 87 percent in 1990) when buying a product for the first time. The FMI/Prevention Magazine survey of grocery shoppers did not show this downturn, finding that significantly more shoppers used the nutrition label (83 percent said "almost always" or "sometimes" in 1993, up 6 percent since 1992) (20). But FMI/Prevention Magazine data covered only a one-year period, and the prevalence for 1993 was lower than that found in the similar FMI Trends survey in

1990. In both studies, the prevalence of nutrition label use among grocery shoppers was higher than that found for adults in general.

While consumers report high levels of label use on surveys, some evidence suggests that actual use is less frequent. In a 1993 observational study in the cereal aisle of supermarkets, shoppers looked at the nutrition panel on only five percent of the cereal packages they inspected (29). Breakfast cereal is a frequently purchased and well-known food product, so it is likely that many of these shoppers had examined the nutrition label previously. Also, some label reading occurs at other places. Nevertheless, this observational finding suggests that in the absence of promotional and educational activities to draw their attention to the new food label, many consumers may not notice it.

B. Who Uses Food Labels?

A trend analysis of the HDS from 1982–1988 showed that demographic patterns associated with label use have been maintained over time (28). Throughout this time period, people who used both the nutrition and the ingredient label were more likely to be female and on self-initiated low-sodium or cholesterol-lowering diets. People who used neither the nutrition nor the ingredient label were more likely to be older, male, less educated, and nonwhite.

Although different studies classify age groups in different ways, results generally agree that adults in middle age groups are more likely than either young adults or the elderly to report using food labels. Middle age is often the time when adults first confront the effects of diet on their own health, providing a strong motivation for looking at food labels.

Older adults should be at least as motivated as middle aged adults, given their greater likelihood of health problems affected by diet, but they reported using food labels less often. An explanation that has been offered is that the ability to extract relevant information from food labels declines with age. A recent study of label use in both a supermarket setting and a laboratory experiment found age differences in the ability to use the label to make nutritious product selections (29). The study involved choosing a breakfast cereal that met two nutritional criteria, low sodium and high fiber. Older shoppers (aged 60 and above) found it difficult to use the nutrition label to select cereal based on the two criteria, even when they were able to write down the relevant information rather than having to retain it in memory. The difficulty seemed to reflect information processing problems associated with aging that made it difficult for the older participants to integrate two decision criteria simultaneously. These findings have important implications for label education for older consumers (29).

Recent data from USDA add an important element to this picture of label users—actual food consumption. An implicit assumption in studies of label use and in discussions of food label education is that if consumers use the label, they are likely to make healthier food choices. Data from the 1989 DHKS and CSFII compared food label users and nonusers in terms of attitudes, knowledge, and dietary intake. Results showed that label users tended to be female and white and to have at least a high school education. Compared with nonusers (29 percent), label users (71 percent) were more likely to be aware of diet-disease relationships, to be knowledgeable about foods, and to report healthful food preparation habits, such as using skim or low-fat milk in cooking (30). Label users more often cited taste, food safety, and nutrition as high priorities in selecting foods, while nonusers more often cited price and ease of preparation.

These encouraging differences notwithstanding, when label users and nonusers were compared in terms of their actual food consumption, few differences were found in consumption of key nutrients. No differences were found in fat or saturated fat intake, and only modest differences were found in salt and cholesterol intake (both of which fell within dietary recommendations among label users and nonusers). Label users consumed slightly more dietary fiber than nonusers, but both groups fell far short of dietary recommendations for fiber.

In an analysis of the 1990 HDS, consumers were classified into one of three label interest groups (high, average, or low) based on their responses to questions that tapped their nutrition label experiences and their interest in improved labels (31). This consumer segmentation provided a different perspective on label use than that gained by comparing label users and nonusers because it differentiated the large label user group.

To be classified as high label interest, consumers had to say yes to three questions: they were willing to pay more for foods in order to have improved food labels, they had changed a purchase or use decision based on label information, and they had looked for nutrition information on a label and not found it. Frequency of label use was not one of the criteria by which the groups were defined. Consumers were classified as low interest if they were not willing to pay for improved nutrition labels, regardless of their responses to the other two questions listed above. Those classified as average interest were willing to pay more for improved labels, but said no to one or both of the questions about changing a purchase decision because of label information and looking for nutrition information but not finding it. The low and high label interest groups represent 20 percent of consumers each, and provide a revealing contrast. The average label interest consumers shared some characteristics of both the high and low groups.

Virtually all consumers in the high interest group were regular food label users and tended to be female, middle-aged, and more educated. They also reported greater awareness and knowledge of diet-disease relationships and more healthful behaviors, including healthful eating habits. In contrast, almost half of the low interest consumers said they used food labels rarely or never. They tended to be the least aware and least knowledgeable consumers, and fewer of them practiced healthful behaviors. Low interest consumers were more likely to be male, older, and less educated.

Like the high label interest consumers, most of those with average interest said they used food labels. However, the average interest consumers tended to resemble those with low interest on many awareness, belief, knowledge, and self-reported behavior questions.

An interesting pattern emerged in the way these label interest groups rated the accuracy of various parts of the food label. No differences occurred in their ratings of the accuracy of three types of label information: ingredients, vitamins and minerals, and statements such as "low cholesterol." However, the high label interest consumers were less likely than average or low interest consumers to believe in the accuracy of health claims or of claims such as "light." In other words, those consumers who were most likely to use the label and to want more information on labels were not more likely to think that the information was accurate. And, for some information, they were less likely to think so. This suggests that the lower level of interest in labels shown by average and low interest consumers was not because of greater skepticism regarding label information.

For highly motivated consumers, like those in the high label interest group, educational strategies that alert them to the presence of new food labels and provide a little guidance on how to take full advantage of the greater amount of nutrition information to achieve healthier diets will probably be sufficient.

Consumers with less interest in food labels tended to be less informed about diet and health, but since they were also older, they are likely to have as much or more need for the types of information available on the new labels as the high interest consumers. Many low interest consumers said they used food labels, so this weak behavior should be encouraged by ensuring that they see the value in the new label information. Educators will need to consider motivation as much as label reading skills when dealing with these consumers. They will likely require more guidance, given generally lower educational attainment, to understand and use label information.

Most consumers who fall in the middle group seem predisposed to use the new food labels, but they are currently not well-informed. Their incipient interest and frequent casual use of the labels provides a foundation for educators to build on. They may need more information on the health and diet reasons for label use to help motivate them.

C. Consumer Confidence in Labels

The level of consumer confidence in the accuracy or believability of label information has been shown to be moderate, at best, and the analysis of trends in confidence indicates no change. Nutrition and health claims have been studied more extensively than other label parts in relation to consumer confidence. Considerable skepticism has been reported, offset somewhat by the greater likelihood of purchasing products with such claims.

The HDS has examined consumer confidence in various parts of the label. Consumers were more likely to believe that the ingredient list (37 percent) and vitamin/mineral content (34 percent) on a label were "always accurate" than to believe "no cholesterol" (28 percent) or "light" claims (17 percent). Even fewer believed claims that eating a food might help to prevent disease (10 percent) were always accurate. This skepticism about information on the food label was unchanged from 1988 to 1990 (10).

Serving size information is another potential source of consumer mistrust. In 1992, a plurality of consumers believed the serving sizes shown on food packages were inaccurate, most often in the direction of getting fewer servings than the package listed (32). This perception of serving sizes as inaccurate may explain the finding that, although many food shoppers queried in a focus group study said they were familiar with the serving size information on labels, they either did not pay attention to it or did not find it very useful (22).

The food industry's increasing use of health claims (e.g., Kellogg's fiber/cancer messages on All-Bran cereal) and nutrient content claims (e.g., "no cholesterol") led to one of the most significant controversies that stimulated demand for improved food labeling regulations.[7] Consumer confidence in these types of claims is low.

The 1990 FMI Trends survey (33) reported that few shoppers considered health claims "very believable" (8 percent), although many rated them "somewhat believable" (58 percent). Those who found manufacturers' health claims less than "very believable" cited as reasons omission of information about harmful ingredients, statements that were not always truthful, and use of claims as an advertising gimmick. However, health claims were reported to affect purchase decisions: 73 percent of shoppers said such statements influenced their decision to purchase a food product "somewhat" or "a great deal."

Roper Reports (34) found that, although significantly fewer consumers could recall seeing health claims in advertising or on food packages in 1992 than in 1989, their degree of skepticism did not change. A plurality said they believed "only some" claims were accurate, and few believed "many of them" or "just about all of them" were accurate. Half the sample were asked about claims in advertising and half about claims on food labels. Ratings of accuracy were virtually identical, suggesting that consumers did not distinguish between advertisements and labels. Nearly half the consumers felt there was "not enough" monitoring of health claims by the government.

In FDA focus group studies, participants expressed considerable skepticism toward nutrient content claims, and they were unable to distinguish absolute ("low fat") and comparative ("reduced fat") claims (35). Confidence appeared to be an important issue to many consumers.

The widely found consumer skepticism toward nutrient content and health claims may be explained in part by consumer perception that these claims are more like advertising than other label information (and some of the research reported above did not explicitly limit the focus to claims on labels). The increased use of health and nutrition claims since 1975 has been documented for advertising (36), so consumers are used to seeing such claims in this context. Substantially lower consumer confidence in the accuracy of information presented in advertising versus labeling has been reported from surveys dating from 1973 (37).

A small study of grocery shoppers in Utah examined consumer comprehension of health claims (38). Consumers looked at fiber messages on boxes of Kellogg's All-Bran cereal. Their understanding of the messages was relatively low, especially among the less educated consumers. Nearly 40 percent of the participants were unable to correctly identify the source of a cereal company message, even though the information was present on the package, and instead attributed the message to the National Cancer Institute or some other government agency, such as the FDA.

FDA research on new label formats provided some insight into how consumers use front panel label claims, particularly nutrient content claims (39). Participants evaluated a product more favorably when just the front panel claims were presented. Once they looked at the nutrition label, their judgment was moderated (i.e., the food was given a lower healthfulness rating), and their ratings came closer to the judgments of nutrition experts (40). Consumers were quite competent at evaluating the truthfulness of nutrient content claims, but they also showed a bias toward believing claims. They were more likely to err in saying a false claim was true than in saying a true claim was false (40).

It is hard to estimate the extent to which consumers rely on front panel claims rather than confirm the information by examining the nutrition label. Although they tended to express skepticism in research settings, they also reported greater probability of buying products with claims. Also, the prevalence of nutrient content claims in the marketplace suggests that such information does increase sales. Perhaps consumers discount the information, but at the same time they assume the food is better than similar products without the claim.

D. Label Use Purposes

An understanding of consumer use of labels requires a more detailed look at why and how consumers use food labels. Research has examined the various purposes for which consumers use labels and the information most frequently sought from labels. The nutrients of greatest interest from the label tend to parallel the nutrients which consumers are most likely to associate with chronic disease.

In a food label format study conducted by a food industry consortium (41), shoppers who said they read the nutrition label (81 percent) were asked whether they used the label for each of a variety of purposes. The most common purposes were the following:

- To calculate exactly how high or low a product is in things like calories, salt, vitamins, fat, etc. (71 percent)
- To get a general idea of nutrition content (66 percent)
- To compare different types of food items to each other (60 percent)
- To help decide which brand of a particular food item to buy (58 percent)

The least common purposes were the following:

- To determine whether something said on the package or in advertising is actually true (46 percent)
- To figure out how much of the product you or someone in your household should eat (38 percent)
- To help in meal planning (38 percent)

These findings show that labels were used predominately for simple tasks, such as general information and product comparisons, and less frequently for complex tasks, such as meal planning and quantitative dietary regulation.

Consumers have reported using the food label to make purchase decisions. Recent surveys found that from one-third to one-half of consumers said they changed their decision to buy a product after reading the nutrition label (10, 18,20,32).

In one study, consumers were asked for which products the food label had the most and least influence on their purchase decisions (42). They reported that labels had the most influence on cereal (35 percent) and canned foods (26 percent), and that labels had the least influence on snack foods (15 percent) and fresh foods (12 percent). These figures probably reflect availability of nutrition labels as much as they do consumer interest in the nutrition content of different types of food.

Under NLEA, virtually all packaged foods will have nutrition labels, and the most frequently consumed raw fruits, vegetables, seafood, meats, and poultry will have nutrition information available in most supermarkets (e.g., on posters, brochures, or shelf labels). This increased availability of nutrition information will provide consumers the opportunity to learn about the nutrient profiles of most of the foods they eat. Interest in the new label may be heightened by emphasizing this increased availability, particularly for food products that consumers have previously had less information about.

Some insights can be gained about how consumers will be affected by new food labels by examining the results of experimental shelf labeling studies. A grocery store program initiated in 1981 highlighted food products with special nutrition characteristics (low sodium, cholesterol, fat, or calories) with prominent shelf markers. Moderate increases (or lack of decreases) occurred in market share for these products compared with control stores (43). This Special Diet Alert program was more effective than in-store health promotion programs that depended on posters or educational handouts that did not tie the relevant information directly to specific brands of products. Although the new nutrition labels will not call attention to products that meet certain dietary criteria, they will contain information about specific brands, and through this information, they may improve the market share of products consumers perceive as meeting their own dietary criteria. Also, the nutrient content claims on some products will serve to call attention to products with desirable levels of some nutrients.

Several surveys have asked consumers what label information they look for or find useful, with consistent results. When asked what specific information they look for on the food label, consumers most often cited fat content (59 percent), followed by calorie content (32 percent), sodium content (31 percent), general ingredients (22 percent), cholesterol content (21 percent), general nutrition content (21 percent), and sugar content (21 percent) (8). A parallel listing of nutrients was found when consumers were asked what information on the label was "most useful" (18).

IV. RESEARCH BASIS FOR THE NEW LABEL FORMAT

A. Format Elements and Label Use Tasks

The major features of the new nutrition label format are based on experimental research conducted by the FDA. This research was designed to identify a format that provided nutrition information in a way that was easy for consumers to use. Focus group research on food labels was also conducted. The findings provide valuable insights into consumer abilities to use the food label for a variety of purposes. The research demonstrated that, even in the absence of any education about the new food label, consumers could use the new label as well as or better than the old label for most tasks. In addition, the new food label (specifically the Percent Daily Value information) was the most effective for the NLEA objective of helping consumers put the food into the context of their total daily diet.

The FDA conducted two label format experiments to inform the agency's choice of a format for the new nutrition label (39, 44). Format Study 1 had only one label use task, product comparison, and evaluated several label format elements.[8] The results showed that the labels with the least information were easier to use. However, consumers preferred the labels that had more information, and they wanted interpretational aids, such as adjectives (44). Older participants and those with less education had the greatest difficulty discriminating product differences and were more likely to perceive differences where none existed.

The format elements for Format Study 2 were selected on the basis of the results of Format Study 1. Format Study 2 added new format elements and expanded the tasks to measure other important label uses beyond product comparison (39).[9] The following five design elements were tested:

- Daily Value listing
- Declaration of nutrient amounts in metric units or percentages
- Adjective descriptors
- Highlighting nutrients meeting dietary recommendations
- Grouping of nutrients into two categories, those to eat more of and those to eat less of each day

The following five label use tasks were examined:

- Product comparison—to compare two food labels to identify differences between them
- Evaluation of nutrition claims—to evaluate the truthfulness of nutrient content claims about the food
- Dietary management—to place the food in the context of a daily diet by identifying the nutrients to eat more of and to eat less of in a day, after eating three servings of this food
- Use of daily value concept—to determine the contribution of a food to the daily diet by estimating how many servings of the food would be needed to get all of the daily requirement for a certain nutrient by eating only this food
- Judging healthfulness—to evaluate the overall healthfulness of the food, first based only on the front panel claims and again after examining the nutrition panel as well

Table 12.1 shows the main tasks, measures, and results of Study 2. The tasks are ordered according to the amount of prior knowledge required to successfully perform them. Prior knowledge includes both nutrition information, such as knowledge of dietary guidelines, and practical skills, such as computation. Only the answers to the first four tasks could be clearly classified as correct or incorrect; the fifth was evaluated by comparing the participants' ratings with those of an expert panel. The participants performed best on product comparison; most people accurately identified 78 percent of the product differences in less than one minute. The most difficult objective task, use of the Daily Value to estimate the number of servings needed in a day, was performed poorly (only 20 percent answered correctly). Somewhat better performance (30 percent cor-

Table 12.1 Label Use Tasks, Measures, and Summary Performance

Task	Definition	Performance
Product comparison	Identify nutrient differences between 2 products	Accuracy—78% of differences Mean task time—58 seconds
Evaluation of nutrition claims	Ratings of front panel claims as true or false	Number true and false claims correctly evaluated—58% Number false rated as true—45% Number true rated as false—18%
Dietary management	Identify nutrients to seek or avoid after eating the food	More often correct (45%) than incorrect (30%)
Use of Daily Value concept	Calculate servings needed to meet Daily Value	Majority unable to correctly estimate servings (20% correct)
Judging healthfulness	Ratings of healthfulness —based first on front panel claims; then after adding nutrition information	Ratings moderated by nutrition information (lower ratings); Consumer ratings closer to expert ratings after looking at nutrition label

Source: FDA Format Study 2. Fein, Derby, and Levy (Ref. 39).

rect) was found in a related industry study that asked a similar question. In that study, consumers were only asked to describe how they would estimate the number of servings needed; they did not have to do any calculations (41).

The experimental design used for the FDA format studies allowed the effects of practice to be examined. The participants performed each label use task four times, with different formats. They did not receive feedback to indicate if they were answering questions correctly, so these findings reflect performance changes due to use alone, not education. Participants improved over the course of the four trials for three tasks: product comparison, evaluation of nutrition claims, and dietary management. But they did not improve on the use of the Daily Value task. This suggests that if consumers use the new food label, they will quickly become more competent at many common label uses, with the main barrier being motivation. Other uses, such as calculating servings needed, require specific skills and knowledge and will not improve through practice alone.

Performance improved the most for those tasks participants performed the best initially, with improvement greatest for the easiest task, product comparison. Tasks that are easy to perform and learn provide a foundation for educators to build upon. The more complex tasks that require putting foods into a broader dietary context will not improve without direct intervention by educators to help consumers learn how to do them.

Half the sample received a brief introduction about Daily Values at the start of the session. The introduction had a significant impact on the ability to use the prototype food labels; participants who received the introduction were more accurate in the task that required using the Daily Value. This finding is noteworthy because it suggests that even cursory information about the new elements will enhance consumers' ability to use the label.

The critical role of the Percent Daily Value in helping consumers correctly interpret food label information was demonstrated in special analyses of the Format Study 2 results (45). For the key macronutrients, response to the dietary management question (identifying nutrients to eat more or less of after eating a particular food) was examined for different levels of the nutrient. The results showed that the size of the number affected the response when only metric information (grams, milligrams) was provided. Good examples are the responses for saturated fat, with a Daily Value of 20 grams, and sodium, with a Daily Value of 2400 milligrams. For a label that displayed five grams of saturated fat, fewer participants listed saturated fat as a nutrient to eat less of when they saw only the metric amount. For sodium, even a small amount nutritionally (e.g., 115 milligrams) resulted in more participants listing sodium as a nutrient to eat less of because the number appeared large compared with the other nutrient amounts on the label. However, when interpretive information (adjectives or Percent Daily Value) was included on the label, participants were much better at recognizing the nutritional significance of the nutrient amounts.

B. Consumer Characteristics Associated with Performance Level

The characteristics that differentiated those who performed well from those who did not are similar to the characteristics that differentiated label users from nonusers, reported in the previous section. Those consumers who most needed to use food labels (those with a diet-relevant health condition or on doctor-initiated diets) did not perform well on many of the label use tasks. The better scores of label users is consistent with the finding that practice of the tasks improves performance.

The following characteristics were associated with better performance on the label use tasks (40):

- Demographics—female, middle aged, white, more educated
- Health status—those *without* a diet-relevant health condition, such as high blood pressure or heart disease

- Diet status—those on self-initiated low fat, low cholesterol, low sodium or low calorie diets; those on doctor-initiated diets did better than those not on a diet
- Label use—regular label readers, particularly for tasks affected by product characteristics (nutrition claims, dietary management)

V. PRACTICAL SUGGESTIONS FOR EDUCATION

Research consistently shows that awareness of diet-disease relationships, nutrition knowledge, healthful eating habits, and food label use are related to consumer characteristics such as age, education, and relevant health conditions. In addition, it shows how consumers currently use labels and which types of label use tasks are hard and easy for them. Educational initiatives, from advertisements to intensive programs, will have their greatest impact when the messages and materials are tailored to the target audience. The findings here can provide a foundation by identifying what consumers already know, common areas of confusion or misunderstanding, and concerns and motivations that predispose consumers to pay attention to new messages.

However, nutrition education is a complicated activity, and each initiative must consider specific characteristics of the audience and the message in much greater detail than general consumer research can provide. In addition, the nutrition education literature addresses the complex issues of providing information to consumers with various characteristics and special needs, and this material cannot be summarized here. The following section offers some tips related to informing consumers about the new food label that derive from the consumer research.

A. Consumer Characteristics

1. Demographic Characteristics and Health Conditions

To date, most materials on the new food label focus on awareness of the new label, and they target general audiences. While this is a critical first step, consumers will be better prepared to make effective use of the information available on the new food labels if they receive specific, personally relevant information.

Age is one characteristic by which educational programs might be targeted. Older consumers are less likely to use the food label, yet they are more likely to have health conditions which might be treated by diet. In addition, they are more likely to have several characteristics that indicate a need for specialized programs: a lower average education than younger consumers; physical limitations, such as poor vision; and information processing limitations that make the presentation and amount of information critical considerations.

Tips:

- Limit the number of messages presented at any one time.
- Emphasize simpler label uses, such as product comparison and low-high evaluations. When appropriate, provide rules of thumb to guide low-high evaluations.
- Encourage use of nutrient content or health claims to find products with desired nutrient characteristics.
- Use large print with high contrast for written materials.
- Develop simple memory aids to enable older consumers to take more than one nutrient into account when selecting foods.

Consumers with low education levels may also benefit from specially targeted programs. Education is closely associated with label use, and it is also related to the skills consumers have available to use the label in different ways. Some research has found that less educated consumers are more likely to acquire nutrition information from television than from print sources (46).

Tips:

- Use simple words, sentence structure, and concepts.
- Use broadcast media, such as videotapes, to illustrate basic concepts and uses.
- Recognize that background skills may need to be explicitly taught.
- Use tips listed for older consumers (except for large print), since these would be effective with less educated consumers as well.

Consumers on restricted diets will benefit from specific messages targeted to them because their immediate need to make effective use of new labels is greater, and they may use the label in more complex ways than the average consumer. The research showed that people on restricted diets generally used food labels more often than others, but they were not better at understanding label information or more knowledgeable about diet and health.

Tips:

- Provide specific advice on foods to increase as well as foods to limit, since many consumers lack background knowledge about relevant food sources of nutrients.
- Identify target levels for specific nutrients in terms of the Percent Daily Value (key points to include are described in Section V.B).

- Consider barriers related to age and limited education as well as health condition (see preceding tips) that may be appropriate to the target audience.

Some materials and programs are underway to target special groups or special messages, although gaps remain. The FDA, in cooperation with other federal, state, and local agencies, is sponsoring food label education pilot projects at the local level (47). These projects target elderly and low-income consumers as well as those from diverse ethnic backgrounds. Some steps have also been taken toward targeting label education to people on specific diets. The FDA plans to publish articles in the *FDA Consumer* geared to specific health conditions. The American Heart Association collaborated with the FDA to develop a brochure introducing the new food label, with an emphasis on how the label relates to preventing heart disease. The National Food Processors Association, in consultation with the FDA, USDA, and major health and consumer organizations, developed a set of consumer "fact sheets" on specific nutrients (fat, saturated fat, salt/sodium, calories, cholesterol, carbohydrates, dietary fiber, vitamins, and minerals) and on particular health conditions (diabetes, heart disease, cancer). These materials are part of a larger educators guide that includes technical information and food label use exercises (48).

2. Awareness and Concern

American consumers have been responsive to public health campaigns relating specific nutrients to serious health consequences. Impressive gains have been made in awareness of diet-disease relationships over the past 20 years, and, although education differentials remain, improvements in awareness have occurred in all education categories. Currently, the nutrient of greatest concern is total fat, but a relatively high level of concern also exists for cholesterol and sodium. Less well known nutrients, such as saturated fat and fiber, may be more easily considered in daily diets now that the nutrient information is listed on most products.

Tips:

- Alert consumers that the nutrition label will now be on almost all foods
- Alert consumers that nutrients of public health concern (such as saturated fat, fiber, and carbohydrate) are now listed on the label
- Highlight the improvements to the label, such as its new format that makes it easier to compare products
- Link concerns about specific nutrient-disease relationships to the information on the new label (i.e., show how the label provides personally relevant information)

- Tailor the content of messages and the tips on using the label to the particular concerns of the target group

3. Knowledge and Beliefs

Misperceptions and knowledge gaps persist, even though consumers have become more knowledgeable about diet and health issues generally. Consumers often lacked the information necessary to translate their nutrient concerns into healthier diets. Many consumers did not know the recommended amounts or major sources of nutrients they wished to add to their diets (such as fiber) or to limit in their diets (such as fat).

Tips:

- Demonstrate how the Percent Daily Value provides an easy to use nutrition profile of foods
- Encourage consumers to refer to the Nutrition Facts panel to learn which foods they eat are good sources of nutrients they want to increase in their diets and which foods are high in nutrients they want to limit
- Explain how the Percent Daily Value incorporates important dietary advice, so the consumer does not have to remember specific guidelines
- Alert consumers that nutrient content claims can be used with confidence to help find products with desired nutrient characteristics

4. Dietary Management and Label Use

Important changes have occurred in the American diet over the past 30 years, suggesting that public health messages do affect eating behavior. The research showed that most consumers use the food label at least some of the time, but some of this use is probably perfunctory. The experimental studies showed that consumers are able to use the label accurately for some purposes. But in these studies, participants were motivated to cooperate and to pay attention to the labels. In the real world, time pressures and distractions may limit label use. Even minimal instruction is likely to help many consumers understand new elements of the label, as the experience in the format study showed with a brief description of Daily Values.

Tips:

- Show consumers how to extract relevant label information quickly and easily
- Develop guides for using the label to follow particular diets or dietary recommendations

5. Purposes for Which Consumers Use Labels

Survey research showed that consumers tend to use labels for simple tasks, such as product comparison or determining how low or high a product is in a certain nutrient. The format research showed that virtually all consumers are able to use labels for product comparison, and many can make low-high judgments quite accurately.

Quantitative dietary management (i.e., keeping running totals for specific nutrients over the course of a day or other period of time) is an infrequent use of current labels and is likely to be a rare use of the new labels as well. Consumers may be discouraged from using the new labels if they believe such time-consuming and difficult tasks are its major purpose.

Tips:

- Emphasize easy, familiar uses of the label, such as product comparison and low-high judgments
- Provide practical hints on how to use the label for familiar tasks (e.g., comparing the Percent Daily Value columns)
- Reserve suggestions for quantitative dietary management to those consumers who truly need this level of complexity

B. Label Format Elements

Consumers may need assistance to effectively use the information on the new food label, particularly new or changed information. Suggested points to include about each of the key format elements are discussed in this section.

1. Serving Size

Interpretation of almost all the nutrition information on the label depends on referencing the serving size. Consumers' previous skepticism about serving size information may keep them from recognizing that it has changed or appreciating why they will be better able to use the other label information if they first examine the serving size. For example, consumers need to know that if they eat all the food in a package that says it contains two servings, the nutrient amounts need to be doubled.

Points:

- New labels have new serving sizes. The familiar serving sizes from the old labels may change. This information is particularly important for consumers on restricted diets.

- Serving size is based on the amounts customarily consumed, so they should better reflect the amount consumers actually eat.
- The serving size is now set by regulation so that similar products have the same serving size, making them easier to compare. (But some apparently similar products have different serving sizes; examples are different types of breakfast cereals and snack versus non-snack crackers. Consumers still need to check the serving size before comparing nutrients in different foods.)
- Serving size must be declared in common household units, so consumers should find it easier to relate the serving size to the amount they eat.
- Serving sizes on new food labels are *not* recommended amounts to consume for good health. Other consumer information, such as the Food Guide Pyramid and Five-a-Day program, is based on different serving sizes and provides dietary advice. (A consumer brochure and educator's guide that link the new food label with the Dietary Guidelines for Americans and the Food Guide Pyramid is being developed by the USDA).
- Those on special diets who rely on systems such as the American Diabetes Association/American Dietetic Association Exchange Lists for Meal Planning need to pay careful attention to label serving sizes, since they differ significantly from these guides.

2. Percent Daily Value

The Percent Daily Value presents product specific information in relation to dietary guidelines. For example, the Percent Daily Value for total fat shows how much the grams of fat in one serving contribute to the total day's intake, based on a reference calorie amount of 2000 calories and the dietary recommendation that fat contribute 30 percent (or less) of daily calorie intake.

Points:

- The Percent Daily Value information is similar to the information on the old labels for the vitamin and mineral content (expressed as percent of U.S. RDAs).
- The Percent Daily Value information, combined with the number of servings consumed, contains all the information required for the most common label use purposes.
- The Percent Daily Value information eliminates the need for consumers to remember specific recommended amounts (e.g., how much fat to eat in a day).
- The Percent Daily Value eliminates the need for consumers to do any computations or pay attention to how nutrients are measured (e.g., in grams or milligrams) to relate the amount of a nutrient in the food to the amount needed daily.

- For product comparisons, it does not matter if the consumer eats more or less than the 2000 calorie reference amount.

Although the recommended amounts of some nutrients change with different calorie needs, the relative amounts listed as the Percent Daily Value accurately portray the nutrition profile of the food and can be used without adjustment for these common purposes. For example, a food with 20 percent of the Daily Value for total fat has twice as much fat as a food with 10 percent of the Daily Value for total fat. This is true regardless of the individual's calorie needs.

- Low-high evaluations can be made without adjusting the percentages if consumers eat a little more or less than 2000 calories, because the Percent Daily Value changes are not great for even a 400 calorie difference from the reference 2000 calories.

For example, a product that contains 10 grams of total fat in one serving has a Percent Daily Value of 15 percent. For a diet of 2400 calories, this product has a Percent Daily Value of 13 percent, and for a diet of 1600 calories a day, it has a Percent Daily Value of 19 percent. For people who must monitor their diets closely, these differences might be significant as they add up nutrient amounts in the foods eaten over the course of the day. But most consumers will use the Percent Daily Values as estimates to decide if the food has a lot or a little of the nutrient in question. To get a general sense of whether a nutrient is high or low, the Percent Daily Value is adequate.

- People who need to follow dietary recommendations precisely and who do not eat 2000 calories a day can easily apply the Percent Daily Value information by adjusting the percent targets.

For example, Percent Daily Values can be used to follow a quantitative diet that meets the recommended amounts for nutrients but limits calories to 1600. The label shows gram amounts that could be tallied for each nutrient, but the consumer would need to remember different gram amount targets for each nutrient. An easier strategy is to use the Percent Daily Value column and to calculate an adjusted percentage target for all nutrients that vary with calorie level. For this example, the calorie goal is 80 percent of the label reference amount (1600 divided by 2000 = 80 percent). The dieter would therefore strive for a daily target of about 80 percent rather than 100 percent of the Percent Daily Values for fat, saturated fat, total carbohydrate, and dietary fiber. A caution with lower calorie diets is that the NCI recommends a minimum of 20 grams of fiber a day (80 percent of the reference Daily Value). For cholesterol and sodium, which do not vary by calorie intake, the target for the summed Percent Daily Values would continue to be no more than 100 percent, respectively. Similar adjustments could be made for those who eat more than 2000 calories per day.

- Consumers who need to restrict a particular nutrient can keep a daily tally more easily by adjusting the total percent, so they do not need to search for gram amounts or handle more difficult figures. For example, if a person needs to restrict sodium intake to 1200 milligrams per day, the adjusted Percent Daily Value for sodium would be 50 percent (1200 divided by 2400, DV).
- Consumers need to be aware that the Percent Daily Value shows how much of the nutrient the food contributes to the *total daily diet*; it does not show how much of the nutrient is in the food. Misperceptions may arise, especially with consumers who are familiar with percentages in terms of the contribution of a nutrient to the calories in the food, such as the percent of calories from fat. Many consumers recognized the advice to limit calories from fat to 30 percent of total calories, and they tended to overgeneralize by assuming it referred to individual foods as well as to the total diet. In addition, some pre-NLEA food labels highlighted the percent of calories in the food that came from fat.

3. Daily Values

The listing of Daily Value metric amounts for two calorie levels is displayed in an expanded footnote, available only on larger food packages. Unlike the other information on the Nutrition Facts panel, the footnote is not product-specific. Rather, it provides general dietary guidance, based on current scientific knowledge concerning diet and health.

Points:

- The Daily Values listing identifies nutrients of public health concern, by virtue of including them in the footnote.
- The Daily Values listing illustrates that some nutrients vary by calorie intake while others do not.
- The Daily Values listing shows that some guidelines represent maximum amounts (those that say "less than") while others are goals or minimums.
- The Daily Values listing displays the metric amount on which the Percent Daily Values are based for each macronutrient. Although metric information has limited use for most consumers, those following a quantitative diet can use it to calculate an adjusted percentage daily target for the nutrients on which they keep a tally, such as that shown in the preceding example for sodium.

4. Nutrient Content and Health Claims

Many consumers are skeptical of nutrient content and health claims, probably in part because they do not distinguish claims in advertisements from claims on food labels.

Points:

- Consumers can use claims on labels with greater confidence than in the past because nutrient content claims and health claims must meet regulated definitions. The definitions for some claims include reference to the amount of related nutrients in the product; for example, a product may not claim "low cholesterol" if it has more than two grams of saturated fat (which also raises blood cholesterol levels).
- Claims on food labels can provide a shortcut to identify a product with desired characteristics.
- Claims are voluntary; the absence of a claim does not mean the food has less desirable nutrition characteristics than another product that has a claim.
- Education on interpreting claims (comparative versus absolute claims, for example) and the continuum from "no" or "free" to "low" to "reduced" would be useful for those who use claims to identify products with specific characteristics.
- Consumers should not rely only on claims to choose foods; they should check the nutrition facts panel on the label to obtain more complete information about the nutrition profile of a food product.

C. General Guidance and Rules of Thumb

Managing a total daily diet for good health requires complex decisions for which consumers may need guidance or a rule of thumb. Such guidance is easier for some tasks and nutrients than others.

Points:

- Better low-high judgments might be made if consumers know that, on average, people eat 15 to 20 different foods in a day. A Percent Daily Value of five percent or less for any nutrient represents a small contribution to the day's total.
- A rule of thumb for "high" is more difficult to state because what is high varies for different nutrients, and for some nutrients depends on how the food is used in the day's diet.

For example, fiber is a nutrient that most Americans underconsume and that is found only in selected foods, generally in small amounts; a Percent Daily Value of 10 percent is considered significant. In contrast, setting a rule of thumb for total fat is more complex. Fat is a nutrient that consumers should limit, but it is pervasive in foods and easily over consumed. Fat is also often relatively high in protein foods that make an important nutritional contribution to main meals. Therefore, a relatively high Percent Daily Value for a main dish may be nutritionally necessary since it represents an eating occasion that will occur only once or twice a day. The same level of fat for a snack is more likely to lead to overconsumption of fat, and, in fact, the fat contribution of most foods eaten in a day will need to be relatively low to balance a high level for the main meal.

- The nutrition profile of some foods will present difficult trade-offs for consumers. For example, one food might have less total fat but more sodium than a comparison food, or one food might have less total fat but more saturated fat than another. Consumers trying to adhere to special diets, in particular, may need guidelines to help them make nutrient trade-offs.

VI. CONCLUSIONS

The implementation of NLEA marks a beginning. American consumers will have unprecedented access to accurate and detailed nutrient information on most of the foods they purchase. The new food labels will provide an opportunity to renew and reinforce consumers' interest in diet and health, and they will give consumers a versatile tool they can use to make healthier, more informed choices. Changing behavior is difficult, and although lack of information presents a barrier to change, the availability of information does not produce change, but simply makes it possible. Nutrition education can provide the impetus needed to help consumers use the newly available information to improve their diets.

VII. ENDNOTES

[1] A detailed review of food labels, their deficiencies, and ways to improve them to help consumers adopt healthier diets, is provided in a report sponsored by the Food and Drug Administration and the U. S. Department of Agriculture: D. V. Porter and R. O. Earl (eds.), *Nutrition labeling: Issues and directions for the 1990s*, National Academy Press, Washington, D.C. (1990).

[2] Public Law 101-535, November 8, 1990, 104 Stat. 2357.

[3] The various studies cited in this chapter have sampled different populations. These sample differences probably account for some of the differences in results that were found. The three main population groups sampled were American adults, main food preparers or planners, and food shoppers. Different results may also arise because the studies worded questions differently (see footnotes 5 and 6 for examples).

[4] Descriptions of the HDS, DHKS, and CSFII are available in a government publication that describes a wide variety of federal and state activities related to nutrition monitoring. Information includes the sponsoring agency, purpose, target

population, dates of data collection, design and methods, key variables, selected publications, and contact persons: Interagency Board for Nutrition Monitoring and Related Research, J. Wright (ed.), *Nutrition Monitoring in the United States: The Directory of Federal and State Nutrition Monitoring Activities*, Public Health Service, Hyattsville, Maryland (1992).

5 These percentages represent "top of the mind" awareness and thus include an element of saliency. Consumers were asked, "Have you heard anything about [health condition] being related to things people eat or drink?"

6 In the DHKS, the household member identified as the main meal planner or preparer was asked, "Have you heard about any health problems that might be related to how much [nutrient] a person eats?"

7 Chapter 13 by Gardner discusses this topic. For more information, see P.B. Hutt, Government regulation of health claims in food labeling and advertising, *Food Drug Cosmetic Law Journal*, *41*: 3-73 (1986) and R.M. Cooper, R.L. Frank, and M.J. O'Flaherty, History of health claims regulation, *Food Drug Cosmetic Law Journal*, 45:655-691 (1990).

8 Five formats were tested in Study 1: (a) Control—a metric declaration of nutrient values similar to the old format; (b) Control/Daily Value—a metric declaration of nutrient values with a list of Daily Values for key nutrients; (c) Adjective—a metric declaration and the list of Daily Values, with adjectives (low, medium, high) describing each nutrient; (d) Numeric—a metric declaration and the list of Daily Values with a separate panel that listed the percent of Daily Value; and (e) Bar Graph—a metric declaration and list of Daily Values with a separate panel that expressed the percent of daily value both numerically and as a bar graph.

9 Seven formats were tested in Study 2, three from Study 1 with slight modifications, and four new ones. The three from Study 1 were the following: (a) Control; (b) Control/Daily Value; and (c) Adjective. The four new formats were the following: (a) Percent/No Daily Value—a metric and percent of Daily Value listing of nutrient values, but no listing of the Daily Values themselves; (b) Percent/Daily Value—a metric and percent of Daily Value listing, with the list of Daily Values for key nutrients; (c) Grouping—a metric declaration and the list of Daily Values for key nutrients, with nutrients grouped under two headings ("Choose a diet low in" or "Choose a diet high in"); and (d) Highlighting—a metric declaration and the list of Daily Values, with bold type to emphasize nutrients at a level consistent with dietary recommendations.

VIII. REFERENCES

1. F.R. Shank, The Nutrition Labeling and Education Act of 1990, *Food and Drug Law Journal*, 47, pp. 247–252 (1992).
2. United States Department of Health and Human Services, Public Health Service, *The Surgeon General's report on nutrition and health*, Government Printing Office, Washington, D.C. (1988).

3. National Research Council, *Diet and health: Implications for reducing chronic disease*, National Academy Press, Washington, D.C. (1989).
4. United States Department of Agriculture and United States Department of Health and Human Services, *Nutrition and your health: Dietary guidelines for Americans*, Third edition, Home and Garden Bulletin, N. 232, Washington, D.C. (1990).
5. A.S. Levy, *Quantitative research at FDA and USDA: Key findings for nutrition education*, Paper presented at the National Exchange for Food Labeling Education Conference, June 11, 1993, Washington D.C. [Copy of talking points and slides available from the author. (Food and Drug Administration, Consumer Studies Branch HFS-727, 200 C St., SW, Washington, D.C. 20204).
6. A.S. Levy and J.T. Heimbach, *Recent public education efforts about health and diet*, Food and Drug Administration, Center for Food Safety and Applied Nutrition, Consumer Studies Branch, Washington, D.C. (1989).
7. Division of Consumer Studies/National Exchange for Food Labeling Education, *Summary of consumer research on health and diet attitudes and knowledge and use of food labels*, Food and Drug Administration, Center for Food Safety and Applied Nutrition, Washington, D.C., September (1992). [Copies available from Brenda Derby, Food and Drug Administration, Consumer Studies Branch, HFS-727, 200 C Street, SW, Washington, D.C. 20204].
8. Food Marketing Institute/Opinion Research Corporation, *Trends in the United States: Consumer attitudes and the supermarket 1993*, Washington, D.C. (1993).
9. Interagency Board for Nutrition Monitoring and Related Research, *Nutrition monitoring in the United States. Chartbook 1: Selected findings from the National Nutrition Monitoring and Related Research Program*, B. Ervin and D. Reed (eds.), Public Health Service, Hyattsville, Maryland (1993).
10. Food and Drug Administration, *Health and Diet Survey*, Center for Food Safety and Applied Nutrition, Consumer Studies Branch, Washington, D.C. (1990).
11. A.S. Levy, *Causal modeling of dietary intake behavior*, Paper presented at the meeting of the Association for Consumer Research, Nashville, TN (October, 1993).
12. U.S. Department of Agriculture, *Diet and Health Knowledge Survey*, Human Nutrition and Information Service, Hyattsville, Maryland, (1989).
13. B. Schucker, J.T. Wittes, N.C. Santanello, S.J. Weber, D. McGoldrick, K. Donato, A.S. Levy, and B.M. Rifkind, Change in cholesterol awareness and action: Results from national physician and public surveys, *Arch Intern Med*, 151, pp. 666–673 (1991).
14. Produce for Better Health Foundation, More Americans think 5!, *5 A Day News*, *3*(1), Newark, Delaware (December 1993).
15. National Institutes of Health, *Understanding benefits and barriers to fruit and vegetable consumption: Findings from focus groups conducted as part of the 5 a day for better health program*, Office of Cancer Communications, National Cancer Institute, Bethesda, MD: (1992).
16. International Food Information Council/The American Dietetic Association, *Gallup survey of public opinion regarding diet and health*, Princeton, New Jersey, (1990).
17. A.S. Levy, S.B. Fein, and M. Stephenson, Nutrition knowledge levels about dietary fats and cholesterol: 1983-1988, *J Nut Ed*, 25:60-66 (1993).

18. Wirthlin Group/Kraft, *Survey of American dietary habits*, American Dietetic Association, (September, 1993).
19. S.A. Cremer and L.G. Kessler, The fat and fiber content of food: What Americans know, *J Nut Ed*, 24:149-152 (1992).
20. Food Marketing Institute/Prevention Magazine (1993). *Shopping for health, 1993: A report on diet, nutrition and ethnic foods*, Washington, D.C. (1993).
21. D.S. Putler and E. Frazao, Assessing the effects of diet/health awareness on the consumption and composition of fat intake, In *Economics of Food Safety*, J.A. Caswell (ed.), New York: Elsevier 1991, pp. 247–270.
22. American Meat Institute/The Roper Organization, *Nutrition labels and meat: Consumer knowledge, usage, and attitudes—Report of a national consumer survey*, New York: Roper (1992).
23. C. Lenfant and N. Ernst, Daily dietary fat and total food-energy intakes—Third National Health and Nutrition Examination Survey, Phase 1, 1988–91, *Morbidity and Mortality Weekly Report*, Feb. 25, pp. 116–125 (1994).
24. Food Labeling News, *Americans using low-fat foods; cholesterol concerns wane*, Washington, D.C. pp. 6-7 (April 1, 1993).
25. Calorie Control Council, Reduced-fat, low-calorie eating more popular than ever, *Calorie Control Council Commentary*, Spring 15(1):1–2 (1993).
26. A.S. Levy, and R.C. Stokes, Effects of a health promotion advertising campaign on sales of ready-to-eat cereals, *Public Health Reports*, 102:398–403 (1987).
27. The Roper Organization, *Roper reports: Food and nutrition* (91-2, pp. 54–61), New York (1990).
28. M.M. Bender and B.M. Derby, Prevalence of reading nutrition and ingredient information on food labels among adult Americans: 1982–1988, *J Nut Ed*, 24:292-297 (1992).
29. C.A. Cole and S.K. Balasubramanian, Age differences in consumers' search for information: Public policy implications. *J Consumer Research*, 20:157–169 (1993).
30. J. Guthrie, *Using USDA data for food label education research*, Nutrition Label Users. Who Are They? Paper presented at the National Exchange for Food Labeling Education Conference, June 11, 1993, Washington, D.C.
31. B.M. Derby, *Consumer use of the food label: A market segmentation*, Food and Drug Administration, Division of Market Studies, Consumer Studies Branch. Summary prepared for National Exchange for Food Labeling Education Conference, June 1993. Paper in preparation.
32. The Roper Organization, *Roper Reports: Food labeling*, (92-4, pp.2–9), New York (1992).
33. Food Marketing Institute/Opinion Research Corporation, *Trends: Consumer attitudes and the supermarket 1990*, Washington, D.C. (1990).
34. The Roper Organization, *Roper Reports: Food and health*, (93-1, pp. 34–39), New York, (1993)
35. Market Facts, *Summary report: FDA nutrition labeling focus groups*, Food and Drug Administration, Center for Food Safety and Applied Nutrition, Clinical Nutrition Branch, Washington, D.C. (February, 1992).

36. B.W. Hickman, G.E. Gates, and R.P. Dowdy, Nutrition claims in advertising: A study of four women's magazines, *J Nut Ed*, 25(5):227–235 (1993).
37. A.S. Levy, *Consumer beliefs about the credibility of food advertising and food labeling*. Food and Drug Administration, Center for Food Safety and Applied Nutrition, Clinical Nutrition Branch, Washington, D.C. (April, 1987).
38. S. Fullmer, C.J. Geiger, and C.R.M. Parent, Consumers' knowledge, understanding, and attitudes toward health claims on food labels, *J Am Dietetic Assoc*, 91(2):166–171 (1991).
39. A.S. Levy, S.B. Fein, and R.E. Schucker, *Evaluation of nutritional label formats: FDA Study 2*, Food and Drug Administration, Center for Food Safety and Applied Nutrition, Consumer Studies Branch, Washington, D.C. (1992).
40. S.B. Fein, B.M. Derby, and A.S. Levy, *Consumers' abilities to use food labels: Implications for education,* Paper presented at the meeting of the American Psychological Association, Washington, D.C. (August, 1992).
41. National Food Processors Association/M/A/R/C, *Food industry labeling study: An assessment of alternative label formats*, Washington, D.C. (1992).
42. National Food Processors Association/Opinion Research Corporation, *Food labeling and nutrition . . . What Americans want*, Washington, D.C. (1990).
43. A.S. Levy, O. Mathews, M. Stephenson, J.T. Tenney, and R.E. Schucker, The impact of a nutrition information program on food purchases, *Journal of Public Policy and Marketing*, 4:1–16 (1985).
44. A.S. Levy, S.B. Fein, and R.E. Schucker, More effective nutrition label formats are not necessarily preferred, *J Am Dietetic Assoc*, 92(10):1230–1234 (1992).
45. A.S. Levy, *The advantage of percent declaration formats on dietary management tasks*. Food and Drug Administration, Center for Food Safety and Applied Nutrition, Consumer Studies Branch, Washington, D.C. (July, 1992).
46. American Heart Association/Product Development Services, *Food label use needs assessment*, Washington, D.C. (1991).
47. FDA/USDA Food Labeling Education Information Center, Pilot programs target elderly, low-income, and ethnic consumers, *Update*, Fall/Winter 1993, pp. 1,7 (1993).
48. M.B. Browne, *Label facts for healthful eating: Educator's resource guide*, National Food Processors Association, Mazer Corporation, Dayton, OH (1993).

13

Consumer Communications: Informing, Educating, and Persuading the Public to Use the New Food Label

Lauren Swann

Concept Nutrition, Inc., Bensalem, Pennsylvania

I. INTRODUCTION

"It is estimated that the new food label will cost FDA-regulated food processors between $1.4 billion and $2.3 billion over the next 20 years. The benefits to public health—measured in monetary terms—are estimated to well exceed the costs"(1). This assertion hinges on the greatest speculations of the NLEA: that consumers can and will utilize the new label to make healthful food choices; that these choices will subsequently contribute to lower risks associated with degenerative disease; and that the ultimate result will be lowered health care costs from reduced medical care and treatment for chronic health conditions.

Introduced into Congress in 1990, the new food label—after years of controversy over NLEA—is finally a reality and an integral part of public health goals. "The proposed Year 2000 Objectives for the Nation, provide a number of nutrition objectives that relate to consumer knowledge of foods, nutrition, labeling, and availability of improved food products. Objectives specific to labeling are to: increase to at least 80 percent the proportion of people age 21 and older who use food labels to make food selections; increase nutrition labeling that provides information to facilitate choosing foods consistent with the Dietary Guidelines for Americans to at least 80 percent of processed foods and 40 percent of fresh meat, poultry, fruits, vegetables, baked goods and ready-to-eat foods" (2).

It is the "E" in NLEA which addresses the crucial link between informed consumers and healthful food choices. "The NLEA education role is to enable the public to readily observe and comprehend required information and to understand its significance in the context of a total daily diet" (3). It is also the "E" in NLEA which entails a complex process to meet the challenge of successfully influencing the public to select a balance of foods which create a healthy diet. "Before any food choices can be made, consumers, must first have the information available (exposure), notice appropriate parts of the nutrition panel (attention), understand the terminology and how it is being presented (comprehension) and finally, be able to use their knowledge in various food-decision situations (retrieval)" (4).

II. THE NEED FOR COMMUNICATIONS SUPPORT FOR FOOD LABEL EDUCATION

New food labels are, at best, questionably effective without public education for their use. "It would be a terrible waste if all the thought, planning and effort that went into the new food label went for naught because consumers either didn't know there was something new or didn't bother to use it in planning their daily diet" (5).

The challenge of a public education campaign about the most sweeping and dramatic changes ever experienced by the food industry creates an enormous task. Fred Shank, Ph.D., director of the FDA's Center for Food Safety and Applied Nutrition, and other nutrition experts acknowledge that the task of teaching consumers how to use the new label will be formidable. "It's been a long haul," Shank said about efforts to revamp the label. "But the greatest challenge lies ahead—in educating consumers" (6).

"NLEA calls for activities which educate consumers about the 1) availability of nutrition information in the label or labeling of food, and 2) the importance of that information in maintaining healthful dietary practices" (7). Because the nation's food supply and its regulation are a convergence of private industry; local, state and federal government; not-for-profit trade associations; and professional organizations, the government is justified in seeking and encouraging support and cooperation from public, private, and community sectors in this effort. "Neither FDA nor USDA by themselves has adequate resources to effectively inform and educate the entire American public about the new food label and how to use it to plan a healthy diet. Therefore, the agencies are encouraging consumer groups, industry associations, health professional associations, voluntary health organizations, and other federal agencies to work together to provide consumers with what they need to know to make the most of the new food labels" (7).

Methods typically utilized by private industry are recognized as a valuable means of meeting these goals. "Goals of the NLEA include consumer education and improved diet. Industry, government, consumer groups and academics can promote healthier diets and better nutrition through effective marketing and advertising" (7).

There is evidence of a mutual gain for industry and consumers from a cooperative effort. "Consumers read nutrition labels on food products for a variety of reasons; mostly to determine whether a product is high or low in some specific nutrient or to compare similar products. Consumers generally find it easy to extract information from the nutrition label for these purposes. Reading nutrition labels on food products is a very desirable practice—but only motivated people do it. We need to motivate more people to read nutrition information on food products" (8). Besides the intended health benefit for consumers, the need to motivate more people to read labels has value for food manufacturers as well. Expanding consumer use of food labels beyond simple information extraction of isolated nutrients for product comparison could lead to enhanced perception of a branded product and strengthen loyalty or win new market share.

"People do not go to the grocery store to read and gain nutrition knowledge. They want to use the label for quick reference to aid in food selection. Should I buy this or that, yes or no" (9). A well rounded communications campaign can increase product awareness prior to the point of selection, be reinforced by label influence at the point of purchase, and strengthen repeat purchase to build brand loyalty. Appropriate use of the communication process is key for effective education.

III. COMMUNICATIONS COMPONENTS

Communication is defined as the act or means of making something known; an interchange of thoughts; the act of giving, receiving, sending, or exchanging information, ideas or messages. "Communication can increase awareness of a health issue, problem or solution; affect attitudes to create support for individual or collective action; demonstrate or illustrate skills; increase demand for health services; remind about or reinforce knowledge, attitudes or behavior" (10).

A. The Communications Umbrella

In a sense, communications encompasses a continuum including *information, education*, and *persuasion.* This continuum begins with information. Information is defined as facts, news, or knowledge told; or the act of giving knowledge of something, telling, or acquainting. Therefore, information, by definition, is heavily based in objectivity. Messages are sent with minimal intended influence on the recipients thought process relative to subsequent action.

Education can be considered the midpart of the umbrella. It is defined as instruction or training by which people learn to develop and use their mental, moral, and physical powers; a gaining of experiences; the process of developing knowledge, skill, or character; or the act of giving/developing knowledge, skill, mind, or character. With education there is an intended effect or consequence as a result of the communications process. It is noteworthy that while education is generally considered a predisposition to learning—defined as the acquisition of knowledge, skill, or ability by study, instruction, practice, or experience—learning is measured by a permanent change in behavior. Therefore, education and learning are not necessarily synonymous.

At the opposite end of the continuum is persuasion, defined as the act of inducing by argument or advice to do something, the act of convincing to cause or do especially by urging, coaxing, or reasoning; to cause to do something especially by reasoning, urging, or inducement; to induce by argument or advice; or to convince. Persuasion, then, is that aspect of communications which has the maximum intent of subsequent action as a result of the communication process, but is based least in objectivity.

Every communications process involves a message originating from the send- er. It is the recipient's *perception*—the act of becoming aware by comprehension—that is the ultimate result of the communication process. For example, "In response to the perceived concern that some consumers have over the use of monosodium glutamate, food flavor potentiators have been sought that are capable of replacing MSG in foods" (11). Clearly, knowing consumer perception of food composition technicalities is far more valuable than simply knowing technical nutrient facts. It is these different components of the communications continuum which form the basis for the major methods of applied communications.

IV. MARKETING COMMUNICATIONS VEHICLES

Marketing communications is the practical application of these defined communications processes. It encompasses advertising, public relations, and consumer education and utilizes mass media as a channel of dissemination. Marketing communications campaigns evolve from product marketing goals. which begin with product positioning.

A. Product Positioning

Described as "the most ambitious and comprehensive revisions of food labeling rules in history" (12), the vital disclosure aspects of nearly every food product sold in the United States are affected by NLEA, creating a new era for food marketing. All required and optional food label declarations are an integral part of product positioning. The label functions as a comprehensive disclosure message. Requirements include product identity statements or common names, which are determined by product composition; ingredient declarations and Nutrition Facts. Optional fanciful names and vignettes contribute additional descriptors and perceptions. Ingredient and nutrient content claims, also optional, can create a specific marketplace position.

Determination of product positioning begins with marketing research to evaluate current product perception, key appealing product characteristics, other influential characteristics, nutritional promotion of competitive products, and the influence of brand equity and loyalty in market share. For food labeling, evaluation of nutrient characteristics relative to individual health needs, general public health goals, and population variability, evaluation of normal or average product consumption (i.e., meal or recipe usage) is warranted to determine how to effectively position the product through labeling messages. Most importantly, the fundamental role of taste must be considered. While studies indicate that "a growing number of shoppers are reading food labels and using them to make purchasing decisions, taste remains the primary factor in food selection" (13).

Examination of these issues enables identification of aspects of the current label which influence purchase and review of nutrient and ingredient qualities with appeal. The roles that brand names, vignettes, and graphic design play in product positioning and competitive choice must also be considered along with a determination of whether ingredients, quality, processing, or basic composition are stronger characteristics than nutrient profile.

The food label is a comprehensive disclosure message which is the bridge between regulations and point of purchase promotion providing product information for purchase. The impetus for new regulations is truthful, not misleading product promotion. By building on a solid premise, not a fad, for lasting equity in product positioning, the label becomes a marketing communications tool. For maximum effectiveness, the food label is part of a total product communications effort. Once product positioning is established, marketing communications plans can be developed.

B. Advertising

Advertising is based on persuasion. Brand advertising is built on product attributes as they reflect perceived needs and desires of the intended audience. Because advertising is regulated by the Federal Trade Commission (FTC), food product claims in advertising are not necessarily subject to FDA labeling requirements. However, "the agency (FTC) is responsible for balancing consumer protection with free commercial speech" (14). "Since the new label guidelines are aimed at reducing confusion in the marketplace, helping consumers make informed food choices, and providing incentives for the food industry to produce innovative, healthful products, advertising policy will be largely aimed at the same thing. The issue facing regulators is to what extent advertising content will be specified and mandated, and just how consistent ads and labels must be" (15). Chapter 4 by FTC Commissioner Deborah Owen and Joyce Plyler describe the agency's current position on NLEA nutrition labeling and advertising.

Because the ultimate goal of advertising is based on persuasion, it is different than the FDA-mandated disclosure, which is objective information about product composition. "It can be shown again and again that the pure recital of technical claims about a product leaves the reader cold and does not arouse the emotions necessary to make him change to a new product" (16). These are the limitations of strictly using strictly required disclosure as the sole basis for an ad. Nor is there the same opportunity or need in advertising, as there may sometimes be in labels, to ensure that all of the minutiae related in some way to a particular claim are simultaneously communicated to the consumer. Rather, the FTC as has noted, with advertising the opposite is generally true. If the contents of an effective and entirely accurate label were converted into an advertisement, that advertisement could be ineffective: not only failing to generate interest in the product, but missing the opportunity to educate consumers with respect to nutrition and health" (15). Advertising has a different modus operandi than labeling. Claims and declarations can be expanded, contextual and dramatized to become relative to consumer situational circumstances. Advertising can function to create product awareness in support of positioning and draw attention to the label. To sum up advertising goals in print ads, for example: "Copy is intended to shape motives and desires, to build believability and to provide a reason for selecting a particular brand over all others" (17). Advertising techniques such as this are crucial to effectiveness. Persuasive ad techniques expand beyond the label and need to relate to the audience to be effective.

Because advertising is heavier in persuasion, unproven or minimally supported research can be easily abused when it is introduced this way. Advertising has limited time in broadcast media, and limited space in print media, which often precludes lengthy educational messages. Whenever possible, advertising and labeling can and should reinforce and support each other to minimize consumer confusion. Ads can and do inform and educate but persuasion to sell the advertised product is the primary goal. Ads educate only enough to sell the product. Therefore, advertising is best utilized to create product awareness, particularly with new product packaging which bears the new label.

"Some product beliefs may be derived from explicitly stated features or attributes that are conveyed in the advertisement. In contrast, inferences are defined as beliefs that the individual derives as a result of elaboration and abstraction process. Therefore product beliefs that the consumer stores in memory may result from both explicitly stated descriptive advertisement information and individuals inferences which occur as the result of the advertisement processing. In sum, product beliefs that result from advertising may go far beyond what is explicitly conveyed in the ad, and an advertisement may be capable of generating residual beliefs well after exposure. Advertising should be targeted to an audience that is highly involved with the product or service being advertised. This can be achieved through ad placement or timing in media where the target audience is likely to be highly involved (18). Advertising to promote the new label is probably most effective within the context of a comprehensive campaign including public relations and consumer education.

C. Public Relations

"Public Relations relies on information. Public Service Announcements (PSAs) educate and inform. Public relations responds to inquiries" (10). Intended primarily to generically promote a product or image and strengthen company visibility, public relations is based on the information aspect of the communications umbrella. "We know now that there is a straight line relationship between product purchase and excellence of company image" (19).

Public relations programs can range from simple mailings of a single brochure to a targeted audience to elaborate multi-level campaigns targeting several interrelated audiences and utilizing all forms of media over an extended length of time. Public relations can involve events, sponsorships, and partnerships or focus heavily on concentrated publicity in print and broadcast media. Public relations differs from advertising because advertising is purchased print space or air time to promote a specific brand with the express purpose of increasing sales. The true public relations approach avoids overt brand name promotions, although the ultimate goal is company or brand name mention or visibility as a result of the effort. Additionally, when advertising is part of a public relations campaign, the ad is usually promoting a concept or issue as a opposed to a branded product.

Two public relations vehicles are applicable to consumer food label education:

1. Consumer Education

This is the aspect of the communications continuum based on education. Consumer education programs rely heavily on print materials to communicate straightforward, unbiased facts about a product-related issue or to generically support product characteristics. Materials should offer accurate complete and relevant messages; be appropriate for target audience in format, style, and readability; and be available and affordable (10).

Consumer education materials are an excellent way to provide straightforward information and facts about the new food label to consumers, using branded company products as the label example. Materials can be appropriately tailored for intended audiences and distributed through direct mailings to multiplier audiences; used in support of large scale public relations programs; or sent to consumers primarily upon request.

Consumers are eager for information about nutrition issues but are confused by conflicting messages. To cope with the confusion, skepticism, and negativism, practitioners must provide far more information than ever before. That information must be more scientific—and thus presumably more reliable. Yet it must also be clear and understandable and, above all, the source must be credible, untainted by vested interests" (20).

Consumer information can include product literature, recipe brochures, toll-free numbers, booklets, pamphlets, fact sheets and posters.

The success of the new food label depends on the ability of consumers to use it. If they are unaware, unmotivated or confused, the label is useless. Labels are complex and consumers need training and background information in order to interpret them. Education programs must address the embedded assumptions—the skills necessary to use the label—as well as the content of the label itself.

To avoid losing consumer acceptability, consumer affairs programs should maintain public relations as a minimally branded method of marketing communications. Great care should be taken to ensure that the materials are free of either explicit or implicit bias and that no general approval of sponsoring industries products is implied (21).

2. Mass Media

The mass media, particularly newspapers and news programs, are based primarily on the information end of the communications continuum. Utilizing the mass media for food label education is part of the government plan. "Among the activities FDA and USDA have agreed to carry out are: Working with the print and electronic media—both news and entertainment—to involve them as much as possible in increasing public awareness of the new food label" (7).

Entailing print, broadcast, and electronic media, the mass media is undoubtedly the most powerful and broadest-reaching means of communicating today.

Mass media—whether print or audiovisual as in radio or television—are best used to create awareness or expand knowledge about a particular new concept or program. Mass media can be effectively used to introduce a new concept or piece of information to target populations and thus create a better knowledge base about the meaning of that new concept or idea (22).

Use of the mass media, however, must be carefully leveraged to avoid sensationalism and potential bias.

> Under the glare of intense media attention, specific consumer food safety concerns can erupt into widespread alarm. A recent example occurred when use of the chemical daminozide (sold under the brand name Alar) on apples was questioned. Although there was no clear scientific consensus on the health risk posed by Alar, the public reacted strongly (23).

Additionally, different channels of the mass media hold different values. Television is visual yet intrusive, requiring use of two senses for effective message comprehension; radio is a background and portable media; print media requires a minimum educational base and mental concentration. One must determine the audience they want to reach and then carefully select the appropriate media channel.

Mass media can transmit news quickly to broad audience but cannot be expected alone to motivate people to change their behavior. the mass media are generally the public's primary source of information but may be less trusted than more intimate sources of information and are constrained by time, space and newsworthiness to the extent they can explain complex information properly. Media may focus too much attention on new information or information affecting limited segments of the population; may increase the chances for miscommunication of complex or controversial news; may communicate incomplete information. Loss of control over how the information is communicated may be a trade-off for broad and rapid transmission. The purpose of mass media is to inform and entertain, not educate. Therefore, if the complete message is too complicated or simply not considered interesting enough, you must redesign for media professionals and their perceptions of audience desires. Working with media professionals will help assure that messages are interesting as well as accurate (10).

V. MARKETING COMMUNICATIONS CAMPAIGN PLANNING

With the new labeling mandate, manufacturers benefit from encouraging consumers to use the new label as well as taking responsibility for the nutrients that their products contribute to consumer diets. This responsibility forms the basis for a credible marketing communications background. "Personal or direct influence strategies are based on providing information or applying educational, persuasion, and behavior modification techniques directly to individuals or small groups" (22). The specifics of campaign planning begin with basic determinations. In consumer based health communications, a good strategy includes four core questions (24):

- What is the purpose?
- Who is the target?
- What is the promise?
- What is the support?

The communications strategy statement should include (10):

- The program objectives, the primary and secondary audiences.
- The target information to be communicated.
- The benefit as perceived by audience.

Successfully communicating credible news to the consumer today is not without challenges. "The largest issue for food and beverage public relations as we see it is maintaining and building consumer confidence in the food supply" (20). When developing messages, the goal must reflect current consumer perception. The message must be relevant to consumer interests and concerns. It must focus on what consumers want to know. The message must be backed up with simple but accurate explanation and cues to action. The consumer must be drawn into making conscious decisions regarding purchases and eating behavior. Repetition is an essential ingredient (25).

A. Need for Integrated Campaign Planning

Using health messages to sell products without a balanced perspective brought us to NLEA—stricter government regulation to knock down the "Tower of Babel" as former DHHS Secretary Louis Sullivan described the food labeling state of affairs when reform was initiated. Another challenge resulted from unbalanced nutrition messages: "Consumers are confused about what foods are healthy and unhealthy, largely because of conflicting reports coming from nutritional experts. The result is a decline in those who say nutrition is important to them or that they are very concerned about nutrition" (26).

"People are becoming turned off by information overload and confusion" (13). A classic example is a magnified "cholesterol free" claim on foods such as margarine or vegetable oil. Although the claim is technically correct, a perceived health benefit is potentially implied to consumers understandably confused by the complexities of cholesterol versus fat reduction versus fatty acid versus calorie control recommendations.

> In order to facilitate dietary changes, consumers must understand the contributions that specific foods and food types make to the overall diet. From this perspective, science-derived diet and disease messages must be supplemented by information about the nutritional

> characteristics of specific foods and food types, and how to buy and prepare appropriate foods and meet appropriate quantitative goals. From the perspective of helping consumers make intelligent food selections, one difficulty with diet and disease messages is that these messages usually are based on food components whose scientific role is not well understood by consumers. Consumers may need to acquire knowledge about specific nutrients, food components or food and nutrition processes in order to implement the generic dietary advice implied by such messages. Appropriate dietary behavior depends on consumers' ability to recognize foods with desirable properties, to understand relevant terminology and apply it to food choices, to critically evaluate claims, and to assess the relative benefits of possible dietary changes in their own diets (22).

The preceding can only be achieved through utilization of information, education and persuasion models via advertising, public relations and consumer education in the mass media.

The information consumers find most difficult to understand on food labels are chemical names, technical terms and unfamiliar ingredients (27,28). When developing an integrated program, different aspects of the label may need a different emphasis and therefore a specific vehicle of consumer communications in the integrated process. Complex messages or explanations require a suitable medium for effective comprehension.

For example, the USDA Food Guide Pyramid offers the manufacturer of grain based foods an excellent opportunity for combining nutritional promotion and education with marketing. On the other hand, while many bread companies include the pyramid in print advertisements and on the label, their use of dough conditioners to improve the quality, handling and processing characteristics of the food product, health-conscious consumers may be intimidated by the string of chemical compound names they read on the label (11). Clearly, a consumer education component of this public relations campaign must support and strengthen the effectiveness of the primary theme about bread. Health Communications programs cannot produce behavior changes without supportive program components to address and prioritize the issues and desired message(s) (10).

The effectiveness of an integrated approach is well supported by successfully implemented efforts. During 1989 and 1990, Phil Sokoloff, a heart attack victim and founder of National Heart Savers, a non-profit organization funded primarily by Sokoloff to increase cholesterol consciousness, implemented "poisoning of America" advertisements in several major newspapers in an effort to "bring about change in the food industry and increase public awareness of the potentially harmful effects of dietary cholesterol and saturated fat." During one phase of the campaign, he held a press conference and distributed releases to local television networks the same day the ads ran, successfully generating heavy coverage on the evening news. "Sokoloff demonstrated the efficiency with which the mass media can be used to induce and accelerate change in the food industry"(29). In September 1994, celebrating NLEA and the new nutrition labels he rented a billboard in Times Square with the message "Cut Fat Intake and Live Longer" together with a typical nutrition label format. In addition, he offered $1 million in a supermarket sweepstakes promoting better nutrition. Both presentations attracted widespread media coverage. One article was entitled "Times Square on a Diet" (30).

B. Role of Partnerships

Private industry, government agencies, and not-for-profit organizations continuously contribute to the shape and direction of the food system by leveraging and balancing their interests and motives. It is the conscious and concerted initiative to join forces for a specific issue which creates partnerships. "An effective education campaign requires carefully constructed communication strategies, the help of partnerships and networks, and the support of a body of research. The task of educating the entire American public is too resource-intensive for any one group to manage alone."

"Partnerships and networks divide the workload and multiply the results, of a public education program. Partnerships take time to arrange, but are worth it because they can accomplish so much" (31). With integrated marketing communications planning, each sector is suited to one of the marketing communications vehicles: private industry is experienced in and financially equipped for advertising; community and public health organizations are compatible with grass roots consumer education; not-for-profit trade associations and professional organizations have a clearer advantage for public relations methods. Cooperative ventures can help nutrition education and marketing programs work and reach a broader audience (10).

> The benefits of an association between nutrition educators and industry—provided the linkages are well chosen and credible—can be enormous for both sides. There are those who believe that in some respects it would be irresponsible for the corporate sector not to be active in educational partnerships. After all, what's the benefit of manufacturing and marketing more health-conscious food products to a consumer caught in a nutrition maze? Linkages with corporate partners can provide far more than the obvious financial contribution (21).

Five major types of relationships can exist (21):

- Product endorsements.

- Endorsements of health messages endorsements of educational or informational materials to which a professional society is willing to affix its name or logo or seal of approval.
- Certification of food products whose nutrient composition meets designated standards of the society/association so that its name can be displayed on the products.
- Allowing advertising of food products in an association's publications and the promotion of food products at its meetings or exhibits.

Project Lean, a component of the Henry J. Kaiser Family Foundation campaign to reduce the nation's risk levels for five preventable major morbid health conditions, has among its objectives: "to increase availability of low fat foods from supermarkets, restaurants, cafeterias and increase cooperation and collaboration among national and community organization committed to reducing dietary fat levels." Targeting consumers and those who influence choices—food processors and outlets—"Project Lean is a collaborative and interdisciplinary undertaking advised by 'Partners for Better Health' a 30 member coalition of the nation's leading health organization. Program materials, concepts and strategies are developed and exchanged with and through these member organizations" (29). Project Lean also demonstrated the effectiveness of utilizing opinion leaders as gatekeepers. "Partnerships take longer but are worth it, if sources are credible and you have mutual goals" (31).

C. Professional Opinion Leaders—Gatekeepers To Consumers

"Multipliers such as classroom teachers and physicians are important as well because they function as information gatekeepers. If you expect them to carry your message to consumers, you must address their needs and limitations when designing your program. Include multiplier audience representatives in the planning, advisory and review stages of product development and marketing" (32). For nutrition messages, opinion leaders include dietitians, nutritionists, nurses, physicians, fitness instructors, and home economics teachers. This role can even expand to celebrities such as fashion models, performers, athletes and high profile fitness experts. Gaining their confidence and acceptance first increases the chances of effectively communicating with consumers because consumers will seek their opinions, advice, and approval and be influenced by their endorsements.

Gatekeepers also have the multiplier effect of passing materials on to their publics. A single brochure with a request form for additional copies sent to a selected professional audience can measure interest prior to executing a questionably effective mass mailing effort. Requests for feedback can also be a means of gaining valuable insight on the appeal of the materials. Newsletters and multi-faceted educational kits are two examples of materials developed for professional audiences.

D. Targeted Audience Campaign Planning

During the 1980s, marketers moved away from simply using demographics—age, race income, education—to using psychographics for analyzing and developing marketing plans. The American population has become so diverse, psychographics takes values such as socioeconomic status and lifestyle choices into consideration, supporting the importance of recognizing and understanding different segments of the population and, most importantly, their perceived needs and motivational factors. "Increasingly, food companies need to target products at particular segments" (23). Less educated consumers, in particular those with less than a high school education, are more likely to agree that there are so many recommendations about healthy ways to eat (29), "I don't know what is good or bad" and "what I eat won't make much difference in my chances of getting heart disease" (33). "Elderly consumers and those with less education are the most vulnerable population of label readers. They have greater difficulty using labels to discriminate product differences and tend to perceive differences where none exist" (27,34). "Less educated shoppers are more likely to acquire nutrition information from television than from print sources such as newspapers and magazines" (27,35).

The need to reach less affluent, literacy-compromised populations is emphasized because it is often these populations that have the greatest need for public health intervention and assistance to improve health habits. "The greatest challenge of the education campaign is reaching low income, low literacy, unmotivated and disinterested populations. These 'should knows' need nutrition education the most, as compared with the 'must knows' who have food allergies the 'need to knows' who have health risks and the naturally curious 'would like to know'" (4).

Campaign planning with sensitivity to populations is essential to achieve the ultimate NLEA goal—lowered health care costs.

> The importance of segmenting your audience cannot be overstated, because there are many separate publics. Beware of trying to target too many consumers at one time. Low literacy, for example, is too broad a group to target. You can narrow an audience by defining specific cultural or ethnic differences. Develop your message, tailoring it every step of the way to for your specific audience (36).

VI. SUMMARY

NLEA is a government mandate for required disclosure. It behooves manufacturers to educate consumers so labeling is a meaningful way to promote their products and gain consumer confidence. Used properly, these regulations for non-deceptive labeling information can help regain consumer confidence in the food system. It is a step toward educating the consumer, ultimately promoting the contributions each product can make toward a healthful diet. "Realize that the availability of information on the food label is of little use if consumers are not aware of it or if it is not clearly communicated in understandable terms" (26). As part of simple promotion and publicity plans or elaborate full-scale campaigns, the food label is intended to serve as consumer education for making wise food choices which will help minimize degenerative disease risk and ultimately reduce health care costs. Utilizing public relations, consumer education and advertising in the mass media is vital to the success of NLEA.

VII. REFERENCES

1. *FDA Backgrounder* HFI-40, BG92-4. Food and Drug Administration, December 10, 1992.
2. U. S. Department of Health and Human Services, *Promoting Health/Preventing Disease: Year 2000 Objective for the Nation*, Public Health Service Washington, D.C. (1989).
3. B. Derby, Highlights from FDA food label research: Implications for education, Proceedings of the National Exchange for Food Labeling Education, Washington, D.C., 1992.
4. L. Sims, An organizing framework for using information on the food label. , Proceedings from Exchange '92 The Communication Process: Research, Materials and Audience, Washington, D.C., 1992, pp. 16-17.
5. Taking the guesswork out of good nutrition, FDA Consumer Special Report; *Focus On Food Labeling, Read the Label Set a Healthy Table*, (1993).
6. P. Kurtzweil, Good reading for good eating, FDA Consumer Special Report; *Focus On Food Labeling, Read the Label Set a Healthy Table*:13 (1993).
7. FSIS and FDA Backgrounder, Public education campaign on the new food label, National Exchange for Food Labeling Education, February 1992, pp. 1–2.
8. A. Levy. Quantitative research at FDA and USDA: Key findings for nutrition education. Proceedings from the National Exchange for Food Labeling Education, Washington D.C., 1993.
9. E. Tuckermanty, Approaches to educating special populations about the food label: What does and doesn't work and lessons learned. Proceedings from Exchange '92

The Communication Process: Research, Materials and Audience, Washington, D.C., 1992, pp. 48.
10. *Making health communication programs work*, Department of Health and Human Services, National Cancer Institute and National Institutes of Health, Publ. No. 92-1493, 1992, pp. 1,13,59.
11. D. Pszczola, Friendly Labels: Responding to consumer desires, *Food Technology 5*:124 (1993).
12. J. Villforth, Proceedings of the New Food Label: An Authoritative Seminar on the New Labeling Rules, Washington, D.C., 1993.
13. Shopper behavior verifies need for nutrition education, *Milling & Baking News* 19:1 (1993).
14. Advertising guidance likely to differ from food label rules, *Milling & Baking News* 29 (1993).
15. P. Hollingsworth, Food advertising at the crossroads, *Food Technology 7*:74 (1993).
16. E. Dichter, A Psychological view of advertising effectiveness, *The Journal of Marketing 7*:61-66(1949).
17. A. Politz, The Dilemma of creative advertising, *The Journal of Marketing 10*:1-5(1960).
18. S. F. Gardial, D. W. Schumann, E. Petkus, R. Smith, Processing and retrieval of inferences and descriptive advertising information: The effects of message elaboration, *Journal of Advertising 1*:25 (1993).
19. J. Plachta, The art of gentle persuasion, *Marketing* 9/6:25 (1990).
20. R. Skolnik, A full plate: consumer concerns challenge food and beverage practitioners, *Public Relations Journal 10*:24 (1993).
21. C. Wilson, Strange bedfellows? (Cooperative relationships between the food industry and nutrition groups), *Food in Canada 5*:S12-S13 (1993).
22. *Presentation of Nutrition Information on Food Labels, Nutrition Labeling Issues and Directions for the 1990s* (D. V. Porter and R. O. Earl. , eds), National Academy Press, Washington, D.C. , 1990, p. 266.
23. B. Senauer, E. Asp, J. Kinsey, Major trends in a consumer-driven food system, *Food Trends and the Changing Consumer*, Eagan Press, St. Paul, MN, 1991, p. 7.
24. S. Sutton, Consumer based health communications, Proceedings from the National Exchange for Food Labeling Education, Washington D.C., 1993.
25. O. Matthews, Public and private partnerships, food label education and the supermarket. Proceedings from Exchange '92 The Communication Process: Research, Materials and Audience, Washington, D.C., 1992, pp 56.
26. B. Gatty, FMI presentation highlights consumer concerns, trends, questions about food safety, nutrition remain, satisfaction, economic indicators on the rise, *Grocery Marketing 6*:6 (1993).
27. *Summary of consumer research on health and diet attitudes and knowledge and use of food labels*. Division of Consumer Studies Center for Food Safety and Applied Nutrition Food and Drug Administration. Washington, D.C., 1992.

28. *Food Labeling and Nutrition What Americans Want*, National Food Processors Association, Washington, D.C., 1990.
29. R. J. Adams, K. M, Jennings, Media advocacy: A case of study of Philip Sokolof's cholesterol awareness campaign, *Journal of Consumer Affairs Summer 1*:145 (1993).
30. "Times Square on a Diet," New York Times, Sunday, September 4, 1994, Business Section, p. 20.
31. K. Durham, S. Sachs, Food label education process, Proceedings from Exchange '92 The Communication Process: Research, Materials and Audience, Washington, D.C., 1992, pp. 18.
32. *Health and diet survey*, Food and Drug Administration, Division of Consumer Studies, Center for Food Safety and Applied Nutrition, Washington, D.C., 1990.
33. A. Levy, S. Fein and R. Schucker, *A study of nutritional label formats: performance and preference*, Division of Consumer Studies, Center for Food Safety and Applied Nutrition, Washington, D.C., 1991.
34. Food label use needs assessment, American Heart Association. Washington, D.C., 1991.
35. P. N. Anderson, Low Literacy Guidelines, Proceedings from Exchange '92 The Communication Process: Research, Materials and Audience, Washington, D.C., 1992, p. 21.

14

Lost in the Supermarket: Consumer Confusion and Marketing Mania

Stephen Gardner

Southern Methodist University, Dallas, Texas

I. INTRODUCTION

Those of us who do the family shopping know from painful experience that the Saturday morning rush through the supermarket aisles is not the ruminative experience we might wish it were. Our goal cannot be a studied contemplation of the contents and import of food labels. Instead, we are reduced to the more-restricted goal of getting out of the store with most of what we came there to get and little of the things the three-year-old grabs and throws into the cart.

American consumers have so little leisure time that they welcome any way to improve their use of that time. This likely accounts for the fact that so many working parents opt for fast food instead of a home-cooked meal for dinner. On the other hand, they also express a desire to learn about and purchase healthy foods.

Where does this leave us? At a point where we have decided as a society to mandate improved food labels and to restrict some of the claims that can be made with respect to the benefits of the foods bearing those labels. The Nutrition Labeling and Education Act of 1990 (NLEA) is the manifestation of that societal decision.[1]

This chapter will trace the genesis of NLEA from the unrestrained—and insupportable—claims made by some marketers in the 1980s, through both the predictable and the unanticipated regulatory responses, ending with the adoption of the NLEA.[2]

The purpose of this chapter is to discuss food labeling issues from an unabashed, though not uncontrolled, consumer perspective. This chapter therefore relies to a great extent on—and is shaped to a significant degree by—the author's experiences as an assistant attorney general for the States of New York and Texas, two of the states that were at the forefront of state law enforcement initiatives in the 1980s. It is a view from the front of food labeling reform.

II. HOW MANY GRAMS IN AN INCH?

A. Consumer Needs and Desires for Food Labels

Before embarking on the somewhat disheartening journey through the depths to which food marketing sank in the 1980s, it is first appropriate to start with the person who is the target of concern of consumer advocates and consumer products companies alike—the consumer.[3]

After all, it is the consumer that all the labeling hullabaloo is about. It is for the consumer that we consumer advocates advocate. It is for the consumer that consumer products companies produce. It is for the consumer that Congress congregates. And so on, through the whole feeding chain of interests that have been involved in food labeling issues over the past ten years.

Repeatedly, consumers express preferences for healthier foods. Consumer concerns with nutrition remain high, although demands for greater amounts of information have apparently peaked.[4] This peaking of the demand for more information could presage a peaking in consumer interest in all nutritional information.[5] It could also portend simply that consumers have received the maximum level of information they find useful.

One problem with predicting actual consumer behavior based on polls of their expressed needs and desires is that sometimes consumers give in to the natural tendency (familiar to priests, psychiatrists, and police) to admit to somewhat higher aspirations than they in fact have.[6] That is, consumers may indicate a preference for a low-sodium, non-fat hamburger in response to a mall-intercept pollster with a clipboard. When faced with an actual choice, however, they opt instead for two all-beef patties, special sauce, lettuce, cheese, pickles, onions, on a sesame-seed bun—in all its salt-soaked, fat-dripping, tasty glory.

A second problem with predicting consumer behavior with respect to diet and health is that there is a considerable gap between expressed consumer desires and actual consumer knowledge of the relative minutiae of nutrition. Thus, though most consumers report that health concerns have caused a major change in their diets[7] and that they use food labels in their search for healthier foods,[8] they are also lacking in some of the most basic information necessary to make any significant change in their diet—such as the relationship between HDL and LDL cholesterol,[9] the saturated-fat level of coconut oil,[10] and what a complex carbohydrate is.[11] (One suspects that many of the readers of this book—certainly themselves far above the curve on nutritional issues—would hesitate to volunteer certain knowledge of these same bits of information if any sizable amount of money rested on it.)

Perhaps this is the reason that consumers tend to express what would at first glance appear to be mutually-exclusive desires. First, they want simple means of conveying information. Second, they want enough information to make an informed decision.[12]

The challenge to marketers, regulators, and consumers alike as we move forward in these post-NLEA days is how to walk the tightrope between sufficient information presented in an understandable manner and too much information, which will lead to information overload and a shutdown of consumer interest. For example, the oat bran bubble was beginning to deflate of its own accord when it was fully burst by one study that said that oat bran was not proven to be effective in reducing cholesterol.[13] This shows that consumers are willing to listen to only so much when it comes to claims for the beneficial effects of foods. Once their credibility is strained past the breaking point, they give up pretty quickly. It also would seem to indicate that consumers were only grudgingly eating foods containing oat bran and were all too happy to abandon the stuff without a backward glance as soon as just one study debunked the benefits of oat bran.[14] Yet they remain willing at the present to attempt to make the journey to more healthful eating.

To start down that road, consumers indicate that their four primary sources of information regarding diet and health are:

- News stories
- Health organizations
- Physicians and other health professionals
- Food labels[15]

But for the fourth, these various sources have never been seriously challenged as viable and trustworthy sources of such information. And whatever doubts we may have had about labels should have been effectively laid to rest by NLEA and the subsequent Food and Drug Administration (FDA) and Department of Agriculture (USDA) regulatory response. So far, so good.

However, to the intense regret of all of us on the consumer side of the fence and to great joy of marketers and advertisers, consumers also report that they find informational statements by food companies—advertising—to be believable.[16] Presumably, if consumers believe a statement in an ad, they'll use it in their search for the truth.

If we could rely on the kindness of advertisers, we would probably be able to provide consumers with exactly what they say they want—information that is sufficient and simple to understand. However, we cannot.

Marketers are not out to inform the public. Instead, they are out to sell a product. One primary way they sell their product over all others is positioning their product by creating a point of difference.[17] This point of difference may well be created out of the whole cloth where no perceived difference had existed and where no meaningful difference does exist.[18] Perhaps the best expression of the ethics of advertisers in this regard is by that grand old man of advertising, David Ogilvy, who bragged that:

> I could have positioned Dove as a detergent bar for men with dirty hands, but chose instead to position it as a toilet bar for women with dry skin. This is still working 25 years later.[19]

Jockeying for positioning by food companies has been the prime cause of the excesses of health-related advertising in the past decade. What was once just oatmeal, or margarine, or soup became the number one ingredient in the battles against heart disease and cancer. Or so some companies wanted consumers to believe, when they positioned their products as just that.

It is in fact advertising, far more than labeling, that was the bone of contention in all of the food fights between the state attorneys general and the food manufacturers discussed in Section IV, below. And it is likely to be in the area of advertising that we will see the future issues affecting health-related claims in the future, as discussed in Section VI, below.

III. REGULATORY GRIDLOCK AND DEAD-END STREETS

A. The Cause and Effect of Federal Non-Response to Burgeoning Deceptive Health Claims

"Deregulation" was a byword of the Reagan Administration.[20] Conservative ideologues within the government firmly believed in the principles of new federalism—getting the federal government out of the business of regulating the lives of Americans and American business in particular, and leaving the business of regulation up to the individual states, to act, as described by United States Supreme Court Justice Brandeis, as laboratories of democracy.[21] Each state was free to experiment with differing manners and methods of governing, without interference from the federal bureaucracy.[22]

So it went. The architect of President Reagan's transition team at the Office of Management and Budget, dedicated to dismantling the federal system as rapidly as possible, was James Miller, who was subsequently appointed chairman of the Federal Trade Commission (FTC) in 1981. The FTC under Chairman Miller was the antithesis of activist, fulfilling the role of deregulation by enforcement marked more by avoidance than by observance.[23]

Cynics, unhappy with the prevailing winds at the White House during the 1980s, saw this shift from enforcement less as a true ideological shift than as an intellectually-supportable denial of protection to the average consumer in favor of corporate America.[24] The deflated FTC took its place alongside other agencies, such as the Environmental Protection Agency and the Equal Employment Opportunity Commission, all of whose enforcement activities slowed to a standstill or went careering into reverse.

Whatever the true motivation of the Administration, the reaction of much of the marketing community was unhesitating and unequivocal. They took federal deregulation as a "Get Out of Jail Free Card" and as an uncategorical imperative to go forth and profit by deception at the expense of consumers, who were left unprotected.

The prime example of the results of deregulation fever was the burgeoning growth of unfounded and illegal claims for the health and nutritional benefits of foods. As with many floods, this began with a chink in the dam. The Kellogg Company developed an understated and mild campaign that promoted one of its cereals for use as part of a diet that could be used to help prevent some forms of colon cancer.[25] Kellogg's campaign was carefully developed and had the collaboration and endorsement of the National Cancer Institute.[26] It was also thoroughly illegal.[27] The Food, Drug, and Cosmetic Act as it existed at the time strictly prohibited promotion of a food for prevention of disease without approval by the Secretary of Health and Human Services.[28] This is precisely what Kellogg did, with the National Cancer Institute as its (perhaps) unwitting accomplice. The FDA, which enforces federal food and drug laws, took exception and began enforcement steps that would have stopped the claims made by Kellogg.[29]

It would have done so, that is, had FDA had the chance to do its job. Instead, the deregulation mavens stepped in. Officials at the Office of Management and Budget effectively muzzled the FDA and prevented it from enforcing the law.[30]

Once Pandora's cereal box had thus been opened a crack, pandemonium ensued. Companies of every ilk and repute began making a variety of disease-based claims, all without the scientific support Kellogg had amassed and without the cooperation and oversight of the National Cancer Institute or any other regulatory or non-profit body that did not have a financial stake in the truthfulness and legality of the claims.

The synchronous apex and nadir of these claims was probably oat bran beer. The very idea of promoting beer to Bubba as a way to fight cholesterol without having to do more than pull a ring-tab caused even some marketers to stop short. And consumer advocates stopped a lot shorter still.

IV. THE CHOW HOUNDS

A. The Inception and Activities of State Attorneys General in the Health Claims Area

The result of this free-for-all market was a call by consumer advocates and marketers alike for renewed federal activity. Unfortunately, this call fell on plugged ears—there continued to be an enforcement vacuum at the federal level.[31]

Among the other forces of nature that abhor a vacuum are the state attorneys general. Before the 1980s, the attorneys general had focused their consumer protection efforts on problems in their own states, leaving most national consumer protection enforcement to their federal counterparts at the FDA and FTC. But with the advent of deregulation at the federal level came a rise in activity at the state level. The attorneys general had already come together to deal with deception in automotive repair,[32] in discount airline advertising,[33] and in rental car practices,[34] among other areas. As they worked together, they learned that they could have a significant impact on the practices of major national companies that deceived the citizens of their states. Even if the federal agencies charged with consumer protection were out of commission, the state attorneys general were willing to pool their resources to protect their own citizens.[35]

Well before the adoption of NLEA, several state attorneys general banded together to bring enforcement actions against a number of companies, including the makers of Campbell's soups, Sara Lee pastries, and Nabisco's Fleischmann's margarine, for a variety of health-related but deceptive claims for foods.[36]

Industry, which had reacted so positively to the new federalism when it meant no law enforcement at the federal level, began to raise all manner of objections to the several states that fulfilled the promise of the new federalism by enforcing their own consumer protection laws when necessary.[37] Charges of preemption, commerce clause problems, and First Amendment infringement began to be leveled against the states that chose to act against deceptive claims for foods.[38] They all proved fruitless.[39] The state attorneys general didn't go away. The cumulative effect of the rise in state activities was a renewed cry by marketers for the renascence of the FDA, the FTC, and other federal agencies.[40]

The Chicago-school economic theories that had fueled the deregulation fever on the Potomac in the 1980s had been running on empty for some time. In its simplest form, the Chicago-school hypothesis, applied to marketing practices, was that information is good, the more the better. If the information contains falsehoods, that is bearable, because the marketplace will step in to correct the falsehoods. This hypothesis was rejected. In fact, the marketplace did adjust to deceptive claims. But it adjusted down—honest marketers sank to the level of their dishonest competitors just to be able to compete.

V. THEY FOUGHT THE LAW AND THE LAW WON: FREE SPEECH AND COSTLY LIES

In the heat of the battle over deceptive and illegal health and nutrition claims for foods, many in the food industry decided that the First Amendment was a handy sword and shield with which to fight to protect their right to deceive with half-truths. However, First Amendment law does not avail these arguments, as can be seen from a brief overview of the relevant cases.[41]

> Obviously, much commercial speech is not provably false, or even wholly false, but only deceptive or misleading. We foresee no obstacle to a State's dealing effectively with this problem. The First Amendment, as we construe it today, does not prohibit the State from insuring that the stream of commercial information flow cleanly as well as freely.[42]

NLEA only regulates speech relating to marketing specific products.[43] The fact that these claims address health and nutrition concerns that are the subject of public debate does not take them out of the definition of commercial speech. As one federal court succinctly put it: "Kellogg's invocation of the protections of the First Amendment is misplaced."[44]

NLEA does not make it impossible to sell a product without making a health or nutrition claim. Nor does it make it impossible to inform consumers about health and nutritional issues without endorsing a particular product. NLEA thus does not inextricably intertwine commercial speech, which may be regulated, with non-commercial speech deserving of full First Amendment protection.[45]

Therefore, the claims regulated by NLEA are solely commercial speech. As commercial speech, these claims are entitled to lesser protection than other constitutionally guaranteed expression.[46] Most importantly, the Constitution does not protect false, misleading or deceptive commercial speech.[47] Thus, the government may prohibit or restrict it, as Congress has done in this instance.

From a consumer standpoint, the excesses of the 1980s prove that a reasonable amount of government regulation is essential to protect consumers from questionable or groundless claims made by food marketers. Health, nutrition, and disease-prevention claims for foods are not such as can be tested by average consumers to determine their accuracy.[48] Therefore, the role of regulating these claims for truth and accuracy must fall to government.

The courts have repeatedly upheld government prohibitions on deceptive advertising and labeling of foods and drugs.[49] More significantly, courts have also upheld prohibitions against the use of specific terms in commercial speech unless required governmental standards are met.

In what is perhaps the most on-point decision relating to the right of government to control the content of a food label without running afoul of the First

Amendment, a federal district court held that the State of Texas could prohibit labeling claims made by Kellogg Company for a new cereal. Kellogg had introduced a cereal that it chose to call Heartwise, which contained a significant level of psyllium, a high-fiber grain that is the active ingredient in laxatives such as Metamucil.[50] Texas health officials detained shipments of Heartwise because of labeling statements linking consumption of Heartwise and prevention of heart disease.[51] Recognizing that the state could impose an outright ban on the sale of food products containing psyllium, the court held that it was a less restrictive alternative for the state instead to insist that its labeling laws be met when a company chose to lace its cereal with laxative, without imposing on Kellogg's First Amendment rights.[52] Drawing an important distinction between speech and marketing, the court noted that "there is nothing about the detentions that restrains Kellogg's right to say whatever it wants about the health benefits of a high-fiber diet in general or psyllium in particular."[53]

In another case, a different federal district court rejected a First Amendment challenge to labeling regulations. As with the NLEA, those FDA regulations required foods to meet a specific standard before using a defined term on the label.[54] The trade group that brought that case specifically complained of the FDA regulation of "frozen heat and serve dinners" because the permitted use of that term required a listing of components that had to be in the product in order for that term to be used. The court noted:

> The obvious objective was to provide to consumers sufficient information on the labels of food products so that reasoned and informed shopping decisions could be made.[55]

Rejecting a First Amendment challenge, the court concluded:

> These regulations constitute the conclusion by the Commissioner that labeling which fails to meet the requirements of the regulations is misleading or otherwise not in compliance with the Act.[56]

The potential to mislead consumers is even greater when consumers have difficulty in evaluating the claims independently. One federal court of appeals considered "the difficulty for the average consumer to evaluate such claims through personal experience, and the consequent tenacity of advertising-induced beliefs about superiority" in deciding to hold claims to a high level of substantiation. The court stated:

> Because consumers cannot accurately rate the products for themselves, advertising, and the expectations which it engenders, becomes a significantly more influential source of consumer beliefs than it would otherwise be.[57]

The Grocery Manufacturers of America (GMA) has recognized the value of NLEA. Regarding the FDA regulations issued pursuant to the NLEA regarding nutrient descriptors (e.g., "low fat," "light," and "high fiber"), the GMA explained:

> Prior to the NLEA, analysis of the meaning of these terms depended on the meaning that consumers would attach to them, but because different consumers could interpret the claims differently, there was no universal standard. The purpose of FDA's rulemaking was to arrive at consistent, objective measures to describe the nutrient characteristics of the labeled foods. This the rulemaking accomplished.[58]

Thus, even the GMA conceded the value of regulating claims that consumers do not have the scientific expertise to evaluate independently.[59]

The question now remains as to what form that regulation will take in the future.

VI. THE RETURN OF THE FDA AND THE FTC TO ENFORCEMENT ACTIVITY

It is incontrovertible that both FDA and FTC activity has increased over the past few years. The FDA activity in the enforcement area has been largely put on hold in favor of the mammoth task—now concluded—of coming up with the implementing regulations necessary to make NLEA work. The FTC on the other hand has certainly increased its individual case activity, as evidenced by recent investigations of health-related claims by various liquid diets, Nestle for Coffee-Mate non-dairy creamer, Bertolli for claims about olive oil, and Kraft for claims for the milk content of its Singles slices.

However, there is some reason to doubt the depth of the FTC activity, despite its seeming breadth. Although the FTC's enforcement numbers are indeed up, most of the cases are brought against small companies that operate on the fringe of health claims—bee pollen, baldness cures, diet patches, weight loss creams, and the like. Few of the companies that become the subject of investigations are exactly household names, nor do their activities seem likely to influence the purchasing decisions of any but the ignorant, the unthinking, and the credulous. Granted, even those unfortunates deserve help from the Government, but it appears that the FTC efforts have been and will continue to be focused on the fringe and not on those who set the standards.

The state attorneys general learned in the 1980s that the quickest and most effective way to have an effect on deceptive advertising was not to scratch away at the fringe but instead to go to the root of the problem—the prominent, well-heeled national companies that had chosen to engage in deceptive and illegal behavior in order to position their products as the best thing since sliced, high-soluble-fiber, low-fat, no-cholesterol bread. Whatever those in the industry who were engaged in these activities might have thought about the propriety of these actions, there was no doubting their effectiveness.

Unfortunately, the FTC, either willingly or ignorantly, has refused to profit by this experience and continues to focus its efforts on those companies operating on the edge. Nonetheless, it appears that its efforts will continue along this road, with a significant level of activity and an occasional strong action against a company that matters.

Even more unfortunately, it appears that the opposite effect has happened—the FTC's regulatory approach has been largely adopted by the state attorneys general. Increasingly in the past few years, the attorneys general have become a force to be ignored, just as they had been before the heady days of the 1980s. Whether this is due to a desire to shift national labeling and advertising regulation to the FDA and the FTC, to the loss of several leading attorneys general,[60] to a simple change in philosophy, or to more base political reasons, the end result is the same.

VII. CONCLUSION

Many of the battles of consumer advocates for better regulation of health-related claims for foods were won with the passage of NLEA, which contained most of the specifics and all of the precepts that lay behind the complaints of advocates about marketing activities in the 1980s.

The past is always prologue. The industry practices, regulatory policies, and law enforcement prosecutions discussed above will serve as a basis for what we see going forward. But the most important part of this puzzle is yet to be found—how consumers react when they find that they have to a large degree gotten their wishes.

VIII. ENDNOTES

1. The author asks the reader's willing suspension of disbelief to the point that Congress is deemed to have some degree of credibility as the expression of our goals as a society. Otherwise, put down this book and go watch TV.
2. Although this chapter purports to be a consumer perspective on these issues, it is not a summary of NLEA's specific responses to the issues. For details on the specific requirements of NLEA and the regulations adopted pursuant thereto, look to Chapter 7, "A Comprehensive Review of the NLEA Regulations," by Ralph Shapiro.

3 For another viewpoint on consumer perceptions and desires, see Chapter 12, "Meeting the NLEA Education Challenge: A Consumer Research Perspective," by Brenda Derby, Ph.D., and Sarah Fein, Ph.D. Drs. Derby and Fein are with FDA and bring a federal regulatory approach to bear on this issue.

4 Center for Food Safety and Applied Nutrition, Food and Drug Administration, *Summary of Consumer Research on Health and Diet Attitudes and Knowledge and Use of Food Labels* 7 (1992). This study, written by the same Brenda Derby, Ph.D., to whose chapter the reader was referred in the preceding note, is an excellent summary of research on consumer preferences, knowledge, and behavior.

5 Center for Food Safety and Applied Nutrition, Food and Drug Administration, *Summary of Consumer Research on Health and Diet Attitudes and Knowledge and Use of Food Labels* 8 (1992). An excellent summary of research on consumer preferences, knowledge, and behavior is available from the Center for Food Safety and Applied Nutrition, Food and Drug Administration.

6 Robert B. Settle and Pamela L. Alreck, *Why They Buy: American Consumers Inside and Out* 34-35 (1989).

7 Center for Food Safety and Applied Nutrition, Food and Drug Administration, *Summary of Consumer Research on Health and Diet Attitudes and Knowledge and Use of Food Labels* 22 (1992).

8 Center for Food Safety and Applied Nutrition, Food and Drug Administration, *Summary of Consumer Research on Health and Diet Attitudes and Knowledge and Use of Food Labels* 8 (1992).

9 Center for Food Safety and Applied Nutrition, Food and Drug Administration, *Summary of Consumer Research on Health and Diet Attitudes and Knowledge and Use of Food Labels* 21 (1992).

10 Center for Food Safety and Applied Nutrition, Food and Drug Administration, *Summary of Consumer Research on Health and Diet Attitudes and Knowledge and Use of Food Labels* 20-21 (1992).

11 Center for Food Safety and Applied Nutrition, Food and Drug Administration, *Summary of Consumer Research on Health and Diet Attitudes and Knowledge and Use of Food Labels* 21 (1992).

12 Levy, Alan S. et al., *More Effective Nutrition Label Formats Are Not Necessarily Preferred,* 92 J. Amer. Dietetic Assn. 1230, 1234 (1992).

13 Swain, *Comparison of the Effects of Oat Bran and Low-Fiber Wheat on Serum Lipoprotein Levels and Blood Pressure*, 322 New Eng. J. Med. 147 (1990); Liesse, *America's Oat Bran Craze on the Wane*, Advertising Age, Jan. 22, 1990, at 1, col. 2.

14 It is also worthy of mention that the study that caused the bubble to burst—while an adequate and well-controlled study—was probably no better as studies go than the many tests funded by industry grants that tended to show the opposite. One conclusion to be drawn from this clear consumer reaction is that people will only go so far to keep healthy and will abandon all efforts in that direction at the least pretext. On the other hand, one could equally conjecture that this wholesale rejection of oat bran precipitated by the one study instead demonstrates that consumers had reached their limit of credibility and simply didn't know whom to

trust. Faced with informational overload (at least some of it of questionable scientific validity), they just threw up their hands. And very likely their oat bran, as well.

15 Center for Food Safety and Applied Nutrition, Food and Drug Administration, *Summary of Consumer Research on Health and Diet Attitudes and Knowledge and Use of Food Labels* 7 (1992).

16 Center for Food Safety and Applied Nutrition, Food and Drug Administration, *Summary of Consumer Research on Health and Diet Attitudes and Knowledge and Use of Food Labels* 19 (1992).

17 Levitt, T., The Marketing Imagination 85 *et seq.* (Expanded Ed. 1986).

18 *See, e.g.,* Theodore Levitt, The Marketing Imagination 86 (Expanded Ed. 1986).

19 Ogilvy, D., Ogilvy on Advertising 12 (1983) [emphasis added].

20 *See* Exec. Order No. 12,612, 52 Fed. Reg. 41,685 (1987). This section is derived in large part from a law review article written by the author. Stephen Gardner, *How Green Were My Values: Regulation of Environmental Marketing Claims,* University of Toledo Law Review, Volume 23, No. I (1991), to which the author wisely retained copyright. There is indeed nothing new under the sun.

21 *New State Ice Co. v. Liebmann,* 285 U.S. 262, 311 (1932) (Brandeis, J., dissenting).

22 Exec. Order No. 12,612, 52 Fed. Reg. 41,685 (1987).

23 *Report of the American Bar Association Section of Antitrust Law Special Committee to Study the Role of the Federal Trade Commission,* 56 Antitrust & Trade Reg. Rep. (BNA) No. 1410, at S-11 to S-12 (Apr. 6, 1989).

24 Burros, M. *Eating Well,* N.Y. Times, Feb. 27, 1991, at C3, col. 1.

25 U.S. House Comm. on Governmental Relations, Disease-Specific Health Claims on Food Labels: An Unhealthy Idea, H.R. Rep. No. 561, 100th Cong., 2d Sess. 2-3 (1988). An article by Bruce Silverglade entitled "Pre-emption—The Consumer Viewpoint," 45 FOOD DRUG COSM. L.J. 143, 146 (1990) contains an excellent overview of the causes and effects of the mania for health claims, written from the perspective of the legal director of the Center for Science in the Public Interest, the consumer-interest group that played perhaps the most significant role in getting NLEA through Congress relatively undiluted.

26 At least one person in the consumer-activist fold has posited that it was indeed Kellogg's careful adherence to many criteria advanced by health professionals, consumer advocates, and regulatory officials that made its advertising campaign both informative and effective, although unfortunately unique in this regard. Bruce Silverglade, *A Comment on Public Policy Issues in Health Claims for Foods,* 10 J. Public Policy & Marketing 54, 55 (1991).

27 U.S. House Comm. on Governmental Relations, Disease-Specific Health Claims on Food Labels: An Unhealthy Idea, H.R. Rep. No. 561, 100th Cong., 2d Sess. 2-3 (1988). *See also* Bruce Silverglade, *Preemption—The Consumer Viewpoint,* 45 Food Drug Cosm. L.J. 143, 146 (1990).

28 Under then-existing 21 U.S.C. § 321(g)(B) any "article[s] intended for use in the diagnosis, cure, mitigation, treatment, or prevention of disease in man" is a "new drug." 21 U.S.C. § 355 prohibited the introduction of "new drugs" into commerce

without prior approval. 21 U.S.C. §§ 321(f, p), 343 and 355. *See also* U.S. House Comm. on Governmental Relations, Disease-Specific Health Claims on Food Labels: An Unhealthy Idea, H.R. Rep. No. 561, 100th Cong., 2d Sess. at 6-11.

29 U.S. House Comm. on Governmental Relations, Disease-Specific Health Claims on Food Labels: An Unhealthy Idea, H.R. Rep. No. 561, 100th Cong., 2d Sess. 2-3 (1988).

30 U.S. House Comm. on Governmental Relations, Disease-Specific Health Claims on Food Labels: An Unhealthy Idea, H.R. Rep. No. 561, 100th Cong., 2d Sess. at 22-26.

31 One consumer advocate commented that the problem was that the FDA and FTC were "understaffed, underfunded, and under Reagan."

32 *Big Suits. Texas v. AAMCO*, Texas Lawyer, Mar. 2, 1987, at 12, col. 1.

33 National Association of Attorneys General, *Report and Recommendations of NAAG Task Force on Air Travel Industry: Guidelines for Air Travel Advertising, 53* Antitrust & Trade Reg. Rep. (BNA) No. 1345 (special supp.) (Dec. 17, 1987). This particular state enforcement effort came for naught when the United States Supreme Court held that states were preempted from enforcing state laws against deceptive airline advertising. *Morales v. Trans World Airlines, Inc.*, 112 S.Ct. 2031 (1992).

34 National Association of Attorneys General, *Final Report and Recommendations of the Task Force on Car Rental Advertising and Practices,* 56 Antitrust & Trade Reg. Rep. (BNA) No. 1407 (special supp.) (Mar. 16, 1989).

35 *Report of the American Bar Association Section of Antitrust Law Special Committee to Study the Role of the Federal Trade Commission,* 56 Antitrust & Trade Reg. Rep. (BNA) No. 1410, at S-12 (Apr. 6, 1989).

36 Sugarman, *The New Chow Hounds. States Join Forces to Monitor Product Claims,* Wash. Post, Sept. 21, 1988, at El, col. 3; Burros, *Eating Well,* N.Y. Times, Feb. 27, 1991, at C3, col. 1.

37 Probably the finest example of these arguments can be found in an instance of the FTC carrying the industry's water. John E. Calfee and Janis K. Pappalardo, How Should Health Claims for Foods Be Regulated? An Economic Perspective, Bureau of Economics, Federal Trade Commission (1989). As an artifact of a failed regulatory approach, this piece is must reading. One way for those opposed to consumer protection efforts in the 1980s to advance an intellectual justification for their inactivity was to use a cost-benefit analysis. Because they found any inconvenience to industry a major cost and found no benefit to a deception-free marketplace, the cost-benefit battles were over before they began, as far as this breed of economists was concerned. Calfee and Pappalardo's attempt to quantify that which is essentially metaphysical reached its charmingly nutty peak when they put forth the proposition that the best justification for FDA failing to act on illegal health claims was derived from the formula: $EV = P_tB_t - (1 - P_t)C_f$ where EV is the expected value of allowing a health claim, P_t is the probability that the claim will turn out to be true, B_t is the estimated net benefit of allowing the claim if it turns out to be true, and C_f is the estimated net cost of allowing the claim if it turns out to be false. *Id.*, at 39-44. Got it?

38 For good discussions of the evils of preemption, and the lack of legal underpinnings for it, *see* Charles P. Mitchell, *State Regulation and Federal Preemption of Food Labeling*, 45 Food Drug Cosm. L. J. 123 (1990); and Richard L. Cleland, *The Regulation of Food Labeling: An Effective, Uniform National Standard Without More Preemption*, in James E. Tillotson (Ed.), America's Foods: Health Messages and Claims: Scientific, Regulatory, and Legal Issues (1993).

39 *See, e.g., Kellogg Co. v. Mattox*, 763 F. Supp. 1369 (N.D. Tex. 1991), *aff'd* 5th Cir. 1991. Kellogg sued Texas Attorney General Jim Mattox claiming several constitutional grounds—including the Commerce Clause and the First Amendment—for its right to violate state food labeling laws. The court denied Kellogg's motion for preliminary injunction, in a strongly-worded opinion.

40 *FTC's Welcome Return*, Advertising Age, Feb. 6, 1989, at 16, col. 1; Saddler, *FTC, Under Industry Pressure, Shows New Life in Backing Deceptive Ad-Laws*, Wall St. J., Apr. 17. 1989. at B4. col. 3.

41 This section is based in large part on a legal brief written by the author and Sharon Linden, an attorney with the Center for Science in the Public Interest. The nice thing about being a lawyer is that this sort of thing is called "precedent" rather than "plagiarism." Those readers fortunate *not* to be lawyers are granted permission to skip this section.

42 *Friedman v. Rogers*, 440 U.S. 1, 9-10 (1979) (citations omitted).

43 *Board of Trustees v. Fox*, 109 S.Ct. 3028 (1989); *Bolger v. Youngs Drug Products Corp.*, 463 U.S. 69 (1983); *Virginia State Board of Pharmacy v. Virginia Citizens Consumer Council, Inc.*, 425 U.S. 748 (1976).

44 *Kellogg Co. v. Mattox*, 763 F. Supp. 1369, 1381 (N.D. Tex. 1991). *Accord Central Hudson Gas & Electric Corp. v. Public Service Comm'n of New York*, 447 U.S. 557, 563 n.5 (1980) (emphasis added); *Fox*, 109 S.Ct. at 3032; *Bolger*, 463 U.S. at 67-68.

45 *Board of Trustees v. Fox*, 109 S.Ct. at 3031; *Riley v. National Federation of the Blind of North Carolina*, 487 U.S. 781, 796 (1988).

46 *United States v. Edge Broadcasting Co.*, 125 L.Ed. 345 (1993); *City of Cincinnati v. Discovery Network, Inc.*, 113 S.Ct. 1505 (1993); *Board of Trustees of State Univ. of New York v. Fox*, 492 U.S. 469, 477 (1989); *Central Hudson Gas & Electric Corporation v. Public Service Commission*, 447 U.S. at 563; *Ohralik v. Ohio State Bar Ass'n*, 436 U.S. 447, 455-56 (1978).

47 *Central Hudson*, 447 U.S. at 566; *Friedman v. Rogers*, 440 U.S. at 9-10.

48 Indeed, one basis for opposition to improved nutrition labeling by food marketers was that the testing was too difficult and expensive even for them to perform.

49 *See, e.g., Kraft, Inc. v. F.T.C.*, 970 F.2d 311 (7th Cir. 1992) (upholding FTC ban on deceptive calcium claims for processed cheese products); *Bristol-Myers Company v. F.T.C.*, 738 F.2d 554, 762 (2d Cir. 1984) (upholding FTC prohibition against certain advertising claims for analgesics); *United States v. General Nutrition, Inc.*, 638 F. Supp. 556, 562 (W.D.N.Y. 1986) (upholding Food and Drug Administration [FDA] prohibition of certain nutritional claims on the product label); *F.T.C. v. Pharmtech Research*, 576 F. Supp. 294, 303 (granting preliminary injunction against deceptive advertisements for dietary supplements); *United States*

v. Articles of Food, Etc., 67 F.R.D. 419, 424 (D. Idaho 1975) (rejecting First Amendment defense to forfeiture action because "freedom of speech does not include the freedom to violate the labeling provisions of the Federal Food, Drug, and Cosmetic Act"); *United States v. Articles of Drug*, 32 F.R.D. 32 (S.D. Ill. 1963) (upholding FDA ban on false and misleading medical guidance in pamphlets and other literature sold with supplements); *United States v. 8 Cartons, Etc.*, 103 F. Supp 626 (W.D.N.Y. 1951) (upholding FDA ban on unapproved health claims in literature distributed with product).

50 *Kellogg Co. v. Mattox*, 763 F. Supp. 1369, 1372 (N.D. Tex. 1991), *aff'd* 5th Cir. 1991.

51 *Kellogg Co. v. Mattox*, 763 F. Supp. at 1376-1377.

52 *Kellogg Co. v. Mattox*, 763 F. Supp. at 1381.

53 *Kellogg Co. v. Mattox*, 763 F. Supp. at 1381.

54 *American Frozen Food Institute v. Mathews*, 418 F. Supp. 548 (D.D.C. 1976) *aff'd* 555 F.2d 1059 (D.C. Cir. 1977).

55 *American Frozen Food Institute v. Mathews*, 418 F. Supp. at 551.

56 *American Frozen Food Institute v. Mathews*, 418 F. Supp. at 555 [emphasis added]. *See also National Nutritional Foods Association and Solgar Company, Inc.*, 504 F.2d 761, 808 n. 11 (2d Cir. 1974) (dismissing First Amendment challenge to FDA regulations defining standards of identity and prescribing label statements for foods); *62 Cases, Etc. v. United States*, 340 U.S. 593 (1951) (upholding FDA's authority to require foods bearing a particular name on the label to meet certain standards); and *U.S. v. Dakota Cheese*, 906 F.2d 335 (8th Cir. 1990) (upholding food standard for "mozzarella cheese").

57 *American Home Products Corp. v. Federal Trade Commission*, 695 F.2d 681, 698 (3rd Cir. 1982),

58 May 25, 1993 letter from C. Manly Molpus, President and Chief Executive Officer of GMA to Janet D. Steiger, Chairman of the Federal Trade Commission, p. 19.

59 Despite its recognition of the value of NLEA's restrictions on deceptive labeling, GMA curiously continues to oppose similar regulations for food advertising, which is not covered by the NLEA.

60 For example, the attorneys general of California, Iowa, New York, and Texas—all of whom were leaders in the fights against deceptive food labeling in the 1980s—have all left office, mostly due to unsuccessful bids for the governorships of their respective states.

15

The Impact of the Nutrition Labeling and Education Act—An Industry Perspective

Guy H. Johnson

The Pillsbury Company, Minneapolis, Minnesota

I. INTRODUCTION

The Nutrition Labeling and Education Act of 1990 (NLEA) requires that virtually all food products sold at retail in the United States bear nutrition labeling. The FDA (1) and USDA (2) have estimated that more than 595,000 labels will be directly affected by this new law.

The regulations implementing NLEA (3) specify a very detailed graphic format ("Nutrition Facts") which will provide nutrition information on a uniform basis. The regulations also provide detailed guidelines for determining the serving size of food products, and strictly control the use of both nutrient content claims (e.g., "Low fat," "Lite") and health claims (e.g., "May reduce the risk of certain forms of cancer").

The FDA established the following objectives for NLEA:

- Provide consumers with the information necessary to make informed dietary selections.
- Eliminate confusion (particularly with respect to claims), and restore credibility to the food label.
- Provide food manufacturers with an incentive to develop and market "healthier" products.

The ultimate success of these objectives will be judged by NLEA's effect on the dietary intake and health status of American consumers. Unfortunately, such an assessment cannot be made until consumers are exposed to the new system and learn to apply it.

The food industry, however, is in a position to offer a more immediate assessment of NLEA. The process of implementing these new regulations has given food manufacturers a detailed, practical understanding of how they apply to an extremely diverse range of products. In addition, many companies are qualified to make an informed prediction of how consumers will respond to the new labels based on a rapport they have developed over the years.

The survey reported in this chapter was conducted among major food companies doing business in the United States. These companies were asked to rate how well they believe the food labeling regulations published in the January 6, 1993, *Federal Register* will address FDA's objectives.

These data were collected to provide some early insight on the impact of this important nutrition initiative and to identify its most significant benefits and potential problems.

II. METHODS

During June of 1992, an attempt was made to identify persons knowledgeable and/or accountable for the implementation of NLEA at the 50 largest food companies (based on total 1991 food sales) doing business in the United States.

Thirty-five such individuals were identified and sent a short survey soliciting information about NLEA. The majority of these individuals had technical backgrounds and were affiliated with scientific or regulatory affairs functions. The companies surveyed were given an assurance of complete anonymity. The Pillsbury Company did not participate in the survey.

The survey asked respondents to rate how well the NLEA regulations will accomplish FDA's objectives on an overall basis and in the specific areas of label format, serving size, nutrient content claims, and health claims. Each topic was subdivided into two or more specific areas. The survey also asked respondents to identify the components of NLEA that were expected to be the most and least effective in accomplishing its objectives, and solicited any other open-ended comments.

The following five-point scale was provided to assess how well NLEA objectives would be met:

1	=	Not at all	4	=	Good
2	=	Poor	5	=	Excellent
3	=	Fair			

III. RESULTS AND DISCUSSION

A. General Observations

1. Sample Composition

As noted above, only large food companies participated in this survey. Most of these companies were very experienced with nutrition labeling under the voluntary system and had closely monitored the NLEA regulations as they evolved.

No attempt was made to solicit the opinion of smaller companies. The results of this survey, therefore, should not be interpreted as a reflection of the food industry as a whole.

2. Study Design

This survey was conducted as a quick, informal poll of major food manufacturers for a presentation at the 1993 annual meeting of the Institute of Food Technologists in Chicago. The study was not intended to adhere to a strict scientific design, and it should not be interpreted as such.

B. Response Rate

Twenty-five of the 35 surveys distributed were returned (71%). Therefore, one-half of the 50 largest food companies doing business in the United States were represented.

C. Overall Rating of NLEA

The distribution of responses for the overall rating of NLEA is presented in Figure 15.1. In general, NLEA was expected to do a fair job of accomplishing its objectives. The overall rating was 2.7 (0.77) [mean (± S.D.)].

As noted in Figure 15.1, there was a wide range of responses to this question. This diversity of opinion was evident throughout the survey.

The data obtained from this survey should be interpreted in the proper context. NLEA poses an enormous challenge to food manufacturers. The regulations implementing NLEA are extremely complex (more than 800 pages in the *Federal Register*), and the time permitted for implementation was very short (less than 18 months).

The survey was distributed approximately six months after publication of the final regulations, when efforts to implement NLEA were at full intensity. During this period companies were faced with the need to quickly understand and interpret the new regulations, facilitate adequate internal communication, redesign computer systems and data bases, determine new or revised label declarations based on chemical analysis or data bases, redesign packaging, prepare

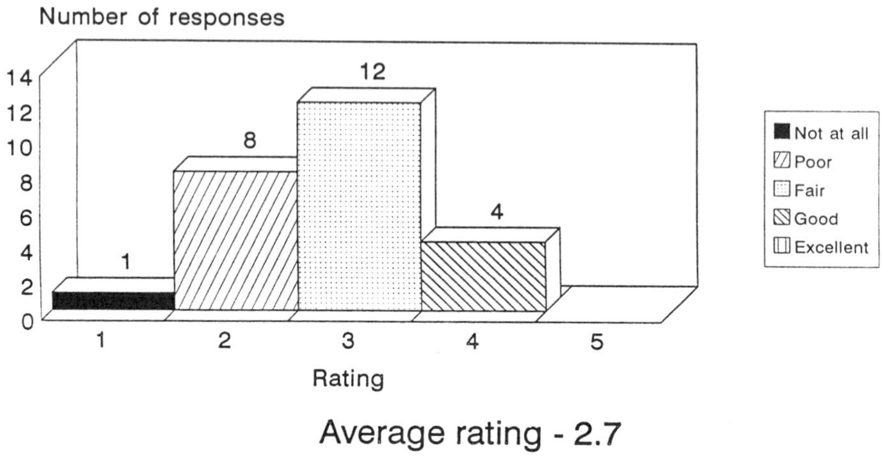

Figure 15.1 How well will NLEA meet its objectives?

products to determine new serving sizes, address marketing issues (including possible reformulation, repositioning, or discontinuation of products that could no longer make claims), and a host of other activities. The task of coordinating these activities for hundreds or thousands of product labels was an enormous challenge.

The pressures associated with this challenge may have had a negative impact on the data, and its interpretation should be tempered accordingly.

D. Format

The survey addressed four aspects of the NLEA format: graphic presentation, nutrients included, use of the Percent Daily Value (DV), and information in the footnote. Respondents were asked to rate these four areas for each of the FDA's three objectives using the five point scale described above. This information is presented in Figures 15.2 and 15.3.

Food manufacturers gave relatively high ratings to the graphic presentation and the nutrients included in the new label with respect to providing information to consumers. The fact that the label will appear on virtually all foods contributed to this optimism. However, respondents were less confident that the new label would reduce consumer confusion in these areas.

Not surprisingly, the graphic format was not expected to provide an incentive to industry. On the other hand, the nutrient content of the label was regarded as a modest incentive to develop healthier products. The fact that the nutrient content of virtually all foods will now be disclosed may explain this response.

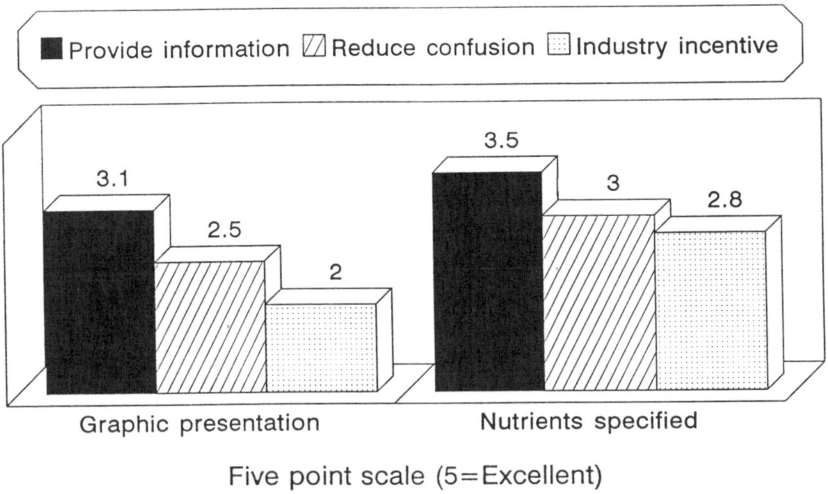

Figure 15.2 Format features (graphic presentation and nutrients specified).

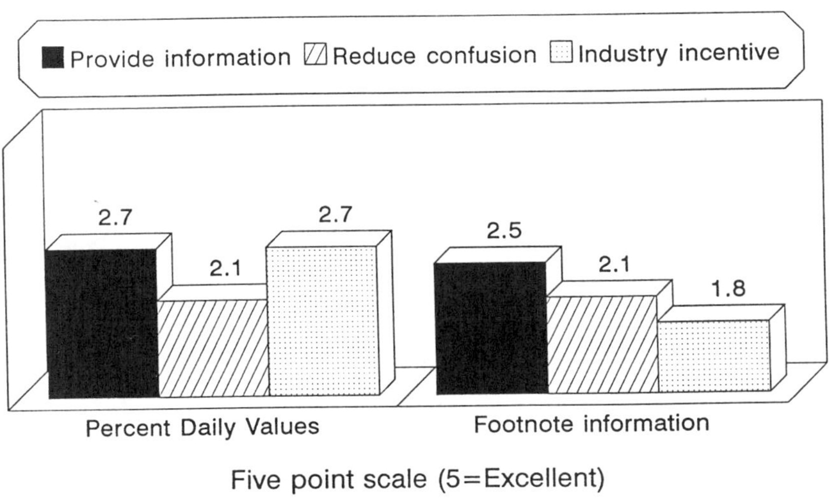

Figure 15.3 Format features (percent daily values and footnote information).

Use of the Percent Daily Value and the requirement to provide the DVs for 2,000 and 2,500 calorie diets in a footnote on many labels received much lower ratings. Of particular note, respondents were skeptical of the ability of DVs to reduce consumer confusion. These data reinforce the need for consumer education to facilitate understanding and appropriate use of the new label.

E. Serving Size

Data related to serving size are presented in Figure 15.4. Although some companies expressed concern about the appropriateness of individual Reference Amounts Customarily Consumed (RACC), the ratings for this area were similar for all three of FDA's objectives, and identical to the overall rating for NLEA.

The ability of the serving size declaration to provide useful information and reduce consumer confusion received higher marks. Standardization of serving sizes was noted as the most important advantage in this regard—particularly for the purpose of making product comparisons. This advantage overshadowed concern expressed by some companies about confusion for products that require preparation. The serving size of many such products must be expressed in the common household measure of the unprepared product (e.g., 1/3 cup pancake mix), rather than in the more recognizable prepared form (e.g., 3 pancakes).

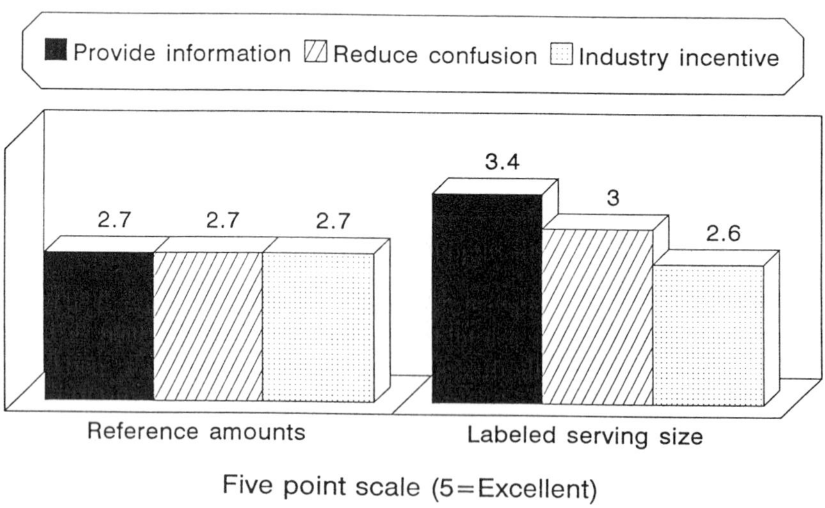

Figure 15.4 Serving size.

F. Nutrient Content Claims

Survey results pertaining to nutrient content claims are presented in Figures 15.5 and 15.6. Respondents generally believed that the scope of claims (i.e., the range of terms available) is adequate to describe most products, and that a manageable number of such terms will reduce consumer confusion.

In addition, the availability of consistent quantitative definitions for claims contributed to above average ratings in this area. Respondents expect that this uniformity will add to the credibility of the label and of products that bear such claims.

This area also received a higher rating for providing an incentive to industry than any other category. Although this rating is still only fair, it suggests that manufacturers will be able to apply the quantitative definitions to a range of existing and newly formulated products.

The use of referral statements (e.g., "See side panel for nutrition information.") was rated the lowest of all claim-related areas. Of particular note is the low rating (2.2) for reducing consumer confusion. Some companies expressed the concern that consumers are likely to ignore referral statements—particularly after they become commonplace.

Industry reaction to the use of reference foods for comparative claims was similar to the overall average for NLEA and did not differ among the FDA's three objectives.

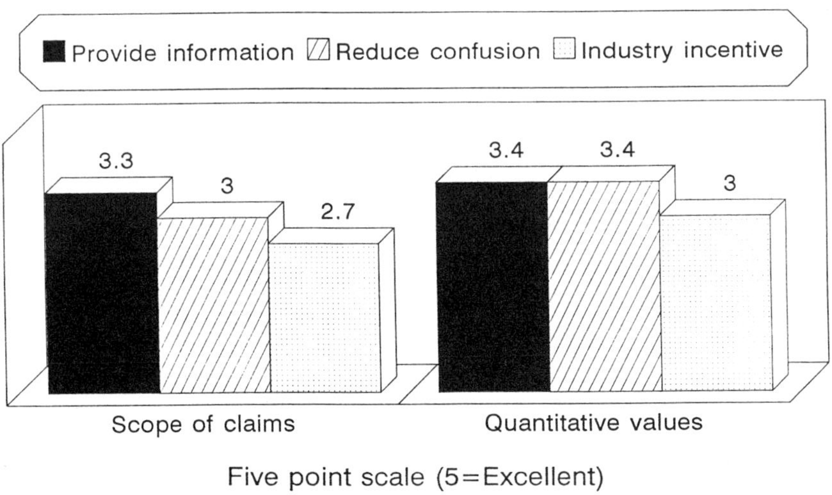

Figure 15.5 Nutrient content claims (scope of claims and quantitative values).

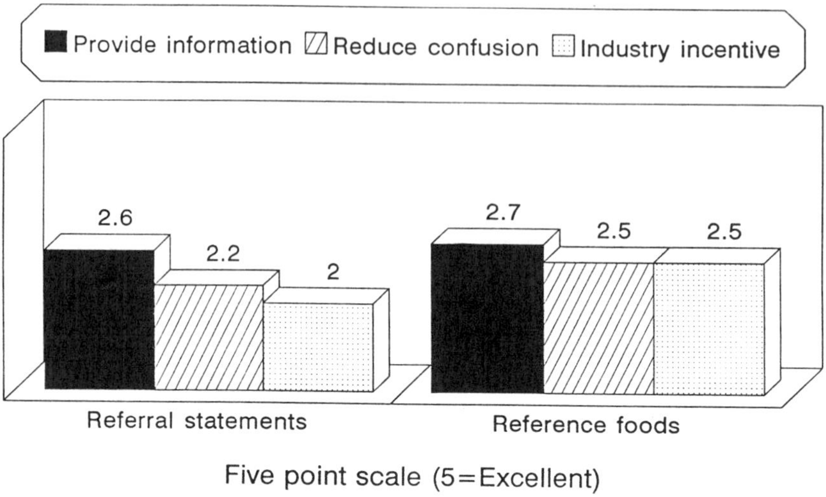

Figure 15.6 Nutrient content claims (referral statements and reference foods).

G. Health Claims

The survey results relating to health claims are presented in Figure 15.7. These data suggest that the greatest potential is to provide information and reduce confusion in the seven areas originally authorized by FDA:

- Dietary saturated fat and cholesterol and heart disease
- Dietary fat and cancer
- Sodium and hypertension
- Calcium and osteoporosis
- Fiber containing foods and cancer
- Soluble fiber containing foods and heart disease
- Fruits and vegetables and cancer

However, the opportunity to provide useful information in "new" health claim areas was regarded as poor. Manufacturers cited the lengthy nature and high cost of the petition process as serious concerns. In addition, the complexity of the health claim messages required by the NLEA regulations was noted to provide little incentive for industry to develop healthier products.

H. Most/Least Effective Aspects of NLEA

Table 15.1 lists the most common responses to the question, "What aspect of NLEA do you feel will be most effective?"

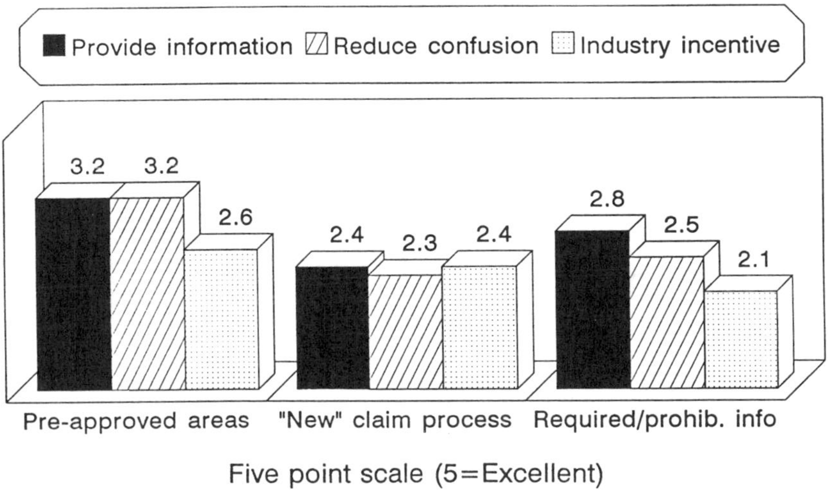

Figure 15.7 Health claims.

The positive attributes of NLEA noted by large food manufacturers generally related to an anticipated increase in uniformity of the new food label. Uniformity of claims, serving sizes, and the nutrition label format were among the top responses. More consistency in these areas could enhance the overall credibility of the food label, and benefit the industry and its consumers as a whole.

The aspects of NLEA that were cited as least effective by the survey respondents are reported in Table 15.2. The leading concerns expressed about NLEA by industry representatives related primarily to consumer education.

Adequate education is necessary for consumers to understand the new nutrition label and the products that bear it. A thorough understanding of a product's benefits is absolutely essential to its commercial viability—particularly for products with a nutrition rationale.

Table 15.1 The most effective aspects of NLEA

Response	Percent of sample
Uniform definition of claims	38
All products will be labeled	23
Standard serving sizes	12
New format easier to read	12
Label uniformity	8

Table 15.2 The least effective aspects of NLEA

Response	Percent of sample
Percent DVs	31
Lack of resources for education	19
Claims provide disincentive	15
Health claims awkward	12
Serving size confusion	12
Lack of graphic flexibility	8

Areas cited as being particularly difficult to explain were the percent DV (because it depends on individual energy needs), health claims (because of their length and complexity) and serving size declarations (because they may be in unfamiliar units and/or reflect unprepared product).

IV. SUMMARY AND CONCLUSIONS

As previously mentioned, the results of this survey should be interpreted with caution. Design limitations and a stressful environment at food companies during the data collection period may have skewed the results. Nevertheless, the need to implement NLEA quickly provided the food industry with a detailed, practical understanding of the new regulations. This experience, coupled with an ongoing understanding of the consumer, uniquely qualified food companies to provide an early assessment of NLEA.

The data collected in this survey support the following conclusions:

- The overall rating of how well NLEA is expected to accomplish its objectives was near the mid-point of a five-point scale.
- There was a broad range of responses to all questions in the survey. These differing responses reflect many factors, including the great diversity among commercially available food products.
- Respondents tended to believe that NLEA will do a better job of providing information to consumers than giving industry incentives to develop healthier products.
- The most effective aspects of NLEA generally related to providing uniformity among food labels.
- The greatest concerns about NLEA were generally related to the need for consumer education.

V. REFERENCES

1. Department of Health and Human Services, Food and Drug Administration, Regulatory Impact Analysis of the Proposed Rules to Amend the Food Labeling Regulations, *Federal Register 56*:60856 (1991).
2. U.S. Department of Agriculture, Food Safety and Inspection Service, Supplement to the Preliminary Regulatory Impact Analysis of Proposed Regulations for Nutrition Labeling of Meat and Poultry Products, *Federal Register 57*:5956 (1992).
3. Department of Health and Human Services, Food and Drug Administration, Food Labeling; General Provisions; Nutrition Labeling; Label Format; Nutrient Content Claims; Health Claims; Ingredient Labeling; State and Local Requirements; and Exemptions; Final Rules *Federal Register 58*:2066 (1993).

16

NLEA and the Calorie Control Food and Beverage Industry

Lyn O'Brien Nabors

Calorie Control Council, Atlanta, Georgia

I. INTRODUCTION

The use of low calorie, reduced fat, and "light" products has grown substantially over the past decade. In 1993, 81 percent of adult Americans, age 18 and over, consumed low calorie, sugar free, and/or reduced fat foods and beverages (1). That projects to 152 million consumers—a significant increase from the 141 million consumers of these products in 1991 (2).

Low calorie and reduced calorie processed foods were initially made by replacing sugar with a low calorie sweetener. With today's emphasis on dietary fat reduction, the number of products reduced in fat has increased substantially and numerous fat replacers are now available making these products possible.

In 1992, for the first time, more new reduced/low fat products (1,257) were introduced than reduced/low calorie products (1,130) (3). The number of products bearing labeling indicating a reduction in the fat, sugar, and/or caloric content, therefore, represents a significant portion of the processed food market.

The new U.S. food labeling regulations provide both opportunities and challenges for the many companies making products reduced in calories and/or fat. For the first time, nutrition labeling is required on most processed foods. The calorie and fat content of products will be displayed for all to see. Consumers will no longer have to guess or be able to fool themselves about which products are low in calories and/or fat.

The use of descriptors has changed. Some have been redefined and some defined for the first time. And, if it is not defined, you cannot use it unless approved by petition. Food processors can no longer choose the serving size they deem appropriate. The use of portion control to adjust the calorie and/or fat content of a serving is, consequently, largely a thing of the past.

The U.S. Food and Drug Administration now provides defined claims for individual foods, main dishes, and meals. Descriptors of interest to the calorie control industry include "free," "low," "reduced," and "light." Synonyms also have been provided for these terms. There are a number of restrictions concerning the use of these claims. Referral statements and other declarations are often required.

II. CALORIE CONTENT CLAIMS

Calorie content claims include "calorie free," "low calorie," and "reduced calorie." Historically, "reduced calorie" has been much more frequently used than the more restrictive "low calorie." "Reduced calorie" is now even less restrictive but has less appeal to the marketer than the now more restrictive "light." Previously, a one-third reduction in calories was required for reduced calorie claims (4). The decreased requirement, to 25 percent, may facilitate the development of additional reduced calorie products, such as reduced calorie candy.

A. Calorie Free (5)

"Calorie free" may be used if the individual food contains less than five calories per reference amount. Synonyms for "calorie free" are "free of calories," "no calories," "zero calories," "without calories," "trivial source of calories," "negligible source of calories," and "dietarily insignificant source of calories." If a food contains less than 5 calories per reference amount without special processing, the claim must be accompanied by a qualifying statement (e.g., "cider vinegar, a calorie free food").

B. Low Calorie (6)

The terms "low calorie," "few calories," "contains a small amount of calories," "low source of calories," or "low in calories" may be used on an individual food containing no more than 40 calories per reference amount (except sugar substitutes), or per 50 g, if it is a small serving size food. A small serving size food is a food with a reference amount of 30 g or less or 2 tablespoons or less. Again, if no special processing is required, a qualifying statement is needed (e.g., "celery, a low calorie food").

A "low calorie" meal or main dish product must contain 120 calories or less per 100 g and meet the other requirements for low calorie individual products.

C. Reduced Calorie (7)

The term "reduced calorie" may be used if an individual product contains a minimum reduction of 25 percent fewer calories per reference amount than the appropriate reference food. "Reduced in calories," "calorie reduced," "fewer calories," "lower calories," and "lower in calories" are considered synonyms for "reduced calorie."

The "reduced calorie" claim must be accompanied by explanatory information required for all relative claims, including a referral statement (e.g., "see side panel for nutrition information"), the reference food, the percent or fraction by which the calorie content was reduced and quantitative information comparing the amount of calories in the food and the reference food. The claim is not allowed if the reference food meets the "low calorie" definition (see Figure 16.1).

The 25 percent reduction for meal or main dish products is measured per 100 g as compared to an appropriate reference food. The requirements for an individual food otherwise apply.

III. FAT CONTENT CLAIMS

Fat content is extremely important to the calorie control industry. In recent years, health professionals have emphasized decreasing fat intake for a number of reasons. To the calorie control food manufacturer, reducing fat is the most expeditious method of reducing calories, since fat at 9 calories per gram, has over twice the calories of carbohydrates at 4 calories per gram.

Fat is an asset to product development and greater success is generally achieved from a taste standpoint if all the fat is not removed. Manufacturers, therefore, endeavor to reduce the fat content as much as possible without sacrificing product taste. A gradual reduction of the fat content in the American diet over time would likely improve the long-term success of fat free and low fat products. This, however, cannot be accomplished by individual food processors.

Fat content and, consequently, fat reduction are important from a health claim perspective. Health claims may not be made for an individual food product containing more than 13 g total fat for a main dish product containing more than 19.5 g per label serving size or for a meal product containing more than 26 g per label serving size (8).

"Fat free," "low fat," "reduced fat," and "percent fat free" claims are of particular importance to labeling of calorie control foods. Consumers now ex-

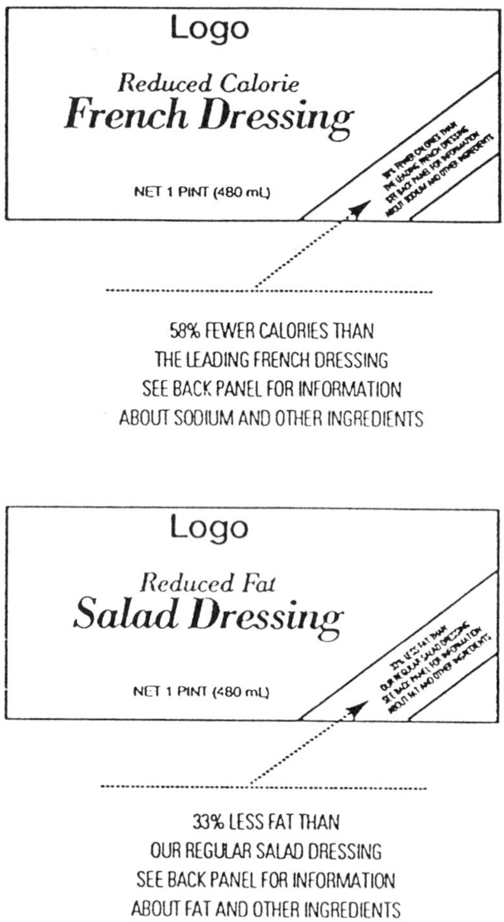

Figure 16.1 Examples of "Reduced Calorie" and "Reduced Fat" labeling.

hibit more interest in products reduced in fat than products reduced in sugar. (Saturated fat and cholesterol content claims which are related more to health than calorie control will not be discussed here.)

A. Fat Free (9)

"Fat free," or its synonyms "free of fat," "no fat," "zero fat," "without fat," "nonfat," "trivial source of fat," "negligible source of fat," and "dietarily insignificant source of fat," may be used on individual foods with less than 0.5 g total fat per reference amount. Meal products and main dish products must contain less than 0.5 g total fat per label serving size. A food qualifying for the term without special processing must include an appropriate qualifying statement (e.g., "broccoli, a fat free food").

If the fat free food contains a fat ingredient or one consumers generally understand to contain fat, the ingredient's listing in the ingredient statement must be followed by an asterisk referring to a qualifying statement below the ingredients list (e.g., "contains a negligible amount of fat").

FDA has been specifically asked if emulsifiers are considered fat-containing substances that must be declared in a footnote. An FDA official responded that, if it is something the consumer would not necessarily recognize as fat, it does not have to be asterisked and highlighted (10).

B. Low Fat (11)

The term "low fat" ("contains a small amount of fat," "low source of fat," and "little fat") may be used if an individual food contains a maximum of 3 g total fat per reference amount or per 50 g for a "small serving." Meal or main dish products must contain 3 g or less fat per 100 g and not derive more than 30 percent of calories from fat. If the food qualifies for the claim without special processing, the claim must be accompanied by an appropriate qualifying term (e.g., "frozen perch, a low fat food").

Permitted synonyms are "low in fat," "contains a small amount of fat," and "little fat."

C. Reduced Fat (12)

"Reduced fat" may be used on individual foods containing 25 percent less fat than the appropriate reference food. For meal or main dish products, the 25 percent reduction is per 100 g. As with other relative claims, a reduced fat claim must be accompanied by a referral statement, reference food, percent or fraction by which the nutrient (fat, in this case) content is reduced, and quantitative information comparing the amount of fat in the food and in the reference food (see Figure 16.1).

D. Percent Fat Free (13)

To be labeled "percent fat free," a food must meet the criteria for "low fat." The term "fat free" and the percent reduction must be in uniform type size. A "100 percent fat free" claim may only be made on products which are "fat free," contain less than 0.5 g of total fat per 100 g, and contain no added fat.

IV. SUGAR CONTENT CLAIMS

Sugar content claims refer to the level of all mono- and disaccharides in a product, not merely sucrose (14). Products in which sugar alcohols have been used to replace sugar, therefore, may be labeled "sugar free." Terms such as "Useful only in not promoting tooth decay" or "Does not promote tooth decay" have long been used on "sugar free" chewing gums and candies sweetened with sugar alcohols. The sugar free candy business is now a substantial one and the noncariogenicity claim is important to the promotion of both sugar free candy and chewing gum. The FDA has now determined, however, that the noncariogenic claim is an unauthorized health claim, although the proposed regulations had provided for such a claim on chewing gum. In addition, the FDA acknowledges that there is substantial evidence that sugar alcohols do not cause tooth decay (15,16).

Industry has filed objections to the FDA change and requested a stay of the earlier regulations allowing the noncariogenicity claim (17). Consequently, until the FDA resolves the issue, both the old regulations and the new ones are lawful. There is no indication as to when the FDA will resolve matter. The agency, however, has indicated that should the noncariogenicity claim ultimately be disallowed, an appropriate transition period would be provided for manufacturers who continue to use the "does not promote tooth decay" language (18).

A. Sugar Free (19)

The term "sugar free" may be used if the food contains less than 0.5 g sugar per reference amount or per labeled serving for meal or main dish products. The food may not contain any ingredient that is a sugar or is generally understood by consumers to contain sugars without a qualifying statement (e.g., "adds a trivial amount of sugar"). Permitted synonyms are "free of sugar," "no sugar," "zero sugar," "without sugar," "sugarless," "trivial source of sugar," "negligible source of sugar," and "dietarily insignificant source of sugar."

Since, according to FDA, consumers would expect "sugar free products" to be low or reduced calorie they must be labeled "low calorie" or "reduced calorie" or bear a relative claim of special dietary usefulness. If the product cannot be so labeled, a qualifying statement must be made each time the "sugar free" claim is made (e.g., "not for weight control").

B. No Added Sugar (20)

The terms "no added sugar," "without added sugar," and "no sugar added" may only be used if the product contains no added sugar or other ingredient that functionally substitutes for sugars; contains no ingredient containing added sugars such as jelly, jam, or concentrated fruit juice; the sugar content has not been significantly increased by some means, such as the use of enzymes; the food for which it substitutes normally contains sugars; and, if appropriate, bears a statement that the product is not a low or reduced calorie food and directs the consumer to the nutrition panel for sugar and calorie content.

"No added sugar" claims are not prohibited on products (e.g., juices) that have an inherent sugar content, are unsweetened, or contain no added sweeteners.

C. Reduced Sugar (21)

A reduced sugar (or "reduced in sugar," "sugar reduced," "less sugar," "lower sugar," or "lower in sugar") claim may be made if the product contains 25 percent less sugar per reference amount as compared to the reference food. Meal and main products must have a 25 percent reduction per 100 g. The claim, like other relative claims, must provide a referral statement to the nutrition panel, the reference food, the percent or fraction of sugar reduction, and quantitative information comparing the amount of sugar in the food with the reference food.

V. LIGHT CLAIMS

The term "light" was not previously defined by the Food and Drug Administration and was liberally used by industry. The FDA initially proposed that products bearing a light claim must be reduced in calories and fat by 50 percent (22). This proposal greatly concerned the calorie control industry. In comments to the FDA, the industry pointed out that in several product categories it is extremely difficult to reduce both fat and calories by 50 percent and provide good tasting products. When fat is removed it must be replaced and this replacement does not always result in a 50 percent reduction in calories (23).

Subsequently, the FDA revised its original proposal and light products must now be reduced in fat by 50 percent or in calories by one-third. If, however, the food to which it is compared derives 50 percent or more of its calories from fat, a 50 percent reduction in fat is required. The food label must bear a statement comparing the grams of fat and calories in the light food and the reference food (e.g., "light cheese cake—200 calories, 4 g fat; regular cheese cake—300 calories, 8 g fat per serving").

For light claims the reference food must be similar to the product bearing the claim (e.g., potato chips for potato chips). The nutrient value for fat or calories in a reference food that is used as a basis for the "light" claim may be determined in a number of ways. It may be the average value of the top three national or regional brands; a representative value from a valid data base; a market basket norm; or, where its nutrient value is representative of the food type, an individual food such as a market leader. The reference food may not be low calorie and low fat.

The light claim also must be accompanied by information on the identity of the reference food and the percentage (or fraction) by which the nutrient has been modified (e.g., "1/3 fewer calories and 50 percent less fat than our regular cheese cake"). This comparison must appear on either the information or principal display panel (see Figure 16.2).

If the product contains more than 13 g of fat, 4 g of saturated fat, 60 mg of cholesterol, or 480 mg of sodium per reference amount and per labeled serving, or if a food has a reference amount that is 30 g or less (or 2 tablespoons or less), the referral statement must include the nutrient(s) exceeding the maximum level.

Light claims also can be made for the sodium and organoleptic characteristics of food. A "light" claim referring to at least a 50 percent reduction in sodium may be made if the reference food contains 40 calories or less (low calorie) and 3 g of fat or less (low fat); otherwise the sodium reduction must be "light in sodium." Light can continue to be used to refer to organoleptic characteristics of food, so long as the characteristic (e.g., color) is stated in the same style and color of type in at least half the height of "light." However, names that have a long history of use, such as "light brown sugar," may still be used without qualifying information (24).

VI. DESCRIPTORS USED IN BRAND NAMES (25)

Food processors may no longer use brand names containing nutrient content claims that are not defined in the new regulations or approved via a petition submitted by the company. An exemption is provided for an undefined claim in a brand name in use on a specific food product before October 25, 1989. This exemption does *not* apply to product line extensions. Claims that are part of the product name as a result of a standard of identity are not subject to nutrient content regulations.

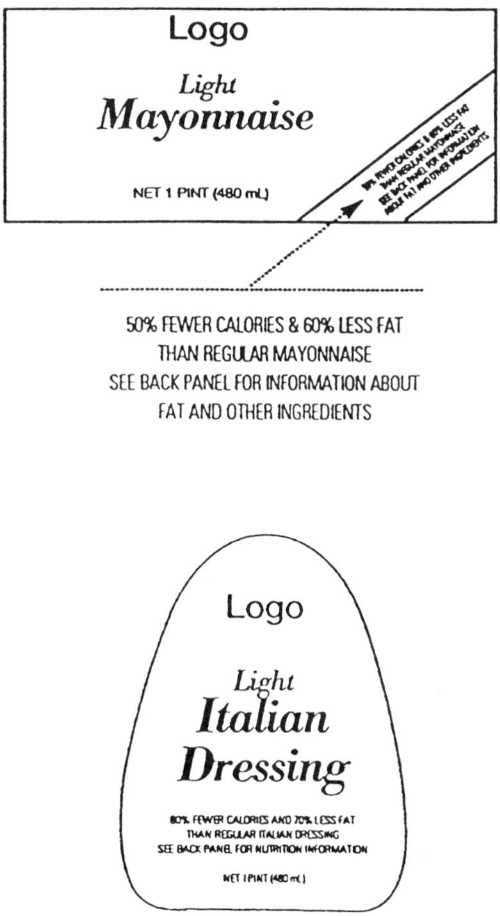

Figure 16.2 Examples of "Light" labeling.

A specific exemption is provided for the use of the term "diet" in the brand name of a soft drink used prior to October 25, 1989. Referral and disclosure statements are not required. Soft drinks marketed after October 25, 1989 may use "diet" provided that use complies with the 1993 final rules for special dietary use.

The labeling of "diet" soft drinks has been complicated by the FDA August 1993 technical amendments. Diet soft drinks provide no nutrients and, therefore, qualify for the simplified format label (26). The FDA technical amendments clarified that the use of a claim (e.g., "diet") on the label is a criterion for necessitating the statement, "Not a significant source of __________" (with the blank to be filled in by the names of nutrients present in insignificant amounts) (27). In discussions and correspondence with the National Soft Drink Association, the FDA has reportedly agreed to allow a simplified statement, "Not a significant source of other essential nutrients," on diet soda labeling for non-existent nutrients if the manufacturer has determined that its product does not contain significant amounts of other nutrients (28).

VII. MODIFIED STANDARDIZED FOODS (29)

New regulations concerning standardized foods allow for additional calorie control products. The new regulations do not amend existing food standards but provide a general definition for modified versions of traditional, standardized foods named by combining a nutrient content claim with the name of a standardized food (e.g., "low fat sour cream").

The FDA has determined this approach more efficient than issuing new standards for each new modified version of a standardized food or issuing temporary marketing permits (TMP's) for such products (30). A number of temporary marketing permits were issued for modified versions of standardized dairy products during the late 1980s and early 1990s (e.g., "no fat sour cream," "nonfat cottage cheese").

Any nutrient content claim must meet the applicable definition. The modified food must not be nutritionally inferior to the standardized food. Nutritional inferiority includes any reduction in the content of an essential nutrient that is present in a measurable amount in the standardized food, but does not include a reduction in caloric or fat content (31). Essential nutrients removed in the process of removing a component of the standardized food may be restored so the product is not nutritionally inferior.

Certain ingredients also may be added, under particular circumstances, so that the product is not inferior in performance characteristics. For example, fat replacers may be added to replace part of the fat in accordance with the regulations. Ingredients mandated by the standard must, however, be present in a significant amount. If a standard, therefore, calls for fat, a fat free version (e.g., "fat free mayonnaise") is not possible because a significant amount of fat must be present.

If the modified food has a significant difference in performance characteristics as compared to the standardized food, the difference must be disclosed on the label. For example, a "reduced fat margarine" might require a label statement such as "not recommended for frying."

Details on the impact of NLEA on the labeling of standardized foods are discussed in Chapter 17 by Thomas Balmer and Kimberlee Carson.

VIII. SPECIAL DIETARY PRODUCTS FOR CALORIE CONTROL

The FDA 1993 revisions in labeling requirements for foods for special dietary use in reducing or maintaining weight are not substantive in nature. They are, moreover, to align the special dietary regulations with those for nutrient content claims.

The FDA has advised that it would generally regard a brand name such as "Weight Watchers," when accompanied by information on the product label that suggests that the product can be useful in an overall weight control diet plan, to be a claim solely concerning its usefulness for a special dietary purpose. Such a claim would, therefore, be subject to special dietary regulations and not those for nutrient content claims. Consequently, the FDA has stated that a brand name, such as "Weight Watchers," may continue to be used irrespective of whether a specific product under that brand name was introduced into the marketplace before October 25, 1989 (32).

A. Label Statements Relating to Usefulness in Reducing or Maintaining Body Weight (33)

Any food that purports to be or is represented for special dietary use because of usefulness in reducing or maintaining body weight must bear 1) nutrition labeling, unless exempt, and 2) a conspicuous statement of the basis upon which the food claims to be of special dietary usefulness. If the food achieves its special dietary usefulness by use of a non-nutritive ingredient (i.e., one not utilized in normal metabolism), the label must contain a statement that discloses the non-nutritive ingredient and its percentage by weight, unless the ingredient is a non-nutritive sweetener, in which case the percentage by weight does not have to be declared.

Foods purporting to be "low calorie," "reduced calorie," or otherwise containing fewer calories than a reference food must comply with the nutrient content claim regulations. A food may be labeled with terms such as "diet," "dietetic," "artificially sweetened," and "sweetened with a non-nutritive sweetener" only if the claim is not false and misleading and the food is labeled "low calorie," "reduced calorie" or bears another comparative claim according to the nutrient content claim regulations. Exemptions to this include the use of a term that is specifically authorized by regulation governing a particular food, or, unless otherwise restricted by regulations, to any use of the term "diet" that clearly shows that the food is offered solely for dietary use other than regulating body weight (e.g., for low sodium diets); the use of a term on a formulated meal replacement or other food that is represented to be of special dietary use as a whole meal; and the use of such terms on soft drinks as authorized by regulation.

IX. CONCLUSION

The full impact of the new labeling on the calorie control food and beverage industry is, of course, yet to be seen. The descriptors that may be used and how they may be used is now clearly defined by the FDA. Food processors will have to work even harder to make good tasting products with attractive labeling. The development of "fat free," "low fat," and "reduced fat" products is, for example, still in its infancy. The calorie control industry expects to meet the challenges put before it by the NLEA and to continue to develop excellent products low in calories, sugar, and fat.

X. REFERENCES

1. Dieting and low-calorie/reduced-fat products survey, Booth Research Services, Inc., Atlanta, GA (Unpublished), 1993.
2. Dieting and low-calorie/reduced-calorie products survey, Booth Research Services, Inc., Atlanta, GA (Unpublished), 1991.
3. Lyn Dornblaser, New Products, Prepared Foods, p. 39, March, 1993.
4. Foods for special dietary use, Code of Federal Regulations [21 CFR 105.66 (d)(i)], Revised as of April 1, 1991.
5. Covington & Burling, A Compilation of Current Food Labeling Regulations of the Federal Food and Drug Administration Principally Implementing the Nutrition Labeling and Education Act (FDA Regulations), Title 21, Sec. 101.60 (b)(1), September, 1993.
6. Covington & Burling, FDA Regulations, Sec. 101.60 (b)(2), September, 1993.

7. Covington & Burling, FDA Regulations, Sec. 101.60 (b)(4), September, 1993.
8. Covington & Burling, FDA Regulations, Sec. 101.13 (h), September, 1993.
9. Covington & Burling, FDA Regulations, Sec. 101.62 (b)(1), September, 1993.
10. FDA debates use of "light" on standardized food labels, Nutrition Labeling Watch, p. 3, October 18, 1993.
11. Covington & Burling, FDA Regulations, Sec. 101.62 (b)(3), September, 1993.
12. Covington & Burling, FDA Regulations, Sec. 101.62 (b)(4), September, 1993.
13. Covington & Burling, FDA Regulations, Sec. 101.62 (b)(6), September, 1993.
14. Covington & Burling, FDA Regulations, Sec. 101.9 (c)(6)(ii), September, 1993.
15. Rules and Regulations, Federal Register, Vol. 58, No. 3, p. 2326, January 6, 1993.
16. Rules and Regulations, Federal Register, Vol. 56, No. 229, p. 6037, November 27, 1991.
17. Melvin S. Drozen, Keller and Heckman, Petition to FDA for Stay and Reconsideration by Working Group of Sugar Alcohol Manufacturers, Docket No. 91N-3847, February 5, 1993.
18. Eugene Lambert, Covington & Burling, Communication to Calorie Control Council, November 16, 1993.
19. Covington & Burling, FDA Regulations, Sec. 101.60 (c)(1), September, 1993.
20. Covington & Burling, FDA Regulations, Sec. 101.60 (c)(2), September, 1993.
21. Covington & Burling, FDA Regulations, Sec. 101.60 (c)(4), September, 1993.
22. Rules and Regulations, *Federal Register*, Vol. 56, No. 229, p. 60472, November 27, 1991.
23. Calorie Control Council, Comments to FDA on Food Labeling, Docket No. 91N-0384, February 24, 1992.
24. Covington & Burling, FDA Regulations, Sec. 101.56, September, 1993.
25. Covington & Burling, FDA Regulations, Sec. 101.13 (q)(1), September, 1993.
26. Covington & Burling, FDA Regulations, Sec. 101.9 (f), September, 1993.
27. Rules and Regulations, *Federal Register*, Vol. 58, No. 158, p. 44069, August 18, 1993.
28. FDA bends a bit on diet soda labeling; new proposals due, Food Labeling News, p. 3, November 25, 1993.
29. Covington & Burling, FDA Regulations, Sec. 130.10, September, 1993.
30. Rules and Regulations, *Federal Register*, Vol. 58, No. 3, p. 2432, January 6, 1993.
31. Covington & Burling, FDA Regulations, Sec. 101.3 (e)(4), September, 1993.
32. Rules and Regulations, *Federal Register*, Vol. 58, No. 3, pp. 2427–2430, January 6, 1993.
33. Covington & Burling, FDA Regulations, Sec. 101.56, September, 1993.

17

Labeling Dairy Foods Under NLEA: New Products, New Opportunities

Thomas Balmer and Kimberlee J. Carson

International Dairy Foods Association, Washington, D.C.

I. NUTRITION LABELING OF DAIRY FOODS—OVERVIEW

Although the Nutrition Labeling and Education Act (NLEA) of 1990 mandates nutrition labeling for many foods for the first time, the dairy industry has had substantial experience in the area for over twenty years. Because of their nutrient-dense nature and well-developed system of standards of identity (based largely on milkfat content), many dairy foods have long borne nutrient content claims on their labels and carried full nutrition labeling on their packaging.

While compliance with the regulatory aspects of NLEA remains the main focus of dairy processors in regard to food labeling, the new regulations also offer the dairy industry a new opportunity to develop and market many new products not permitted under the old regulations. Products such as "lowfat ice cream," "reduced fat cheddar cheese," "nonfat sour cream," and "light butter" may now be marketed without the burdensome and confusing "substitute" or "alternative" nomenclature previously required for these types of products.

Enforcement of labeling regulations for most food products rests with the Food and Drug Administration (FDA). However, under NLEA, state regulatory agencies also have the authority to enforce the food labeling provisions of the federal Food, Drug and Cosmetic Act (FD&C Act). Because of the historical involvement of state departments of agriculture and/or health in marketing dairy foods, the labels of these products are expected to receive more attention at state level than those for other categories of foods.

This chapter describes the dairy industry's approach to compliance with NLEA and its efforts to market innovative new products in a new regulatory environment.

A. Standards of Identity

Section 401 of the FD&C Act provides statutory authority for the establishment of standards of identity for foods. While the FDA has established standards of identity for many food products, no segment of the food industry is more affected by their use and enforcement than the dairy industry. Approximately 40% of all federal standards of identity fall into one of the following dairy categories: milk and cream, cheese and related cheese products, or frozen desserts. In addition, two other dairy products, butter and nonfat dry milk, have standards of identity which were established directly by Congress. With the exception of butter and nonfat dry milk, all federal standards of identity for dairy products are referenced in Title 21, Code of Federal Regulations (CFR), Parts 100–169.

Food standards are established to promote fair competition among manufacturers and to eliminate consumer confusion. The federal standards of identity basically address the following areas:

- Process of producing the food, required and optional ingredients.
- Methods of analysis for various components.
- Finally, product nomenclature and required and voluntary label declarations.

Microbiological specifications and quality grades are not included in the federal standards of identity for dairy foods.

In addition to the federal standards of identity, many states have also promulgated standards of identity for dairy products. In most cases, state standards have been identical to the federal standards, but significant differences have existed for some products (e.g., standards of identity for fluid milks in California). As of November 8, 1990, NLEA established that where federal and state standards exist simultaneously, the federal standard preempts the state version. However, in the event no federal standard exists for a specific dairy product and a state standard has been promulgated, the state standard is in effect.

Many of the dairy products subject to a standard of identity are required to bear a nutrient content claim (descriptor) in the name of the food (e.g., lowfat milk, lowfat cottage cheese, nonfat yogurt, light cream). The use of such descriptors in dairy product nomenclature predates the establishment of definitions for nutrient content claims as mandated by NLEA. In recognition of the historical use of descriptors in the labeling of certain dairy foods subject to standards of identity, NLEA exempted such products from the labeling requirements normally required for foods bearing nutrient content claims (21 CFR §101.13(q)(6) *Nutrient content claims—general principles*).

B. Nonstandardized Foods

While the vast majority of dairy foods are subject to a standard of identity, several well-known products are either completely nonstandardized or are subject to a state standard. Examples of such products are frozen yogurt and cheeses such as ricotta and brie. In the case of frozen yogurts that bear nutrient content claims (e.g., lowfat or nonfat frozen yogurt), the product may be subject to a state standard in terms of physical composition and production method. However, because of the preemptive authority established by the NLEA in the area of nutrient content claims, the product is subject to meeting the federal definitions for the terms "lowfat" or "nonfat."

Because of the inconsistencies existing between not only federal and state definitions, but also among state standards of identity, the International Ice Cream Association (representing manufacturers of frozen desserts) has petitioned the FDA to establish a standard of identity for frozen yogurt.

II. NUTRITION LABELING OF DAIRY FOODS: HISTORICAL PERSPECTIVE (1973–1993)

As mentioned in the chapter overview, most dairy foods have borne full nutrition labeling for many years. While nutrition labeling was voluntary for most foods prior to the passage of NLEA, many dairy foods were required to bear full nutrition labeling due to either the inclusion of a vitamin, mineral, or protein (e.g., "Vitamin D Added" or "Protein Fortified") or the use of nutrition claims on the food's label (e.g., "Lowfat Milk").

In the early 1970s, the dairy processing industry recognized that industry compliance with the new nutrition labeling regulation would be extremely difficult and costly without the use of a nutrient database for dairy ingredients.

In the absence of a nutrient database, dairy processors would have faced the expensive prospect of ongoing nutrient analyses of their finished products. As a result, two dairy trade associations, the Milk Industry Foundation (MIF), representing processors of fluid milk products and cultured dairy foods, and the International Ice Cream Association (IICA), representing manufacturers of ice cream and frozen desserts, developed a nutrient database for the two major components of dairy foods, milkfat and nonfat milk solids. The values developed by MIF and IICA were submitted to the FDA and accepted for use by the dairy industry beginning in 1973. These nutrient database values provided the basic information for manufacturers to calculate nutrition labeling information for fluid milk and creams, cultured dairy foods, and ice cream and frozen desserts over most of the pre-NLEA period.

Nutrition labeling on cheese and butter products was rather limited through most of the 1970s and 1980s. In many cases cheese and butter labels did not bear any label statements that would have triggered mandatory nutrition labeling. In addition, unlike fluid milk products and frozen desserts, these products did not lend themselves as readily to the calculation approach to nutrition labeling since these categories of foods experience some degree of nutrient concentration and loss during processing.

III. INDUSTRY COMPLIANCE WITH NLEA: 1994 AND BEYOND

With the NLEA bringing all dairy foods under mandatory nutrition labeling, the cost of complete nutrition analysis on finished products is higher to dairy processors than ever before. As a result, many companies will employ nutrient databases or a combination of nutrient databases and analysis to develop nutrition labels.

The U.S. dairy processing industry, through its major trade association, the International Dairy Foods Association (IDFA),[1] has developed a nutrient database for a variety of dairy foods. This database has been submitted to the FDA for acceptance and use by the dairy processing industry.

As described above, the manufacturing processes and inherent nature of some dairy foods like fluid milks, creams, and ice creams are quite conducive to employing nutrient databases based on milkfat and nonfat milk solids content. Other products, like butter and standardized natural cheeses (i.e., cheddar, swiss, and monterey jack), may also employ industrywide nutrient databases based on analytical values since their composition does not vary significantly among different manufacturers.

Other dairy foods, like process cheeses and yogurt will vary in composition and manufacturing procedures among processors to the extent that industrywide databases are impractical. Consequently, dairy processors can be expected to develop their own sources of nutrition information for these products.

The IDFA nutrient database submission contains nutrient profiles for milkfat and nonfat milk solids in addition to values for nineteen cheeses and butter.

IV. USE OF NUTRIENT CONTENT CLAIMS ON STANDARDIZED DAIRY FOODS

For many years, dairy processors who desired to market lower fat versions of traditional dairy foods were prohibited from doing so if they also wished to

Figure 17.1 Presentation of sugar-free ice cream under the new regulations.

market the products using the standard name of the food (e.g., ice cream, sour cream, cheddar cheese). Since, in most cases, the products no longer contained sufficient milkfat to use the name as specified by the standard of identity, the lower fat versions were required to bear such confusing and unappealing terms as "substitute" or "alternative." As a result, dairy processors were not compelled to market lower fat versions of traditional dairy foods since these products could not be presented to consumers in terms which accurately described both the physical and organoleptic properties of the food.

Figure 17.2 Presentation of lowfat ice cream under the new regulations.

Figure 17.3 Presentation of reduced fat, reduced calorie sour cream under the new regulations.

Figure 17.4 Presentation of reduced calorie butter under the new regulations.

Figure 17.5 Presentation of reduced fat, reduced calorie cheddar cheese under the new regulations.

However, the Food and Drug Administration has recognized this dilemma and, with prompting from the dairy foods industry, issued a new regulation, 21 CFR §130.10 *Requirements for foods named by use of a nutrient content claim and standardized term*, to help remedy the situation.

A. Temporary Marketing Permits

Prior to the passage of NLEA, dairy processors wishing to market a product which deviated from an established federal standard of identity were required under 21 CFR §130.17 *(Temporary permits for interstate shipment of experimental packs of food varying from the requirements of definitions and standards of identity)* to file a request with the FDA for a temporary marketing permit (TMP). Although sometimes difficult to obtain from the FDA in a timely fashion, TMPs have been employed by the dairy industry successfully over the past several years.

A TMP allows a manufacturer to market a product which deviates from the standard of identity in order to test the consumer acceptance of a product which has been modified by food technology. Technically speaking, the sole purpose behind the granting of a TMP is to allow a manufacturer the opportunity to collect the necessary data to support, on reasonable grounds, a petition to amend a food standard. It also allows a manufacturer to complete a study of the product deviating for the standard of identity and to assure that the interest of consumers is adequately protected.

TMPs have been used extensively in the dairy industry for products such as light sour cream, light eggnog, and nonfat cottage cheese. Although in the past TMPs have been granted primarily to dairy companies whose products deviate from the required milkfat level, TMPs have also been used to market products differing from the standard in other ways.

While TMPs may continue to be employed in the future by the dairy industry for certain products, the cumbersome and time-consuming procedure involved in securing a TMP from FDA and the recent creation of a more efficient procedure (21 CFR §130.10 *Requirements for foods named by use of a nutrient content claim and standardized term)* would seem to limit their widespread use in the future.

B. 21 CFR §130.10—The Generic Standard

While not actually NLEA-initiated, 21 CFR §130.10 *Requirements for foods named by use of a nutrient content claim and standardized term* is a new regulation that essentially creates a generic standard of identity which allows products with a federal standard of identity to bear an appropriate descriptor claim. Products with a standard of identity may deviate from the prescribed standard of identity in order to meet the requirements of a nutrient content claim that has been defined by the FDA. The product must comply with the nutrient content claim definition and, with exception, the parameters of the respective standard of identity.

A product with a standard of identity bearing a nutrient content claim is considered a type of substitute food, in that it substitutes for the product that it resembles. The agency requires a substitute product to be nutritionally equivalent to the unaltered product of the same standard of identity, minus the component (nutrient) for which the claim is being made.

A product with a standard of identity bearing a nutrient content claim may use safe and suitable ingredients to improve the acceptability and functionality of the substitute food.

In addition, products marketed under a TMP in order to make a nutrient claim, such as light eggnog, light sour cream and nonfat cottage cheese, must comply with the nutrient content claim regulations, regardless of the approval which was given to the product label under the TMP.

A nutrient content claim made on a product with a standard of identity is subject to all the nutrient content regulations, including type size, location, referral statement, and quantitative modification and comparison statements. As mentioned earlier, products with a standard of identity that contain a nutrient content claim as part of the common or usual name of the product (e.g., nonfat milk, lowfat yogurt, low sodium colby cheese are unaffected by this regulation.

1. Mandatory Compliance with the Generic Standard

A modified version of a standardized food that qualifies for use of a nutrient content claim and meets the other parameters of the standard must be labeled according to the provisions of the generic standard regulation. The agency states that a food is misbranded if it purports to be, or is represented as, a food for which a standard of identity has been established unless it conforms to such standard and its label bears the name of the food specified in the standard. Therefore, altered foods that conform to the definition and standard established in the generic standard (21 CFR §130.10 *Requirements for foods named by use of a nutrient content claim and a standardized term*) must be labeled with the name provided under the generic standard regulation or it is considered misbranded.

For example, sour cream is required to contain not less than 18 percent milkfat and sour half and half is required to contain not less than 10.5 percent, but less than 18 percent milkfat (21 CFR §131.60 *Sour cream* and 21 CFR §131.185 *Sour half and half*, respectively). If the total fat in sour cream has been reduced from 18 percent to 13.5 percent in order to qualify for a "reduced fat" claim, the product falls under the standard of identity for sour half and half. The product may **not** be labeled "reduced fat sour cream," but must be labeled "sour half and half." However, if the total fat was reduced in the sour cream product from 18 percent to 7 percent, the product falls below the minimum milkfat level for sour half and half and may be appropriately labeled as "reduced fat sour cream" (60 percent total fat reduction), "light sour cream" or, alternatively, "reduced fat sour half and half" (25 percent total fat reduction). In such cases, manufacturers have the option of determining the parent product to which modification is made (sour cream or sour half and half) and indicate this as the comparative food on the label.

2. Performance Characteristics

As is the case with the performance characteristics requirement for non-standardized products bearing a nutrient content claim, a standardized product bearing a nutrient content claim must perform similarly to the product for which it substitutes. In addition to the performance characteristics listed in the general principles for nutrient content claims (physical properties, functional properties, and shelf life), the generic standard specifically requires that the standardized food bearing a nutrient content claim be similar in flavor characteristics.

The generic standard also requires that the substitute food bearing a nutrient content claim perform one of the principal functions of the standardized product in a manner equivalent to the standardized product. FDA believes that consumers should be able to depend on using a substitute food in the same manner for which they use the traditional standardized food or, at the very least, for one of the principal functions of the standardized food. Specific performance standards for each product have not been established.

For example, a reduced fat sour cream must, at a minimum, be suitable for eating directly from the package as a condiment or for baking and cooking.

Performance characteristic differences that functionally limit the use of the product must be disclosed in easily legible print or type and in a size no less than that required for the net quantity of contents statement, except where the nutrient content claim is less than two times the required size of the net quantity of content statement, in which case, the performance characteristic difference must be no less than one-half the size of the claim, but no smaller than 1/16 of an inch. The disclaimer must be located adjacent to the most prominent claim.

a. Nutritional Inferiority

The performance characteristics include nutrition equivalence as a requirement. Products that are not nutritionally equivalent to the traditional food must be labeled as an "imitation" food. A manufacturer may avoid "imitation" labeling if the altered product is nutritionally equivalent to the product for which it substitutes and resembles (i.e., reduction of 2 percent or more of the RDI) (formally U.S. RDA of protein, vitamins, and minerals).

The addition of vitamins or minerals may be required to ensure that the altered product is nutritionally equivalent to the traditional product with a standard of identity. For example, reduced fat ice cream must be nutritionally equivalent to regular ice cream. This may require the addition of vitamin A to replace that which was removed with the milkfat. Nutrients added to restore nutritional equivalency must be included in the ingredient statement.

The agency is also permitting noningredient deviations from a product with a standard of identity bearing a nutrient content claim in order for the substitute food to maintain similar performance characteristics. Noningredient deviations include moisture content, food solids, processing conditions, etc. These deviations must be to the minimum level or amount necessary for the product to qualify for the nutrient content claim. Standardized foods bearing a nutrient content claim that excessively deviates from the standard are considered misbranded.

For example, a cottage cheese product may increase the moisture content in the substitute food bearing a nutrient content claim in order to qualify for a "nonfat" claim.

3. Ingredients

Within the limitations discussed herein, the ingredients used in products deviating from the standard of identity in order to make a nutrient content claim may differ from those used in the product conforming to the standard. The substitute product may differ by the addition of safe and suitable ingredients, the addition or removal of ingredients specified in the standard, and ingredients used in excess of the level provided for in the standard.

a. Safe and Suitable Ingredients

A substitute product deviating from the standard in order to make a nutrient content claim may use safe and suitable ingredients to improve texture, add flavor, prevent syneresis, extend shelf life, improve appearance, or add sweetness, so that the product is not inferior in performance characteristic to the traditional product. For example, the regulation specifically indicates that water and fat analogs may be added to replace fat and calories to maintain performance characteristics, including improving texture and replacing an ingredient or component that is specifically required by the standard (e.g., milkfat).

b. Ingredients Required in the Standard

Ingredients that are required by the standard of identity may not be replaced or exchanged with a similar ingredient from another source unless the standard provides for the addition of such an ingredient (e.g., nonfat milk substituted for milk). For example, vegetable oil may not be substituted for the milkfat in sour cream.

c. Ingredients Not Addressed in the Standard

Ingredients or components which are not specifically addressed in the standard of identity may be altered in order to produce a product which qualifies for a nutrient content claim. For example, cholesterol may be extracted from lowfat milk in order to make a "reduced" nutrient content claim for cholesterol.

d. Ingredients Not Permitted in the Standard

Ingredients not specifically permitted in the standard of identity may be used in a standardized product bearing a nutrient content claim. Additionally, ingredients may be used in excess of the level provided for in the standard. Ingredients not provided for or used in excess of those levels provided for in the standard must be flagged in the ingredient statement. An asterisk (*) or some other symbol must be placed adjacent to the ingredient and the statement "*Ingredient(s) not in regular ________"(the blank is filled in with the traditional standardized food) or "* Ingredient(s) in excess of amount permitted in regular __________" (the blank is filled in with the traditional standardized food) or both statements, as appropriate. The ingredient disclosure statement must immediately follow and appear in the same type size as the ingredient statement. The ingredient disclosure statement is not required if the ingredient was added to restore the level of a nutrients which was altered in order to produce the substitute food making a nutrient content claim.

e. Ingredients Prohibited in the Standard

Ingredients or components which are specifically prohibited in the standard of identity may not be added to those products with a standard of identity making a nutrient content claim under the "generic standard" regulations. For example, vegetable oil may not be substituted for the milkfat in "milk," because the standard of identity for milk (21 CFR §131.110 *Milk*) requires that the product "contain not less than 3.25 percent milkfat."

f. Optional Ingredients

Optional ingredients or components allowed for in the standard of identity may be added to a product to meet a nutrient content claim. For example, nonfat yogurt may add granola to make a "source" nutrient content claim for fiber.

4. Existing Standards of Identity Bearing a Nutrient Content Claim

"Nonfat milk" (21 CFR §131.143)
"Acidified nonfat milk" (21 CFR §131.144)
"Cultured nonfat milk" (21 CFR §131.146)
"Lowfat milk" (21 CFR §131.135)
"Acidified lowfat milk" (21 CFR §131.136)
"Cultured lowfat milk" (21 CFR §131.138)
"Light cream" (21 CFR §131.155)
"Light whipping cream" (21 CFR §131.157)
"Lowfat yogurt" (21 CFR §131.203)
"Nonfat yogurt" (21 CFR §131.206)
"Lowfat cottage cheese" (21 CFR §133.131)
"Low sodium cheddar cheese" (21 CFR §133.116)
"Low sodium colby cheese" (21 CFR §133.121)

As mentioned previously, foods that were subject to a standard of identity on November 8, 1990, that contain a nutrient content claim as part of the common or usual name of the food, are exempt from complying with the requirements of expressed or implied claims, including the referral statement and other disclosure statements (21 CFR §101.13 *Nutrient content claims—general principles*).

Products with a standard of identity that contain a nutrient content claim include the following:

Nutrient content claims, in addition to those specified in the standard of identity, may be used in conjunction with the above products, so long as they meet the requirements of the nutrient content claim. For example, if a lowfat cottage cheese product is formulated so that it meets the requirements of a reduced sodium product, then the sodium nutrient content claim may be used in conjunction with the name of the food, such as "reduced sodium lowfat cottage cheese."

The type size of the nutrient content claim, when it is part of the name of the food with a standard of identity, is not subject to the nutrient content claim type size requirements or any other general nutrient content claim requirements, such as a referral statement, quantitative modification, or comparison statement. Claims which are not part of the name of the food are subject to the nutrient content claim regulations.

V. SUMMARY

The dairy industry has been involved with nutrition labeling since its inception, over 20 years ago. In the early 1970s, the Milk Industry Foundation (MIF) and the International Ice Cream Association (IICA) established a nutrient data base which provided dairy processors basic information for developing nutrition labels. More recently, the International Dairy Foods Association (IDFA) has submitted a data base for a variety of dairy foods to the FDA for approval and use by the dairy processing industry.

Since the vast majority of dairy products have Federal standards of identity, this chapter discusses the rules applicable to the use of nutrient content claims (descriptors) in the name of these foods and the resulting opportunities for developing and marketing new products that previously were not permitted under the "old regulations."

VI. ENDNOTE

1 The constituent organizations of the International Dairy Foods Association are the Milk Industry Foundation (MIF), The National Cheese Institute (NCI), and the International Ice Cream Association (IICA). The American Butter Institute is also affiliated with IDFA.

18

The Impact of Mandatory Nutrition Labeling Rules on the U.S. Meat and Poultry Industry

James L. Marsden

American Meat Institute, Washington, D.C.

I. INTRODUCTION

The American Meat Institute (AMI) represents packers and processors of meat and other animal protein products. Our members produce the vast majority of meat and meat food products sold in the United States. From this perspective, AMI has a vital interest in new nutrition labeling rules at the FDA and USDA, and played a major role in developing the new requirements.

AMI has long supported the overall principle of uniform, mandatory nutrition labels for most processed foods. When consumers can see nutrition profiles of meat products compared in similar formats to other processed foods, we believe they will be reassured that meat products are nutrient-dense and not excessively high in fat, sodium, cholesterol, or calories. In fact, in January, 1991, the AMI called on the Secretary of Agriculture to revise USDA's nutrition labeling rules to establish mandatory nutrition labels on processed meat and poultry products and a comprehensive nutrition information program for fresh meat and poultry products.

As the meat and poultry industry worked with the USDA to develop labeling requirements, the AMI stressed that the first priority in implementing the new rules must be to assure that nutrition information is accurate, informative and useful to consumers. The AMI believes that the format chosen by the USDA and FDA is overly complex and may not be understood by most Americans.

In February 1992, AMI submitted to both theFDA and USDA results of a national consumer survey, "Nutrition Labels and Meat: Consumer Knowledge, Usage and Attitudes," which was conducted by The Roper Organization, Inc. This research provided extensive information regarding the ability of consumers to comprehend nutrition information presented in different formats and to use that information to make informed food choices.[1]

II. CONSUMERS AND NUTRITION LABELS

The consumer survey conducted by The Roper Organization in late 1991 was the first national survey to measure consumers' understanding and usage of current nutrition label information on meat and poultry products. Among the survey's key findings:

- Nearly as many Americans do not read, understand, or use nutrition label information (33 percent) as those who do (39 percent.) Only 6 percent of Americans are "committed" nutrition label readers who always read, understand, or apply the label information to their dietary choices. More than twice that amount, 15 percent, never bother.

  This raises an interesting question: If one-third of all Americans rarely or never read, understand, or use today's short nutrition label, and another third find it useful only sometimes, how will they react to the new nutrition labels that will display nearly triple the information that has appeared on most nutrition labels for the past 20 years?
- Those who read, understand, or use nutrition labels at least some of the time tend to be female, affluent, and educated. Whites tend to find nutrition labels more useful than blacks or hispanics.

  This finding suggests that some Americans may need help using the new nutrition labels. If the traditional label appears to serve primarily affluent, educated American women, how will the USDA and FDA ensure that the new nutrition label is understood and used by *all* Americans?
- If more nutrition information equates to higher food prices, most consumers choose lower-priced brands of food with less nutrition information. If additional nutrition information is free, most consumers choose food brands with more nutrition information on the label.

  It is notable that most consumers say they are not willing to pay for a lengthier nutrition label. How will they respond to higher food prices once the new labels are in place? Unlike the hypothetical situation in our survey, consumers will not be able to choose products with more or less nutrition information.

- Three out of four Americans do not know what percent of their calorie intake should come from fat. Calories from fat, and saturated and unsaturated fat are among the six nutrients consumers most often say they don't understand. Only one in five consumers claims to understand the term DRV.

If consumers lack understanding of basic nutrition information today, how do the USDA and FDA propose to help them learn enough to make use of the new expanded nutrition labels?

Consumer education will be an important component of the success of the new nutrition labels. Meat and poultry companies will do their part to educate consumers, but the USDA and FDA must fulfill their mandates to educate consumers as well.[2]

III. COSTS OF NUTRITION LABELING TO THE MEAT AND POULTRY INDUSTRY

The AMI estimated that the cost of the original proposed rule for mandatory nutrition labels of processed meat and poultry products ranged between $2.5 and $3 billion in the first year alone. This figure was the equivalent of twice the processing industry's annual payroll for its 67,000 production workers. It was 10 percent of the industry's annual sales and three times the industry's annual profits.

According to an AMI survey of meat and poultry processing companies, the first year implementation costs would have outweighed USDA's estimated first year benefits by a factor of 27.5 to one.

But AMI worked with tje USDA to neutralize the excessive costs of the original proposed rule through two important modifications:

- Allowing processors to use government-approved data bases instead of costly laboratory analysis to create nutrition labels.
- Phasing in the implementation of mandatory nutrition labeling to allow time for both industry and government to teach consumers how to use the new labels, and to give industry sufficient time for an orderly and cost-efficient manufacturing change.

One element of the final USDA nutrition labeling rule that significantly reduces industry compliance costs is allowing industry to rely upon government-approved data bases to calculate nutrient values for foods. A large segment of nutrition labeling compliance costs to the meat and poultry industry would have been laboratory analysis to calculate nutrients in food products.

Databases, however, allow industry to use respected, existing resources to calculate accurate nutrient composition of foods. Lab analysis need only be used for monitoring and verification, only, thus substantially lowering industry costs without diminishing the accuracy of the nutrition information on the label. This method has been used successfully in other parts of the food industry regulated by FDA.

Without the use of data bases, it has been suggested that the cost of laboratory analysis to confirm the nutrients on the new nutrition label would have been $1,500 per label. Based on this figure, the AMI estimates that the use of data bases could effectively reduce compliance costs to the meat and poultry industry by $1.5 billion.[3]

The cost of laboratory analysis to confirm the nutrients required on the new nutrition label would have been approximately $1,500 per label. The data base allowance effectively reduces industry compliance costs by $1.5 billion.

Meat and poultry industry compliance costs were also reduced significantly when the industry was given more time to implement the final rule. The originally proposed six month time frame was extended to 18 months.

Extending the compliance date to an 18 month period gave companies an opportunity to deplete existing supplies of packaging and labeling materials and order new materials with new information to replace existing stocks. In addition, an 18 month period allowed companies to spread the costs of laboratory analysis, package redesign and possible product reformulation over time. This has produced a more orderly and economically feasible transition to new labels.

IV. BENEFITS OF NUTRITION LABELING RULE

The primary benefit of uniform, mandatory nutrition labels on processed meat and poultry is that more accurate information will be available to all consumers who choose to use it for managing their diets. Research shows that many Americans will not or can not use the information to make dietary choices. But educational efforts over time should help more Americans to derive benefit from the new labels.

In initiating the rulemaking process, the federal government engaged in an interesting but controversial study of the potential benefits of the new label. Using research commissioned by the FDA, USDA reported in its Preliminary Regulatory Impact Analysis (PRIA) that nutrition labeling of meat and poultry products would benefit the United States at a value of $2 billion over a 20-year period. This equates to approximately $100 million per year, or $.40 each year for every American.

The value of the benefit was calculated by assigning dollar values to human lives—lives which presumably would be extended or saved as a result of nutrition labeling reform. This striking assertion was supported by an analysis which made several assumptions.

First, it assumed that the presence of more nutrition information on more food product labels will result in men and women decreasing their intake of fat, saturated fat and cholesterol. It seems a giant leap of faith to assume that more information automatically begets behavioral changes. If that were the case, speed limit signs would replace highway patrolmen as sufficient motivation for drivers not to speed, and the calorie information that has been on most food packages for the past 20 years would be sufficient motivation for people not to overeat. Unfortunately, these are both unrealistic assumptions.

Second, the analysis further assumed a direct link between the consumption of fat and cholesterol and the development of heart disease and cancer, without accounting for any other contributing dietary, genetic, behavioral, or environmental factors. The assumption that excess fat and cholesterol consumption is the primary cause of these diseases ignores a myriad of other factors and distorts the relationship of diet to health. For example, Americans' longer lifespan, which has doubled during this century, is clearly a factor in the increased incidence of these age-related ailments.

Third, using a computer model that did not account for any other factors, the analysis asserted that 5,818 lives would be saved over a 20-year period by nutrition labeling. These lives would be saved, according to the analysis, because these people would read and understand nutrition labels on meat and poultry products, which would lead them to decrease their fat and cholesterol consumption, thereby decreasing the incidence of fatal cancer and heart disease.

It is noteworthy that a similar analysis published by Dr. Warren S. Browner in the June 26, 1991, *Journal of the American Medical Association* found that by lowering total fat intake to 30 percent of calories, Americans could add a total of three to four months to their lives—and only after they had survived to age 65.

And finally, perhaps the most remarkable of all of the assumptions contained in the analysis was the value assigned to a human life: $1.5 million. It was this value, pro-rated by the anticipated extended years of life expectancy and multiplied by the number of lives presumed to be saved as a result of nutrition labeling of meat and poultry products, that produced the $2 billion value of USDA's initial proposed nutrition labeling rule.

V. CONCLUSION

The American public and the food industry deserve a realistic assessment of the costs and benefits of this major public policy change. While it is reasonable to suggest that more nutrition information may benefit those consumers who wish to make more informed food choices based on the nutritional attributes of foods, it is misleading and counter-productive to over-sell the benefits of nutrition labeling.

The meat and poultry industry—together with other food industry segments—will bear an enormous cost to provide new nutrition labels on all processed foods. Time will tell whether the labels actually benefit consumers. And education will be the most important factor influencing whether the labels are read and understood by most Americans.

Industry, government, and consumer and public health organizations have a major educational challenge ahead to maximize the benefit of the new nutrition labels. The meat and poultry industry stands ready to do its part.

VI. ENDNOTES

1 An extensive review of consumer knowledge, usage, and attitudes regarding nutrition and nutrition labels is presented by Brenda M. Derby and Sarah Beck Fein in Chapter 12.

2 Consumer education and FDA approach and programs for educating consumers about nutrition labeling are discussed by Naomi Kulakow, Chapter 11, and Brenda M. Derby and Sarah Beck Fein, Chapter 12.

3 The current analytical costs for nutrition labeling varies between $395–$752 per label. Many meat and poultry products do not require analysis for all the compounds required on the label because they occur in insignificant amounts. This could further reduce the cost. See Chapter 3 for details.

19

The Impact of the Nutrition Labeling and Education Act on the Frozen Food Industry

Leslie G. Sarasin

American Frozen Food Institute, McLean, Virginia

I. INTRODUCTION

With its passage of the Nutrition Labeling and Education Act of 1990 (NLEA), the U.S. Congress began the process of implementing the most sweeping reforms to the country's food labeling regulations in decades. By the end of 1992, when the Food and Drug Administration (FDA) completed its development of regulations to implement the provisions of the NLEA, this tremendous reform effort affecting the food industry coincided with the 50th anniversary of an industry segment which is considered to be the most evolutionary and innovative—the frozen food industry.

Primarily a cottage industry in 1942, frozen foods have since exploded into a multibillion dollar industry. Frozen products in 1942 generally were commodity-type vegetables and fruits, but products currently available to consumers in the freezer cases of their local supermarkets include such upscale and innovative items as hors d'oeuvres, gourmet items, breakfast foods, baked goods, prepared vegetables, sandwiches, pizza, pot pies, entrees, and dinners. Growth in the frozen food industry continues to be fueled by the never-ending evolution and growing diversity of new frozen food products available both for retail and foodservice uses, as well as by increases in grocery store freezer space, packaging advancements, changing consumer and advertising trends, and emerging global opportunities.

While this variety of product selection has been beneficial to consumers and to the industry itself, the NLEA-imposed requirement of including detailed nutrition information on the labels of virtually every food product sold in the United States has presented myriad and, at times, unique challenges to the frozen food industry. These special labeling considerations for the frozen food industry are worthy of further exploration.

II. DEVELOPMENT OF A NUTRIENT DATABASE FOR FROZEN FRUITS AND VEGETABLES

The FDA refers to a database as a "collection of related information," and historically has indicated strong support for the development and use of nutrient databases for commodity-type products. At least as early as 1979, the FDA, the U.S. Department of Agriculture and the Federal Trade Commission encouraged the concept of using industry-wide databases to reduce the costs of developing nutrition labels (1). In that regard, the FDA pledged to work with industry to develop valid databases and to resolve any compliance problems that might arise for foods labeled on the basis of a database reviewed by the agency.

The FDA reiterated its support for industry databases in a database guide it issued in 1993, in conjunction with the final regulations to implement the NLEA (2). In the guide, the FDA stated that "expansion of mandatory nutrition labeling to nearly all foods regulated by the agency has heightened interest in the use of industry-wide databases for some food products" (3). The agency stated further that it has had nearly 20 years of experience assisting industry in the development of nutrition label information.

The frozen food industry, specifically as represented by the American Frozen Food Institute (AFFI), also has had experience with nutrient databases. In fact, AFFI developed one of limited scope in 1980, which still was in use by some companies when the new nutrition labeling regulations were issued. AFFI is the national trade association which represents manufacturers of frozen food products, as well as their suppliers and marketers. AFFI's more than 550 member companies are responsible for the sale and distribution of frozen food products throughout the United States, representing more than $60 billion in sales annually.

In the months following the passage of the NLEA, AFFI considered the possibility of updating its 1980 nutrient database for frozen vegetables, or creating a new database (4). By mid-1991, it became clear that such a project would not only be desirable, but for some segments of the frozen food industry, including frozen fruits and vegetables, would be necessary if those segments were to continue to operate efficiently and also meet the statutory compliance deadline.

Several factors contributed to the growing importance to the frozen food industry of FDA's encouragement for the development of industry-wide nutrient databases. Perhaps foremost for the companies participating in the development of the nutrient database for frozen fruits and vegetables was the agency's statement that if a manufacturer uses a database for nutrition labeling in lieu of its own resources, the agency

> . . . will not take immediate action against a company for a product that does not contain the labeled level of nutrients, provided the Agency has reviewed and accepted the database for the product and provided the company has followed good manufacturing practices in producing the food. The Agency would instead work with the developers of the database to resolve the regulatory issue. Action would be taken if non-compliance is a result of failure to follow good manufacturing practices (5).

The FDA statement makes clear its intention, through its thorough database review process, to analyze the testing protocols and the quality of the data submissions from database participants. Through this process the agency may determine that representative samples have been included for testing from each participating company and that the testing methodologies have been followed correctly. Following acceptance of a database proposal, FDA has committed to working with database participants in the event a discrepancy is discovered in its nutrition information, rather than taking enforcement action against a company following testing of a product in the marketplace.

This flexibility has proven instrumental in encouraging the development of industry-wide databases, especially in the frozen food industry. Both FDA and the industry agree that a well-constructed database should yield average nutrient values which, in a large majority of instances, accurately convey the nutrient value of the food as packaged. Given the inherent variability of nutrient content of foods and the associated complexity of establishing a representative database, the food product will not provide the specific/declared amount of the nutrients 100 percent of the time. A flexible approach by FDA, therefore, is critical to providing adequate incentives for the development of and reliance on such databases.

During the early stages of development of the nutrient database for frozen fruits and vegetables, some industry representatives argued against the necessity of participating in an industry-wide nutrient database, indicating that a company merely could copy the nutrition information from any package label of a participating company it chose in the supermarket. While technically that was and continues to be true, subsequent input from FDA made it clear the data generated by nutrient databases generally would be proprietary to database participants. As a result, if the nutrient information on a product label were found to be inaccurate by FDA, the agency would seek documentation from the database developer, in this case AFFI, to support the label information. Non-participating companies would not have the documentation necessary to support the label values, nor would the developer's documentation be available to the non-participating companies. The result would be no exemption from compliance for the non-participants and possible enforcement action by FDA.

For many frozen fruit and vegetable packers, it was difficult to imagine a more important reason for participating in an FDA-accepted nutrient database than exemption from compliance. For those requiring more encouragement, however, there were at least two other important issues for consideration. One of these was cost. Several frozen food companies indicated an interest in developing their own nutrient databases to receive an exemption from compliance, without having to participate in a group project such as the one undertaken by AFFI. While this certainly was an option, at least for larger companies, the sharing of responsibility for performing nutrient analyses by all participating companies proved to reduce significantly the costs incurred by each of the companies, making participation in the industry-wide effort a worthwhile and cost-efficient endeavor.

Moreover, FDA's guidelines dictate an extremely complex procedure for reviewing the nutrient literature, performing nutrient analyses, and verifying the data (6). By participating in a group database, companies have been able to receive assistance in their sampling procedures and have been able to rely on a collective effort in compiling the data, writing the proposal and submitting the information for review and acceptance by the agency. The result has been a more cost-effective project for each of the participating companies. It also produced better quality data than any one company could have developed on its own.

For companies involved in packing private label products, databases address a unique set of issues. Retail operations which own their own store labels buy product for those labels from many different processors. This certainly is true in the case of frozen fruits and vegetables.

These packers of frozen fruits and vegetables, in turn, purchase product for freezing from various sources, giving rise to possible extreme variations in nutrient values associated with the product. These variations may be due to a host of factors, including the seasonality of the pack, the geographical distribution of the product sources, the condition of the soil in which the product was grown and the weather conditions in the growing region.

Without an acceptable nutrient database which has factored in these variables, and one which is available for use by packers of private label fruits and vegetables destined for many of the same consumer outlets, discrepancies on the nutrition panels of different packages of the same commodity placed next to each other on supermarket shelves would necessarily occur. The disparate nutrition information among various suppliers to private label retailers would create immense confusion among consumers and could result in increased scrutiny by FDA.

The only apparent alternative to the use of group nutrient databases in this situation would be for retailers with private label products to print a number of different labels for the same product, depending on the source. Retailers then could select the label that best represents the particular qualities of the product being packed in any given order. Not only would this be a highly unlikely scenario due to its costly nature, but the option would present an administrative nightmare with unpredictable results for consumers. The nutrient database for frozen fruits and vegetables, therefore, became inevitable.

Since AFFI's original consideration of the development of a follow-up to its 1980 nutrient database for frozen fruits and vegetables effort, the association, on behalf of 27 processors of frozen fruits and vegetables, has coordinated the development of proposals for databases for 26 commodity items and submitted those to FDA for review and acceptance. Within a few months of its submission, FDA accepted formally the first proposal. This proposal constituted the development of a nutrient database for frozen broccoli to be completed over a ten-year period, making AFFI's the first *major* database to be recognized by the agency. The FDA's approval allows a two-year provisional use by AFFI database participants of the nutrient information contained in the proposal, based on a multi-year study to gather additional data points to verify the nutritional profile established.

The remaining 25 proposals, including asparagus, green beans, carrots, cauliflower, brussels sprouts, corn, okra, spinach, green peas, mushrooms, celery, onions, sweet peppers, strawberries, fordhook limas, pea pods, summer squash, water chestnuts, southern peas, turnips, potatoes, lima beans, leafy greens, sweetened strawberries, and winter squash subsequently were accepted on a provisional basis by the agency.

While the collection of the relevant data has at times been a painful process for the database participants and has required tremendous levels of planning and coordination, this has been far outweighed by the benefits accruing to the participating companies, as well as the benefits which will accrue to the consuming public in the form of accurate and timely nutrition information on the labels of frozen fruits and vegetables.

III. USE OF NUTRIENT CONTENT CLAIMS IN FROZEN MEAL-TYPE PRODUCTS

Increasingly throughout the 1980s and continuing into this decade, many U.S. consumers have revised their eating habits such that the relative healthfulness of a particular food product has been elevated significantly on the list of food selection criteria. Overall diet has begun to play a significant role in the move to healthier lifestyles. In that regard, the U.S. dietary guidelines recommend reducing the consumption of fat, especially saturated fat, as well as cholesterol, and increasing the consumption of low-fat, high-fiber foods, such as fruits, vegetables, and grains.

The food industry has responded to consumers' increasing desire to follow these recommendations, as expressed by their demand for healthier prepared-food items. No segment of the industry has been more responsive to this demand than the frozen food category.

Total sales of low-fat, low-cholesterol foods equaled $32 billion in 1991, up over 10 percent from $29 billion in 1990. Approximately $1.5 billion of those sales were low-fat, low-cholesterol frozen dinners and entrees. This sales figure constituted a compound annual growth rate of 11.7 percent from 1986. In that same period, sales of all frozen dinners and entrees increased at an annual rate of only 5.3 percent (7).

Furthermore, the market for low-fat, low-cholesterol frozen dinners and entrees is expected to continue to grow at an annual rate of 10.5 percent through 1996, bringing sales to $2.4 billion, which will represent a 39 percent share of the frozen food dinner and entree market (8). It is anticipated that sales of low-fat, low-cholesterol frozen desserts, which include ice milk, frozen yogurt, and sherbert, will climb from $2 billion in 1991 to $8.7 billion in 1996. At that time, low-fat, low-cholesterol frozen desserts are expected to comprise 63 percent of the total frozen desserts market (9).

To accomplish this goal of meeting consumers' increasing demand for healthier frozen food products, manufacturers have invested heavily in the background work necessary for new product development. They have dedicated entire corporate research and development departments of staff talent and expertise to meeting the demands of the marketplace. They have revised and reformulated their products, including healthier ingredients as substitutes for high-fat, high-cholesterol ones. Through this process, the manufacturers have sought products with nutrient profiles which are consistent with the U.S. recommended dietary guidelines.

Most important, they have sought food products which taste good. As indicated by the sales figures expressed previously, they appear to have been successful.

Much of the success of these new products, and the resulting health benefits derived by the consumers who purchase them, can be attributed to the fact that manufacturers have been able to communicate the food products' healthful attributes to their customers. Although some of this has been accomplished through advertising, primarily it has been achieved through the inclusion of messages on food labels.

The frozen food industry has long believed the food label should serve as an accurate means for providing nutrition information to assist consumers in selecting foods consistent with individual dietary preferences. It is essential to consumers' ability to make informed food selections that manufacturers of food products have every opportunity to provide truthful and non-misleading information on food labels. In fact, because of the frozen food industry's dedication to providing healthy frozen food selections for consumers, its manufacturers have long been leaders in providing meaningful nutrition labeling information to consumers. Indications are, however, that this trend of providing the most truthful and non-misleading information possible to consumers may inadvertently have been put in jeopardy by the new food labeling regulations.

In its final food labeling regulations, FDA adopted rules establishing parameters for use of nutrient content claims, as well as other claims in food labeling. As a result, a food label may not bear an express or implied claim that characterizes the level of a nutrient in a food unless the term has been defined by regulation. Thus far, FDA has established definitions for relatively few terms, thereby excluding the use of any others by manufacturers in their label communications with consumers. The defined terms include "free," "low," "reduced"/"less," "light"/"lite," "lean," "extra lean," "good source," "high," and "more" (10).

As one might expect, many qualified individuals disagree about the appropriateness of the provisions contained in the specific definitions which have been established by FDA for these terms, and that is unlikely to change. The critical issue, however, is that by refusing to allow any label statements regarding the nutrient content of a particular food product except those specifically defined by regulation, the agency effectively is minimizing, if not eliminating, the incentive for industry segments which have made a long-term and successful commitment to the development of healthful food products, such as the frozen food category, to continue doing so. The unfortunate victim in this situation is the consumer, who is deprived of information which would be useful in making personal dietary choices.

The only option available to manufacturers wishing to use an undefined term in labeling involves a petition process. Petition procedures are established for interested parties that seek to establish a regulation authorizing a nutrient content claim, a synonymous term, or use of an implied claim in a brand name (11). While the NLEA dictates certain time frames in which FDA initially must act on the petition, many in the frozen food industry view the detailed requirements concerning the content and form for each type of petition as being so unduly burdensome as to constitute no opportunity for the use of previously undefined claims at all. By virtually eliminating the prospects of obtaining approved use of new claims, due to the elaborate information required to support a petition, the petition provision serves as a significant disincentive for the development of new healthful products.

Many also have pointed out that the information which would be required to support a petition is far more detailed than any of the information upon which FDA relied in issuing its proposed or final nutrient content regulations. For example, a petition would have to include an analysis of the effect of the use of a claim on corresponding changes in nutrient intake, which most view as an unrealistic requirement.

The unfortunate ramifications presented by this section of the new food labeling regulations are exacerbated by the fact that they excessively specify not only the type of information which may be included on a food label, but also the location and size in which the information must appear, as well as the prominence which must be given the information. Furthermore, the requirements associated with these labeling mechanics are complicated and difficult to understand and apply.

For example, on the one hand, the type size and style of a nutrient content claim may be no larger than two times that of the statement of identity of the food for which the claim is made. Disclaimers, disclosures, referral statements, and principal display panel information accompanying relative claims, on the other hand, in most cases must be in a typesize equivalent to that required for the net content statement. Referral statements required to disclose the presence of a nutrient in a food product exceeding an established level, however, must include a statement as follows: "See [appropriate panel] for information about [nutrient requiring disclosure] and other nutrients" (e.g., "see side panel for information about total fat and other nutrients"). This statement must be placed in a location specified in the regulations (12).

Not only are these requirements extensive, complicated, and difficult to apply and understand, but most manufacturers believe them to be unnecessary and inconsistent with FDA's long-standing regulatory approach. Moreover, they severely limit a manufacturer's ability to provide information in a creative, meaningful manner.

In addition to the investment companies have made in new product development, manufacturers have dedicated tremendous resources to obtain a special recognition of their brand names by consumers which evokes quality and value. To accomplish this goal, they have spent not only enormous sums of money, but also many hours of creative talent to design packaging for their products which provides consistency in the presentation of their brand name across many product lines and which is recognizable immediately to the consuming public. Many manufacturers have found that there simply is insufficient room on most frozen food packages to accommodate the information required to be included, along with the brand name and artwork in the form with which consumers are familiar. The inclusion of the FDA-required additional information, although instructional and useful to consumers, may not be possible due to the extensive requirements of referral statements, disclosures and other information. The result of its inclusion would be an illegible, unattractive, and probably unpurchased food package.

The other ongoing debate in this context is the extent to which the food label should be considered the nutrition educator for consumers, as well as the extent to which food manufacturers should be responsible for educating consumers through their food labels. There is a balance to be achieved between requiring that consumers be provided the information necessary to evaluate a particular food product's appropriateness for their dietary needs and requiring that the food label be the absolute public nutrition educator. Many in the frozen food industry believe that while FDA has made a laudable attempt in this area, this balance may not yet have been achieved through the final labeling regulations.

While the use of nutrient content information is made extremely difficult by the lack of the FDA-defined terms and the extensive typesize, placement and disclosure requirements, perhaps the most troubling challenge for manufacturers of frozen food products is determining that which constitutes a nutrient content claim. Under the regulations an express nutrient claim is "any direct statement about the level (or range) of a nutrient in the food" (e.g., "cholesterol free" or "contains only 100 calories") (13). This definition seems to have proved useful and comprehensible to most in the industry.

That which constitutes an implied nutrient content claim, however, is not as clear and has presented difficulties in interpretation for many in the frozen food industry. Implied nutrient claims, according to the final regulations, include any claim which 1) "describes the food or an ingredient therein in a manner that suggests that a nutrient is absent or present in a certain amount" (e.g., "high in oat bran" implies high dietary fiber), or 2) "suggests that the food, because of its nutrient content, may be useful in maintaining healthy dietary practices and is made in association with an explicit claim or statement about a nutrient" (e.g., "healthy, contains 2 g of fat") (14).

While the agency has attempted to clarify this confusing category of claims by identifying label statements pertaining to the nature of a product that would not be regulated as nutrient content claims, FDA has been careful to note that even those included in its list would be considered implied nutrient content claims if "such statements are made in a context that would make them implied nutrient content claims." The agency also has stated that factual statements that do not characterize nutrient level explicitly or implicitly are not regulated, nor are statements of pure authoritative dietary guidance that do not place the product in a nutritional context. Further, FDA advised that in certain ingredient claims (e.g., "contains oat bran" or "corn oil margarine") which may imply nutrient content, it will determine on a case-by-case basis whether to regulate ingredient claims as nutrient content claims, focusing on whether the ingredient claim explicitly or implicitly identifies a nutrient and whether it indicates that the nutrient is absent or present in a certain amount (15). Without clear direction from FDA, this, of course, leaves the industry in the difficult position of taking a risk on whether its interpretation of that which constitutes an implied claim will match that of the agency.

Many in the frozen food industry have been critical of this treatment by the agency. Critics argue that the definitions established by FDA are far broader than that required by the NLEA and are sufficiently vague as to provide little guidance to the food industry in determining the statements which would be considered by FDA to constitute implied claims. The NLEA states that a food is misbranded if a claim expressly or by implication characterizes the level of any nutrient in the food, unless the claim is permitted by regulation. The industry has maintained, therefore, that only a statement which characterizes the level of a specified nutrient in a food product, such as "contains no saturated fat," or "contains no cholesterol," could be regulated by FDA as a nutrient content claim.

Perhaps in no area has this potential for variability of interpretation been more critical or more controversial than in its application to brand names. While the NLEA addressed this area to some extent by providing an exemption from definitional criteria for undefined nutrient content claims in brand names in use before October 25, 1989, much has been left to FDA's discretion for determining those terms in brand names which constitute implied claims.

During the rulemaking process, the frozen food industry developed a position that a brand name which includes a descriptor defined by regulation must comply with the criteria established in such regulation if the brand name, in the context in which it is used, explicitly or implicitly characterizes the level of a nutrient in the food or if an explicit claim is made in reference to the level of a nutrient in the product. The industry determined further its view that a brand name which includes a descriptor that has been defined by regulation need not comply with the criteria set forth in such regulation provided the brand name, as used in its context, neither explicitly nor implicitly characterizes the level of a nutrient in the food because in these cases, the term should not be regulated as a nutrient content claim. Finally, for purposes of the regulation of brand names, including descriptors used to characterize the level of nutrients in food, the industry determined that single-component products should comply with different criteria than meal-type products.

While this appeared to the frozen food industry to be a responsible and reasonable approach to the issue of brand names, the agency thus far has failed to endorse such an approach. On May 10, 1994, FDA published a final rule addressing a definition for one of the most controversial terms used in a brand name—"healthy" (16). AFFI has objected to the final regulation by petitioning the Agency to reconsider its position that raw fruits and vegetables deserve a special exemption for the use of the term "healthy" that the Agency declined to extend to frozen fruits and vegetables. AFFI has met with FDA representatives and presented extensive documentation to support the use of this term on labels of frozen fruits and vegetables, but as of this writing the Agency has not yet responded to AFFI's position.

The other unknown factor relating to the consideration of brand names as claims is the extent to which the companies owning the brand names in question will be willing to go to battle with the Agency to maintain their usage. FDA received a number of comments from trade associations and individual companies which challenged the restrictions on brand names and other labeling statements under the First Amendment (which protects commercial speech) and the Fifth Amendment (which bars unlawful takings of private property by the federal government).

In its final regulations, FDA provided extensive legal justification for its position that the restrictions on brand names do not violate food manufacturers' rights to protection of their commercial speech. The agency determined that the governmental purpose served and the potential misuse of nutrient content claims warrant the limitations established. Similarly, FDA rejected arguments that prohibiting continued use of brand names containing nutrient content claims that are not in compliance with a definition constitutes an unconstitutional taking in violation of the Fifth Amendment. The agency found that the character of its regulatory action, the interference with expectations of continued use of brand names, and the resulting economic impact do not rise to a "taking" level (17).

Some industry attorneys have opined that the agency's position may not withstand a court challenge should one become necessary.

IV. PROSPECTS FOR THE FUTURE IN FOOD LABELING

Since publication of the final food labeling rules, representatives of food manufacturing companies, including frozen food processors and packers, have poured over the approximately 4000 pages of the FDA regulations, as well as those issued by the U.S. Department of Agriculture, to determine the manner in which the regulations apply to their specific products. Companies have been forced to consider whether further nutrient analyses were required, undertake the necessary analyses, develop nutrition panels based on the analyses, ensure the nutrition panels comply with the format requirements established by FDA, order and begin using new packaging that bears the revised labels, and, finally, hope consumers are equipped to read and understand all the new information.

Far from a one-time label change process, this effort promises to be an evolving one, with many permutations over the coming years. In a period of only a little more than two years, the agency issued proposed regulations, final regulations, several sets of technical amendments, a question-and-answer document, and more proposed regulations, all of which affect food manufacturers who are trying diligently to place labels on their products which are in compliance with the law. To the agency's credit, much of this publication effort has been the result of its attempt to respond to manufacturers' input regarding the difficulty in applying the regulations to their particular types of operations and products.

For example, the final FDA regulations contain specific requirements of the format in which the mandatory nutrition information must be presented. In its original form, the regulation provided little if any flexibility for manufacturers, regardless of whether their packaging was in a box, tubular, or bag form, or whether it contained seams or other intervening details which would prevent inclusion of the required format in a legible manner. With almost no exceptions, the required format was vertical.

This became a particular problem for the frozen food industry because many frozen products, particularly frozen fruits and vegetables, are packaged in plastic polybags. These plastic polybags are produced in a flat roll for the frozen fruit or vegetable processor, and the bags are, in effect, formed around the fruit or vegetable content and then sealed. For this reason the seals are not created exactly in the same location on each product. The seal which runs horizontally along the back of the polybag would make it impossible to print information in a manner which covers a portion larger than one half of the package. As a result, the vertical format mandated by FDA for nutrition labeling would have been impractical for these packages.

Following the industry's extensive discussions with FDA representatives, as well as the submission of detailed comments and samples of frozen fruit and vegetable packaging, the agency revised the previous regulation in its technical amendments of August, 1993. As a result, the use of a tabular format on packages greater than 40 square inches will be allowed when there is insufficient continuous vertical label space (i.e., approximately three inches) to accommodate the required components of the nutrition label up to and including the mandatory declaration of iron (18). In fact, in his press conference to announce publication of the technical amendments, FDA Commissioner David Kessler utilized the example of a frozen vegetable product packaged in a polybag to demonstrate the need for this particular amendment.

The frozen food industry is well established in its resolve to continue in its role as the most innovative and dynamic segment of the food industry. Being the first to go through the process of developing a nutrient database in this new food labeling environment has created unique issues for its participants. Addressing the other frozen food-specific issues which have arisen, and which will continue to surface, has and will continue to require creativity, diligence, and above all, patience.

V. REFERENCES

1. 44 *Federal Register* 75,990 (December 21, 1979).

2. *FDA Nutrition Labeling Manual: A Guide for Developing and Using Databases,* Washington, D.C., 1993.
3. *FDA Nutrition Labeling Manual: A Guide for Developing and Using Databases,* Washington, D.C., 1993, p. 7.
4. *Nutrients in Frozen Vegetables,* American Frozen Food Institute, McLean, VA, 1980.
5. *FDA Nutrition Labeling Manual: A Guide for Developing and Using Databases,* Washington, D.C., 1993, p. i.
6. *FDA Nutrition Labeling Manual: A Guide for Developing and Using Databases,* Washington, D.C., 1993.
7. *The Market for Lowfat and Low-Cholesterol Frozen Dinners and Entrees,* FIND/SVP, New York, NY, 1992, p. 99.
8. *The Market for Lowfat and Low-Cholesterol Frozen Dinners and Entrees,* FIND/SVP, New York, NY, 1992, p. 108.
9. *The Market for Ice Cream and Other Frozen Desserts* and *The Frozen Yogurt Market,* FIND/SVP, New York, NY, 1992.
10. 58 *Federal Register* 2410 (January 6, 1993).
11. 58 *Federal* Register 2423 (January 6, 1993).
12. 58 *Federal Register* 2411 (January 6, 1993).
13. 58 *Federal Register* 2411 (January 6, 1993).
14. 58 *Federal Register* 2411 (January 6, 1993).
15. 58 *Federal Register* 2423 (January 6, 1993).
16. 58 *Federal Register* 2944 (January 6, 1993).
17. 58 *Federal Register* 2392 (January 6, 1993).
18. 58 *Federal Register* 44,093 (August 18, 1993).

20

NLEA and the Foodservice Industry

Guy E. Livingston
Food Science Associates, Inc., Dobbs Ferry, New York

I. THE FOODSERVICE INDUSTRY IN THE UNITED STATES

A. Size and Importance of the Industry

The foodservice industry in the United States has shown a steady and dramatic growth during the last several decades. According to the National Restaurant Association (NRA), foodservice sales increased from $42.8 billion in 1970 to $238.8 billion in 1990 (1). NRA's analysis of the 1991 Bureau of Labor Statistics survey of consumer expenditures revealed that the annual expenditure for all households in the United States for food away from home reached $1,620 per household or $623 per capita. Per capita spending was influenced by a variety of factors, including location, age of the household head, and household income. For example, per capita spending on food away from home in Miami—the highest in any metropolitan statistical area—attained $1,150 in 1991 (2).

The growth in foodservice, of course, means that an increasing proportion of the consumer food dollar is being spent eating away from home. From 33.6% in 1972, this percentage grew to 43% in 1990 (1). In fact, in justifying its June 15, 1993 publication of its proposal to amend its food labeling regulations by removing the provisions that exempt restaurant menus from the requirements of how nutrient content claims and health claims are to be made, the Food and

Drug Administration (FDA) stated, "almost half of the American food dollar is spent on food consumed away from home, and perhaps as much as 30 percent of the American diet is composed of foods prepared in food service operations" (58 FR 2302 at 2387). Thus, if FDA is to effectuate its policy objectives under the 1990 amendments, it is vitally important that FDA ensure that claims made in restaurants comply with the act" (3).

B. Structure of the Industry

Although the final FDA regulations on the implementation of NLEA in the foodservice industry have not been published at the time of this writing, it is certain that these regulations will impact differently on the various sectors of the industry.

NRA categorizes foodservice establishments into three groups: Group I, Commercial Foodservice; Group II, Institutional Foodservice—business, educational, governmental or institutional organizations that operate their own foodservice; and Group III, Military Foodservice. The composition of these groups is shown in Table 20.1, which is adapted from the NRA's projected 1993 foodservice industry food and drink sales.

It will be noted that commercial foodservice accounts for 88.52% of the industry sales and that, within that sector, restaurant and commercial cafeteria sales as a group account for a total of 63% of industry sales. Significantly, limited menu restaurants, the category that includes the fast food operations, by itself now accounts for almost one-third of industry sales. Given the product standardization and centralized management which prevail within this industry sector, opportunities for implementing nutrition labeling and education activities are significantly greater than they are in individually owned and operated restaurants.

Commercial food contractors account for 6.1% of industry sales. Within this sector a number of large and sophisticated multi-unit food management companies, with diversified interests in contract feeding, operate in colleges and universities, hospitals, business and industrial facilities, airline catering, amusement parks and vending. Such operators, most of whom employ full-time dietitians, also have the ability to implement FDA requirements with respect to nutrition labeling in the operations which they manage.

Schools, hospitals and nursing homes generally require the services of dietitians for menu planning and should, therefore, be able to substantiate health or nutrient claims which might be made for menu items served.

Even allowing for the stated exceptions, it would appear that over 55% of industry sales are generated in operations which currently lack the technical expertise to deal knowledgeably with FDA requirements which may eventually be promulgated.

Table 20.1 Projected 1994 Foodservice Industry Food and Drink Sales (1)

	Sales ($ 000)	Percent
Group I—Commercial Foodservice		
Eating Places		
Fullservice restaurants	$85,510,450	31.1
Limited-service restaurants	$86,025,561	31.3
Commercial cafeterias	$4,744,230	1.7
Social Caterers	$2,867,922	1.0
Ice-cream, frozen-custard stands	$2,677,794	1.0
Total Eating Places	$181,825,957	66.1
Bars and Taverns	$8,741,976	3.2
Total Eating and Drinking Places	$190,567,933	69.3
Food Contractors		
Manufacturing and industrial plants	$4,014,292	1.5
Commercial and office buildings	$1,397,241	0.5
Hospitals and nursing homes	$1,863,951	0.7
Colleges and universities	$3,756,219	1.4
Primary and secondary schools	$1,511,301	0.5
In-transit foodservice (airlines)	$1,613,164	0.6
Recreation and sport centers	$2,203,072	0.8
Total Food Contractors	$16,359,240	5.9
Lodging Places		
Hotel restaurants	$14,204,520	5.2
Motor-hotel restaurants	$520,469	0.2
Motel restaurants	$738,458	0.3
Total Lodging Places	$15,463,447	5.6
Retail-host restaurants	$11,440,899	4.2
Recreation and sports	$3,236,727	1.2
Mobile caterers	$881,002	0.3
Vending and nonstore retailers	$6,192,974	2.3
TOTAL GROUP I	$244,142,222	88.73

Table 20.1 (Continued)

	Sales ($ 000)	Percent
Group II—Institutional Foodservice—Business, Educational, Governmental or Institutional Organizations that Operate their own Foodservice		
Employee foodservice	$1,925,954	0.7
Public and parochial elementary, secondary schools	$4,265,091	1.6
Colleges and universities	$5,110,379	1.9
Transportation	$1,538,657	0.6
Hospitals	$9,446,688	3.4
Nursing homes, homes for the aged, blind orphans, and the mentally and physically disabled	$4,205,603	1.5
Clubs, sporting and recreational camps	$2,320,880	0.8
Community centers	$1,073,902	0.4
TOTAL GROUP II	$29,887,154	10.86
Group III—Military Foodservice		
Officers' and NCO clubs ("Open Mess")	$732,529	0.3
Foodservice-military exchanges	$376,926	0.1
TOTAL GROUP III	$1,109,455	0.40
GRAND TOTAL	$275,138,831	100.00

(10) Adapted from NRA foodservice industry sales projections (NRA Research Dept., 1993).

II. REGULATION OF NUTRIENT AND HEALTH CLAIMS FOR RESTAURANT FOODS

A. FDA Final Rules Implementing NLEA

On November 8, 1990, the President signed into law the Nutrition Labeling and Education Act of 1990 (Public Law 101-535). Sections 403(q)(5)(a)(i) and (ii) of the act exempted from the nutrition labeling requirement foods that are served in restaurants or similar foodservice establishments that are principally processed and prepared in retail establishments, that are ready for consumption although not necessarily for consumption at the place of sale, and that are not offered for sale outside the establishment (4). The FDA accordingly also exempted from the nutrition labeling requirements foods sold by a distributor who sells principally to restaurants and other foodservice establishments. The agency, however, cautioned manufacturers, packers, and distributors of foods for restaurant use to label their food products if there was a reasonable possibility that the food would be purchased directly by consumers. Although it was not stated, this caution undoubtedly arose from the recent growth in the United States of so-called warehouse club stores that sell large package size food products to the public as well as to foodservice operators.

B. FDA Proposed Rule

The final Food Labeling Regulations Implementing the Nutrition Labeling and Education Act of 1990, published by the Food and Drug Administration on January 6, 1993, stated in §101.10 (Nutrition Labeling of Restaurant Foods) that "Nutrition labeling in accordance with §101.9 shall be provided upon request for any restaurant food or meal for which a nutrient claim . . . or a health claim . . . is made (except on menus)" (3).

Legal challenges by consumer groups and criticism by some members of Congress following the above-cited January 6, 1993, publication and its exemption for menu claims, led the FDA to reverse itself. On June 15, 1993, the agency proposed to amend its food labeling regulations by removing the provisions that exempted restaurant menus from the requirements governing nutrient and health claims. It also proposed to delay the effective date of these regulations for small restaurant firms for one year. "Small restaurant firms" were defined as those consisting of 10 or fewer establishments (5).

III. FDA AND INDUSTRY ASSESSMENT OF NLEA IMPACT

A. FDA Assessment

In its January 6, 1993, publication of the Final Rules for Nutrition Labeling, the FDA estimated the total costs of food labeling regulations to range from \$1.4 billion to \$2.3 billion. Of this total, the agency suggested that "the costs to restaurants and other food service establishments to reprint menus not in conformance with nutrient content or health claim proposed regulations would be \$116 million (6). FDA based its calculations on information received from the foodservice industry, adjusting that information to allow for a longer compliance period, the use of options other than analytical testing to support its claims, and the deferral of enforcement for restaurants with less than 10 individual establishments.

The economic impact of the June 15, 1993 removal of the exemption on menu claims by FDA (5) was estimated by the agency to require approximately 2,800 large and medium restaurants to make simple menu design changes affecting 3,600 menus at a cost of \$500 per menu, and 900 complex changes at $1,700 per menu, for a total cost of $3.4 million. However, if 90 percent of large and medium size restaurants already had a reasonable basis to support menu claims currently being made, the proposal would result in costs of less than $1 million.

Small foodservice firms, with 10 or fewer units, constitute 75 percent of foodservice establishments. Enforcement of compliance with claim definitions had originally been postponed to May 1995. Under the June 15, 1993 proposal, these small firms would need to change their menus and other labeling to comply with nutrient claim regulations within 4 months of the publication of the final rule in the *Federal Register*, i.e. approximately 6 to 8 months earlier than under the previous date. The agency estimated that this change would require approximately 8,500 restaurants to implement 10,800 simple menu changes at $500 per menu and 2,700 complex menu changes at $1,700 per menu, for a total potential cost of $10.1 million. However, if 90 percent of these restaurants already had a reasonable basis to support their claims, the cost could be reduced to about $1 million (5).

B. Restaurant Industry Assessment

The National Restaurant Association is the leading trade association serving the foodservice industry. Started in 1919, its current membership of 25,000 companies encompasses 150,000 of the 721,000 foodservice outlets in the United States. The association has served as the principal industry representative in analyzing the potential impact of the nutrition labeling regulations on foodservice operations.

From the outset, NRA has endorsed the goals of the NLEA but maintained that the proposed regulation of nutrient content claims and health claims would be counterproductive as applied to the restaurant industry in that it would "essentially bar the dissemination of virtually all nutrition information concerning restaurant foods" (7). Recalling that NLEA exempted restaurant foods from the restrictions placed on claims related to cholesterol, fat, and fiber content, the association asserted that the FDA's proposed regulations governing nutrient content and health claims for such foods would "be impracticable, costly, and . . . reduce the availability of nutrition information to consumers." The NRA argued that "restaurant meals are not standardized" and that "there is no control over preparation of seemingly like products given the manner in which foods are prepared in a restaurant kitchen." A single menu item, such as a sandwich, could be ordered with or without any one of numerous condiments or toppings, each combination having a different nutritional profile.

The association contended that FDA erroneously applied definitions developed for frozen meals and entrees to restaurant meals, which in fact cannot be standardized, given seasonal differences in ingredient availability, regional preferences and marketing considerations. By and large, there is no central control over preparation, and portion sizes often reflect the judgment of individual food handlers.

Serving sizes in restaurant meals are conspicuously larger than those in packaged frozen meals. Restaurant operators are motivated to satisfy their customers and "do not, and cannot manipulate serving sizes to 'trick' the customer about nutrient levels."

The association also pointed out that "even among larger and presumably more sophisticated operations, there is no absolute degree of central control over food preparation." As a result of having hundreds of food preparation workers around the country influencing ingredient and serving size decisions, it is "impossible even for the multi-unit operator to guarantee the accuracy of nutrient analysis."

Perhaps the most problematic of the proposed regulations, from the NRA's perspective, were those governing nutrient content claims. Clearly, these regulations were developed by FDA to cover packaged foods, not restaurant foods. The notion that restauranteurs could compare the nutrient level of a "reduced" food described on a menu as having "less" cholesterol, fat, or sodium with that of a reference food implied that the restaurant industry had widely used norms that would permit such a comparison to be made, which in fact is not the case. Other arguments could be made with respect to the problems which the restaurant industry would have with proposed regulations of descriptors such as "light" or "fresh."

Many restaurants, NRA advised FDA, use recipes developed by the American Heart Association and designate the menu items so prepared with a heart-shaped symbol or term such as "heart healthy." Rather than comply with the FDA's requirements governing the use of health claims, most operators would discontinue their present practices, thus denying their customers the opportunity to select nutritionally desirable menu offerings.

The NRA concluded that nutrition labeling should not be required for restaurant foods for which a nutrient or health claim is made.

IV. MENU HEALTH AND NUTRIENT CLAIMS

Restaurant menus have traditionally offered a fertile outlet for the verbal creativity of chefs and restauranteurs. Operating under the marketing aegis that you must "sell the sizzle " in order to sell the steak, menu writers have been known to take considerable license in naming their offerings. Some descriptions are so extravagant that the customer is not likely to be fooled. For example, how much credibility does "Today's Catch: Red Snapper" have when it is featured on the menu of a restaurant in the heart of Des Moines, Iowa? On the other hand, many examples of creative menu writing are clearly deceptive, such as "filet of sole" when the fish served is, in fact, flounder (8) or even cod; "garden fresh vegetables" that came out of the kitchen freezer, or "whipped cream" which turns out to be a mixture of milk solids, vegetable oil, stabilizers, and flavorings in an aerosol can. Modern food technology has compounded the problem by offering an array of excellently engineered, high acceptance, look-alikes that can easily pass for a higher priced food, as in the case of surimi for crab meat, but clearly lack the menu appeal of the original.

It should surprise no one, therefore, that nutrient or health-related claims for menu offerings have also been used loosely within the foodservice industry, with varying degrees of justification, to enhance menu appeal. Health claims for foods are hardly new, as any student of the classics knows. In fact, twenty four hundred years ago, Plato, in Book II of his dialogue, the *Republic*, described a diet consisting of "barley-meal and flour of wheat . . . salt, and olives, and cheese . . . roots and herbs . . . figs and peas, myrtle-berries and acorns" and concluded "And with such a diet they may be expected to live . . . to a good old age, and bequeath a similar life to their children after them."

In its June 15, 1993, proposed rule, the Food and Drug Administration proposed "to include claims made for restaurant foods on menus, as well as those made on signs and placards within the coverage of the food labeling regulations implementing the 1990 amendments. Should the actions proposed in this document be adopted and go into effect, any nutrient content or health claim made for any food sold in a restaurant, in or on any labeling in the restaurant, including menus, signs or placards will be subject to FDA's food labeling rules" (5).

The proposed regulation recognized that the foodservice industry is not as well equipped as the food processing industry to communicate and to control the nutrient content of its products and, therefore, provided that "Nutrient levels may be determined by nutrient data bases, cookbooks, or analyses or by other reasonable bases that provide assurance that the food or meal meets the nutrient requirements for the claim. Presentation of nutrition labeling may be in various forms, including those provided in §101.45 and other reasonable means" (5). The information on the nutrient amounts that are the basis for the claim, may be presented in a simple fashion. For example, in the case of "low fat," "this meal provides less than 10 grams of fat" may serve as the functional equivalent of complete information as described in §101.9.

The NRA has pointed out that the proposed definition for "low fat" is too restrictive for use in the foodservice industry. It cited, as an example, a meal consisting of a lean meat, a starch and a vegetable to be consumed as part of a daily diet of 1,600 calories, meeting the recommended limit of 30% of total calories from fat (7). With a serving size of 10 oz, this particular meal would contribute 18 grams of fat, far more than the allowable limit under §101.62 (b)(3)(i). Similarly, other healthful meals, based on the fact that they provided one-third of the nutrients recommended in a *daily* diet, could not be so promoted by foodservice operators under the proposed regulations.

A. Expressed Claims

Both expressed and implied claims have found their way into restaurant menus and signs. Fisher listed some examples of the descriptive terms used in the industry which would be considered expressed claims by the FDA:

"lower in cholesterol, fat and, sodium"
"lite bits" or "lighter fare"
"light menu"
"light, healthy meals"
"heart healthy"

Some of the descriptors used by foodservice operators, such as "lower" are not defined by the FDA. In its Food Labeling Regulations of January 6, 1993, the agency did not propose a definition for "light" meal-type products because, as it explained, "the agency could not identify appropriate reference foods to permit this use of the claim" (9). But in the same publication, the agency stipulated that terms such as "light fare," "lite bites," or "light entrees" would not be considered nutrient content claims if the sign or the placard on which they appeared made it clear that they were not intended to characterize the level of a nutrient. A term such as "light fare," supported by an explanation that it referred to dishes with smaller portion size than normal would not then be considered a nutrient claim.

Under the proposed regulations, other descriptors, such as "free," "low," "lean," "reduced," or "high," as they would normally apply to calorie, sodium, fat, sugar, or fiber content, when used by foodservice operators would be required to meet the criteria governing the permissible usage of these term as defined elsewhere in this volume (see Shapiro, Chapter 7).

B. Implied Claims

The advent of foods engineered or modified to improve their nutritional quality has resulted in the appearance of implied nutrient claims on restaurant menus. These can result from the highlighting of an ingredient used in the preparation of the menu time, e.g. "made with oat bran," the usage of a well-recognized brand name, e.g. "Egg Beaters® Omelet"[1], or the adoption of a term implying reduced fat and calorie content, such as "spa cuisine." Under the proposed regulations, subjecting restaurant nutrient claims and health claims to the same burdens of proof as required of packaged foods, a restauranteur would have to demonstrate that an "Egg Beaters® Omelet" contained less than 20 milligrams of cholesterol, 19.5 grams of total fat, and 2 grams of saturated fat.

V. FOODSERVICE PROCEDURES REQUIRED TO INSURE COMPLIANCE WITH NLEA

As noted above, the FDA's proposed regulation for nutrition labeling of restaurant foods requires "assurance that the food or meal meets the nutrient requirements for the claim." This simple statement spells out the challenge facing foodservice operators.

Granted that §101.10(a)(3) allows foodservice operators to establish original nutrient levels required to substantiate nutrient or health claims by means of nutrient databases, cookbooks, or other reasonable means—and, thereby, spares the industry the frightful expense of potentially millions of laboratory analyses. The burden of assuring that menu items bearing such claims are consistently prepared according to the recipes and procedures underlying the claims is by no means trivial.

A foodservice operator dedicated to complying with the nutrition labeling regulations would have to address the following issues:

- how to evaluate the permissibility of a claim
- how to ensure that the menu items bearing claims will consistently qualify for them
- how to monitor compliance

In its comments to the FDA on proposed regulations covering health and nutrient content claims for restaurant menus, the Association of Food and Drug Officials (AFDO) wrote

> We would suggest that FDA, in its discussion of the final regulation, advise regulated industry that acceptable recipes and/or data bases must be reinforced by accurate weight and volume measurements, and that specific procedures for preparation must be used. No deviations from the recipes and preparation procedures may be permitted (which would alter the nutrition profile of the food) if claims are to be made. Also, personnel must be trained to follow established procedures. Appropriate records should be kept to document the use of the recipes/data bases in the preparation of the menu foods.

A. Standard Recipes

Standard recipes have, of course, long been a fact of life for larger foodservice organizations, whether of the profit or non-profit type. Standard recipes are a must if consistent quality, yields, cost and customer satisfaction are to be attained. In the case of nutrition labeling, they are a vital link in the chains of controls needed to insure compliance with the regulations.

Therefore, even if restaurant operators currently do not have standard recipes in use, they will have to create them if they wish to make nutrient or health claims on their menus or signs. These standard recipes must spell out the exact ingredients (with sufficient detail to prevent ingredients different in composition from being used), the exact quantities of each ingredient required, the mode of preparation and cooking, and the exact portioning procedures to be used.

B. Nutritional Analysis and Claim Substantiation

Since most recipes in use are created (or modified) by chefs or dietitians associated with the operations in which these recipes are used, nutrient data on prepared foods contained in commercially available data bases are of little value to foodservice operators in determining the nutrient profile of their own menu items. On the other hand, the technique of computerized nutritional analysis, using compositional data on the exact ingredients used in a recipe, which is allowable under the FDA regulations, offers a convenient and relatively inexpensive way to develop the data needed to support nutrient and health claims.

While most computer nutrient analysis programs on the market today are relatively simple to use and accurate in their computations, the validity of the data which they yield is primarily a function of the knowledge and experience in food chemistry and analysis of the individual carrying out the determinations. Because the programs are so user-friendly, there is, unfortunately, a tendency on the part of many persons associated with the industry—including chefs, managers, owners and dietitians—to believe that they are capable of generating accurate data on the nutrient content of prepared foods, a belief that can be shattered only by comparing the results of their calculations with laboratory reports on the same products.

A variety of factors may account for disparities between the two sets of data, including the selection of wrong ingredients from the data base, the failure to consider losses in food preparation and cooking, and failure to allow for post-preparation losses in holding or storage. Unless the analyst is willing to spend the time collecting nutrient data on the specific brands of ingredients being used from the suppliers of these products and adding them to the data base, ingredient selection is frequently made on the basis of an ingredient in the data base which resembles in name the product in use, but which may in fact be quite different in composition. Another frequent cause of error in computerized nutrient analysis is the selection for computation of the raw or frozen ingredients actually used in preparing the menu item, rather than the cooked or canned ingredients which would more accurately reflect in their nutrient losses the nutritive values of the cooked products.

When accompanied by careful weighing of ingredients at the various stages of preparation, calculation of losses due to trimming, cutting, cooking, dripping, rendering, evaporation, oxidation, hot holding, etc., and subsequent adjustment of the nutrient profile generated by the computer program to take all these relevant changes into account, the finished nutrient profile may be highly credible.

An even better practice is to actually analyze the finished product for its moisture, fat and sodium content and adjust the computer-generated nutrient profile based on the laboratory findings. Such a combination of laboratory and computer analyses, which is fairly inexpensive compared to the cost of a complete laboratory analysis, will provide a high degree of reliability in terms of total calories, total fat, fatty acids, cholesterol, and sodium—all issues in nutrient claims.

The complexities of the NLEA regulations virtually require that the interpretation of the data to substantiate nutrient and health claims be carried out by someone thoroughly familiar with these regulations, another point which could be easily overlooked by foodservice operators.

C. Control of Ingredients

Unlike large food manufacturing plants, which purchase raw materials on specification for use in the preparation of a finite number of finished products, restaurant operations generally purchase a wide array of ingredients needed to support the preparation of several meals a day and frequently changing menus. Knowledgeable chefs and stewards procure their ingredients on a competitive-bids basis from a number of approved distributors. Changing market conditions and sources of supply place a premium on a chef's ability to prepare the best possible meals at the lowest possible cost. New products—differing in pack, origin, or composition—are constantly evaluated in a neverending search to reduce food costs, and creativity is encouraged. All of these practices and endeavors mitigate against a stable and unvarying set of ingredients being used in the preparation of restaurant foods. Yet, somehow in this changing environment, foodservice operators will be forced to institute new procedures to insure that menu items for which nutrient or health claims are made are prepared in an unabidingly unchanging manner, identical with the requirements of the recipes upon which the claims were based.

While food manufacturers generally employ trained personnel, working in "ingredient rooms" to ensure that the correct ingredients and quantities are delivered to the work stations, restaurant cooks are most often responsible for securing and measuring their own ingredients. Under the demanding conditions prevailing in most commercial kitchens, weighing procedures, if used at all, tend to be haphazard. For the sake of compliance with nutrient and health claims, restauranteurs will have to set up new procedures to train their personnel, and acquire the requisite equipment, e.g., balances capable of weighing accurately quantities of a gram or less, if they are to ensure that stipulated recipe amounts are respected.

D. Plating Procedures

In most commercial foodservice operations, food is eyeballed rather than weighed as it is plated. Compliance with nutrition labeling will require that the components of a plate be weighed or measured volumetrically to ensure that the required quantities are used. Written weight charts, accompanied by color photographs of the finished platters, should used by the kitchen personnel responsible for assembling the meals for which claims are made.

E. Training of Food Production and Serving Personnel

Because the procedures that will be required to ensure compliance with health claims are so much more demanding than has been previously required of foodservice personnel, a training program must be established in the operations electing to make (or continuing to make) such claims to ensure that everyone concerned—chef, steward or purchasing manager, cooks and serving personnel—understands the absolute need to prevent deviations from the established recipes and procedures. The individuals requiring the most instruction will be those who are made responsible for assembling, weighing and labeling the correct ingredients for use in each recipe involved. Records should be kept to show daily quantities of ingredients assembled so that these may be compared to the number of special meals sold.

F. Monitoring

Management will also have to set up monitoring procedures to ensure, not only to their satisfaction but also to the satisfaction of local or state inspectors, that the standard recipes upon which the claims are predicated are accurately used on an ongoing basis.

The easiest way to accomplish this task will be by means of single use "batch sheets." These sheets will list all of the required ingredients and preparation and portioning steps. Workers should be required to initial each step as it is completed. Workers in the weighing areas will have to attest to the fact that the correct ingredients and amounts were used; cooks will have to corroborate the fact that they followed the prescribed cooking steps and procedures; and portioning personnel will need to sign off on the accuracy of the weights of components served to the guest.

These written records should, of course, be kept on hand for at least six months for the review of inspectors to vouchsafe for compliance with FDA requirements.

VI. EXAMPLES OF NUTRITION INFORMATION PROGRAMS USED BY THE FOODSERVICE INDUSTRY

Nearly all major foodservice operators have nutrition information programs in place to satisfy the public's desire to know the nutritive value of the meals which they are consuming away from home. The sophistication of these programs vary, depending upon the nature of the operations and the clientele to which they cater. Obviously, food management companies operating facilities in schools, colleges, universities, hospitals, business establishments and factories, have a longer-lasting impact on the nutritional well-being of their customers than public restaurants, recreational facilities or airlines.

Fast food operations which, as pointed out at the beginning of this chapter, now represent the single largest sector of the industry, are rising to the responsibilities imposed upon them by their newly-acquired popularity. They are accomplishing this, not only by communicating nutrition information about their products to their customers, but also by introducing new products, more in keeping with current dietary recommendations, than the standard fare of cheeseburgers, milk shakes, and french fries.

A. ARA Services

ARA Services is a diversified foodservice management company, with interests in many sectors of the industry. The company has developed the Treat Yourself Right® program, which is designed to provide its customers with healthy menu choices. To support this comprehensive program, the company has prepared materials for use by their managers, cooks , servers, and customers. An inservice training program has been instituted to instruct company personnel in the dietary guidelines, as well as the procedures required to follow the recipes and portioning instructions provided. Special Treat Yourself Right menus are used to convey information about the program to the company's customers. ARA has had a computerized nutrition information program in place since 1972. The program is maintained by the company's Culinary Standards Department and is sent by disk to 1,200 foodservice facilities for their use. Managers are encouraged to use the nutritional analysis module offered to verify that their local recipes meet the Treat Yourself Right guidelines. The following are excerpts from a Treat Yourself Right pamphlet:

> "Welcome to the world of TYR (that's Treat Yourself Right®), ARA's exciting health-boosting program. . . ."
>
> "TYR is NOT a diet. Diets put tight restrictions on certain foods. With the TYR program, there are no food restrictions, because there are no *good or bad* foods. There are, however, good, bad, or indifferent choices of food people make, day-by-day, week after week, that add up to healthy or unhealthy eating habits."
>
> "Health experts say the biggest and best change most of us can make in our eating habits is eating less fat. Too much fat increases our chance of getting heart attacks or strokes, cancers of the bowel and, for women, breast cancer. That's why the recipes you'll be using for the TYR program are low in fat."
>
> "Another change for the better is cutting down on salt. Salt promotes high blood pressure in some people. And high blood pressure, as you probably know, is another trigger for heart attack and stroke. So TYR recipes are low in salt, but high in flavor."
>
> "Sugar is loaded with calories called *empty* calories, because sugar has no vitamins or minerals, only calories. Too much sugar can make you fat and take away your appetite for healthy foods. So TYR recipes are low in sugar."
>
> "TYR encourages heaping helpings of starchy foods like bread and potatoes, macaroni, dried beans and whole grain cereals. These foods are low in fat and high in the complex carbohydrates and fiber you've heard so much about in breakfast cereal commercials. Other foods that fill you up without filling you out are vegetables and fruits, especially citrus fruits. Help yourself to low fat foods like these."

B. Marriott Health Care Services

Marriott Management Services is the world's leading contract services organization, providing food and services management to over 3,000 locations. In 1986, the company introduced Wellness & You!®, a nutrition and lifestyle program which provides customers with health and nutrition information, and motivation. The program features seven modules: a Basic Wellness & You! Starter Kit; Cancer Awareness; Weight Management; Women and Wellness; Fitness and Exercise; Heart Health; and the Nutrition Train, a special nutritional

Table 20.2 Examples of Marriott's "Wellness and You" Nutrition Information[a]

	Serving size	Calories	Sodium (mg)	Fat (g)	Cholesterol (mg)
Juices and fruits					
Orange	8 fl.oz.	121	3	0.1	0
Grapefruit	8 fl.oz.	111	3	0.4	0
Apple	8 fl.oz.	121	8	0.3	0
Fresh strawberries	½ cup	24	1	0.3	0
Grapefruit half	5.5 oz.	51	0	0.2	0
Banana	3.5 oz.	90	1	0.5	0
Entrees					
Egg, fried	1 each	85	55	6.7	185
Egg, poached	1 each	66	246	4.4	186
Egg, scrambled	4 oz.	170	148	11.8	436
Breakfast meats					
Crisp bacon	2 slices	61	169	5.2	0.1
Frizzled ham	1 oz.	55	320	3.3	13
Sausage links	2 each	96	338	8.1	22
From the grill					
Buttermilk pancakes	2 each	220	555	7.5	62
Waffles	2 each	274	593	12.4	81
French toast	2 halves	138	288	5.1	119
Side orders					
Hash brown potatoes	3 oz.	152	219	8.1	0
Grits	2 oz.	67	300	4	0
Cereals					
General Mills Cheerios	1 oz.	110	290	1.8	0
Nabisco Shredded Wheat Miniatures	1 oz.	90	0	0	0
Oatmeal	6 oz.	100	294	1.6	0
Wheatena	6 oz.	101	297	0.8	0
Cream of Wheat	5 oz.	69	149	0.3	0

[a] Marriott Management Service's Wellness Program Nutrition Information provides customers with information about the foods served in the company managed cafeterias as exemplified by the excerpts presented in this table.

Table 20.3 Example of McDonald's Nutrition Information Program[a]

	Amount (% Daily Value)			
	Total fat	Saturated fat	Cholesterol	Sodium
Breakfast				
Whole-grain cereal (Wheaties® or Cheerios®); 1% Lowfat milk, 8 fl oz; Orange juice, 6 fl oz; 280 calories, 30 calories from fat	3 g (5%)	1.5 g (8%)	10 mg (3%)	370 mg (15%)
English muffin; Strawberry preserves, 1 pkt; 1% Lowfat milk, 8 fl oz; 315 calories, 55 calories from fat	6 g (9%)	2.5 g (13%)	10 mg (3%)	415 mg (17%)
Scrambled egg, 1; Apple bran muffin; 1% Lowfat milk, 8 fl oz; 360 calories, 65 calories from fat	7 g (11%)	3 g (15%)	225 mg (75%)	475 mg (20%)
Egg McMuffin; Orange juice, 6 fl oz; 360 calories, 100 calories from fat	11 g (17%)	4 g (20%)	235 mg (78%)	730 mg (30%)
Hotcakes with syrup; Apple juice, 6 fl oz; 445 calories, 35 calories from fat	4 g (6%)	1 g (5%)	10 mg (3%)	585 mg (24%)
Lunch/Dinner				
McLean Deluxe; Side salad, Lite vinaigrette dressing, 1/2 pkt; Diet Coke, small; 375 calories, 110 calories from fat	12 g (18%)	5 g (25%)	90 mg (30)	845 mg (23%)
Chicken McNuggets, 6 piece; Sweet 'N Sour Sauce; Sprite, small; 470 calories, 135 calories from fat	15 g (23%)	3.5 g (18%)	55 mg (18%)	805 mg (34%)
Regular cheeseburger; Vanilla shake; 615 calories, 165 calories from fat	18 g (28%)	8 g (40%)	75 mg (25%)	895 mg (37%)
Regular hamburger; Small french fries; Coca-Cola Classic, small; 620 calories, 190 calories from fat	21 g (32%)	6 g (30%)	35 mg (12%)	615 mg (26%)
McGrilled Chicken sandwich; Side salad, Lite vinaigrette dressing, 1/2 pkt; Apple juice, 6 fl oz; 535 calories, 130 calories from fat	14 g (22%)	5 g (25%)	110 mg (37%)	845 mg (35%)

[a] McDonald's Corporation provides customers with information based on the NLEA "Daily Value" concept. This table represents exerpts from McDonald's "Menu of the 90's" brochure. Copies may be obtained from McDonald's Nutrition Information Center, McDonald's Corporation, Oak Brook, IL 60521 (708-575-FOOD).

Table 20.4 Example of Hardee's Nutrition Information Made Available to Customers[a]

	Serving (g)	Calories	Total fat (g)	Saturated fat (g)	Cholesterol (mg)	Sodium	Total carbohydrate (g)	Protein (g)
Breakfast								
Rise 'N'Shine™ Biscuit	83	390	21	6	0	1000	44	6
Bacon, egg and cheese biscuit	137	530	31	11	155	1470	45	18
Canadian Rise 'N'Shine™ Biscuit	164	570	32	11	175	1860	46	24
Big Country Breakfast™ (Sausage)	274	930	61	19	340	2240	61	33
Blueberry muffin	113	400	17	4	65	310	56	7
Sandwiches								
Hamburger	107	260	9	4	20	460	33	11
Mushroom 'N' Swiss™ Burger	203	520	27	13	45	990	37	30
Friso™ grilled chicken	244	620	34	10	95	1730	44	35
Regular roast beef	124	270	11	5	25	780	28	15
Hot Ham 'N' Cheese™	201	530	30	9	65	1710	49	18

Fried Chicken/Sides								
Breast	148[b]	370	15	4	75	1190	29	29
Wing	66[b]	200	8	2	30	740	23	10
Thigh	121[b]	330	15	4	60	1000	30	19
Leg	69[b]	170	7	2	45	570	15	13
Cole slaw (4 oz)	113	240	20	3	10	340	13	2
Mashed potatoes (4 oz)	113	70	c	c	0	330	14	2
Gravy (1.5 oz)	43	20	c	c	0	260	3	c
Salads/Fries								
Side salad	140	20	b	b	0	20	3	1
Garden salad	264	190	14	9	40	280	3	12
Chef salad	268	200	13	8	45	910	5	20
French fries (medium)	142	350	15	4	0	150	49	5
Crispy Curls™	85	300	16	3	0	840	36	4
Shakes/Desserts								
Shake (vanilla)	326	370	9	6	25	210	59	14
Cool Twist™ cone (vanilla/chocolate)	118	170	4	3	15	85	29	5
Big Cookie™	57	280	12	4	15	150	41	4

[a] This table represents exerpts from Hardee's "Menu Items." Copies may be obtained from HFS, Inc. Research and Development, 1233 Hardee's Boulevard, Rocky Mount, NC 27804-2815 (800-777-8000). Hardee's Food Systems Inc. makes no nutrient or health claims but provides its customers with nutrition information to assist them in making appropriate menu choices.

[b] Denotes one piece, portion size may vary = Edible portion without bone

[c] Denotes less than 1

education program for school children. Each module includes a series of approved menus and informational materials for customers to take home, including brochures and recipe cards. Daily menu selections follow the latest recommendations with respect to calories, fat, sodium, and cholesterol. Brochures are available at each food station to detail this information and help customers in making their menu selections. The brochures cover breakfast (Table 20.2), salads, the grill, the deli, desserts, beverages, and the "Healthy Traveler."

C. McDonald's Corporation

McDonald's, the world's largest foodservice organization, provides ingredient and nutrition information to its customers about menu items by means of a "Did You Know" wall displays in the lobbies of its restaurants. Additional published materials are available to nutrition professionals to assist them in advising their patients on menu choices (Table 20.3). New materials, printed or revised in line with NLEA requirements, will include all the information required for Nutrition Facts panels. To insure compliance with expressed or implied nutrient claims, the company relies on its suppliers to control the composition of the ingredients which they manufacture under McDonald's specifications.

D. Hardee's Food Systems

The company, a leader in the fast food sector of the industry, makes available to its customers data on the calorie, fat, saturated fat, cholesterol, sodium, carbohydrate, and protein contents of its menu items (Table 20.4). With the exception of several fat-free salad dressings supplied by commercial packers, the company features no offerings for which nutrient or health claims are made.

VII. ACKNOWLEDGMENTS

The author gratefully acknowledges the assistance provided to him by Janna P. Trout of ARA Services, Amelia Catakis of Marriott Management Services, James D. Motzer of Burson-Marsteller (on behalf of McDonald's), and Ruby Parker and Janet LaCombe of Hardee's Food Systems, for supplying information and samples of nutrition information materials in use within their respective organizations.

VIII. ENDNOTES

[1] Egg Beaters® is a regtistered trademark of the Standard Brands Co.

IX. REFERENCES

1. H. Riehle, The big picture: Foodservice sales soared between 1970 and 1990, *Restaurants USA, 12*, 9: 38 (1992)
2. R. Iwamuro, Spending for food away from home hits $623 per person; Miami is tops again, *Restaurants USA 13*: 5, 41 (1993)
3. FDA, 21 CFR Part 101—Food Labeling: § 101.10 Nutrition labeling of restaurant foods, *Federal Register, 58*, 3, 2410 (1993a)
4. FDA, *Federal Register, 56*, No.229, p. 60368, Nov. 27 (1991)
5. FDA, 21 CFR Part 101. Food Labeling: Nutrient Content Claims and Health Claims; Restaurant Foods, *Federal Register, 58*: 113, 33055 (1993b)
6. FDA, 21 CFR Parts 5, 20, 100, 101, 105 and 130—Regulatory Impact Analysis of the Final Rules to Amend the Food Labeling Regulations, *Federal Register, 58*, 3, 2927 (1993c)
7. W. P. Fisher, Letter to Dockets Management Branch, FDA, Feb. 25 (1992); R. Iwamuro, Spending for food away from home hits $623 per person; Miami is tops again, *Restaurants USA 13*: 5, 41 (1993)
8. R. V. Noble, Ritz-Carlton Hotel, *A Guide to Distinctive Dining: Recipes from Famous Restaurants of America*, The Berkshire Publishing Company, Cambridge, MA, 1954, p. 13.
9. FDA, 21 CFR Part 101—Food Labeling: § 101.10 Nutrition labeling of restaurant foods, *Federal Register, 58*, 3, 2383 (1993d)
10. D. Sowards, Jr., Official Comments on Docket No. 93N-153, Health and Nutrient Content Claims for Restaurant Menus, Communication from the Association of Food and Drug Officials to Dockets Management Branch (HFA-305), U.S. Food and Drug Administration, July (1993)

21

The Impact of the Nutrition Labeling and Education Act on Functional Foods

Clare M. Hasler and Richard L. Huston
University of Illinois, Chicago, Illinois

Eve M. Caudill
University of Illinois at Urbana-Champaign, Urbana, Illinois

I. INTRODUCTION AND BACKGROUND

We are entering a new era in the food and nutrition sciences. Although a connection between diet and health is not new, nutrition research has traditionally focused its efforts on the prevention of nutrient-deficiency diseases. The classically held view that, for optimal health, the food supply should be safe and provide sufficient energy, macronutrients (protein, lipids, carbohydrates), and micronutrients (vitamins and minerals) has shifted to an examination of the role of diet, particularly nonnutritive components of diet, in reducing the risk of chronic disease. Diet-based disease prevention is increasingly seen as the wave of the future and there is currently a revolution of interest and effort in the formulation of diets or the development of foods tailored to meet specific health needs, so-called functional foods. The Nutrition Labeling and Education Act of 1990 (NLEA) will be the engine that powers the second revolution in the dietary prevention of disease in this country and will have a significant bearing on the future of functional foods.

A. Definition of Functional Foods

There is currently lack of agreement on the precise meaning of the term functional foods. Indeed, it is thought to be a concept which is still evolving (1). The term was first introduced in Japan to describe a class of food products defined by the Japanese Ministry of Health and Welfare (MHW) as "processed foods containing ingredients that aid specific bodily functions in addition to being nutritious" (2). Functional foods were developed in response to efforts by the MHW to reduce Japan's rising health care costs in the face of an aging population. In this country, the term "designer foods" was coined in 1989 by Dr. Herbert Pierson, then director of the National Cancer Institute's $20 million Designer Foods Program (3), to describe foods which naturally contain or are enriched with nonnutritive, biologically active chemical components of plants (phytochemicals) that are effective in reducing cancer risk. Likewise in 1989, the term "nutraceutical" was coined by Dr. Stephen DeFelice, chairman of the Foundation for Innovation in Medicine, to refer to any substance that may be considered a food or part of a food and provides medical or health benefits, including the prevention and treatment of disease (4).

Since 1989, a variety of other terms have been used to describe foods which may have beneficial effects in reducing the risk of chronic disease, including pharmafoods, phytofoods, phytogenic substances, performance foods, smart foods, therapeutic foods, genetically engineered foods, genomic foods, and phytochemical sources (5–9). A recent book published by the Institute of Medicine, *Opportunities in the Nutrition and Food Sciences* (10), has defined functional foods as those that "encompass potentially healthful products," including "any modified food or food ingredient that may provide a health benefit beyond the traditional nutrients it contains." We will utilize the term functional foods throughout this chapter to refer to this exciting and rapidly evolving field of technology within the food and nutrition sciences.

B. Factors Contributing to the Recent Interest in Functional Foods

Several factors have contributed to the current explosion of interest in functional foods. First, there is overwhelming evidence supporting the vital role of nutritional factors in health and disease (11,12). Diet has been implicated in 6 of the 10 leading causes of death in the United States (heart disease, cancer, stroke, diabetes, atherosclerosis, and liver disease), as well as in several nonfatal but disabling disorders, such as osteoporosis (13). More specifically, consumption of a plant-based diet high in fruits, vegetables, whole grains, and legumes is thought to protect against the development of chronic disease, particularly cancer (14). Persons with low fruit and vegetable intake have twice the risk of cancer at most sites compared to individuals with high intake (15). It is likely that the increased concentration of biologically active phytochemicals in a plant-based diet contributes to a reduction in cancer risk (Table 21.1) (16,17). This has led to intense research interest in the role of physiologically functional foods in disease prevention and health promotion.

Second, America has a rapidly aging population. Americans aged 65 and above made up 12.3% of the population in 1988, and that figure is expected to increase to 22% by the year 2030 (18). Food and health are also critically linked for many aging baby boomers who believe the right foods will prevent or counter chronic disease (19). Improved nutrition intervention strategies could dramatically increase the quality of life in older Americans.

Third, emphasizing good nutrition in all age groups could reduce skyrocketing health care costs. In 1987, health care and disability costs of four major chronic diseases (coronary heart disease, cancer, diabetes, and osteoporosis) in which diet plays a key etiological factor approached $145 billion (20). Despite this fact, funding for nutrition-related research remains appallingly low. In 1988, the federal expenditure in support of nutrition research and training was less than 0.3% of the nation's health care budget. If improved nutrition reduced health care costs by 10%, a savings to the country of $14.6 billion would be realized (21).

Finally, recent technological advances in the food industry have made it possible for the field of functional foods to evolve. The challenge will be how to incorporate products made from these fast moving, multifaceted technologies into the existing regulatory framework, which radically changed with the enactment of the NLEA (22).

II. IMPACT OF THE CURRENT REGULATORY FRAMEWORK ON FUNCTIONAL FOODS

A. The NLEA and Explicit Health Claims

The NLEA is the most comprehensive revision of the United States food labeling laws since the Federal Food, Drug and Cosmetic Act (FDCA) was passed in 1938 (23). This legislation provides, for the first time in the history of the Food and Drug Administration (FDA), a specific statutory base for explicit health- or disease-related claims on the label, or in the labeling, of conventional foods.

As defined by the NLEA, a health claim is any claim that expressly, or by implication (including third party references, written statements, a brand name, or symbols such as a heart), characterizes the relationship of any substance to a disease or health-related condition (24). The FDA views third-party references as "references, made through a name or logo, to a person or organization such as a professional society or association that is independent of the product's manufacturer or distributor, on product labeling or advertising, to promote the organization's approval of a product." A third-party reference does not automatically constitute a health claim unless the information can be understood to characterize the relationship between a substance and a disease or health-related

Table 21.1 Naturally Occurring Substances that May Fight Disease

Active food component	Possible desease fighting properties	Found in the following foods
Allylic sulfides	Inhibits cholesterol synthesis and protects against carcinogens	Aged garlic extract
Alpha-linolenic acid	Reduces inflammation and stimulates the immune system	Flaxseed, soy products, purslane, walnuts
Carotenoids	Antioxidants that protect against cancer and may help reduce the accumulation of arterial plaque	Parsley, carrots, yams, winter squash, sweet potatoes, cantaloupe, apricots, spinach, kale, turnip greens, citrus fruits
Catechins	Studies have linked catechins to low rates of gastrointestinal cancer; may aid the immune system and lower cholesterol	Green tea, berries
Coumarins	Prevents blood clotting and may have anticancer activity	Parsley, carrots, citrus fruits
Flavonoids	Block receptor sites for certain hormones involved in cancer promotion	Parsely, carrots, citrus fruits, broccoli, cabbage, cucumbers, squash, yams, tomatoes, eggplant, peppers, soy products, berries
Gamma-glutamyl allylic cysteines	May have role in lowering blood pressure and elevating immune system activities	Aged garlic extract
Indoles	Induce protective enzymes that deactivate estrogen	Cabbage, brussels sprouts
Isothiocyanates	Powerful inducers of protective enzymes	Mustard, horseradish, radishes

Table 21.1 (Continued)

Active food component	Possible desease fighting properties	Found in the following foods
Limonoids	Powerful inducers of protective enzymes	Citrus fruits
Lycopene	Powerful antioxidant that helps the body resist cancer and its progression	Tomatoes, red grapefuit
Monoterpenes	Cancer-fighting antioxidants that inhibit cholesterol production and aid protective enzyme activity	Parsley, carrots, broccoli, cabbage, cucumber, squash, yams, tomatoes, eggplant, peppers, citrus fruits, whole grains, berries
Phenolic acids	May help the body resist cancer by inhibiting nitrosamine formation and affecting enzyme activity	Parsley, carrots, broccoli, cabbage, tomatoes, eggplant, peppers, citrus fruits, whole grains, berries
Phthalides	Stimulate the production of beneficial enzymes that detoxify carcinogens	Parsley, carrots
Plant sterols	Block estrogen promotion of breast cancer activity; help block the absorption of cholesterol	Broccoli, cabbage, cucumber, squash, yams, tomatoes, eggplant, peppers, soy products, whole grains
Polyacetylenes	Protect against certain carcinogens found in tobacco smoke and help regulate prostaglandin production	Parsley, carrots, celery
Triterpenoids	Prevents dental decay and acts as anti-ulcer agent; binds to estrogen and inhibits cancer by supressing unwanted enzyme activity	Citrus fruits, licorice-root extract, soy products

condition. Use of symbols (such as a heart) without some clarification of their meaning in context is likely to cause the food to be misbranded. Some symbols (e.g., the USDA Food Pyramid, the American Heart Association symbol, or the American Cancer Society logo) may or may not constitute a health claim, depending on whether the labeling contains a reference to a substance or a disease or health-related condition.

Thus, in order to constitute a health claim, two essential elements must be present: (a) the reference to a "substance" and (b) characterizing the relationship of that substance to a disease or health-related condition.

B. Currently Allowed Health Claims

The NLEA specifically directed the FDA to consider 10 topics associating substances with a disease- or health-related condition. FDA's final regulations permit health claims for food labeling in 7 of the 10 specific areas reviewed by the agency: calcium and osteoporosis; sodium and hypertension; saturated fat and cholesterol and risk of coronary heart disease; dietary fat and cancer; fiber-containing grain products, fruits, and vegetables and cancer; fruits, vegetables, and grain products that contain fiber, particularly soluble fiber, and risk of coronary heart disease; and fruits and vegetables and cancer (25). Recently, a claim for folic acid and neural tube defects was also proposed (26), after initially being denied by the agency. Health claims were not authorized for zinc and immune function in the elderly or omega-3 fatty acids and coronary heart disease. In addition, a claim for antioxidant vitamins and cancer was also not allowed, a decision which has sparked significant controversy and was the subject of a recent public conference (27).

C. Standard and Rationale for Use of Health Claims

The NLEA has three basic objectives: (a) to encourage product innovations through the development and marketing of nutritionally improved foods, (b) to eliminate consumer confusion by establishing definitions for nutrient content claims, and (c) to make available nutrition information that can assist consumers in selecting foods that can lead to healthier diets (28). Although it is estimated that the new food label will cost FDA-regulated food processors $1.4–2.3 billion and the federal government $163 million over the next 20 years (29), public health benefits may well exceed the costs. A recently completed study estimated that the NLEA may prevent up to 300,000 deaths over the next 20 years and reduce medical costs as much as $12 billion (30).

A primary goal of the NLEA is to protect consumers from unfounded health claims (31). This is reflected in the rigorous, dual standard now required for permitting health claims on food labels: (a) the health claim must be supported by the totality of the publicly available scientific evidence, and (b) there must be significant scientific agreement among qualified experts that this support exists. Although the FDA contends it will "not require that health claims be supported by a full consensus among scientists," the issue of what constitutes significant scientific agreement is currently the subject of much debate. With regard to functional foods, the concern is that if standards for health claims are too high, few companies will be willing to accumulate the evidence needed to meet them. Health claims will have a profound effect on the food industry and will be a central issue in the research and development of functional foods. In addition, the issue of how to effectively regulate functional foods is critical to the future discovery, development and availability of innovations in this area of food technology.

D. Functional Foods and the Existing Regulatory Food Categories

According to Wrick (32), functional foods exist at the "food-drug interface" because of their potential in the prevention of chronic disease (Figure 21.1). Products at the food-drug interface include (a) consumer products regulated as foods that are purchased for a health benefit, (b) food ingredients that affect the nutritional properties of foods, (c) pharmaceuticals used for their nutritional or dietary properties, and (d) compounds in foods that have a physiological benefit once consumed (Figure 21.2).

A major issue with the functional foods paradigm is that no universally accepted provision exists in current food regulations for foods consumed with the intention of disease prevention. Substances with this type of physiological effect are normally classified as drugs; these are "articles intended for the diagnosis, cure, mitigation, treatment, or prevention of disease," as defined by the FDCA (23). A substance fulfilling criteria as a drug must meet very stringent premarket approval requirements or be subject to regulatory action as an unapproved drug. The current research and development cost for a successful new drug application, which must be reviewed and approved by FDA prior to marketing, typically exceeds $200 million and requires ten years or longer from the start of preclinical testing to final approval (33).

Given that functional foods exist at the food-drug interface, should they be classified as foods or drugs?

Complicating this scenario further still is the fact that functional foods, depending on their identity, use, processing, or claims, may be categorized in several different ways: (a) conventional foods, (b) GRAS (generally recognized as safe) substances, (c) food additives, or (d) drugs, as explained in a recent paper by Hathcock (5), using garlic as an example. Garlic has recently received significant attention as a functional food, both in the popular press (34) and peer-review scientific journals (35,36). Whole, unprocessed garlic (without

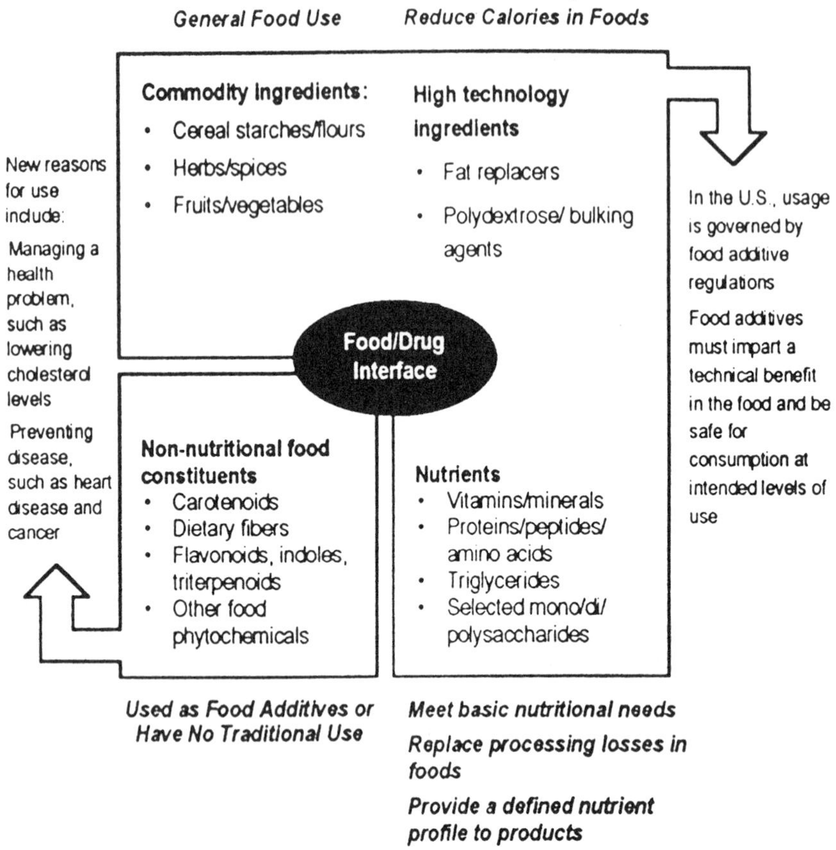

Figure 21.1 Functional foods exist at the food-drug interface.

a health claim) is marketed as a conventional food, while dehydrated garlic and natural garlic extracts have affirmed GRAS status. According to Hathcock however, if a particular garlic phytochemical were isolated, purified, or concentrated and added to a food, it may or may not be affirmed as GRAS. In addition, if the garlic phytochemical were determined to be a food additive, it would require the approval of a food additive petition, which can be nearly as rigorous as the approval process for a new drug. Furthermore, if a health claim were made on the label or in the labeling of any garlic or garlic-derived product, it would be classifiable as a drug unless the claim had been authorized by the FDA under the current rules of the NLEA.

Products regulated as foods, purchased for a health benefit
- High-fiber cereals
- Vitamin/mineral supplements
- Enteral nutrition solutions

Food ingredients that affect nutritional properties of foods
- High-performance sweeteners
- Fat substitutes/ replacements
- Dietary fibers

Food constituents that have a physiological benefit once consumed
- β-Glucans
- Tocotrienols
- β-Carotene

Pharmaceuticals used for nutritional/ dietary properties
- Psyllium-based laxatives
- Appetite suppressants
- Total parenteral nutrition solutions

Figure 21.2 Products at the food-drug interface.

E. Product Categories and Label Claims Exempt from NLEA Health Claims Regulations

The NLEA imposes significant restrictions on the use of health claims. However, according to Stephen McNamara of the Hyman, Phelps & McNamara law firm in Washington, D.C., there are certain product categories or classes of label claims that are outside the scope of NLEA health claims regulations (and therefore are not subject to their restrictions), into which functional foods may fit: (a) medical foods, (b) foods for special dietary use, (c) truthful and non-misleading information about the "physiological" effect of a food, (d) nutrient content claims, and (e) dietary supplements (37).

Medical foods (38) are a class of "special dietary foods intended for use solely under medical supervision to meet nutritional requirements in specific medical conditions." They are exempt from NLEA health claims regulations. These foods are regulated by the FDA as foods, not drugs—even though such products bear extensive medical-type labeling. Medical foods have also received statutory recognition as a valid type of food in the Orphan Drug Amendments of 1988. If a company chose to market a functional food product as a medical food, the market for such a product would clearly be smaller than if it were marketed as a conventional food. However, it may be large enough to generate significant return on investment, given the fact that this product could bear a health claim, and the current consumer interest in diet and health (S. McNamara, personal communication).

Foods for special dietary use include foods "supplying particular dietary needs which exist by reason of a physical, physiological, pathological, or other condition, including but not limited to the conditions of diseases . . ." The FDA has stated that, "any statement that appears on the label or in the labeling of a food intended for 'special dietary use' that is consistent with provisions of the regulations. . . . will not be regulated as a health claim." An example of such a food would be one designed for individuals with lactose intolerance. However, if such labeling resulted in an implied health claim, the food would be misbranded. However, if such labeling resulted in an *implied* health claim, (e.g., ". . . statements, symbols, vignettes [see discussion on page 3] or other forms of communication that a manufacturer intends, or would be likely to be understood, to assert a direct beneficial relationship between the presence or level of any substance in the food and a disease or health-related condition"), the food would be misbranded.

Truthful and non-misleading claims about the "physiological effect of a food when consumed" is another type of claim which could be used to market a functional food product without triggering NLEA health claims or drug regulations. Examples of such claims would be the "laxative effect of prune juice" or the "stimulant effect of caffeinated coffee." Such information would not fit the definition of a health claim if it did not link a nutrient to a disease or health-related condition (39).

Claims made solely about the nutrient content of a product would also not constitute a health claim if they did not link that nutrient to a disease. An example would be the nutrient content claims now appearing on cereals which state that the product "contains β-carotene." Such a nutrient content claim leaves it up to the consumer to make the link between β-carotene, antioxidant function, and potential cancer risk reduction, depending on their knowledge of this issue—which could be considerable given recent media attention. Nutrient content claims for other physiologically functional nutrients could likewise be effective in the marketing and promotion of functional foods.

With respect to supplements, the Dietary Supplement Act of 1992 placed a one-year moratorium on the application of the NLEA's final rules for health claims to dietary supplements of vitamins, minerals, herbs, or other similar nutritional substances (40). The moratorium expired in December of 1993, thus allowing the NLEA to go into effect for dietary supplements beginning in July,

1994. However, a new bill designed to keep dietary supplements exempt from NLEA regulations has recently been passed. The Dietary Supplement Health and Education Act of 1994—an amended version of S784—was signed into law on October 25, 1994. This new legislation could significantly reduce FDA's authority for regulating dietary supplements and, conceivably, functional foods.

Thus, there are certain product categories that are exempt from NLEA health claims regulations into which functional foods may fit. In addition, certain claims that link food products to health-related conditions either directly or indirectly (by the consumer) may not fit the definition of a health claim, and thus may not trigger FDA sanctions.

These product categories and claims may be useful in the marketing and promotion of functional food products, given that there is no appropriate mechanism to regulate health claims or intellectual property rights for functional foods in this country, despite the fact that these foods have entered the mainstream of science (41), medicine (42), and the consumer marketplace (43). If U.S. companies decline to invest in functional foods research, foreign countries may take the lead in functional foods technology. Indeed, this is already thought to be the case, as discussed in the folowing section.

F. Regulations for Functional Foods Outside the United States

The United States is significantly behind Europe and Japan in the development and marketing of functional food products (33). In Japan, functional foods are seen as an alternative to drug therapy (44), and in July of 1991, a new regulatory category for functional foods was established, *Foods for Specified Health Use* (FOSHU) (45). The Japanese government considers 12 classes of ingredients as physiologically functional, including sugar alcohols, peptides, glycosides, alcohols and phenols, isoprenoids and vitamins, cholines, lactic acid bacteria, minerals, and simple sugars (oligosaccharides) (2). At least 200 companies are involved in functional foods research and development in Japan, resulting in the plethora of such products currently on the market. Stamina drinks have become immensely popular, as well as beverages like Fibe-Mini (a high-fiber soft drink), Daily Bottle (a soft drink containing collagen and vitamin C), and Zixe (a beverage produced by fermenting whey with a mixture of 38 strains of bacteria) (46). In June of 1993, Shiseido became the first company in Japan to win approv- al from the MHW to sell a physiologically-active functional food, a rice product from which an allergenic protein globulin has been removed (47). MHW approval will allow Shiseido to claim medical benefits on the label of this product.

Although the European market for functional foods is more difficult to classify than the market in Japan, there is a growing trend in most Western

European countries toward products purported to make people healthier. Contributing to the potential for a strong functional foods market in Europe is the fact that many traditional foods, such as yogurt and high fiber cereals, are consumed routinely by much of the population.

Most countries in the European Community now have systems that grant marketing exclusivity for health claims for functional foods lacking patents based on proprietary research. In France, for example, the government grants a type of license for products not under patent, entitling companies to make health claims with ten years of market exclusivity. Under this system, Laboratories Lincoln (Paris) is marketing Maxepa, a fish oil product containing omega-3 fatty acids that lower triglycerides. Competitors may not make this health claim if they sell a similar product (47).

Unlike Japan and Europe, where functional foods now constitute 10% or more of the over-the-counter drug market, the regulatory climate in the United States has been essentially antagonistic to functional foods research and development (48). Economic and regulatory reform for functional foods is needed in this country, not only from a public health standpoint, but to prevent U.S. food and drug industries from losing their global competitiveness in food technology.

III. THE IMPACT OF TECHNOLOGY ON FUNCTIONAL FOODS

A. Introduction

Over the past decade, the food industry has clearly taken an aggressive stance toward the use of innovative technologies in order to satisfy consumer needs (49). High technology applications in the food industry have resulted in a vast array of techniques, process applications, and product possibilities. The market has been flooded with so-called healthy fare, appealing to consumers coming to grips with their own mortality and concerned about reducing their risk of chronic disease. This is reflected by escalating sales of low-fat processed foods, which reached $32 billion in 1991, and is expected to climb to $55.5 billion in 1996 (50). This new focus on technology is unusual for the food industry; unlike the pharmaceutical and computer industries, the food industry is traditionally not technologically driven and is not structured to diversify the investment risks of technology development (51).

B. Biotechnology in the Food Industry

One particular technology which has had enormous impact on the food industry is biotechnology. Products of classical methods in food biotechnology (e.g., food fermentations and conventional agricultural breeding programs) are centuries old. However, it is now possible to manipulate genes in plants, animals, and microbes to produce unique raw materials and new food ingredients (52). Genetic engineering techniques are now used to achieve a wide variety of agronomic effects (tolerance to herbicides, pesticides, viruses, salt, and drought; enhanced yields; decreased time to maturity) as well as technical effects (new phytochemical and amino acid profiles; improved fruit ripening and shipping properties) in plant-derived foods. Nearly 800 genetically modified edible plants have received environmental release permits since 1986 (53) and by the late 1980s, scientific advancements spawned transgenic plants for over 30 varieties used for human food (54–56). Although techniques of biotechnology are "useful in enhancing the quality, nutritional value, and variety of food available for human consumption" (57), applications of biotechnology to new food production processes may be of concern because these techniques can impact food safety.

C. Food Safety Concerns

The FDA is on record as stating that (a) adequate provisions are contained in the FDCA to ensure the safety of food products incorporating biotechnology procedures, and (b) any issue relevant to regulating foods or food ingredients developed via new technologies will be considered under the proper regulatory or statutory provisions of the FDCA (58).

Most safety issues relating to products developed with new technologies will involve two components of food law: (a) the adulteration provision of the FDCA, which allows FDA to condemn foods containing poisonous or deleterious substances injurious to health, or (b) the Food Additives Amendment (FAA), which applies to substances intentionally added to foods for a technical effect (e.g., preservation, texture, sweetness) (59). Interpretations of comments and questions about FDA's food policy were recently outlined by Flamm (60).

The adulteration provision of the FDCA applies to natural components of foods (aflatoxin or other natural toxicants) that may ordinarily render the food harmful to health. Post-market aspects of food and drug law places the responsibility for demonstrating food safety with food manufacturers and distributors (60). Consequently, if a functional food is produced containing elevated levels of phytochemicals, the FDA can question the safety of that practice. The potential for adverse effects from phytochemicals does exist, depending on the specific phytochemical, the biological characteristics of the exposed person, interaction with other food components, and the exposure level. According to Hathcock (5), increased exposure to phytochemicals in plants foods can occur through (a) increased dietary intake, (b) conventional plant breeding techniques to increase their concentration, (c) biotechnology techniques that genetically increase their content, (d) various processing techniques that selectively enhance their concentration, and (e) isolating and selectively adding them to foods to enhance their concentrations.

The second component of food law relevant to elevated levels of functional food phytochemicals the FAA covers is substances intentionally added to foods that may render them injurious to health. In this case, any substance that does not occur naturally as a component of food is viewed as an added substance, and manufacturers must prove to FDA that those added substances are safe. Hence, the FDA can take action if the substance in question is (a) a natural component of food that would ordinarily render the food injurious to health, or (b) a substance added to a food that may render the food injurious to health. The FDA need only show that a reasonable possibility exists for the food to be harmful if consumed. Thus, under this provision, the FDA can apply the "may render" standard to foods containing elevated levels of phytochemicals that stem from some technological change (e.g., biotechnology) as previously discussed. Therefore, a food modified by genetic engineering that contains elevated levels of a phytochemical could be in violation of the law (61).

On May 29, 1992, the FDA announced its oversight policy for new plant varieties and bioengineering processes used in developing new food products (62). This policy addresses the issue of whether consumers should be informed when foods are derived from genetically engineered plants. According to the FDA, the agency's policy that information on a food label must be "truthful and not misleading" does not apply to plant breeding techniques because it provides no useful information about food composition or properties per se. Also, it would be impractical to demand for example, that wheat varieties developed via different techniques be segregated during production, processing, storage and distribution. According to FDA regulations, whether technology or bioengineering is used to develop the food product is not a relevant issue. Rather, food manufacturers must ensure that newly marketed foods contain substances within acceptable ranges for that food whether they come from plants developed via conventional breeding techniques or genetic engineering. If it is the latter, plant breeders may choose to test those foods more rigorously than warranted from a safety viewpoint, due to public concerns about the use of biotechnology and genetic engineering in food production (63).

Certainly everyone agrees that a safe food supply is a primary concern, on the part of the public as well as the FDA. If functional foods are developed that are fortified extensively with phytochemicals, the general public intake of these physiologically active compounds could increase dramatically, raising safety concerns. Although phytochemicals may have great benefit in disease prevention, as shown in Table 21.1, concerns about their toxicity have also been raised (64).

Additional questions that may impact safety have yet to be resolved. How will genetically modified additives be defined? Will they be viewed as food additives, GRAS substances, biological residues, or unavoidable substances? Are foods prepared from transgenic plants adulterated if foreign genes are overexpressed at levels greater than that needed to achieve intended technical effects? What about the biological and chemical properties of gene products? Are marker genes safe for humans? What about gene products and their secondary protein metabolites that are generally considered nontoxic and do not bioaccumulate? Until these issues are resolved, developing data to answer safety questions for functional foods will be difficult.

IV. EFFECT OF THE NLEA ON THE DEVELOPMENT OF FUNCTIONAL FOODS

A. Intellectual Property Rights and Patenting Issues

The creation and effective use of intellectual property (patents, trade marks, service marks, copyrights, and trade secrets) is the driving force of economic development and the ticket to technology transfer exchange. Unfortunately, patent strengths in the life sciences are not as strong as those in hard sciences such as engineering. Moreover, because of their very nature, functional foods do not lend themselves to the development of strong patents that work as funding inducements in the pharmaceutical industry.

In the pharmaceutical industry, where exclusive claims are created by patent protection, investment in research and development is robust. However, the current U.S. regulatory structure fails to provide the necessary market exclusivity to food companies willing to invest in exploratory research and development of new functional food products (4). A different strategy is needed to give more exclusivity to industry for functional foods research and development discoveries, comparable to patents in the pharmaceutical industry. Without market exclusivity, there will be no incentive for the private sector to invest in functional foods research and development.

B. Effect of the NLEA on Incentives to Invest in Functional Foods Research and Development

The current regulatory system discourages rather than encourages investment in the research, development, and availability of important advances in functional foods technology. Companies are hesitant to invest in the costly research necessary to support a health claim for a functional food given the stringent requirements for the claims as dictated by the NLEA. As an incentive for research, it has been suggested (4) that companies should be granted seven years exclusivity for health claims based on approved proprietary research as mandated in the Orphan Drug Act (65). Under the Orphan Drug Act, testing requirements for pre-market approval of functional foods would be reduced in comparison to the approval process for drugs or food additives. Since the inception of the Orphan Drug Act in 1983, the number of orphan drugs has increased almost 20-fold, demonstrating how effective proprietary information and intellectual property rights can be in stimulating product development.

It appears that where flexible regulatory requirements and proprietary rights to research investments are in place, private industry has strongly supported research and development. Indeed, in countries such as Europe and Japan, where laws on functional foods are more progressive, the functional foods market is flourishing, constituting 10% or more of the over-the-counter drug market (48). Sales of functional foods are expected to reach $8 billion in Japan within a few years, and $3 billion in Europe by the year 2000. By comparison, one source estimates the current functional foods market in the United States at $495 million (33).

C. Existing Versus New Products

Since the present regulatory environment for functional foods in the United States is inflexible, and health claims are not proprietary, companies will be encouraged to invest in the promotion of existing products, rather than in the research and development of product innovations. This could be viewed positively—as an opportunity to take advantage of currently approved health claims by (a) using basic foods and traditional processing techniques to extend available product lines (66), or (b) adding value to and reposition existing products by utilizing emerging technologies to take advantage of the anticipated increase in demand for food products enriched with phytochemicals and current emphasis on healthy lifestyles (67). For the most part, however, the promotion of existing products in lieu of product innovations will be a negative consequence of the NLEA, potentially preventing U.S. food companies from maintaining their global competitiveness in this rapidly advancing field of technology.

V. EFFECT OF THE NLEA ON THE MARKETING AND PROMOTION OF FUNCTIONAL FOODS

A. Impact of Health Claims on Product Sales

Over the past decade, as consumer interest in health and diet has increased, health claims have become a particularly popular means of promoting nutritional attributes by food companies—particularly because they can dramatically increase product sales, as demonstrated by an effective marketing campaign launched by the Kellogg Company in 1984.

The use of explicit health- and disease-related claims by Kellogg for their All-Bran breakfast cereal was a watershed development that eventually resulted in FDA's reevaluation and current restructuring of their policy toward health claims (68). Kellogg utilized advertisements and explicit package labeling claims to inform consumers that a high-fiber/low-fat diet could reduce the risk of developing certain forms of cancer. Adding to the credibility of this promotion was a third party reference—the National Cancer Institute—which reviewed Kellogg's fiber-cancer claims prior to the launch of its All-Bran cereal promotional campaign. Although Kellogg was successful in increasing consumer awareness of the link between dietary fiber and cancer, it contradicted a long-standing policy by the FDA which had banned the use of health claims to promote food products. The FDA was not consulted prior to this campaign, and an explosion of new health- and disease-related claims resulted. The New York Times documented that in the first half of 1989, 40% of the new food products introduced bore a health claim (69).

With regard to oat bran, success resulted not from the dissemination of information by a food manufacturer, but through a non-commercial information source. The relationship between oat bran and a reduction in blood cholesterol was first noted in an article published in the Journal of the American Medical Association in 1988 (70). Information on the benefits of oat bran was immediately picked up and disseminated through the popular media. The overwhelming consumer demand for oat bran drew interest from food manufacturers such as Quaker Oats, Kellogg, and General Mills, which began to promote the benefits of oat bran in a variety of different products. In addition, there was increased consumer demand for cereal products which were healthier overall. Market analysis indicates that new cereals introduced after 1984 had lower levels of both sodium and fat (71). Thus, food companies looking to differentiate their products through healthy attributes focused on not only increasing the fiber content, but also on lowering the amount of negative nutrients in the product.

B. Impact of Health Claims on Consumer Knowledge and Behavior

Clearly, the ability to make a health claim can dramatically affect the sale and overall development strategies of functional food products. Moreover, health claims have a positive effect on consumer behavior and their knowledge of the link between diet and health. An analysis of the ready-to-eat cereal market during the time period directly before and during the Kellogg's All-Bran cereal campaign found increases in consumer awareness of the relationship between fiber and cancer (in all education levels and income groups), and increases in the consumption of fiber-enriched cereal (71). Thus, health claims not only inform consumers of desirable product features and why those particular features are important, they also provide the motivation necessary to change behavior. This may be hampered, however, by the required wording of the seven health claims currently approved, which is difficult to translate into consumer-friendly language as shown in the following example of a model claim: "Low fat diets rich in fruits and vegetables (foods that are low in fat and may contain dietary fiber, vitamin A or vitamin C) may reduce the risk of some types of cancer, a disease associated with many factors."

The overall effect of the NLEA will be to limit health information disseminated through commercially-generated sources at point-of-purchase locations, not only because of the limited number of currently allowed health claims, but because of their required language. Thus, it will be more difficult to communicate the healthy benefits of specific food products as consumers' purchasing decisions are being made. Although consumers can be reached through other sources (i.e., government agencies, academic researchers, etc.), this information is harder to reach and thus is disproportionately accessed by those consumers motivated to search for it.

VI. FUTURE ISSUES

The NLEA has opened the door for health claims for functional foods. The question is who will invest in the research and development necessary to substantiate these claims, particularly in the current regulatory environment? Will it be the risk-adverse, low-profit-margin food companies, who compete for shelf space in grocery stores emphasizing brand identity awareness for certain product lines, or the "high risk" pharmaceutical companies, who rely on patents and specific claims for higher margin products? Pharmaceutical companies have traditionally spent billions of dollars conducting clinical trials to obtain the market exclusivity for their drug products. Will food companies who traditionally market "me too" or "value added" products with brand identity actively embrace the idea of spending more money to create public sector awareness and to develop markets for patented products? Food companies usually produce and market similar products to create market awareness for that product category and to reduce advertising costs. Will they recognize that shared technology within the competition can establish durable markets for functional foods based on claims currently allowed by the NLEA? One can envision alliances between the pharmaceutical and food companies to capitalize on potential pharmacological actions of food ingredients in developing new product concepts related to functional foods.

Contacts in the food industry have indicated that future success in conceptualizing and developing new products depends on the ability to access and evaluate new information on a flexible and timely basis. Therefore, outside networks (including consortiums, joint ventures, contract research, cross-licensing, technology transfer, and strategic alliances and partnerships) must be utilized. New technologies, ideas, and product development efforts will evolve at a faster rate, with new food products retaining preemptive market positions for shorter time periods (6–9 months), and a significant market share for only 2–3 years. Windows of opportunity will become more specialized with shorter time frames. These trends will demand shorter product development times, greater access to specialized information, and an ability to compete in an expanding market with more flexibility and speed than ever before. In fact, development timeframes will influence product success more than development cost. Thus, sharing development and marketing skills will reduce introduction times into the market, leading food companies to embrace broader scopes of work in order to detect future trends. It has been suggested that hybrid companies or "hyper-links" (72) composed of scores of companies may evolve to research and market new foods and pharmaceuticals with technology closely tied to speed and flexibility.

VII. CONCLUSIONS

Functional foods are foods that contain physiologically-active compounds that may provide a health benefit beyond the traditional nutrients they contain. Thus, functional foods are thought to exist at the food-drug interface. Several factors have contributed to the recent interest in functional foods, including increased knowledge of the link between diet and health, a rapidly aging U.S. population and associated increases in health care costs, and recent technological advances in the food industry.

At present, no appropriate regulatory system exists for functional foods in this country, despite the fact that these foods have entered the mainstream of science, medicine, and the consumer marketplace. Other countries are currently addressing the regulatory issue, making it difficult for U.S. food and drug industries to maintain their global competitiveness in this market. The issue of how to regulate functional foods is critical to their future and has not been adequately addressed by the NLEA. The NLEA does not provide an adequate mechanism for allowing health claims for functional foods, and this will be a central issue in their future research and development. However, functional foods may fall into a number of categories other than conventional foods which are exempt from the NLEA, thus increasing their market potential.

A strategy needs to be developed to give market exclusivity to industry for functional foods research and development discoveries, comparable to patents in the pharmaceutical industry. Without market exclusivity there will be no incentive for the private sector to invest in functional foods research and development which may positively impact public health and reduce health care costs. In order for the field of functional foods to evolve, it is essential that the current food regulatory system *encourage* rather than *discourage* investment in the research, development, and availability of important advances in this new technology.

VIII. REFERENCES

1. D.E. Pszczola, Designer foods: an evolving concept, *Food Technol. 47*:92 (1993).
2. Anon, When food meets medicine, *Food Manufacture 66*:26 (1991).
3. A.B. Caragay, Cancer-preventative foods and ingredients, *Food Technol. 46*:65 (1992).
4. The nutraceutical initiative: a proposal for economic and regulatory reform. The Foundation for Innovation in Medicine (1991).
5. J.N. Hathcock, Safety and regulatory issues for phytochemical sources: "designer foods," *Nutr. Today*, November/December:23 (1993).
6. T.P. Shukla, Nutraceutical foods, *Cereal Foods World 37*:665 (1992).
7. Is the public ready for designer foods? *Food Safety Notebook 4*:35 (1993).
8. J. O'Brien, Health claims and prospects for designer foods, *Trends Food Sci. Technol. 3*:127 (1992).
9. B.F. Haumann, Designing foods. Manipulating foods to promote health. *INFORM 4*:344 (1993).
10. Functional Foods for Health, *Opportunities in the Nutrition and Food Sciences* (P.R. Thomas, R. Earl, eds.), National Academy Press, Washington, D.C., 1994, p. 109.
11. Committee on Diet and Health, Commission on Life Sciences, National Research Council, *Diet and Health: Implications for Reducing Chronic Disease Risk*, Washington, D.C., National Academy Press, 1989.
12. Department of Health and Human Services, Public Health Service, *Healthy People 2000: National Health Promotion and Disease Prevention Objectives.*
13. Department of Health and Human Services, Public Health Service, *The Surgeon General's Report on Nutrition and Health*, Washington, D.C., DHHS {PHS} Publication No. 017-001-00465-1. U.S. Government Printing Office, 1988.
14. K.A. Steinmetz and J.D. Potter, Fruits, vegetables and cancer. I. Epidemiology, *Cancer Causes Control 2*:325 (1991).
15. G. Block, B. Patterson, and A. Subar, Fruits, vegetables, and cancer prevention: A review of the epidemiological evidence, *Nutr. Cancer 18*:1(1992).
16. K.A. Steinmetz and J.D. Potter, Vegetables, fruit and cancer. II. Mechanisms, *Cancer Causes Control 2*:427 (1991).
17. L.O. Dragsted, M. Strube and J.C. Larsen. Cancer-protective factors in fruits and vegetables: Biochemical and Biological background, *Pharm. Toxicol. 72* (suppl. 1):116s (1993).

18. Report of Nutrition Screening 1: Toward a Common View. A Consensus Conference sponsored by the Nutrition Screening Initiative, April 8-10, 1991, Washington, D.C.
19. M. McPhee, Healthy, wealthy & wiser: new products for new markets. *Prepared Foods*, Mid-April:18 (1993).
20. C.M. Weaver, M.K. Schmidl, C.E. Woteki, and W.R. Bidlack, Research needs in diet, nutrition and health, *Food Technol. (suppl) 47*:14s (1993).
21. Food research needs into the 21st century, *Food Technol. (suppl) 47*:3s (1993).
22. Pub L. No. 101-535, 104 Stat. 2353 (1990).
23. Pub. L. No. 75-717, 52 Stat. 1040 (1938).
24. *Fed. Reg. 58*:2066 (1993).
25. The FDA's final regulations on health claims for foods, *Nutr. Rev. 51*:90 (1993).
26. S.L. Nightingale, Proposals for folic acid fortification and labeling of certain foods to reduce risk of neural tube defects, *JAMA 270*:2283 (1993).
27. FDA-initiated public conference: Antioxidant Vitamins and Cancer and Cardiovascular Disease. National Academy of Sciences, Washington, D.C., November 1-3, 1993.
28. Food labeling: nutrient content claims, general principles, petitions, definition of terms; definitions of nutrient content claims for the fat, fatty acid, and cholesterol content of food, *Fed. Reg., 58*:2302 (1993).
29. N.H. Mermelstein, A new era in food labeling, *Food Technol. 47*:81 (1993).
30. G.A. Zarkin, N. Dean, J.A. Mauskopf, and R. Williams, Potential health benefits of nutrition label changes, *Am. J. Pub. Health 83*:717 (1993).
31. F.R. Shank, The Nutrition Labeling and Education Act of 1990, *Food Drug Law J. 47*:247 (1992).
32. K.L. Wrick, The food/drug interface, *Spectrum, Food Industry*, Decision Resources, Inc., Burlington, MA, 1988.
33. L. Miller, U.S. lags in nutriceutical development, *Bio/Technol. 11*:969 (1993).
34. A. Dunkin, Eat nine cloves of garlic, and call me in the morning. *Business Week*, February 15:124 (1993).
35. S. Warshafsky, R.S. Kamer and S.L. Sivak, Effect of garlic on total serum cholesterol. A meta-analysis, *Ann. Int. Med. 119*:599 (1993).
36. E. Dorant, P.A. van den Brandt, R.A. Goldbohm, R.J.J. Hermus, and F. Sturmans, Garlic and its significance for the prevention of cancer in humans: a critical review, *Br. J. Cancer 67*:424 (1993).
37. S.H. McNamara, *Regulatory Affairs 5*:173 (1993).
38. Medical Foods. A Scientific Status Summary by the Institute of Food Technologists' Expert Panel on Food Safety and Nutrition. *Food Technol. 46*:87 (1992).
39. S. McNamara, Esp, *"Regulatory Aspects of Functional Foods,"* presented at the 15th Symposium on Nutrition and Food Technology, March 14, 1994, Chicago, IL.
40. *Fed. Reg. 58*: 2478 (1993).
41. M.S. Kurzer, Planning and interpreting "designer food" feeding studies, *Food Technol. 47*:80 (1993).

42. T. Reynolds, Agreements signed to test foods for cancer prevention, *J. Natl. Cancer Inst. 83*:1050 (1991).
43. K.L. Wrick, L.J. Friedman, J.K. Brewda, and J.J. Carroll, Consumer viewpoint on "designer foods," *Food Technol. 47*:94 (1993).
44. S.S. Harris, Health claims for foods in the international marketplace, *Food Technol. 46*:92 (1992).
45. T. Furukawa, The Japanese system. Presented at the Foundation for Innovation in Medicine Conference. A comparison of the U.S., European and Japanese Nutraceutical Health and Medical Claim Rules, New York City, January, 1993.
46. L. Gilbert, Move over, ginseng, *Prep. Foods*, April:22 (1991)
47. D. Swinbanks and J. O'Brien, Japan explores the boundary between food and medicine, *Nature 364*:180 (1993).
48. T. Orr, Regulatory confusion stymies emerging new health market, *Gen. Eng. News 13*:1 (1993).
49. R.E. Smith, Food demands of the emerging consumer: the role of modern food technology in meeting that challenge, *Am. J. Clin. Nutr. 58*:307s (1993).
50. L. Mancini, Low fat comes of age, *Food Eng.*, June:149 (1993).
51. D. Best, Corporations invest in "back to basics" research. *Prep. Foods*, April:54 (1991).
52. M.E. Sanders, B. Wasserman, and E. Allen Foegeding, Research needs in biotechnology. *Food Technol. 47*:18s (1993).
53. C.I. Beck, Who develops new products and technologies in a recession? *Food Technol.* 47:15 (1993).
54. C.S. Gasser and R.T. Fraley, Genetically engineered plants for crop improvement, *Science 244*:1293 (1989).
55. M.L. De Block, Expression of foreign genes in regenerated plants and their progeny, *EMBO J. 3*:1681 (1984).
56. N.W. Frey, Introduction, In: *Genetic Improvements of Agriculturally Important Crops: Progress and Issues* (R.T. Fraley, N.M. Frey and J. Schell, eds), Cold Spring Harbor, New York, 1988, p. 1.
57. M.E. Kunkel, Position of the American Dietetic Association: biotechnology and the future of food, *J. Am. Dietetic Assoc. 93*:189 (1993)
58. Office of Science and Technology Policy, Coordinated framework for regulation of biotechnology. *Fed. Reg. 51*:23302 (1986).
59. D. Jones and J. Maryanski, Safety considerations in the evaluation of transgenic plants for human food, In: *Risk Assessment in Genetic Engineering* (N. Levin and H. Stranss, eds.), McGraw-Hill, Inc., New York, 1991, p. 67.
60. E. Flamm, Responding to novel foods: Report and conclusions from public response to the FDA, *Science: 99* (1992).
61. D. Jones and J. Maryanski, Safety considerations in the evaluation of transgenic plants for human food, In: *Risk Assessment in Genetic Engineering* (N. Levin and H. Stranss, eds.), McGraw-Hill, Inc., 1991, p. 68.
62. Statement of Policy: Foods derived from new plant varieties. *Fed. Reg. 57*:22984 (1992).

63. T.J. Hoban and P.A. Kendall, Consumer attitudes about the use of biotechnology in agriculture and food production. U.S. Department of Agriculture Extension Service and North Carolina State University, July, 1992.
64. L.S. Gold, T.H. Slone, B.R. Stern, N.B. Manley, and B.N. Ames, Rodent carcinogens: setting priorities, *Science 258*:261 (1992)
65. Federal Food Drug & Cosmetic Act, Food and Drug Administration, HHS, Part 316, Orphan Drugs, 21 CFR Ch. 1 (4-1-93 Edition), p. 160.
66. N. Byal, Are we doing enough? A consumers point of view. *Prep. Foods*, Mid-April:22 (1993).
67. K. Smith, The calcium story: Fortification opens market opportunity, *Prep. Foods*, Mid-April:24 (1993).
68. S.H. McNamara, US FDA rules on health claims for foods, *Trends Food Sci. Technol. 2*:186 (1991).
69. New York Times, February 9, 1990, p. 1.
70. B.P. Kinosian and J.M. Eisenberg. Cutting into cholesterol: cost-effective alternatives for treating hypercholesterolemia, *JAMA 259*:2249 (1988).
71. P.M. Ippolito and A.D. Mathios, Health claims in food marketing: evidence on knowledge and behavior in the cereal market, *J. Public Policy Marketing 10*:15 (1991).
72. P. Lytle, Dancing with the Wallflowers, Presented at the Second Annual Functional Foods for Health Program Retreat, Monticello, Illinois, May 8, 1993.

22

Cereal and Bakery Products: The Nutrition Labeling and Education Act Challenge

Paul Bruns and Cheryl Callen

Nabisco Brands, Inc., East Hanover, New Jersey

I. INTRODUCTION

Since the passage of the Food, Drug and Cosmetic Act of 1938, labeling for retail food products has slowly undergone evolutionary changes. The Nutrition Labeling and Education Act of 1990 (NLEA) and its implementing regulations represented the most revolutionary labeling change in 50 years! For the first time in history, the food industry faced simultaneous changes to ingredient listings, nutrition labeling format, and complex technical criteria for all health-related product claims. Moreover, the original legislation and the implementing regulations triggered mandatory relabeling for virtually every retail food product. The 12–18 month time frame for new label implementation was initially viewed by many as tight but adequate. That soon changed when the extent of label redesign, printing capacity, and the logistics to phase-out the old labels became known. In many cases the expected 12–18 month phase-in of labels had to be reduced by up to 50% to accommodate package redesign and phase-in. This was necessitated by the following circumstances:

- Frequently, labels and packaging for many unique package configurations, especially small retail food packages, had to be totally redesigned to accommodate significant increases in information while retaining brand recognition.
- Health-positioned foods often required extensive reformulation to retain their nutrition-related claims and critical market positioning.

- Products scheduled for market introduction needed an immediate impact assessment, competitive evaluation, claim evaluation, and priority relabeling to assure minimal label write-offs and on-time delivery.
- Multidisciplinary coordination and team work often required accelerated label copy generation to accommodate production schedules and essential inputs from other job functions.
- Label format and claim options had to be quickly identified and leveraged to maintain or enhance competitive advantage.

The impact of NLEA was awesome! In the time it took to go from proposed to final rules (about 18 months), almost 2000 pages of regulations were issued. The new rules represented a blueprint for the future of food product development and marketing. They had to be read, understood, assessed, and implemented. In addition, their technical complexity created a tremendous educational challenge. The NLEA challenge continues to face the entire food industry and may be viewed as either a problem or an opportunity.

For those who embrace the challenge, it represents new product development and marketing opportunities in the growing "healthy foods" marketplace. This chapter is dedicated to helping readers grasp the possibilities confronting the cereal and bakery products industry. Success belongs to those with vision and the courage to see it through to completion. The reader is encouraged to read Chapter 7 for a comprehensive review of the regulations, Chapter 23 for compliance guidelines, Chapter 10 for details on label formats, and Chapter 24 for analytical methods and data bases available for obtaining nutrient content values.

II. SERVING SIZE

Prior to NLEA, manufacturers who provided nutrition information on their labels were free to determine the serving size for their products. Many trade groups and food marketers set reasonable serving sizes for product categories based on consumption data or consumer usage. Occasionally, this freedom led to inconsistencies within a product category or the establishment of artificially low or high serving sizes to satisfy label claims. Addressing this issue, the Food and Drug Administration (FDA) and U.S. Department of Agriculture (USDA) established reference amounts for approximately 160 food categories based on consumption data. The reference amounts serve as the basis for qualifying for nutrient content claims. The reference amounts also serve as the starting point for determining the serving sizes to be used for nutrition labeling. The establishment of reference amounts assures that all foods in a category are labeled in a similar manner and provides for a level playing field with respect to nutrient content claims. Table

Table 22.1 Reference Amounts[a] Common to the Cereal and Baking Industries

Product category	Reference amounts[a]	Label statement
Cereals, dry instant	15 g	cup(s) (g)
Cereals, prepared	110 g	cup(s) (g)
Biscuits, bagels, etc.	55 g	piece(s) (g)
Breads, rolls	50 g	piece(s) (g)
Cookies	30 g	piece(s) (g)
Crackers: snacks	30 g	piece(s) (g)
Crackers: not snacks	15 g	piece(s) (g)
Pies	125 g	piece(s) (g) for distinct pieces
Cakes, heavy weights	125 g	piece(s) (g) for distinct pieces
Cakes, medium weight	80 g	piece(s) (g) for distinct pieces
Cakes, light weight	55 g	piece(s) (g) for distinct pieces
Breakfast cereals		
Hot cereal	40 g dry; 55 g flavored	cup(s) (g)
Ready-to-eat		
20g/cup	15 g	cup(s) (g)
43g/cup	30 g	cup(s) (g)
more than 43/cup	55 g	cup(s) (g)

[a] Reference amount customarily consumed per eating occasion.

22.1 provides some examples of reference amounts common to the cereal and baking industries.

The rules established by the FDA for determining serving sizes are quite complex and mastery of them is essential as the first step in implementing NLEA or in developing new products with a health positioning. Details on reference amounts and the rules governing serving size are presented in Chapter 7.

III. DISCRETE UNITS

The serving size for products sold in discrete units, such as cookies, muffins, or sliced bread, is the number of whole units which most closely approximates the products' reference amount. It is declared in the *Nutrition Facts Panel* as the number of units followed by the metric equivalent in parenthesis. For example, cookies would be labeled as the number of whole cookies closest to the 30-g reference amount followed by the weight of the three cookies in grams: 3 cookies (33 g). If a single unit weighs 67% or more but less than 200% of the reference amount, then the serving size for labeling is one unit. If a single unit weighs more than 50%, but less than 67% of the reference amount, then a single unit may be the labeled serving size. For example, the labeled serving size for a cookie weighing 17 g, may be either one or two cookies. If a unit weighs 200% or more of the reference amount, the labeled serving size may be one unit if it can reasonably be expected that the one unit will be consumed at a single eating occasion.

Products in large discrete units that are typically meant to serve several individuals, such as pies or cakes, are typically labeled with the fractional slice that most closely represents the reference amount. The fractions that can be used include 1/2, 1/3, 1/4, 1/5, 1/6, or smaller fractions that are divisible by two or three.

IV. BULK PRODUCTS

Products such as breakfast cereals, which are sold in bulk and not in discrete units, are labeled with a serving size that is the common household measure that most closely represents the reference amount. "Common household measure" means cup, tablespoon, teaspoon, or other similar household measure. Ounce and fluid ounce are also considered common household measures but are typically recommended for labeling purposes only when other household measures are not applicable. For breakfast cereals that require preparation, such as rolled oats or farina, the package directions may need to be modified so that they correspond to the serving size used in the *Nutrition Facts Panel.*

For nutrition labeling purposes, a tablespoon is 15 mL, a teaspoon is 5 mL, a cup is 240 mL, a fluid ounce is 30 mL, and an ounce is 28 g. Note that these are not exact arithmetical conversions but are rounded to the nearest increment of five.

V. SINGLE-SERVING PACKAGES

The rules for determining the labeled serving size of individual single-serving packages are different than those for discrete units. A product is considered to be a single serving and the entire contents must be labeled as one serving, when it is packaged and sold individually, contains less than 200% of the applicable reference amount, and the reference amount is less than 100 g or 100 mL. A cookie or cracker package containing less than 60 g therefore must be labeled as a single-serving package. If the reference amount is greater than 100 g or

100 mL, as it is for pies and cakes, then the manufacturer may decide whether a package that contains more than 150% but less than 200% of the reference amount is one or two servings. The labeled serving size for single-serving containers does not need to be followed by the gram equivalent since this information is included as part of the Net Weight statement on the principle display panel.

It may prove very useful for cereal and bakery manufacturers to evaluate all package sizes of a product and determine where an increase in net weight would eliminate the need to label a product as a single serving. For example, a cookie package with a net weight of 2 oz. (56.7 g) would need to be labeled as a single-serving container and the labeled serving size would be nearly double that of the same product sold in a multiserving package. If the net weight of this product were increased to 2.2 oz. (62.4 g), the package could then be labeled with a serving much closer to the reference amount. This is especially important when products are labeled with nutrient content claims.

VI. SERVINGS PER CONTAINER

The number of servings per container should be rounded to the nearest whole number except for packages where the number of servings is between two and five. The declared number of servings per container between two and five must be rounded to the nearest 1/2 serving. Since "servings per container" represents rounded numbers, the use of the term "about" is recommended. For example, "about 2 servings" or "about 3.5 servings."

VII. METRIC EQUIVALENTS

As previously noted, the serving size is stated in a common household measure followed by the metric equivalent in parenthesis. The gram declaration should be rounded in whole grams except for quantities less than 5 g. The gram quantity between 2 and 5 g must be declared to the nearest 1/2-g increment. Gram quantities less than 2 g must be declared in 0.1-g increments.

VIII. AERATED PRODUCTS

When the reference amount in grams is modified by the incorporation of air so that the resulting bulk density is lowered by at least 25%, the manufacturer may revise the reference amount for the appropriate food category by adjusting for the difference in density. The new reference amount must be rounded in 5-g increments and supporting data must be available to support the revision. Breakfast cereals and cakes may not use this procedure since density was taken into consideration when their reference amounts were established by FDA. However, it might prove useful for pies containing whipped fillings.

IX. NUTRITION FACTS PANEL

The NLEA and the implementing FDA regulations have changed nutrition labeling from a voluntary program to mandatory for most retail food products. The volume and emphasis of nutrition labeling information was also changed. In this section, we will point out those areas of the *Nutrition Facts Panel* which are of interest to those labeling cereal or bakery products. We encourage readers to read Chapters 7 and 10 for more detailed information on the *Nutrition Facts Panel.*

The overall emphasis of the *Nutrition Facts Panel* has changed from one emphasizing nutrients and food components needed in the diet to those which need to be avoided. The first six components declared on the label are calories, calories from fat, total fat, saturated fat, cholesterol and sodium.

Another significant change brought about by NLEA is the requirement that a food label needs to convey nutrition information in the context of a total daily diet. This was mandated by the Congress and resulted in the establishment of the *Daily Value* chart and *Percent Daily Values* in the *Nutrition Facts Panel.* Section VII of Chapter 7 discusses reference daily intakes (RDI) and daily reference values (DRV) and describes the standard footnote declaring percent daily values that must appear on all labels.

X. CALORIES

Calories are declared in 5-calorie increments up to and including 50 calories and in 10 calorie increments above 50 calories. There are five different ways to determine the caloric content of a food (see Chapter 7, Section IV-A).

The most commonly used general factors of 4, 4, and 9 for protein, carbohydrate, and fat, respectively, may not give you the most accurate or lowest calorie declaration. Maunfacturers may use the calorie calculation method that gives the most favorable results. If it involves the use of factors, apply the factors to unrounded grams of protein, carbohydrate, and fat. It is also useful to analyze products for insoluble fiber, if some insoluble fiber is reasonably expected to be present. This value can then be subtracted from total carbohydrate to reduce the total calories contributed by carbohydrate.

XI. TOTAL FAT

Total fat is declared in 1/2-g increments up to 5 g of fat; less than 1/2 g of fat per serving must be rounded to zero. Total fat is defined as total lipid fatty acids expressed as triglycerides.

XII. SATURATED FAT

Saturated fat is defined as the sum of all fatty acids containing no double bonds. This is significantly different from the previous definition which defined saturated fat as the sum of lauric, myristic, palmitic, and stearic fatty acids. Products which contain a significant amount of butter or other dairy ingredient may see an increase in their declared saturated fat content. Products made with canola oil, which contains saturates at C:20 and C:22, may also see their saturate level rise on a per serving basis.

Saturated fat content is now a mandatory part of nutrition labeling. It is declared in 1/2-g increments up to 5 g; less than 1/2 g of saturated fat must be declared as zero.

XIII. POLYUNSATURATED AND MONOUNSATURATED FATS

Polyunsaturated and monounsaturated fats are an optional declaration unless a claim is made about the fatty acid or cholesterol content of the food. When declared, they are declared in 1/2-g increments up to 5 g; less than 1/2 g is declared as zero. They cannot be declared independently of each other. The declared amount of polyunsaturated and monounsaturated fats includes only the *cis* form of the fatty acids. At this time, the *trans* form of the unsaturated fats is not declared anywhere on the nutrition label. A consumer may be able to determine an estimated *trans* content by subtracting declared polyunsaturated, monounsaturated, and saturated fats from total fat, when all are declared on the nutrition facts panel. The difference is *trans* fatty acid content.

XIV. CARBOHYDRATE INFORMATION

Total carbohydrate is defined as the difference between the total weight of the food and the sum of protein, total fat, moisture, and ash. Sugars are defined as the sum of all free mono- and disaccharides. Ingredients such as food acids and portions of some corn syrups that contain higher-level saccharide units are not considered sugars but other carbohydrate. It is important in products that contain high levels of sweeteners not to overestimate the sugars content of the product by assuming that the sweetener is 100% mono- and disaccharide.

XV. VITAMINS AND MINERALS

The four mandatory micronutrients are vitamin A, vitamin C, calcium, and iron. These were selected based on their continued importance in the diet and their potential relevance to certain diseases. Other vitamins and minerals may be declared as long as they have an established RDI. If a product is fortified, such as a breakfast cereal, or if fortification is required by the standard of identity, such as enriched farina, then the fortified nutrients must be declared in the *Nutrition Facts Panel.* If the product contains as an ingredient a standardized product which must be fortified (e.g., enriched flour or margarine) then the fortification nutrients do not need to be declared as part of the *Nutrition Facts Panel.*

XVI. PERCENT DAILY VALUES

Percent daily values may be determined in one of two ways. They may be derived by dividing the declared value (or rounded value) by the daily value or by dividing the actual value (unrounded value) by the daily value. It is especially important when you are declaring zero grams of fat or saturated fat that the percent daily value also be declared as zero. Otherwise it is confusing to the consumer and may jeopardize any nutrient content claim.

XVII. FORMATS

The *Nutrition Facts Panel* is fairly rigid with respect to type size and format. Flexibility is provided for small package sizes and for products which contribute insignificant amounts of most nutrients. We recommend consulting the regulations to see if your product or package qualifies for format relief. Examples of formats and a discussion of the regulations governing them are presented in Chapter 10.

XVIII. INGREDIENT LABELING

The revised ingredient labeling rules have a significant impact on cereal and bakery products. While ingredients continue to be declared in order of predominance by weight, changes in ingredient nomenclature have the potential to alter consumer perception. Moreover, when choices exist for naming ingredients, it is frequently important to choose label terms which are consumer friendly.

A summary of the new ingredient labeling provisions follow:

- All ingredients in foods with standards of identity must be declared (e.g., margarine [soybean oil, water, vitamin A palmitate, etc.]).
- All certified food colors and their lakes must be declared (e.g., Yellow 5, Blue 1, Red 40, Red 40 lake).
- Sweeteners defined in 21 CFR 168.110 and 168.111—dextrose anhydrous or anhydrous dextrose, and dextrose monohydrate or dextrose, respectively—may also be declared as corn sugar anhydrous and corn sugar monohydrate in addition to the names already provided by their standards of identity.
- All protein hydrolysates must be identified by their food source (e.g., hydrolyzed corn protein).
- Caseinate must be functionally labeled as milk protein derivatives whenever they are used as ingredients in foods which claim to be non-dairy foods (e.g., sodium caseinate [a milk derivative]).
- Sulfites (e.g., sodium bisulfite or sulfur dioxide) must be declared in all standardized foods whether present as direct or indirect additives whenever they are present at more than 10 ppm.
- The addition of water to standardized foods to make other standardized foods requires declaration of the added water (e.g., tomato puree [tomato paste, water]).
- If the percentage of an ingredient is included optionally in the statement of ingredients, it must be declared in whole units to the nearest 1% in parentheses following the name of the ingredient [e.g., menhaden oil (8%)].
- Highly hydrolyzed protein hydrolysates and yeast autolysates due to the presence of high levels of monosodium glutamate (MSG) must be immediately followed by the parenthetical phase "contents glutamate" (e.g., hydrolyzed soy protein [contains glutamate]). An exception is provided for partially hydrolyzed proteins which contain insignificant levels of MSG.

Of the changes listed here, the ones with the most impact are those which cover declaration of all certified colors and ingredients in standardized foods used as ingredients in other foods.

While there are other minor ingredient labeling rules, they do not significantly affect cereal and bakery products.

XIX. NUTRIENT DESCRIPTORS

Nutrient descriptors, such as "low sodium," "fat free," and "low cholesterol," are an aspect of NLEA that will undoubtedly have the greatest impact on the food industry and consumers. Such nutrient descriptors are also called *Nutrient Content Claims* to distinguish them from *Health Messages*. Health messages will be discussed later in this section.

Nutrient content claims are claims based on the nutrient composition of food products. The reader should refer to Section IX, Chapter 7 for a list of permitted claims and their detailed requirements. Both cereal and bakery products lend themselves to a wide variety of potential claims. Such claims constitute the crucial admission price to the growing "healthy foods" category in today's marketplace. No health-related product claims are possible under NLEA rules unless FDA or USDA has authorized such a claim. Nutrient content claims are based on comparisons of reference amounts. Occasionally, a comparison is also mandated on a per 50- or 100-g or labeled serving size basis.

In addition, the food must meet complex technical criteria established by the government agencies before a claim may be used in labeling. When foods are capable of meeting the established claim criteria, the same claims may be extended to advertising. While food labeling is regulated by FDA and USDA, advertising is under the jurisdiction of the Federal Trade Commission (FTC). Although there have not been any test cases to date, FTC has gone on the record as a supporter of FDA and USDA rules. The reader is encouraged to consult the advice of an experienced legal counsel prior to implementing any health-related product claims in advertising. Chapter 4 and Appendix 2 provide information regarding FTC's position on nutrition labeling.

At the present time, FDA has authorized compositional criteria for more than 80 potential nutrient descriptors. While technically complex, these government-sanctioned claim definitions provide the food industry with a level playing field, credible health-related claims, and a powerful stimulus to develop healthier alternatives to traditional food products. At the same time, these highly restrictive claim definitions potentially limit product development and nutrition research in support of new claims.

While a petition process has been approved by FDA for establishing new health-related claims, it is cumbersome, time consuming and costly. From a business standpoint, it may also be viewed as a mechanism for establishing new generic claims which lack competitive advantage. This is due, in part, to the fact that government-established health claim criteria may be used by anyone. Unfortunately, this makes investment in new nutrition claim substantiation research of little interest to the business community. Unless ways can be found to say something unique about a new healthier product (with principal display panel claims and associated promotion), how will consumers recognize its value? A less prescriptive proprietary approval process may be needed once existing claim criteria have been leveraged and implemented. Additionally, we should encourage government agencies to find effective ways for fast-track approval for new health-related claims without misleading consumers. This would stimulate innovation and competitiveness while improving our knowledge and health status.

Since cereal and bakery products are grain-based, they frequently present opportunities for using nutrient descriptors in labeling and advertising. Grains typically contain little fat, saturated fat, cholesterol, or sodium. Frequently, they also serve as sources of dietary fiber, minerals, and water-soluble vitamins. Other considerations include claim opportunities associated with mandatory or voluntary fortification and enrichment of grain-based food to meet legal requirements.

An example would be mandatory addition of thiamin (vitamin B_1), riboflavin (vitamin B_2), niacin (a B-vitamin), and iron to flour, in order to call it "enriched flour". While mandatory enrichment by itself does not trigger mandatory disclosure in the *Nutrition Facts Panel*, a nutrient content claim based on enrichment or naturally occurring nutrients would mandate disclosure in nutrition labeling. This could be a small price to pay for claims of nutritional superiority versus a similar competitive product.

As previously noted, NLEA labeling and claim regulations provide a blueprint for the future development of healthier food products. The challenge is to design nutrient content claims and/or health messages into the process of food product development. This requires product development scientists and marketing experts to become far more knowledgeable about nutrition than in the past. This educational effort will be essential to the process of building better nutritional profiles and claim potential into a competitive advantage while avoiding the competitive disadvantages associated with an inability to match or exceed competitor claims and label values.

Today's new rules require: (a) expertise in the nutrient composition of ingredients; (b) knowledge of competitive product composition; (c) early identification of potential claims; (d) avoidance of negative nutrients; (e) easy access to quality nutrient data bases; (f) mastery of FDA and/or USDA regulations; (g) strategic partnerships with legal, product development and marketing resources; (h) an ability to credibly sell technical solutions, concepts and ideas rather than hype to advertising executives and other nontechnical leaders; and (i) access to ingredient specifications, formulas, and computer literacy.

Nutrient descriptors must be built into new products beginning at the initial development stage. Moreover, it is advisable to develop all potential health-related claims at the same time as the ingredient labels and *Nutrition Facts Panels* are being developed. The reason for this is simple. Information required for claim identification is typically available when ingredient statements and nutrition labels are being developed. Making the labeling process three simultaneous steps rather than separate steps may save considerable time and resources, as will computerizing the technical criteria for nutrient content claims and health messages. As mentioned previously, there are more than 80 claim possibilities. More importantly, a specific claim may be based on meeting up to 15 separate requirements. The calculations can take hours if done manually. Moreover, they are open to human error and associated regulatory, legal, and competitive risk.

XX. HEALTH CLAIMS

Health claims or health messages are claims which characterize the relationship between a food or nutrient and a disease or health-related condition. Such claims may be expressed or implied. At this time there are seven approved health claims or messages related to:

- fat and cancer
- saturated fat, cholesterol and coronary heart disease
- calcium and osteoporosis
- sodium and hypertension
- fiber-containing grain products, fruits and vegetables and their role in cancer
- fruits, vegetables, and grain products that contain fiber, and their role in coronary heart disease
- fruits and vegetables and their role in cancer

FDA is also considering future approval of an additional health claim related to folic acid and its role in prevention of neural tubule birth defects. Approval is expected.

The criteria for health claims are highly technical and voluminous. Since all approved health claims are associated with multifactorial disease states, it is unlikely that any single food or ingredient provides a "silver bullet" cure. Moreover, there is a high likelihood that we may never fully understand the relationship between food and disease. In spite of this situation, the evidence for these seven relationships is generally accepted by the scientific community as compelling enough to warrant preventive action. The reader is referred to Chapter 7, Section XI for detailed information and specific rules.

In the opinion of the authors, the rules for making health claims are generally so complex that they are unlikely to be widely applicable to food labels or advertising. The permitted health messages are also primarily educational in nature. Since the wording of the message is strictly prescribed, they offer little competitive value or consumer attractiveness. However, since few foods will be able to meet the complex criteria, health claims will probably be relatively rare. For this reason alone, they could prove useful as a point of uniqueness and, therefore, offer some competitive advantage for flagship "health brands." While the jury is still out, identification of candidates for health messages certainly makes business sense as an enhancement to potentially greater marketing opportunity. The decision to implement health claims once they have been identified and substantiated is best left to marketing functions.

XXI. PRODUCT DEVELOPMENT

New and reformulated food products are moving rapidly onto retail store shelves. The NLEA implementing regulations have resulted in further segmentation of traditional food categories, and manufacturers are rushing to fill vacant niches for healthier alternatives. The new labeling and claim rules offer significant opportunity to build additional sales volume, not just cannibalize existing brands. Cereal and bakery product manufacturers are increasingly taking advantage of consumer awareness and interest in new or reformulated foods which meet the needs of their healthier lifestyles. Success in this growing market is the result of critical partnerships between product development and the legal and regulatory professions. Teamwork is essential! Manufacturers can no longer afford to relegate their regulatory expertise to traditional compliance functions. Their knowledge of food composition and claim rules must be leveraged to identify new claim opportunities as an integral part of today's product development cycle.

New products must be formulated to meet strict compositional criteria, and existing products can often be reformulated to retain existing market positioning or as line extensions.

NLEA will continue to have a significant impact on new product development for many years to come. As consumers become more knowledgeable regarding the labeling and claim rules, the credibility and usefulness of health-related claims will continue to rise. The use of computerized nutrient data bases will become commonplace as a cost-control measure. Moreover, product formulation will be performed increasingly in a computerized environment where changes in product composition can be estimated as ingredients are manipulated to achieve nutrient contact claims and/or health messages.

XXII. NLEA COMPLIANCE

The new FDA rules became effective for most foods labeled on or after May 8, 1994, but there are exceptions (see Title 21 Code of Federal Regulations Part

101.9 [j]). USDA rules typically affect bakery products containing meat and/or poultry ingredients.

Following these effective dates, USDA and FDA will be obligated to conduct a credible enforcement effort. Clearly, these agencies will be concerned initially about whether firms are making a reasonable effort to comply with the requirements for changes mandated by the act. While detailed information regarding priority enforcement areas is not available at this time, past experience makes an educated guess possible. In our opinion, the following areas will probably receive the most attention:

- nonexempt products that fail to bear nutrition labeling
- products bearing health claims or nutrient content claims that have not been authorized by FDA or USDA
- products that fail to bear information required by NLEA that was not previously required (e.g., ingredient declaration for standardized foods and names of specific certified colors)
- foods bearing approved claims which fail to qualify for those claims
- food labels that provide nutrition labels with major format deviations

XXIII. CONCLUSION

Obviously, it would not be feasible to cover all the NLEA-related rule changes in one chapter. Moreover, cereal and bakery products constitute a very broad food category. Instead of summarizing more than 900 pages of new rules, we chose to provide practical examples of typical business issues associated with the revolutionary labeling changes affecting this industry. However, the reader is reminded that there is no substitute for knowledge or practical experience. They are equal partners for success. For this reason, we encourage you to consult those sections of the 21 CFR as well as other chapters in this book which have been cited in this chapter. We wish you success in your attempt to master the details, as well as the broad principles. Additionally, we are reminded that the meaning of the Chinese symbol for "problem" translates into "danger and opportunity".

23

Compliance with Nutrient Content Declaration Requirements

James T. Heimbach and S. Kathleen Egan

TAS, Inc., Washington, D.C.

I. INTRODUCTION

The objective of this chapter is to give the reader an overall understanding of the Food and Drug Administration's (FDA) requirements regarding nutrient content declarations as well as a tutorial in basic procedures for bringing product labeling into compliance. This chapter is not and cannot be a comprehensive treatise covering all aspects of this subject. For most products, the simple procedures described here will be adequate, but there are many situations that may require the reader to seek information beyond that available here.

Further, there is a strategic decision to be made by the food manufacturer regarding labeling, whether the benefits of carrying out the extensive program of nutrient analysis seemingly called for by FDA are worth the costs. FDA does not mandate a specific approach to determining the nutrient content of food; the agency simply requires that the values declared on the label be "correct" within the meaning of the compliance policy as described below. In determining how best to assure accurate food labels, manufacturers might ask such questions as:

- How much confidence do I have that the nutrient data already available for my product are correct (regardless of whether their origin is consistent with FDA recommendations)?

- Do I plan to make health claims or nutrient content claims (recognizing that this will increase the chances of FDA scrutiny of the basis for the claims)?
- Generally speaking, how risk-averse am I? Do I want to wear a raincoat and carry an umbrella, or do I prefer to take chances and plan to run for cover if necessary?
- How likely is it that FDA will select my product for a compliance review? How likely is it that my product will be found out of compliance if it is reviewed?
- What will actually happen if my product is found out of compliance? Will FDA necessarily exercise the more stringent of their regulatory options, such as product seizure? What are the chances of adverse publicity?
- What will it cost me to assure that my product is in compliance?

This chapter will address many of these issues. It is our goal to provide information that will help in making the strategic decisions of whether to proceed with a program of nutrient analysis as well as technical information that will help readers design such a program.

II. UNDERSTANDING FDA REQUIREMENTS

FDA requirements regarding nutrient content declarations are conceptually simple: declarations must accurately reflect the nutrient content of the food so labeled. If a package declares a sodium content of 100 mg per serving, for example, this number must be sufficiently accurate to allow an interested consumer to make an informed decision whether to purchase the product.

There is thus a fundamental difference in meaning between a label declaration as prescribed by FDA regulations and a food-composition table value such as is reported in USDA's *Composition of Foods: Raw, Processed, Prepared* (1), more familiarly known as *Handbook 8*. While the *Handbook 8* value is chosen to represent the average expected nutrient content of the listed food, the label declaration must refer to the specific food in that package or container and, in essence, "guarantee" its nutrient content. An analogy might be an "average" shelf life for a product vs. the shelf life the manufacturer is willing to promise via a "sell-by" or other expiration date.

A. Classes of Nutrients

What is meant by "accurately" reflecting the nutrient content of the labeled food? After all, nutrient content may vary somewhat from one package to another of the same food. FDA recognizes this and allows some deviation from the declared nutrient content, based generally on how much control FDA judges that the manufacturer possesses. In order to establish precise requirements for accuracy, FDA has formally defined two nutrient classes for compliance purposes. In addition to the two defined classes, there is a third group of nutritional components, those usually regarded as being present in excess in the American diet. These will be referred to as "Third Group" nutrients.

Class I nutrients are those added in fortified or fabricated foods. Since nutrient addition is regarded as being completely under the control of the manufacturer, FDA requires that these nutrients be present at 100% or more of the declared value. Nutrients in Class I may include vitamins, minerals, protein, dietary fiber, and potassium.

Class II nutrients are naturally occurring (indigenous) nutrients. This includes nutrients indigenous to ingredients added to a food. Since the indigenous nutrient content of foods may vary—even when good manufacturing practice is followed in processing—FDA requires only that these nutrients be present at 80% or more of the declared value. Nutrients in Class II may include vitamins, minerals, protein, total carbohydrate, dietary fiber, other carbohydrate, polyunsaturated or monounsaturated fat, and potassium. If a nutrient is both indigenous to a food and added, its labeling is subject to Class I requirements.

Third Group nutrients include calories, sugars, total fat, saturated fat, cholesterol, and sodium. Again, FDA recognizes the variability of nutrient content in products and requires that the actual content of these nutrients be 120% or less of the declared value.

FDA regulations (21 CFR 101.9(g)(4) & (5)) state that a food with a label declaration of a Class I, Class II, or Third Group nutrient "shall be deemed to be misbranded under section 403(a) of the Federal Food, Drug and Cosmetic Act unless it meets [the requirements specified above]."

The requirement that Class II nutrients be present at 80% or more of the declared value, and Third Group nutrients not exceed 120% of the declared value, is popularly known as the "80/120 rule," and will be referred to as such in the remainder of this chapter.

Note that there is a directionality difference in the degree to which FDA allows discrepancies between label declarations of nutrient levels and actual package content. For Class I and Class II nutrients, the concern is with "overdeclaration"—claiming a higher nutrient content than actually exists. For Third Group nutrients, on the other hand, FDA's compliance requirements focus on "underdeclaration"—claiming a lower content than is actually to be found in the product. This directionality reflects a judgment by FDA that the societal costs of underdeclaration of Class I and II nutrients, or of overdeclaration of Third Group nutrients, exceed those of deviations in the opposite directions.

The FDA's compliance policy (21 CFR 101.9(g)(6)) regarding underdeclaration of Class I and II nutrients states merely that "reasonable excesses . . . are acceptable within current good manufacturing practice." Similarly, "reasonable deficiencies" in the content of Third Group nutrients are acceptable. These policies are so vague that it is difficult to conceive of FDA taking enforcement action based upon them. It is likely that the agency regards this matter as largely self-policing. It is very much against a manufacturer's own interest to overdeclare Third Group nutrients or to underdeclare Class I nutrients (which, after all, have been added at some cost). It is generally against the manufacturer's own interests to underdeclare Class II nutrients, although there may be room for judgment here. If a Class II nutrient is expected to be declared only at a low level, perhaps 2% or 4% of the Daily Value, it may not be worth the cost of analysis to support that declaration.

B. The FDA's Approach to Compliance Testing

Although it is not strictly necessary to know how FDA determines compliance in order to develop food labels that meet the requirements, a general awareness of FDA's compliance evaluation procedure will aid in understanding the rationale behind the methods described in this chapter.

The FDA's compliance evaluation procedure, as described in 21 CFR 101.9(g)(1) & (2), is as follows:

- Select 12 consumer units, 1 each from 12 different randomly chosen shipping cases from a single lot of the product. (A lot is defined as "a collection of primary containers or units of the same size, type, and style produced under conditions as nearly uniform as possible, designated by a common container code or marking, or . . . a day's production." FDA also notes elsewhere that actual practice may require some deviation from this definition.)
- Mix these 12 samples together, forming a single composite.
- Analyze the nutrient content of the composite using methods approved by the Association of Official Analytical Chemists or other FDA-accepted methods.
- Compare the analyzed nutrient content with that declared on the product label.

If any nutrient declared on the label is out of compliance, within the limits of analytical accuracy (i.e., the analyzed nutrient content is less than 100% of the declared value for Class I nutrients, less than 80% of the declared value for Class II nutrients, or more than 120% of the declared value for Third Group nutrients), then the product is out of compliance and shall be deemed misbranded.

C. Protection from a Charge of Misbranding

As was noted earlier, FDA has no requirements at all regarding how label values for nutrient content are determined. This is completely the responsibility of the manufacturer, and the manufacturer is under no obligation to reveal to FDA the source of its nutrient content information. All FDA requires is that the declarations meet the compliance criteria described above.

It is only if the product is tested and found to be out of compliance that the question arises of how the nutrient content declarations were determined. At this point, the best—indeed the only—defense against a charge of misbranding is to be able to demonstrate that the label declarations have a sound basis.

If a product is found to be out of compliance, it is not necessarily misbranded. The regulations state (21 CFR 101.9(g)(8)) that: "Compliance . . . may be provided by use of an FDA approved data base that has been computed following FDA guideline procedures and where food samples have been handled in accordance with current good manufacturing practice to prevent nutrition loss." Although a database is not regarded as "FDA approved" unless the agency has stated its approval in writing, there is a clear advantage to the food manufacturer in following FDA guideline procedures. These procedures are set forth in the *FDA Nutrition Labeling Manual: A Guide for Developing and Using DataBases, 1993 Edition* (2).

FDA recognizes that it is not possible to develop reasonable nutrient declarations that provide 100% certainty that every single lot of a product will be in compliance, even with the deviation allowed by the 80/120 rule, and thus such certainty is not required. Rather, FDA guidelines require that:

> Nutrient content declarations must be determined such that the product has at least a 95% probability of being in compliance with nutrient labeling regulations.

An alternative way of stating this requirement is that not more than one compliance check out of twenty (5%) should find the product out of compliance. Or, to restate the requirement positively, it is permissible for up to one product sample in twenty to be out of compliance without the product being deemed misbranded.

However, FDA is not going to analyze twenty separate samples of a product; it is going to analyze only one composite of 12 units drawn from a single lot. From its single analysis FDA has no way of determining whether its findings are representative of the usual nutrient content of the product, or whether the lot selected for compliance testing is unusually high or low in its content of one or more nutrients. What will happen if the product is found out of compliance?

Absent other information, FDA may regard its compliance analysis as definitive and deem the product misbranded. The agency then has a range of actions at its disposal up to initiating a product seizure. In addition to any regulatory penalty, an adverse action by FDA always poses a threat of negative publicity to the manufacturer.

The best defense against such an eventuality is to have available thorough documentation of the basis for the nutrient content declarations made for the product. Any sort of documentation is probably better than none in demonstrating some level of diligence in attempting to comply with labeling regulations. Clearly it is best if the documentation shows that proper procedures, as described in FDA's guidelines, were indeed followed in calculating label values, and this chapter describes the basic methods to follow in order to achieve this.

A final note on the subject of defending against a charge of misbranding: it may possible to submit your nutrient content documentation to FDA for review. FDA has stated that approval will be granted "where a clear need is presented." There is obviously some expense in doing this, and FDA requires a specific format for the submission, but it offers the maximum in protection. FDA has stated that:

> . . . [FDA] will not take immediate action against a company for a product that does not contain the labeled level of nutrients, provided the agency has reviewed and accepted the database for the product and provided the company has followed good manufacturing practices in producing the food.

D. The Likelihood of Noncompliance

At this point, readers may be asking themselves, "Is it really all that likely that my label will be out of compliance?" The answer is that it is quite likely. If all food manufacturers precisely follow FDA's requirements for determining nutrient declarations for all of their labels, about HALF of them will be expected to fail an FDA compliance check.

The discussion to this point has followed FDA's own description of its guideline of 95% confidence that a product's nutrition label be in compliance. However, in all of its discussions of this issue, the agency has ignored the fact that nutrition labels contain many declarations, not just one. In reality, FDA's guidelines do not require that the product, that is to say its entire nutrition label, must be in compliance 95% of the time. Rather, they require that each individual nutrient declaration must have a 95% certainty of being in compliance.

Unlikely events do occur, given enough repetitions. Someone always wins the lottery eventually. This piling up of small probabilities until they become a large probability happens with nutrient declarations also. If two nutrient contents are declared, each with a 95% probability of compliance, the probability of both being in compliance is $0.95 \times 0.95 = 90.25\%$. Declaration of a third nutrient, again with 95% confidence of compliance, multiplies this by 0.95 again, reducing the chances of all three declarations being in compliance to 85.74%.

Estimating the probability of a full nutrition label being in compliance is complicated by the facts that the number of nutrient declarations varies across products and not all of them are independent of each other (e.g., the total fat and saturated fat contents do not vary independently). However, labels with fourteen or more independent declarations are not unusual, and—assuming that each nutrient is declared with 95% confidence—these labels have a probability of being in compliance of only 48.8%. Thus they are more likely than not to fail an FDA compliance review.

Note also that this discussion assumes that proper procedures were followed in calculating label values. Violation of these procedures may increase the probability of noncompliance significantly. To take an extreme case, suppose a manufacturer adds ten vitamins and minerals to a product, and in each case declares the average content on the label. This means that about half of containers of the product would contain somewhat more than the declared value of a given nutrient, and about half would contain somewhat less. These would be Class I nutrients, which must be present at 100% of declared values. The probability of this label being in compliance (ignoring for the moment the question of analytical precision) would be about 0.5^{10}, or less than one chance in a thousand.

It is not possible to state at this time what the chances are of a product being selected by FDA for a compliance review. Given the limits of the agency's resources, FDA will probably instigate very few such reviews on its own initiative. However, it seems likely that there will be considerable "finger pointing" as manufacturers notice competing products with more favorable nutrient content declarations than their own and draw the FDA's attention to them, and it is not clear how the agency will respond.

III. APPROACHES TO DETERMINING NUTRIENT CONTENT

The term "nutrient database," as applied to nutrition labeling, generally refers to an organized collection of nutrient content values that includes foods or ingredients not unique to a single organization. For example, a nutrient database might be constructed for "frozen carrots" with the understanding that the values in the database are valid for most or all processors of generic "frozen carrots," absent some unique feature in a company's source of raw carrots or its handling of them.

On the other hand, a collection of nutrient content values for "brand XXX broccoli with cheese sauce" would not commonly be referred to as a database.

This usage seems excessively narrow. The principles, and indeed the application, of such data collections are identical. It is only in their breadth of application that they apparently differ, and it is not clear that application across several organizations differs in any fundamental way from application within an organization across distinct times or production sites. Consequently, the goals and procedures described in this chapter are regarded as being those for databases in the broad sense of the term (i.e., an organized collection of analytical values) even if the database under consideration is for a single brand.

In classifying types of databases, then, the primary division is between ingredient composition databases and what we will call "finished-food databases."

A. Ingredient Composition Databases

In ingredient composition databases, nutrient content data are associated with each ingredient used in formulating the food (e.g., wheat flour, milk, shortening, salt, sugar, etc.); the nutrient contents of the ingredients are then added to obtain the nutrient content of the food. FDA has had a history of opposition to the use of ingredient composition databases, but more recently has indicated a willingness at least to consider them.

Ingredient composition databases have several attractions:

- There are far fewer ingredients than there are formulated foods, thus the analytical task is greatly diminished.
- If a food is reformulated, its nutrient content can simply be recalculated rather than requiring new laboratory analyses.
- There is greater opportunity for sharing the costs across companies, because many food manufacturers use basic ingredients such as wheat flour, while only one makes brand XXX of product YYY.

At the same time, many challenges face those who would use ingredient composition databases to support nutrition labeling:

- The quality of the data required for an ingredient composition database must be higher than is currently available for most ingredients, including specification of sources of nutrient variation and probability sampling of each ingredient, and calculation of measures of variation in nutrient content for each ingredient. This likely requires a far greater initial effort than is required simply to analyze finished foods.
- Ingredients must be specific to the finished food products in which they are used. For example, if a food manufacturer uses primarily produce grown in a single geographical region, nutrient data based on national averages for that produce would not be appropriate.
- Moisture or fat gain or loss and nutrient loss during processing and handling must be accounted for, either by using nutrient data for processed forms of the ingredients or by adjusting the data for changes in nutrient content. Given the immense variation in processing methods and applications, neither of these solutions is fully satisfactory.
- It must be possible to calculate label values for the finished food such that the requirement for having a 95% probability of passing an FDA compliance

review is demonstrably met. This probably requires the ability to calculate not only the average nutrient content of the formulated food, but also the variability in nutrient content. Allowance must be made for both the variability in the nutrient content of the ingredients and for variability in the formulation (i.e., what proportion of each ingredient is added to each lot of a product across different production sites and over time) which requires dealing with a multivariate density distribution.

Issues in using ingredient composition databases are addressed in another chapter of this book ("Databases," by Jack Smith) and will not be further discussed here.

B. Finished-Food Databases

In finished-food databases, the nutrient data are associated with the food as manufactured, based on laboratory analyses of the finished food. A key question here is how broadly drawn the definition of the food may be. When FDA refers to finished-food databases, the agency generally means databases intended to be used by a number of different manufacturers. Examples are nutrient databases for commodity products such as fresh produce and minimally processed foods (such as frozen or canned produce) for which the intrabrand variability equals or exceeds the interbrand variability—that is, there is as much difference in nutrient content between lots of the same brand as there is between brands.

In such cases, published data such as those available from USDA's *Handbook 8* may be a valuable resource, although FDA has been quite clear that it will not accept data simply lifted from *Handbook 8* without further documentation. As is the case with ingredient databases, this type of finished-food database is discussed in Jack Smith's chapter on databases.

The remainder of this chapter deals with finished-food databases developed and maintained by a single manufacturer, intended for use in determining nutrient label values for that company's products only. Published data such as *Handbook 8* may still be of some use if they are available for the subject product or a similar product, particularly in estimating nutrient variability, but there is no alternative to carrying out laboratory analyses of the finished food if label values are to be developed with confidence that they will be in compliance with FDA regulations.

IV. CHARACTERIZING YOUR PRODUCT

The first step in determining the values to be declared on the nutrition label is to characterize your product in terms of nutrient content and nutrient variability.

This information is critical for determining the nutrients for which laboratory analysis should be performed and the number and source of the samples to be analyzed. In *Food Labeling Questions and Answers* (3), FDA stated:

> It is the responsibility of the manufacturer/packer/distributor to determine the variability of their product(s) and the number of samples needed to provide accurate nutrient data.

Characterization of the product will accomplish several objectives:

- Determining the nutrients present in the product and the approximate levels for each.
- Estimating variability of nutrient content of the product.
- Identifying potential sources of nutrient variability (e.g., ingredients, processing methods, and post-processing storage and handling).
- Identifying nutrients of special significance for the product (e.g., Class I nutrients and/or nutrients for which nutrient content or health claims will be made).

The product description should incorporate general information about the product or similar food (e.g., average nutrient content, general information about cultivation of the product or major ingredients) as well as information specific to the product (e.g., ingredient sources, production and packaging information, processing methods, average shelf life).

A. Review of Extant Data

A review of existing nutrient composition data for the product or similar products is the logical place to begin to characterize nutrient content and nutrient variability. Data compiled specifically for the product from previous laboratory analyses are an excellent resource. Other examples of extant data include food composition tables such as USDA's *Handbook 8* or *Bowes and Church's Food Values of Portions Commonly Used* (4). A review of the scientific literature will identify additional data sources.

The data review should incorporate an evaluation not only of average nutrient values but also other information that will assist in assessing the validity of the nutrient values. Measures of variability (e.g., standard deviation or coefficient of variation) and the number of samples upon which the nutrient values are based are necessary to accurately characterize the expected nutrient variability of your product.

For some nutrients, a number of analytical methods are available to measure nutrient content, oftentimes with different results. Ideally, when evaluating nutrient composition data, one would want to know the basis (i.e., analytical method used) for determining nutrient content. In reality, this information is rarely reported; this is particularly true of food composition tables. Note that while knowledge of the analytical method is not essential for reviewing extant data to characterize the nutrient content, such information is essential if the nutrient composition data are to be incorporated into the calculation of values to be declared on the nutrition label.

B. Measures of Nutrient Variability

In characterizing the nutrient profile of a food product, it is important to know not only the amount of a nutrient present but also how much variability in nutrient content might be expected. Several measures of variability are generally associated with analytical data:

- Standard deviation
- Variance
- Coefficient of variation
- Standard error of the mean

The significance of the variability—and, in fact, the reliability of average nutrient values—is related to the number of samples on which the measures are based. The greater the number of samples, the more dependable are the estimates of average nutrient content and variability in nutrient content.

Measures of nutrient variability are useful in developing a plan to collect samples for nutrient analysis. These measures will assist in determining the number of samples needed to assure accurate representation of the nutrient content of the food. Application of these measures to the development of a sampling plan will be addressed in Section VI.B of this chapter.

In our discussion of sampling, as well as in addressing statistical analysis of analytical results, we will consistently use the standard deviation (abbreviated "SD" or "s") as the measure of variability. Should the reader encounter one of the other measures in reviewing extant data, following are directions for converting them to standard deviation.

1. Variance

The variance is the square of the standard deviation. To obtain the standard deviation, take the square root of the variance.

2. Coefficient of Variation

The coefficient of variation is 100 times the standard deviation divided by the mean (average). To obtain the standard deviation, multiply the coefficient of variation by the mean, then divide by 100.

3. Standard Error of the Mean

The standard error of the mean is actually a measure of the precision of the value of the mean given in the database rather than the variability of the nutrient content of the food. It is equal to the standard deviation divided by the square root of the number of samples. To obtain the standard deviation, take the square root of the number of samples on which the data are based and multiply by the standard error of the mean.

C. Identifying Sources of Nutrient Variability

A number of factors may contribute to the nutrient variability of a food. The nutrient content of the ingredients and the relative amount of each ingredient in the product formulation will contribute to the overall nutrient composition—and variability—of the product. External factors such as conditions under which foods are processed or stored may also affect nutrient content of the final product. Major sources of nutrient variability in foods include the following:

1. *Inherent variability*

- cultivar or variety
- species

2. *Environmental factors*

- regional origin (which may include non-U.S. origin)
- season in the growing region, including early vs. late in the growing season
- growing conditions and agricultural practices
- maturity at harvest

3. *Processing*

- pre-processing storage
- nutrient loss or concentration due to heat processing, freezing, drying or other processing method
- nutrient variability due to differences in product formulation or processing methods

4. *Post-processing handling and storage*

- packaging materials
- storage temperature and humidity
- shelflife

Before developing a plan to select samples of the product for laboratory analysis, the manufacturer should examine the potential sources of nutrient variability in that food. Those factors which may be reasonably controlled for in the sampling plan should be distinguished from those over which the manufacturer has no control. If, for example, carrots are used as an ingredient in the product, beta-carotene content may be affected by several factors: cultivar, season, region, and growing conditions. Extended storage time, both pre- and post-processing, as well as the processing methods, may result in further reduction in levels of beta-carotene. It may be unlikely that the manufacturer will have complete information regarding the regional source or cultivar of the carrots or any information regarding soil characteristics or growing conditions. The manufacturer will, however, have information regarding processing factors and average shelf life of the processed food, and may be able to control for these factors in selecting the samples and analyzing the food for nutrient content. The application of such information to the development of the sampling plan is discussed in Section VI.A.

D. Production Information

A review of information related to production of the food will identify factors expected to contribute to nutrient variability in the final product. Ideally, this information would be incorporated into the design of the plan to collect samples for nutrient analysis. The information may be specific to a production site or common to all products.

Examples of site-specific information would be differences in processing, such as salting methods by individual production facilities that result in varying sodium levels in the packaged food. Another example of site-specific information is reliance of production facilities on regional ingredient sources when regional variations in nutrient content are expected.

An example of information common to all products would be seasonal variability of nutrient content of ingredients. This will affect nutrient content of all products, regardless of the production site. In such cases, the sampling plan should call for collection of samples throughout the year to obtain truly representative nutrient values.

Information about storage of ingredients as well as the final product can be useful for predicting nutrient loss. Storage conditions (temperature and humidity), packaging, and length of storage are factors which may result in decreased levels of labile (unstable) nutrients.

Additional information that is useful in the design of the sampling plan is production volume by facility. If some variability in nutrient content is expected by production site, the manufacturer may want to develop a sampling plan that calls for collection of more samples from sites that account for greater proportion of total production, or alternatively, to apply post-hoc adjustment factors to the analytical data to give greater weight to the samples from the higher-volume sites.

Table 23.1 Daily Values for Mandatory and Voluntary Nutrients

Nutrient	Daily Value (DRV[a] or RDI)
Mandatory Nutrients	
Calories	[b]
Calories from fat	30%
Total fat	< 65 g
Saturated fat	< 20 g
Cholesterol	< 300 mg
Sodium	< 2400 mg
Total carbohydrate	300 g
Dietary fiber	25 g
Sugars	[b]
Protein	50 g
Vitamin A	5000 IU
Vitamin C	60 mg
Calcium	1.0 g
Iron	18 mg
Voluntary Nutrients:	
Calories from saturated fat	10%
Polyunsaturated fat	[b]
Monounsaturated fat	[b]
Potassium	< 3500 mg
Soluble and insoluble fiber	[b]
Sugar alcohol	[b]
Other carbohydrate	[b]
BetaCarotene	[b]
Vitamin D	400 IU
Vitamin E	30 IU
Thiamin	1.5 mg
Riboflavin	1.7 mg
Niacin	20 mg
Vitamin B6	2.0 mg
Folate	0.4 mg
Vitamin B12	6.0 μg
Pantothenic Acid	10 mg
Phosphorus	1.0 g
Iodine	150 μg
Magnesium	400 mg
Zinc	15 mg
Copper	2.0 mg

[a] Based on a 2000-calorie diet
[b] No DRV established
Source: Ref. 5

E. Deciding What Nutrients to Test

A review of extant nutrient composition data for the product will identify nutrients present in the product and the estimated levels of concentration. This information is helpful for designing the new label and determining the nutrients for which laboratory analyses will be done.

The labeling regulations stipulate the nutrients for which labeling is mandatory and those which may be declared voluntarily (Table 23.1). Fourteen nutrients have been designated as mandatory, with five of these considered "core" nutrients (calories, total fat, sodium, total carbohydrate, and protein). For all nutrients, FDA has specified what is considered to be an "insignificant" amount. In most cases, content of nutrients present at or below this amount may be declared as zero; for total carbohydrate, dietary fiber and protein, content at or less than the insignificant amount may be declared as "less than 1 g."

Selection of nutrients to be included on the nutrition label is primarily a regulatory issue although, to some extent (e.g., for voluntary nutrients) the selection will be a marketing decision as well. The cost of nutrient analyses for many voluntary nutrients is quite low and may be more than offset by the marketing advantage gained by their inclusion on the label.

If a nutrient is present in very low amounts, or is known not to be present (e.g., cholesterol in vegetable products), a reliable data source may suffice to document the level of that nutrient in the food. However, FDA has stated that manufacturers are responsible for the values declared on the nutrition label. Thus, it is important that nutrient content declarations are accurate and that records are maintained to document the source of the values declared on the label. This is true for values determined by laboratory analysis as well as those derived from the literature or food composition tables, and includes nutrients for which the declared amount is less than the insignificant amount.

Following are several of the major considerations when deciding the nutrients for which laboratory analyses will be done:

1. Class I Nutrients

For nutrients added to a food, it is essential that the amount declared on the label accurately reflect the actual amount of the nutrient in the food. As the compliance guidelines specify, the amount present in the food must be at least 100% of the amount declared on the label or the food will be deemed misbranded. It is advised that the content of all Class I nutrients be confirmed by laboratory analysis rather than relying on formulation records.

2. Third Group Nutrients

Compliance guidelines specify that the actual amount of these nutrients must not exceed 120% of the declared values, with allowances for "reasonable" underdeclaration of nutrient content. Because excessive intake of these nutrients is considered a public health concern, more attention will undoubtedly be focused on declaration of these nutrients. Determination of the content of all Third Group nutrients by laboratory analysis is advised unless the absence of the nutrient from the food can be thoroughly documented.

3. Nutrients for Which Claims Are Made

Special attention should be paid to any nutrient for which a claim is made. The regulations specify the levels of nutrients required to qualify for making such claims.

4. Voluntary Nutrients

The cost of laboratory analyses for many of the voluntary nutrients is quite low and may be weighed against the marketing advantage gained by declaring these nutrients. A review of extant nutrient composition data will indicate expected levels of these nutrients in the food. The manufacturer may want to consider declaration of voluntary nutrients present above a specific level (e.g., 10% or more of the Daily Value) for which costs of analyses are not prohibitive.

V. DESIGNING A SAMPLING PLAN

Broadly stated, the sampling plan includes all aspects of specifying how many and which units of the product will be subjected to laboratory analysis of nutrient content. In the following discussion, the goal is not to teach a course in statistics but rather to offer practical advice to non-statisticians. It must be recognized that this requires some over-simplification of statistical concepts and terminology.

A. Considerations in Sample Design

In planning a sample, four criteria must be balanced:

1. Design the Sample Plan to Meet the Research Objectives

The research objectives are determined by the product characterization discussed earlier, which provides information on sources of nutrient variability and the range of variability across the product domain. On this basis, decisions must be made regarding such issues as:

a. Is the goal a single label or multiple labels?

For example, if the product's nutrient content is likely to vary by region of the country, should there be a single label which allows for this regional variation or should there be separate labels for different regions? In the former case, a single sample will be drawn across all regions; in the latter case, separate samples will be drawn from each region.

b. What sources of nutrient variability must be addressed in the sample plan?

For example, is year-to-year variation in the nutrient content of the product sufficient to require that a multi-year sample of the product be analyzed, or can it be demonstrated that a single-year sample is adequate?

c. What is the anticipated effect of shelf life on labile nutrients?

Will the product's nutrition label include labile nutrients? Is the content of these nutrients expected to decline significantly over the shelf life of the product?

d. Is the amount of variation of different nutrients similar, or does it differ widely?

Will it be appropriate to conduct the same number of analyses for all nutrients or would it be more economical to plan a greater number of analyses for some nutrients than for others?

2. The Plan Must Allow Computation of Valid Estimates of Sampling Variability

As was discussed earlier, compliance with FDA labeling regulations requires knowledge of the variability of nutrient content of the labeled product; mere knowledge of the average nutrient content is not sufficient.

At a minimum, valid estimation of nutrient variability requires that probability methods of sampling be used to select the units of product that are analyzed. A full discussion of probability sampling is beyond the scope of this chapter, but the defining characteristic of a probability sample is that every element in the population has a known non-zero probability of selection. Note that this does not require that the probabilities of selection of all elements must be equal, only that they be known. Using equal probabilities of selection, however, offers many advantages, including direct estimation of the population mean, and this should be the method of choice when it is feasible.

For example, if a product is produced at two facilities, one accounting for 75% of production and the other 25%, a selection of twelve samples would best be drawn with nine from the first location and three from the second, rather than six from each. The latter could still be a valid probability sample, but the necessity for weighting the data and adjusting for the sample design effect would tend to make it a less efficient approach.

3. The Plan Must Be Practical, Simple, and Economical

FDA requires that nutrient declarations reflect the nutrient content of the product as purchased by the consumer, and suggests that product samples should be drawn as close to the point of purchase as possible—e.g., at the retail level. However, for most products practical considerations militate against this approach. It is very difficult and expensive to draw a probability sample at the retail level; sampling at the point of production is usually easier, more economical, and ultimately more likely to result in a valid sample. If nutrient lability is an issue, the samples can be held for an appropriate period of time before being analyzed.

4. The Plan Must Achieve the Level of Precision Required

Just as a coin flipped ten times will not necessarily land heads exactly 50% of the time, so a sample will not necessarily exactly reproduce the population in nutrient content. Precision is a measure of how closely the nutrient content of a sample of a product (e.g., 12 loaves of bread from each of three lots from a bakery) estimates the nutrient content of the "population" of the product (e.g., the entire bread production of the bakery).

The precision of estimates of nutrient content from a sample of units of a product is a function of two factors: the variability of the nutrient content across units of the product, and the number of units in the sample (the "sample size"). If the nutrient variability is high, the precision of estimate will be poorer than if the nutrient content were more consistent. On the other hand, the precision improves as the sample size increases. Thus, in order to achieve a given precision, the sample size must be based on knowledge of the variability of nutrient content of the product: the more variable the nutrient content, the greater the number of samples of the product that must be analyzed.

FDA does not set a precision requirement for compliance. However, as will be apparent when we examine the calculation of label values for nutrient content, the poorer the precision of estimate of nutrient content, the more allowance must be made to assure that the label is in compliance. This means that a poor precision of estimate will result in the declaration of lower content of Class I and Class II nutrients and higher content of Third Group nutrients than would be otherwise necessary.

It is suggested that a precision of 20% be regarded as a minimum target. That is to say, the number of samples should be selected such that one can be 95% confident that the nutrient content of sampled units will not deviate by more than 20% from the "true" nutrient content of the product. This is very much a minimum standard, and a precision of 10% would be preferable. However, the precision improves only as the square root of the sample size, and so improving the precision by a factor of two requires analyzing *four* times as many samples of the product.

B. Determining the Number of Samples Needed

With a given level of nutrient variability, two factors together determine the sample size needed to determine the nutrient content of a product:

1. The Precision Target

It was noted earlier that the number of samples required to achieve a given level of precision is dictated by the nutrient variability of the product. Using the 95% confidence level as the standard, the needed number of samples is calculated using the formula:

$$N = \frac{(1.96)^2 s^2}{M^2 p^2} \qquad \text{(Equation 23.1)}$$

where N is the number of units that must be analyzed
s is the standard deviation of the nutrient content of the product
M is the mean nutrient content of the product, and
p is the precision goal (expressed in decimal form: ± 20% = 0.2, ± 10% = 0.1, etc.)

Once N is calculated, it must be rounded up to the next integral value to determine the actual number of samples required to achieve the precision target.

Example: Determine the number of samples needed to achieve a precision target of ±15% for estimating the sodium content of a product with an estimated mean sodium content of 70 mg and a standard deviation of 20.

$$N = \frac{(1.96)^2(20)^2}{(70)^2(0.15)^2}$$

$$= 13.9 \rightarrow 14 \text{ (next higher integer)}$$

A similar calculation for each nutrient for which a label value is needed will determine the number of analyses that must be performed for each.

2. Practical Considerations of the Sampling Objective

It was suggested earlier that practicality and simplicity are important criteria in sample design. In the above example, if the product were produced at three sites, with equal production at each, it would probably be worth the slight additional analytical costs of using 15 samples rather than 14 in order to be able to select 5 samples from each site.

Similarly, and of wider application, it is likely that calculations made for each nutrient will result in different numbers of analyses being specified for each. Nevertheless, in most cases the more practical approach is simply to select the highest number and use that as the sample size for all nutrients. The complexities of drawing different samples or of providing difficult instructions to the analytical laboratory regarding which nutrients are to be tested in which samples likely outweigh any cost savings from slightly reducing the number of analyses.

The recommended approach, then, is to determine the number of samples needed to achieve the desired level of precision of the most variable nutrient (defined by s/M—the standard deviation divided by the mean). This number should then be adjusted upward as necessary to meet any sampling exigencies (such as seasonal sampling, number of production sites or other sampling points, etc.). Unless very significant cost savings are available by reducing the number of analyses for some less variable nutrients, analyze all nutrients for which label values are needed in all samples.

One example of an exception to this approach may be noted for salted products. If a product is produced in both salted and unsalted (or lightly salted) forms, and if it can be documented that the *only* difference in nutrient content among the different forms lies in the sodium content, then it may be appropriate to treat sodium differently from the other nutrients. For example, all nutrient analyses might be conducted using the unsalted form, with samples of the salted form being tested only for sodium content.

If the reader decides to use different numbers of analyses for different nutrients, against the advice offered above, an exception should be made for the proximates. The same number of analyses must be used for these nutrients because they are combined to estimate other proximates that are calculated rather than directly analyzed. FDA's recommendation is to determine the number of analyses theoretically needed for each proximate and then to use the geometric mean of these numbers (the nth root of the product of the numbers, where n is how many of them there are).

C. Sources of Estimates of Nutrient Variability

In the previous discussion of determination of needed numbers of samples for analysis, it was noted that it is necessary to estimate the mean and standard deviation of the nutrient content of the product. Earlier in this chapter (Section IV.A.) potential sources of such information were described. But what can be done if no data are available—no previous studies of the target product, no studies of similar products, no published data?

The approach recommended by FDA in the *Nutrition Labeling Manual* is to conduct a pilot study, essentially a miniature version of the planned full program of analysis, in order to estimate nutrient content and the variability of nutrient content.

We recommend a simple extension of the pilot study approach. This is to oversample the product, collecting many more samples than are likely to be needed, but then complete an initial set of laboratory analyses (the pilot study) on only a subset of them. On the basis of this first set of analyses, determine how many more samples to analyze. The primary advantage of this approach is that the costs of designing and executing a sample plan are borne only once.

With regard to the number of samples needed for the pilot study, the controlling factor may be the product characterization as it affects the sampling plan. Using the earlier example of a product manufactured at three sites, a reasonable pilot study might be based on two samples from each site—a total of six analyses.

D. Determining the Sampling Point

A key decision in the sampling plan is the point in the food production and distribution chain at which to sample the product. For agricultural commodities, choices include the farm, the wholesale level (distribution warehouses or other points), and the retail level (supermarkets and other points). For manufactured products, choices may include the manufacturer, the packer, the distributor, the wholesale warehouse, and the retail outlet.

FDA notes that nutrition labeling is required to reflect the nutrient content of food at the time of purchase by the ultimate consumer, and further that its compliance evaluations involve samples selected at the retail level. On these bases, FDA guidelines recommend the retail level is the optimal sampling point.

The advice offered here departs from the FDA recommendation. While there is no question that, all things being equal, sampling at the retail level is preferable, all things are usually not equal in the real world. The reality is that most food manufacturers can select samples at the production level quite easily and accurately, while sampling at the retail level is difficult, expensive, complicated, and liable to error. Further, you may wish to control for known or expected variation in nutrient content across production sites, and this goal is greatly simplified by sampling at those sites.

In addition, sampling at points other than the production site may result in severe problems with what is known as the sampling "frame." The frame is a listing of the sampling units for a particular population, and it is regarded as essential that the frame be complete, accurate, and up to date. To sample a product at the retail level, for example, it is necessary to have a fairly complete list of all retail outlets from which the product is sold, along with accurate information regarding sales volume at each outlet. This information is expensive, and inaccuracies or significant incompleteness will seriously compromise the acceptability of the sample.

Finally, it is useful to keep the sample design as simple as possible. Sampling at the retail level involves at least two-stage, and quite likely three- or four-stage, sampling. It usually requires clustered sampling, because it is less expensive to sample several lots of product from a single store than to sample one lot each at several stores. Each additional complexity in the sample design increases the chances of error. Further, it imposes statistical costs in what is known as the "design effect," which is discussed in Section IX.C.

The recommendation offered here is to sample at a point, most likely production or packaging, under the control of the organization responsible for determining nutrient content of the product (referred to throughout this chapter as the manufacturer). Sample at as many sites as are appropriate, weighting the selection of sites proportionally to the volume of the product that passes through them.

If the number of sites is small relative to the number of samples to be selected, the best plan is simply to draw samples from each site. If, however, there are more sites than samples to be selected, the sites must be chosen through a randomized process. There are numerous acceptable procedures; following is a description and example of one method that produces a properly weighted sample:

- Determine the percentage of the total volume of product that passes through each site; develop a cumulative distribution.
- Divide the number of samples to be selected into 100 to determine the "sampling interval."
- Generate a random number between 0 and the sampling interval to provide a starting point.
- Identify the first sampling point as the starting point, the next as the starting point + the sampling interval, the next as the starting point + twice the sampling interval, and so on.
- Select each sample from the site whose cumulative production percentage includes the identified point.

Example: Six samples are to be randomly drawn from 11 production lines, each accounting for the proportion of production shown. Determine the line from which each sample will be taken.

Line	Percentage of production	Cumulative percentage of production
A	5%	0–5%
B	8%	5–13%
C	10%	13–23%
D	5%	23–28%
E	17%	28–45%
F	8%	45–53%
G	12%	53–65%
H	12%	65–77%
I	10%	77–87%
J	5%	87–92%
K	8%	92–100%

The sampling interval for 6 samples is 100/6 = 16.67.
A random number is generated between 0 and 16.67; assume it is 4.52.

Six numbers are generated, and the line from which each sample will be taken is determined as follows:

Sampling Point

Sample	Identification	Line
1st	starting point = 4.52	A
2nd	4.52 + 16.67 = 21.19	C
3rd	4.52 + 2(16.67) = 37.86	E
4th	4.52 + 3(16.67) = 54.43	G
5th	4.52 + 4(16.67) = 71.20	H
6th	4.52 + 5(16.67) = 87.87	J

The six samples will be drawn with one each from lines A, C, E, G, H, and J.

There are many different ways of choosing sample sites. The crucial requirement is that the method selected be one that assures that each lot of product has a known, non-zero probability of being selected. Further, it is advantageous if these probabilities are equal: that is, every lot has the same probability of being selected. The rationale behind the procedure outlined above is that each line is selected with a probability proportional to the number of lots it produces, and so each lot has an equal probability of being chosen. If each line had been selected with equal probability (6 chances in 11), with a single sample to be chosen from each line, then lots of product from the higher production lines would have less chance of being chosen than those from the lower production lines. This would require that the data be weighted in calculating label values, an unnecessary complication.

One reason that FDA suggests sampling at the retail level is to allow for nutrient loss over the shelf life of the product. The recommendation made here is that, assuming there are no other reasons for sampling at retail, the samples of product merely be stored (under appropriate conditions) until the nutrient content has stabilized or until the normal shelf life is reached, whichever comes first.

E. The Sampling Unit

The discussion to this point has not addressed the question of what is considered a "sample." A sample is the product—or combination of products—that represents the entire universe of products for which nutrient composition is to be measured. For the purpose of nutrition labeling, as is true for compliance sampling by FDA, the goal is to determine the expected or average nutrient content of a single production lot of the food.

The sample generally consists of one or more "sample units." One obvious unit is a single container of the food: a loaf of bread, a box of cookies, a can of green peas, a jar of spaghetti sauce, a candy bar.

The sample unit is an appropriate basis for sampling for many purposes, and it is the only correct basis for sampling if an objective is to learn unit-to-unit variation in the nutrient content of the product. This might be part of a quality assurance program, for example. Since some readers may choose to develop their label information using this approach (especially if they are already collecting such information), Section X.A on calculation of label values addresses this possibility.

However, if the determination of nutrient content for labeling use is the only objective of the sampling and laboratory analyses, analysis of single units of the product is not the optimal approach. Rather, it is more efficient to analyze composites of the product, that is, to combine the contents of two or more single units or containers to form a single sample for analysis. In planning composites for nutrition labeling, it is essential to combine equal amounts of homogeneous units or containers (in this case, ones selected from a single lot). These are referred to as "single" composites. Another type of composite—referred to as a "commingled" composite—may be formed by combining heterogeneous units; an example would be a sample composed of sample units from different production lots or facilities. Analysis of commingled composites is an inexpensive method of determining the average nutrient content of a product because it requires fewer analyses. Nevertheless, we advise against the use of commingled composites for nutrition labeling purposes because they conceal variability across lots of the product and consequently cannot produce the information needed to assure compliance with FDA regulations.

If the nutrient data are for use in labeling, one objective is to predict the results of an FDA compliance review. Recall that FDA's policy is to form a single composite based on 12 samples from a single lot of the product. Consequently, the food manufacturer should use this same procedure: conduct nutrient analyses of single composites formed by combining 12 individual units of each lot chosen for analysis. Each of the 12 individual containers is referred to as a "sample unit," while the 12 sample units together are referred to as the "sample set."

If the composite approach is selected, it is imperative that each composite be formed of 12 units, no more or less, otherwise it is not possible to calculate the label value in a way certain to be acceptable to FDA.

VI. SAMPLE COLLECTION, HANDLING, AND SHIPPING

A. Sample Collection and Shipping

Regardless of the site(s) chosen for selection of samples of the product, the actual collection of samples must be essentially random. Note that random does not mean haphazard or arbitrary. Rather, random sampling is an objective process that assures that the product samples selected fairly represent the lot from which they are drawn.

Of paramount importance is the avoidance of a system that may lead to bias in the sample. For example, avoid drawing all samples from the beginning of a production run; if selecting loose product from bins, avoid taking all samples from the same location in the bins.

It is essential that documentation be maintained describing the sample collection procedure. In order to illustrate methods of collecting randomized samples, one example set of instructions is shown on the following page.

B. Documenting Chain of Custody

It is essential that a standardized procedure be established and maintained for recording pertinent details regarding each sampling unit and attaching a copy to the sample. Information should include study identification, name of collector, product identification, date and place of collection, lot or other identifying information, packing and shipping information. A copy of this document should be retained and another included with the sample container when it is shipped to the laboratory.

Example: Collection Instructions for Sampling from a Production Run

The Sample Set will consist of twelve (12) individual samples of XXX.

- Estimate the time of the run in minutes. 4 hours = 240 minutes.
- Divide by 12, 240/12 = 20 minutes. This is your "sampling interval."
- Draw your first individual sample at one half of the sampling interval into the run, 20/2 = 10 minutes.

Draw your remaining individual samples, one at a time, at the sampling interval after each other:

You drew your first sample at 10 minutes, so

Draw the second sample at 10 + 20 = 30 minutes;
Draw the third sample at 30 + 20 = 50 minutes;
Draw the fourth sample at 50 + 20 = 70 minutes;
Draw the fifth sample at 70 + 20 = 90 minutes;

And so on until you have drawn 12 samples.

- If you realize part way through that the run is going to be shorter or longer than you thought, adjust your sampling interval as necessary. Make this adjustment as soon as you can.
- Place each individual sample in one of the plastic bags provided and seal the bag. Each bag is already labeled with the type of product and a code number that includes the Sample Set number and a number from 1 to 12 to designate the individual sample.
- When you are finished, you should have 12 plastic bags, each containing one individual sample from the production or packing run. The 12 bags together are the complete Sample Set representing that run.
- When all individual samples for one Sample Set have been collected and bagged, place the 12 bags in the box provided for that Sample Set.
- If it is necessary to add packing material to fill the box, do not use newspaper. Plastic packing material or any paper without ink is suitable.
- Complete the "Sample Collection and Chain-of-Custody Tracking Form" and make two copies.
- Place one copy of the form in the box with the 12 individual samples. Seal the box with the tape provided and ship it to using the pre-addressed Federal Express shipping label.

Keep the second copy of the sample collection form for your records and send the third copy of the form to TAS, Inc., using the pre-addressed envelope provided.

Example: Sample Collection and Chain-of-Custody Tracking Form

Product type:	________________
Sample set identification:	________________
Company name:	________________
Address where sample was collected:	________________
Name of responsible individual:	________________
Date sample was collected:	________________
Lot identification:	________________
Sampling method used (check one):	___ Production/packing line ___ Bin ___ Case/package
Approximate weight of sample (total of all 12 bags):	________________
Signature:	________________

Please be sure that all 12 plastic bags of individual samples are included and are tightly sealed. Place one copy of this record in the shipping box, keep a second copy, and send a third copy to TAS, Inc., in the pre-addressed envelope.

An example of a sample collection and chain-of-custody document is shown above.

VII. LABORATORY ANALYSIS

Details of laboratory analysis of food are discussed in a separate chapter of this book. However, two points regarding laboratory analysis are relevant to the subject of compliance with labeling regulations (i.e., in assuring accuracy, validity, and reliability of the nutrient values declared on the label).

A. Analytical Methods

There are a number of different analytical methods available for measuring the same nutrients in foods. The methods may, however, differ in what nutrient components they actually measure or in the precision of the measurements. When comparing nutrient content of foods it is important that the values be based upon comparable analytical methods.

FDA recommends that analytical methods validated by the Association of Official Analytical Chemists (AOAC) be used for nutrient analysis for labeling; alternative methods are recommended only when AOAC-validated methods are not possible (2). In practice, many of the non-AOAC-approved methods are more efficient and equally valid for measuring nutrient content. The fact that the method has not been validated by AOAC does not make it an inferior method of analysis. However, if a non-AOAC-validated method is used, documentation is critical. In the event that nutrient values are challenged by FDA or that the database incorporating results of these analyses is submitted to FDA for review, details of the analytical procedures and performance characteristics of the method must be described in sufficient detail to enable evaluation of the method.

B. Quality of the Analytical Laboratory

A second factor to consider regarding laboratory analysis is the quality of the laboratory performing the analyses. The laboratory should demonstrate that it operates under a well-documented Quality Assurance program. This offers assurance that materials are properly logged, stored, and prepared; employees are well trained; equipment is calibrated; analyses are conducted according to standard operating procedures, including the use of validation materials; analytical data are checked for errors; and the integrity of the data is preserved.

VIII. STATISTICAL ANALYSIS OF NUTRIENT DATA

The first steps in calculating label values based on the laboratory data are to compute: 1) the average nutrient content of the product for each nutrient, and 2) the variability of that nutrient content.

If a simple sample design with equal probability of selection methods was used throughout the study (selection of sampling points, collection of samples for analysis, and preparation of composites), the statistical analysis uses the standard formulas found in elementary statistics texts. If not, adjustments have to be made at this stage.

A. The Average Nutrient Content: the Arithmetic Mean

The mean for all samples, whether simple or complex, based on equal probabilities of selection is calculated using the following familiar formula:

$$M = \sum \frac{X_i}{n} \qquad \text{(Equation 23.2)}$$

where M is the mean
$\sum X_i$ is the sum of the nutrient contents measured in each analysis X_i, and
n is the number of analyses

Example: 5 analyses of the sodium content per 100 g of a food obtained the values (in mg): 22, 26, 30, 31, and 36.

$$M = \frac{22+26+30+31+36}{5}$$

$$= 29.0 \text{ mg}$$

If unequal probabilities of selection were used, the calculation of the mean is more complicated, and a full discussion is far beyond the scope of this chapter. Briefly, it is necessary to weight each observation (nutrient analysis) by the inverse of its selection probability. The general formula is:

$$M = \frac{\sum (X_i/p_i)}{\sum (1/p_i)} \qquad \text{(Equation 23.3)}$$

where M is the mean
$\sum X_i$ is the sum of the nutrient contents measured in each analysis X_i, and
p_i is the probability of selection of each observation

Example: 5 analyses of the sodium content per 100 g of a food obtained the same values as in the previous example. However, each sample came from a different production line, and the last production line actually has twice the output of each of the others. Therefore, a given unit had only half the selection probability from this last line as did those from the other lines, and the formula becomes:

$$X = \frac{22/1 + 26/1 + 30/1 + 31/1 + 36/0.5}{1/1 + 1/1 + 1/1 + 1/1 + 1/0.5}$$

$$= 30.2 \text{ mg}$$

B. The Variability of Nutrient Content: The Standard Deviation

If a simple sample design with equal probability of selection methods was used throughout the study, the standard deviation is calculated using the following formula:

$$s = \sqrt{\left(\sum \frac{(X_i - M)^2}{n-1}\right)} \qquad \text{(Equation 23.4)}$$

where s is the standard deviation
M is the mean
$\sum (X_i - M)^2$ is the sum of the squared differences between the nutrient contents measured in each analysis X_i and the mean, and
n is the number of analyses

Example: 5 analyses of the sodium content per 100 g of a food obtained the values (in mg) 22, 26, 30, 31, and 36. The mean has already been determined to be 29.0.

$$s = \sqrt{\frac{(22-29)^2 + (26-29)^2 + (30-29)^2 + (31-29)^2 + (36-29)^2}{5-1}}$$

$$= \sqrt{\frac{112}{4}} = 5.29$$

If unequal probabilities of selection were used, estimation of the standard deviation becomes difficult, and the reader is referred to an advanced text on statistics. (If non-probability methods were used, it becomes impossible: there is no unbiased estimator of the standard deviation for a non-probability sample.)

C. Complex Sampling Plans: The Design Effect

If the reader has followed our advice and kept the sampling plan simple, this section may be disregarded. However, if any of the following designs were used, the material in this section cannot be disregarded in determining label values for nutrient content:

- Multistage sampling designs, in which the population units are divided into groups (e.g., geographical regions, stores) and the groups are first sampled, followed by selection of units within the chosen groups

- Clustered sampling designs, in which samples are chosen in groups
- Weighted samples, in which units are selected with unequal probabilities which then require weighting to compensate

The formulas given in the preceding section are strictly correct only for data collected by means of simple random samples. They are also acceptable for samples based on systematic selection such as was described in Section VI.D. However, the use of more complex sampling designs—particularly those that involve clustering or weighting—often imposes costs in terms of what is called a "design effect."

This cost arises because multistage sampling, clustering, and weighting, although economically efficient, are mathematically inefficient. For example, it is less expensive to collect five samples from one supermarket than to collect one sample each from five supermarkets, but it is less effective statistically. There is "less information" in the five samples from one supermarket because they tend to resemble each other more than would samples from different supermarkets, just as neighbors tend to resemble each other more than would randomly selected people. The result is that the "effective sample size" of a complex sample design is usually less than the number of samples actually collected.

The design effect usually results in inflation of the standard deviation as calculated, assuming a simple random sample, and may be expressed as a multiplier. While calculation of the true design effect is beyond the scope of this chapter, a conservative adjustment for designs that might be used to collect product samples is to multiply the calculated standard deviation by 1.25.

IX. DETERMINING LABEL VALUES

Procedures for determining the values to be declared on the nutrition label are described in the regulations and in the *FDA Nutrition Labeling Manual*. Some aspects of the process, such as determining the serving size, are complex and depend upon the product type. Careful review of the regulations pertaining to calculation of serving size and declaration of nutrient content per serving is advised.

The objective of this section is to describe the procedures for translating the mean content and the standard deviation of each nutrient into label values that are in compliance with FDA regulations.

There are six steps in this process:

- Calculate the 95th percentile prediction interval.

- For Class II and Third Group nutrients, select the prediction interval versus the mean.
- Determine the serving size.
- Calculate nutrient content per serving of the product.
- Round nutrient content values following FDA's rules.
- Calculate percent of Daily Values (DVs) for each nutrient as appropriate.

These steps are discussed in the following sections.

A. Calculating the 95th Percentile Prediction Interval

The 95th percentile prediction interval is calculated in order to confidently bracket the result of a hypothetical FDA compliance review. That is to say, by using this interval to determine label values, the food manufacturer assures a probability of 95% that an FDA compliance sample will obtain a value for each nutrient that is within compliance allowances.

If the standard deviation being used in the equation is based on the analysis of composites (composed of 12 individual sample units), then the first equation below (Equation 5) should be used. If the standard deviation is based on the analysis of individual units, then the second equation (Equation 6) is appropriate.

1. The Prediction Interval Around the Mean for Composites

$$PI_{\mathrm{M}} = t_{(0.95/df)}(1 + 1/n)^{1/2}(s) \qquad \text{(Equation 23.5)}$$

where PI_{M} is the prediction interval around the mean
t is the 95th percentile of the t distribution with df = $n-1$ degrees of freedom (see Table 23.2)
n is the number of composites that were analyzed to calculate the mean and standard deviation, and
s is the standard deviation

Example: Based on analyses of 8 composites of a vitamin-C fortified product (each composed of 12 individual units), the mean vitamin C content per 100 g of the product has been found to be 120 mg, with a standard deviation of 3.2. What is the 95th percentile prediction interval around the mean?

$$PI_{\mathrm{M}} = (1.895)(1 + 1/8)^{1/2}(3.2)$$

$$= 6.43$$

Table 23.2 The *t* Distribution at the 95th Percentile

df	t	df	t	df	t
1	6.314	11	1.796	25	1.708
2	2.920	12	1.782	30	1.697
3	2.353	13	1.771	35	1.690
4	2.132	14	1.761	40	1.684
5	2.015	15	1.753	45	1.679
6	1.943	16	1.746	50	1.676
7	1.895	17	1.740	60	1.671
8	1.860	18	1.734	70	1.667
9	1.833	19	1.729	80	1.664
10	1.812	20	1.725	90	1.662

Example: Based on analyses of 8 composites of a product containing indigenous vitamin C (each composite composed of 12 individual units), the mean vitamin C content per 100 g of the product has been found to be 85 mg, with a standard deviation of 16.6. What is the 95th percentile prediction interval around the mean?

$$PI_M = (1.895)(1 + 1/8)^{1/2}(16.6)$$

$$= 33.37$$

2. The Prediction Interval Around the Mean for Single Units

$$PI_M = t_{(0.95/df)}(1/12 + 1/n)^{1/2}(s) \qquad \text{(Equation 23.6)}$$

where PI_M is the prediction interval around the mean
t is the 95th percentile of the t distribution with df = n–1 degrees of freedom (see Table 23.2)
n is the number of composites that were analyzed to calculate the mean and standard deviation, and
s is the standard deviation

Example: Based on analyses of 20 individual units of a product, the mean sodium content per 100 grams of the product has been found to be 65 mg, with a standard deviation of 19.3. What is the 95th percentile prediction interval around the mean?

$$PI_M = (1.729)(1/12 + 1/8)^{1/2}(19.3)$$

$$= 15.23$$

3. Using the PI_M to Calculate the Lower Limit for Class I Nutrients

For Class I nutrients, the nutrient content of the product must be 100% of the declared value. The calculated value for the lower limit of the 95th percentile prediction interval is determined by subtracting the PI_M from the mean:

$$PI_{\text{Class I}} = M - PI_M \qquad \text{(Equation 23.7)}$$

where M is the mean nutrient content based on the analyses, and
PI_M is the prediction interval around the mean

Example: The vitamin C in the first example is an added nutrient, and therefore is subject to Class I requirements. What is the lower limit of the 95th percentile prediction interval?

$$PI_{\text{Class I}} = 120\,\text{mg} - 6.43\text{ mg}$$

$$= 113.57\text{ mg}$$

4. Using the PI_M to Calculate the Lower Limit for Class II Nutrients

As with Class I nutrients, the lower limit of the 95th percentile prediction interval is calculated by subtracting the PI_M from the mean. However, because a 20% margin of error is allowed for Class II nutrients (that is, the actual content must be at least 80% of the declared value), the prediction interval for these nutrients must be adjusted to provide this allowance. This is done by means of the following formula:

$$PI_{\text{ClassII}} = \frac{M - PI_M}{0.80} \qquad \text{(Equation 23.8)}$$

where M is the mean nutrient content based on the analyses, and
PI_M is the prediction interval around the mean

Example: The vitamin C in the second example is an indigenous nutrient, and therefore is subject to Class II requirements. What is the lower limit of the 95th percentile prediction interval?

$$PI_{ClassII} = \frac{85\text{ mg} - 33.37\text{ mg}}{0.80}$$

$$= 64.54\text{ mg}$$

5. Using the PI_M to Calculate the Upper Limit for Third Group Nutrients

With Third Group nutrients, the compliance issue is underdeclaration rather than overdeclaration; consequently, it is the upper limit of the 95th percentile prediction interval that is calculated. This is done by adding the PI_M to the mean. Since a 20% margin of error is allowed for Third Group nutrients (that is, the actual content must not exceed 120% of the declared value), the prediction interval for these nutrients must be adjusted to provide this allowance. This is done by means of the following formula:

$$PI_{3rd\ Group} = \frac{M + PI_M}{1.20} \qquad \text{(Equation 23.9)}$$

where M is the mean nutrient content based on the analyses, and
PI_M is the prediction interval around the mean

Example: The sodium in the third example is a Third Group nutrient. What is the upper limit of the 95th percentile prediction interval?

$$PI_{3rd\ Group} = \frac{65\text{ mg} + 15.23\text{ mg}}{1.20}$$

$$= 66.86\text{ mg}$$

B. Choosing the Mean or the Prediction Interval

It may occasionally happen, if the standard deviation is small, that the adjustment of the prediction interval for a Class II nutrient results in a $PI_{ClassII}$ that exceeds the mean nutrient content. In such cases, the mean must be used rather than the $PI_{ClassII}$.

Likewise, the adjustment of the prediction interval for Third Group nutrients may produce a $PI_{3rd\ Group}$ that is less than the mean content. As with Class II nutrients, the mean must be used rather than the $PI_{3rd\ Group}$ when this occurs.

Example: Suppose that the vitamin C in the first example had been indigenous rather than added (i.e., was Class II rather than Class I). The calculation of the lower limit of the prediction interval would have been:

$$PI_{\text{ClassII}} = \frac{120\text{ mg} - 6.43\text{ mg}}{0.80}$$

$$= 141.96\text{ mg}$$

It would not be acceptable to base the label declaration on the figure of 142 mg; rather, the actual mean content of 120 mg must be used.

C. Converting to Serving Sizes

One of the goals of NLEA was to establish more realistic and standardized serving sizes upon which nutrition labeling would be based. As a result, standardized serving sizes (Reference Amounts Customarily Consumed or RACC) have been defined by FDA for 139 food product categories, including 11 categories of foods specially formulated for children under four years of age. These reference amounts represent the amounts of foods customarily consumed per eating occasion by persons 4 years of age or older. Refer to the regulations for the tables of food categories and reference amounts.

The serving size declared on the new food label must be based upon—though not necessarily identical to—the reference amounts defined by FDA. Guidelines for determining appropriate serving sizes are quite detailed and depend upon the nature of the product. Examples of the product types for which determination of serving size may differ include:

- Foods in single- versus multi-serving containers
- Meal and main dish products
- Foods requiring additional preparation (e.g., mixes and condensed foods)
- Foods packaged in large discreet units that are usually divided for consumption (e.g., pies, pizzas)
- Variety packs
- Products with separately packaged ingredients intended for consumption together (e.g., cheese and cracker snack packages)
- Products that naturally vary in size (e.g., pickles)
- Non-discrete bulk products (e.g., breakfast cereals and snack mixes)
- Imitation or substitute foods

Most reference amounts equate to foods as consumed. Nutrition labeling, on the other hand, must be based upon the food "as packaged." For foods requiring further preparation (e.g., cake mix or condensed soup), the serving size equals the amount of packaged food required to prepare the reference amount of the prepared food. Dual declarations are permissible, i.e., listing the nutrient content of the food "as prepared" in addition to the required information for the food "as packaged."

Serving sizes must be expressed in common household measures (e.g., cups, teaspoons, pieces) followed in parentheses by the equivalent metric measurement. FDA's tables of reference amounts indicate the appropriate measures for each product category.

The reader is advised to review the regulations carefully to determine the appropriate method for calculating serving sizes. Consult Sections 101.9 and 101.12 of the regulations for guidelines on determining serving sizes. Reference amounts and appropriate measures for each of the 139 food categories are presented in the tables in Section 101.12(b). If no reference amount has been established for the product being labeled, refer to Sections 101.12(c) through (f).

D. Calculating the Nutrient Content per Serving

After determining the appropriate serving size for the product, the next step is to calculate the nutrient content per serving. Using the nutrient values selected above (i.e., mean versus prediction interval) as the basis for determining the nutrient content per serving, calculate the value as follows:

- Divide the serving size to be declared on the label by the amount for which actual nutrient content is reported. The result is the *factor* by which actual nutrient values will be multiplied to determine the nutrient content per serving.
- To determine the nutrient content per serving, multiply the nutrient value selected as the basis for labeling (mean versus prediction interval) times the factor.
- Following this procedure, calculate the nutrient content per serving for each nutrient to be declared.

Example: The nutrient values to be used as the basis for labeling are expressed per 100 grams of food. The serving size to be declared on the label is 120 grams. The sodium content is 225 mg per 100 grams of food.

Divide the serving size by the amount upon which nutrient content is based to determine the factor:

$$\frac{120}{100} = 1.2$$

Multiply the sodium content (225 mg/100 g food) times 1.2 to determine the nutrient content per serving:

$$225 \text{ mg} \times 1.2 = 270 \text{ mg per } 120 \text{ g serving}$$

Table 23.3 Rounding Rules and Increments for Nutrition Labeling

Nutrient	Increment Rounding
Calories	< 5 cal—express as zero
	≤ 50 cal—express to nearest 5 cal increment
	> 50 cal—express to nearest 10 cal increment
Calories from fat	< 5 cal—express as zero
	≤ 50 cal—express to nearest 5 cal increment
	> 50 cal—express to nearest 10 cal increment
Calories from saturated fat	< 5 cal—express as zero
	≤ 50 cal—express to nearest 5 cal increment
	> 50 cal—express to nearest 10 cal increment
Total fat	< 0.5 g—express as zero
	< 5 g—express to nearest 0.5 g increment
	≥ 5 g cal—express to nearest 1 g increment
Saturated fat	< 0.5 g—express as zero
	< 5 g—express to nearest 0.5 g increment
	≥ 5 g—express to nearest 1 g increment
Polyunsaturated fat and Mono-unsaturated fat	< 0.5 g—express as zero
	< 5 g—express to nearest 0.5 g increment
	≥ 5 g—express to nearest 1 g increment
Cholesterol	< 2 mg—express as zero
	2–5 mg—express as "less than 5 mg"
	> 5 mg—express to nearest 5 mg increment
Sodium	< 5 mg—express as zero
	5–140 mg—express to nearest 5 mg increment
	> 140 mg—express to nearest 10 mg increment
Potassium	< 0.5 g—express as zero
	5–140 mg—express to nearest 5 mg increment
	> 140 mg—express to nearest 10 mg increment
Total carbohydrate	< 0.5 g—express as zero
	< 1 g—express as "Contains less than 1 g" OR "less than 1 g"
	≥ 1 g—express to nearest 1 g increment
Dietary fiber	< 0.5 g—express as zero
	< 1 g—express as "Contains less than 1 g" OR "less than 1 g"
	≥ 1 g—express to nearest 1 g increment

Table 23-3 (Continued)

Nutrient	Increment Rounding
Soluble and insoluble fiber	< 0.5 g—express as zero < 1 g—express as "Contains less than 1 g" OR "less than 1 g" ≥ 1 g—express to nearest 1 g increment
Sugars	< 0.5 g—express as zero < 1 g—express as "Contains less than 1 g" OR "less than 1 g" ≥ 1 g—express to nearest 1 g increment
Sugar alcohol	< 1 g—express as "Contains less than 1 g" OR "less than 1 g" ≥ 1 g—express to nearest 1 g increment
Other carbohydrates	< 1 g—express as "Contains less than 1 g" OR "less than 1 g" ≥ 1 g—express to nearest 1 g increment
Protein	< 0.5 g—express as zero < 1 g—expresss as "contains less than 1 g" OR "less than 1 g" ≥ 1 g—express to nearest 1 g increment
Vitamins and minerals	< 2% of RDI—may be expressed as: (1) zero (2) an asterisk that refers to statement "Contains less than 2% of the Daily Value of this (these) nutrient (nutrients)" (3) for Vitamin A, C, calcium, iron: statement "Not a significant source of (listing the vitamins or minerals omitted)" ≤ 10% of RDI—exporess to nearest 2% increment > 10% < 50% of RDI—express to nearest 5% increment > 50% of RDI—express to nearest 10% increment
Beta-carotene	≤ 10% of Vitamin A—express to nearest 2% increment > 10%– ≤ 50% of Vitamin A—express to nearest 5% increment > 50% of Vitamin A—express to nearest 10% increment

Source: Ref. 3.

E. Rounding

For greater consistency in labeling, the regulations specify the increments by which nutrient values per serving should be rounded. Rounding rules for each nutrient are summarized in Table 23.3.

Rounding of nutrient values is the final step in determining the amount of the nutrient to be declared on the label. After determining the nutrient content per serving, consult Table 23.3 for the proper increments by which to round the nutrient values.

F. Converting to Percent of the Daily Value

The new label uses the term "Daily Value" or "DV" as the dietary reference. This term comprises two sets of dietary standards:

- Daily Reference Value (DRV) for fat, carbohydrate, dietary fiber, protein, cholesterol, sodium, and potassium.
- Reference Daily Intake (RDI) for vitamins and other minerals.

Daily Values have been established for most nutrients with the exception of mono- and polyunsaturated fats, sugars, sugar alcohols, and other carbohydrates. The Daily Values for fats, protein, carbohydrates, and dietary fiber are based on a 2,000-calorie diet. As of the publication date of this text, the reference amounts for vitamins and minerals are the same as the 1973 U.S. RDAs; they are listed in Table 23.1. These values may be updated in the future.

When % DV values fall between two whole numbers, rounding shall be as follows:

- For values exactly halfway between two whole numbers or higher (e.g., 2.50 to 2.99), the values shall round up (e.g., 3%).
- For values less than halfway between two whole numbers (e.g., 2.01 to 2.49), the values shall round down (e.g., 2%).

The regulations specify that percent of the Daily Value be declared for all nutrients with the exception of calories, protein, and nutrients for which reference amounts have not been established. In addition to percent DV, the amount of the nutrient must be declared for macronutrients, sodium, and potassium.

Note that a declaration of percent Daily Value for protein is optional unless a claim is made. If expressed as percent DV, the actual amount of protein in grams per serving should be multiplied by its amino acid score corrected for protein digestibility. The manufacturer is advised to review the regulations carefully regarding the calculation of percent DV for protein.

Calculation of the percent DV may be based either on the actual nutrient content per serving or on the rounded value, whichever method results in more consistent labeling.

To summarize, the nutrition information that must appear on the label includes:

- Calories. Declare the number of calories only.
- Macronutrients, sodium, and potassium. Declare the amount of the nutrient present (for all) and the percent Daily Value (with the exception of protein, sugars, sugar alcohols, other carbohydrates, and unsaturated fats).
- Vitamins and other minerals. Declare only the percent DV.

X. MAINTAINING RECORDS AND KEEPING UP TO DATE

In conclusion, it must be noted that all of the effort exerted in following a program such as has been described in this chapter is wasted if complete documentation is not retained and kept readily accessible. A file should be maintained for each product describing the basis for the nutrient declarations for that product. The documentation should include:

- product characterization
- extant data sources consulted
- mandatory nutrients for which analyses were not performed, with substantiation of the rationale for non-testing
- decisions made regarding testing and/or declaration of optional nutrients
- the sampling plan, with rationale
- records of all sampling, collection, and shipping
- laboratory reports, including analytical methods and certification of adherence to written Quality Assurance protocols
- statistical analysis and determination of label values
- date(s) on which the product samples were drawn
- copies of all labels

The next to last item in the above list concerns the date of the study. This is crucial information, because the nutrient content of food products does not remain static. Obviously the manufacturer will be aware of changes in formulation or in processing methods. However, change in other factors with the potential to affect the nutrient content of the product may not be under the manufacturer's control, or even within the knowledge of the manufacturer. For example, weather and soil conditions change, new geographic sources of supply are opened, new cultivars or varieties are introduced.

Consequently, for each manufacturer and for each product there is a unique constellation of determinants of nutrient content. In order to assure that the product remains in compliance with FDA regulations, the nutrient data supporting labeling must be kept current with the changes affecting the product's nutrient content. It is our recommendation that the procedures described in this chapter be seen as building a sound foundation for a continuing program of product compliance.

XI. REFERENCES

1. U.S. Department of Agriculture, *Composition of Foods: Raw, Processed, Prepared.* Agriculture Handbook Number 8, Washington, D.C.
2. U.S. Food and Drug Administration, Center for Food Safety and Applied Nutrition, *FDA Nutrition Labeling Manual: A Guide for Developing and Using Databases*, 1993 Edition. U.S. Food and Drug Administration, Washington, D.C.
3. U.S. Food and Drug Administration, Office of Food Labeling, Center for Food Safety and Applied Nutrition. *Food Labeling Questions and Answers.* U.S. Food and Drug Administration, Washington, D.C., 1993.
4. J.A.T. Pennington, Ed. *Bowes and Church's Food Values of Portions Commonly Used*, Sixteenth Edition. J.B. Lippincott Company, Philadelphia, PA, 1994.
5. *Federal Register 58*:3 (1993).

24

Analytical Methods and Databases for Nutrition Labeling

Ralph Shapiro

Nutrition International, Dayton, New Jersey

I. INTRODUCTION

The determination of the nutrient content of a food product for nutrition labeling may be accomplished by chemical analysis, use of databases, or by a combination of both.

The Food and Drug Administration (FDA) and the U.S. Department of Agriculture (USDA) Food Safety and Inspection Service (FSIS) do not stipulate how nutrient values are to be determined nor do the agencies require disclosure of the source of this information. It is solely the responsibility of the manufacturer to be sure the label is accurate within FDA/FSIS compliance parameters. The manufacturer must resolve issues of cost, speed, label accuracy, and reliability of nutrient composition data for NLEA compliance.

The purpose of this chapter is to assist the reader in addressing these issues by providing an overview on chemical methods and the use of databases for gathering nutrient data for labeling.

II. SAMPLING

The generation of nutrient data for NLEA compliance is a complex task comprised of several steps. Of paramount importance is the development of a well-designed and economically feasible sampling plan. This plan must take into account the nutrient categories (Class I Nutrients, Class II Nutrients, and a third group of foods not specifically defined by class) established by the FDA, which form the basis of regulatory compliance. In addition, the sampling plan must take into consideration those factors responsible for variation in the nutrient composition of foods (e.g. nature, environment, production, storage, handling) as well as analytical variation.

A good sampling plan must have:

- Clearly defined objectives (e.g., nutrients to be measured; required accuracy and precision)
- A well-defined "target" population (e.g., type of product, geographic location, season of harvest, sampling point—farm, wholesale, point of purchase)
- A clear sampling frame which provides complete, accurate, and current information about the specified product population and location of sampling points
- Specific sampling methods based on probability sampling such as simple random sampling (SRS), systematic sampling, stratified sampling, and cluster/multistage sampling which can be used alone or in combination
- Sample size determination which takes into account statistical objectives, margin of permissible error, and the risk of exceeding this error
- A system, schedule, and procedures for sample collection and preparation

Details of sampling design and sample collection for label compliance are presented in Chapter 23.

III. ANALYTICAL METHODS

As noted above, any consideration of the chemical analysis of foods for NLEA must account for factors which influence the composition of foods as purchased. In addition, it is well recognized that variability in food analysis can result from sample selection (collection) and preparation procedures, differences and inherent variability in methodologies, operational performance by analysts, data interpretation, and differences in statistical treatment.

While there are no regulations specifying how nutrient values are to be derived, the reader should be aware that for purposes of evaluating compliance, both FDA and USDA rely primarily on official procedures of the Association of Offical Analytical Chemists (AOAC) International or other agency validated procedures. FSIS also recommends use of their manual entitled *USDA Chemistry Laboratory Handbook.*

Designation of a method as official by the AOAC requires that the method perform satisfactorily in interlaboratory collaborative studies under conditions of intended use. Methods that satisfy this requirement are given official final action by AOAC International for inclusion in their analytical manual *Official Methods of Analysis* (16th ed., 1995). In 1993, the AOAC International also published a manual entitled *Methods of Analysis for Nutrition Labeling*. Both publications are available from AOAC International, 1970 Chain Bridge Road, Dept. 0742, McLean, VA 22109-0742.

Although the analytical basis of food regulation is the validation process developed by AOAC, it should be recognized that official methods are frequently not methods of choice in terms of speed, sensitivity, simplicity, or suitability for automation, and often do not reflect the latest advances in laboratory technology and instrumentation.

Recognizing this, FDA and FSIS appear to have allowed flexibility in the selection of analytical methods for nutrition labeling. However, to date, they have not established criteria for selecting non-AOAC procedures and still rely upon AOAC methods for compliance evaluation. The manufacturer, therefore, should be aware that there remains a need for the improvement, validation, and standardization of certain methods now used for generating nutrient composition data.

A. Criteria for Selecting an Analytical Procedure[1]

When selecting an analytical procedure for nutrient analysis, the following criteria should be considered:

applicability	accuracy
specificity	precision
sensitivity	detectability

1. Applicability

The method under consideration should be applicable to the nutrient of interest over the concentration range of that nutrient in a food. It should also satisfy established standards of accuracy, precision, specificity, sensitivity and detectability.

2. Specificity

Specificity is the ability of a method to respond exclusively to the analyte of interest.

3. Sensitivity

Sensitivity refers to the change in response (analytical result) per unit of nutrient concentration.

4. Accuracy

Accuracy refers to the degree of closeness of an analytical result to the true value of the nutrient in the food. As a practical matter, nutrient values determined by chemical methods are generally considered accurate if the difference between the analytical value and the reference nutrient value is between ± 2 standard deviations for analytical error.

5. Precision

Precision refers to the variability of analytical results when a particular analysis is performed repeatedly on an identical food under specific conditions. The AOAC uses the terms *repeatability* and *reproducibility* to describe the variation in validated methods under two different circumstances of replication.

Repeatability describes the agreement between successive results with the same method on identical test samples under the same conditions (technician [analyst], instrument, laboratory, time).

Reproducibility describes the agreement between individual results with the same method on identical test samples under different conditions (technicians, instrumentation, laboratories, or time). The reproducibility variance is a composite quantity determined by the sum of the components associated with interlaboratory error and the intra-laboratory component[2].

It is important that the chosen procedure does not have an error of measurement (standard deviation) that is out of line relative to the actual variability of the product. In general, the precision of a reasonably accurate method should not be more than one third of the standard deviation inherent in the product, if data quality is to be high.

6. Detectability

Detectability refers to the lowest concentration that produces a measurable analytical response. For purposes of nutrition labeling, the limit of quantitation (LOQ) rather than the limit of detection (LOD) is of primary interest.

The LOQ is the level of analyte (nutrient) in the test sample (food) which produces a measurable response sufficient to permit determination of the amount of analyte in the sample at least 95% of the time. Nutrient levels above the LOQ should be measured with sufficient accuracy to satisfy the compliance guidelines.

B. AOAC Recommendations

Many validated analytical methods are available for determining nutrients and food components. Most were developed and validated over the last century largely through the efforts of the AOAC (now AOAC International) in collaboration with academia, trade associations, and various professional groups. In general, the procedures involve wet chemistry, microbiology, and instrumental analysis.

With the advent of NLEA, the AOAC assembled a task force in 1991 to evaluate the applicability and validity of these methods. This section summarizes their findings and outlines with references currently accepted methods applicable to NLEA[1].

1. Major Findings

The AOAC Task Force identified:

- Currently available official methods that satisfy the analytical requirements for nutrition labeling
- Methods requiring revision or modification
- Nutrient/matrix combinations requiring methods development and validation
- Availability and sources of reference materials for methods validation and quality assurance

It evaluated the official methods available for 1080 analyte/matrix combinations and found 88% adequate for nutrition labeling. A list of acceptable methods as well as matrixes needing methods development and validation are presented in Tables A3.1–A3.3 and Table A3.6 (Appendix 3), respectively. The AOAC Task Force also recognized that many of the official methods are not contemporary with the latest available analytical technology and instrumentation, and it recommended validation of more current and sophisticated methods.

One of its major concerns is the use of validated matrix-specific methods for a broad spectrum of products. In such cases, the inappropriate selection of one of several different official methods for a given nutrient or food component could result in perfectly valid analytical data but may reflect the bias of the method rather than the true measure of the nutrient in the food. For example, it is well recognized that certain official methods for total fat may include non-fat components which could result in erroneously high analytical values. Alternatively, other official methods could underestimate the total fat in a food because of the selection of an extraction system inappropriate to the product matrix.

The AOAC Task Force also recognized similar analytical ambiguities for such nutrition label analytes as moisture, dietary fiber, sugars, and carbohydrates. To address this problem, they suggested the establishment of a simple

Table 24.1 Definitions of Selected Analytes for Nutrition Labeling

Total fat (M)[a]:	The sum of fatty acids expressed as triglyceride equivalents
Saturated fat (M):	Those fatty acids that do not contain a double bond
Polyunsaturated fat (V)[b]:	The sum of all fatty acids containing two or more *cis-cis* methylene interrupted double bonds.
Monounsaturated fat (V):	The sum of all fatty acids containing one *cis* double bond.
Sugar (M):	The sum of all free mono- and disaccharides (e.g., glucose, fructose, lactose, and sucrose).
Sugar alcohols (V):	The sum of additional derivatives of reducing sugars (e.g., sorbitol, xylitol, mannitol, malitol.)
Dietary fiber (M):	Polysaccharides and lignin that are resistant to hydrolysis by human alimentary enzymes, e.g., non-starch polysaccharides; resistant starch, lignin.

[a] (M): Mandatory declaration for nutrition labeling
[b] (V): Voluntary declaration for nutrition labeling

specific definition for each of these respective components. This would help to establish a framework for specific methods across a broad spectrum of product matrixes. The FDA and USDA accepted AOAC's recommendation and established unique definitions for total fat, saturated fat, polyunsaturated fat, monounsaturated fat, total sugar, total dietary fiber, and sugar alcohols (Table 24.1). Specific definitions for moisture and complex carbohydrates are not yet established.

To assess the applicability, ruggedness, and validity of a specific analytical method across a broad spectrum of foods, a subcommittee[3] of the AOAC Task Force developed a food matrix organizational system which indicates how foods vary on a chemical basis and consequently how they might react under specified analytical conditions. They proposed a food triangle scheme where foods can be organized according to their relative levels of protein, fat, and carbohydrate, the components known to have the greatest influence on the utility of an analytical method for foods. The triangle is divided into nine sectors which group foods according to their relative proportion of these nutrients (Figure 24.1). This segmentation facilitates an assessment of the impact that protein, fat, or carbohydrates might have on the performance of a method and hence its suitability to the food(s) in question. Details of the method with examples are presented in *Methods of Nutrition Labeling Analysis* (AOAC International, 1993, pp. 115–120).

The AOAC Task Force also expressed concern about the availability of appropriate standards and reference materials for methods development, valida-

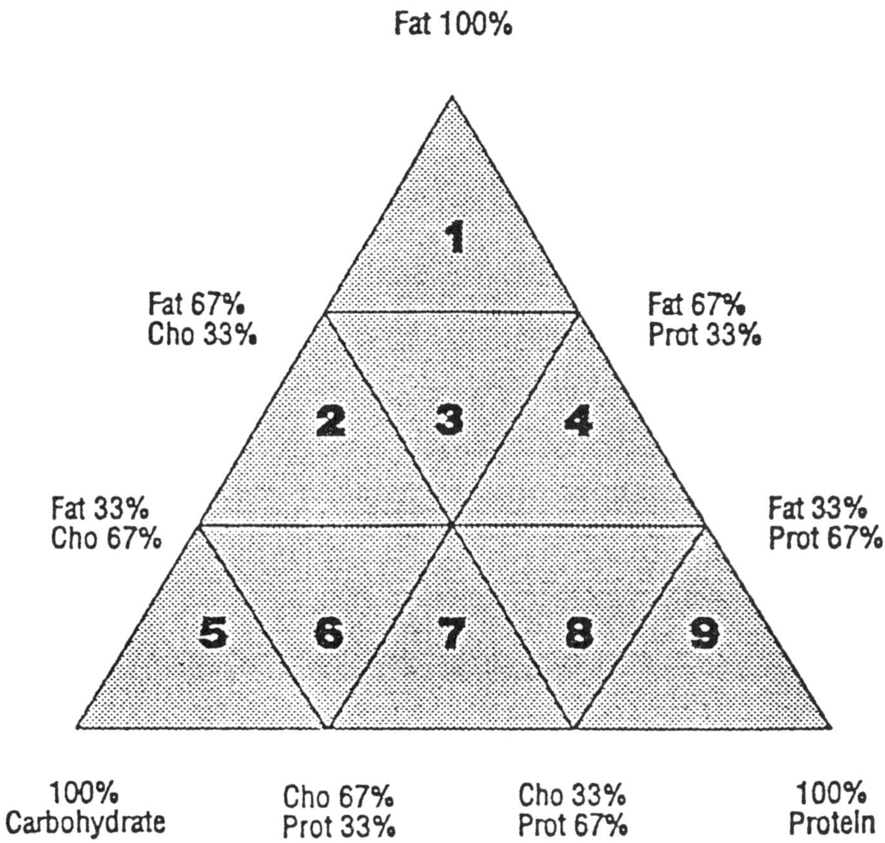

Figure 24.1 Schematic layout of food matrixes suggested for a collaborative study based on protein, fat, and carbohydrate content, excluding moisture and ash. (Reprinted from: *Methods of Analysis for Nutrition Labeling*, AOAC International, 1993.)

tion, and quality assurance. They found that standard reference materials are not available for many of the methods required to support the NLEA and that in the absence of such materials, laboratories use a variety of standards which are not standardized nor certified and are without formalized guidelines for use. To encourage uniformity among laboratories, the AOAC Task Force published a list of certified reference materials available from the National Institute of Standards and Technology (NIST) and from the European Community Bureau of Reference (BCR)[4] as well as a final report on reference materials[5]. Additional information may be obtained from:

- Standard Reference Materials Program, NIST, Gaithersburg, MD 20899; Sales Office; Telephone: 301-975-6766, Fax 301-948-3730
- Community Bureau of Reference (BCR), Commission of the European Communities, Rue de la Loi, 200, B-1049, Brussels, Belgium; Telephone: 32-2-295-3115; Fax: 32-2-295-8072
- The Office of Reference Materials, Laboratory of the Government Chemist, Queens Road, Teddington, Middlesex TW11 OLY, UK; Telephone: 44-81-943-7567, Fax: 44-81-943-2767
- Promochem GMBH, Postfach 1246, D 4230 Wesel, Germany; Telephone: 49-281-530081, Fax: 49-281-89993.

C. AOAC Methods

A list of AOAC approved methods for analyses related to nutrition labeling is presented in Tables A3.1–A3.3 (Appendix 3). The following are some of the salient features and precautions associated with these analyses. The first section discusses sample preparation, the second reviews the procedures for the mandatory nutrients, and the third briefly describes the procedures for the voluntary nutrients.

1. Sample Preparation

Prior to the initiation of any analysis, the matrix to be analyzed must be uniform to ensure representative sampling. Sample preparation and sample homogenicity is essential for accurately determining the nutrient content of a food and for precision. The analytical result is only as good as the sample taken.

At the outset, homogenicity and sample size criteria should be established for each food matrix, and blending procedures should be specified for compositing. After blending, a sufficient amount of sample should be taken to be sure that it is representative of the entire batch, to provide detectable amounts of all analytes, and to allow for repeat analysis if necessary.

In some cases, a food may already be homogeneous (e.g., clear liquids, flour, powders, etc.) and will not require special preparation prior to analysis. For other products, uniformity can be achieved through grinding, milling, homogenization and blending. In some instances, screening may be desired to obtain uniform particle size. At all times, care must be taken to minimize heat generation, exposure to ultraviolet light, and oxidation. All sample preparation equipment must be absolutely clean to avoid cross contamination. After preparation, samples that may be subject to rancidity, microbial activity or other forms of degradation should be refrigerated or frozen; if it is frozen, the entire sample must be thawed and blended before taking samples for analysis. All samples should be thoroughly mixed immediately before being distributed for analysis. Samples should be analyzed as soon as possible after preparation.

2. Mandatory Nutrients

a. Calories

There are four options authorized by the FDA for determining the caloric content of foods.

- Bomb calorimetry with a correction for incomplete protein digestibility by subtracting 1.25 calories per gram protein
- Calculation using the general factors of 4, 4, and 9 calories per gram for protein, total carbohydrate, and total fat, respectively.
- Calculation using the general factors of 4, 4, and 9 calories per gram protein, total carbohydrate less the amount of insoluble dietary fiber, and total fat, respectively
- Calculation using specific Atwater food factors.

Clearly, the accurate determination of the caloric content of food is co-dependent upon protein analysis if bomb calorimetry is the chosen method, and upon protein, fat, and carbohydrate determination if calculations are used.

In the latter case, carbohydrate is generally calculated from a proximate analysis where the sum of protein, fat, moisture, and ash is subtracted from 100%. If applicable, insoluble dietary fiber is then subtracted from the resultant carbohydrate value to obtain digestible carbohydrate. The appropriate general factors of 4, 4, and 9, or Atwater factors, are then applied to obtain the caloric content of the food. Because these components are essential to all caloric calculations, they must be analyzed with extreme care to minimize any cumulative error associated with each analysis. Furthermore, the selection of the proper method is critical to ensure that it is a true measure of the nutrient in the food and not a reflection of the bias of the method.

As previously noted, discrepancies in results as a function of methodology for fat, moisture, and fiber are well known and can be minimized by following the matrix-specific, validated AOAC procedures described in Table A3.3 (Appendix 3) and by using the applicable procedure for the FDA definition of fat (e.g., fatty acid composition). In the absence of a validated procedure, due to a matrix difference, the analyst should objectively select a procedure for a matrix closely representing the product to be analyzed.

b. Proximate Analysis

Protein. As previously noted, a proximate analysis is made up of protein, moisture, ash, total fat, dietary fiber, and total and "other" carbohydrate. Protein is most commonly analyzed by the Kjeldahl method which measures the amount of nitrogen in the food. This procedure involves digestion with sulfuric acid in the presence of a catalyst; subsequent alkalinization releases ammonia which is trapped in acid solution and measured titrimetrically or by colorimeter. The nitrogen concentration is converted to protein, generally using a factor of 6.25

(e.g., protein = N × 6.25). This factor is the reciprocal of 16%, which is the typical amount of amino nitrogen in most proteins. Alternatively, specific factors may be used for those foods which have considerably more (cereals) or less (dairy products) nitrogen. These factors are listed in *Agriculture Handbook No. 8 Series.*

Other AOAC validated procedures for protein analysis include combustion methods for meat products and animal feeds and dye binding and infrared spectroscopy for dairy products[6].

Moisture. Most methods for determining the moisture content of foods use either convection, forced air, or vacuum ovens. The temperatures used depend on the product matrix and fall into three general categories: 60°C for products high in sugar, 70°C for vegetables, and 100–135°C for other foods[7]. The length of drying will depend on the temperature and type of oven used.

It should be noted that these methods measure volatiles rather than water per se. Therefore, it is imperative to use specifically validated AOAC procedures to minimize this potential discrepancy and ensure comparable results. These are listed in Table A3.2 (Appendix 3).

Other AOAC approved methods with limited applications include direct moisture determination (by Karl Fisher titration or by toluene distillation), conductivity, microwave oven drying, and infrared spectroscopy. Direct moisture determination is a true measure of moisture and eliminates the potential errors associated with measuring residual solids or total volatiles.

Ash. The ash content of foods is determined gravimetrically after ignition in a muffle furnace at 525–550°C or higher for certain products, to obtain white ash. Detailed procedures for specific food matrixes are described in *Methods of Analysis for Nutrition Labeling* (AOAC International, 1993, pp. 139–149).

Fat. For the purpose of nutrition labeling, fat is defined as the sum of fatty acids expressed as triglycerides. To date, there are no official AOAC methods that satisfy this definition, but there are a number of official methods that approximate it; still other methods are undergoing collaborative study. Because there are multiple official procedures for fat extraction and determination, selection of the appropriate procedure for a specific matrix is critical. Since the results of fat analyses are dependent on the method used, misleading overestimates or underestimates of the true fat content of a product could be unwittingly (or wittingly) obtained. Of all the analyses required under NLEA, fat poses the most serious problems. Further methods development, validation and standardization are essential. As with moisture determination, to achieve standardization and obtain accurate results, the analyst must be particularly attentive to the method selected for each product matrix. In the absence of a validated procedure for a particular food, a validated procedure for a similar matrix should be chosen to minimize any methods related bias.

Several factors may influence the outcome of fat analyses and should be considered when selecting a suitable method.

- *Particle size*: Fat extraction is frequently improved when the particle size of the matrix is reduced. This could be due to greater surface exposure to the solvent system, thereby leading to more complete extraction.
- *Fat encapsulation*: The use of processing aids and encapsulation per se may render the fat component of a food unavailable to solvent exposure. Such encapsulation usually hinders fat extraction.
- *Moisture content*: High-moisture products often complicate fat extraction and thus necessitate pre-extraction drying to ensure optimum and quantifiable separation of fat.
- *Processing*: Various processing procedures and the use of gums sometimes cause the binding of fats rendering them more difficult to separate.
- *Water solubility*: The presence of emulsifiers in the product may hinder the physical separation of fat from proteins, carbohydrate, and other water-soluble matrix components.

To facilitate analyses, the analyst should be aware of the moisture content and particle size of the product being analyzed, and be informed whenever possible of the use of encapsulation, emulsification, or other special processing procedures.

Pretreatment. Several procedures may be used to make the fat component of the product matrix more readily available for solvent extraction. The pretreatment system selected should be based on the estimated composition of the product: acid hydrolysis breaks emulsions and/or hydrolyzes complex lipids, alkali digestion solubilizes proteins, enzymes hydrolyze proteins and polysaccharides. Products high in sugar should be washed with water to facilitate subsequent solvent extraction. In some instances, pretreatment may not be required and the matrix may undergo immediate solvent extraction. The AOAC enumerates specific pretreatment procedures for various product types.

Generally, increasing the duration and vigor of pretreatment will enhance the availability of the fat component of the matrix to solvent extraction.

Extraction. Selection of the appropriate solvent system is critical to effective and complete fat extraction. In some cases, because of matrix composition, multiple solvent systems and combinations must be used to achieve full separation.

Proper solvent selection is difficult because fats and lipids vary considerably in their structure and polarity. More polar solvents such as chloroform-methanol, extract the more polar lipids in addition to fat. These include phospholipids, sterols, terpenes, waxes, and other non-lipid materials. In general, anhydrous diethyl ether, petroleum ether, chloroform-methanol, and tetrachloroethylene are used to extract triglycerides; ether is used to extract simple glycerides, sterols, and fatty acids.

Increasing the duration and vigor of extraction generally will increase the effectiveness and completeness of fat separation. However, as with pretreatment, in attempting to achieve complete separation of pure fat from its matrix, one must be mindful that excessively vigorous procedures could cause higher than real values due to concurrent removal of other constituents. This can be minimized by appropriate re-extraction and washing to remove any non-fat components.

Most fat determinations are performed gravimetrically. The most commonly used procedures are:

- Chloroform-methanol extraction
- Ether extraction
- Acid hydrolysis
- Alkali treatment (Rose-Gotlieb and Majonnier methods)
- Volumetric analysis (Babcock method)

Table A3.3 (Appendix 3) lists the AOAC official methods for total fat determination in specific matrixes.

Chloroform-Methanol. This procedure is frequently employed with composite foods and foods for which methods are not available. It determines lipids, not fats. The method is not recommended for baked goods because it does not extract all of the lipid material from some of these products.

Because chloroform is mutagenic and carcinogenic, appropriate safety precautions must be employed during its use; the hazardous-waste costs associated with its disposition also must be considered. Table A3.3 (Appendix 3) lists the product matrixes for which this method is applicable.

Ether Extraction. This procedure (also known as the soxhlet method) is typically used with animal feeds, raw cereal grains, meats, nuts, butter and spices. The method requires continuous refluxing of the sample with diethyl or petroleum ether for 12–78 h. (The use of a commercially available Soxtec system substantially reduces the reflux time.) The solvent is then evaporated and the residual fat is weighed. This method extracts mono-, di-, and triglycerides, most sterols and glycolipids. It has three major limitations:

- It underestimates fat in some baked goods.
- Products high in water-soluble carbohydrates, certain sugars, glycol, and lactic acid are believed to interfere with the extraction.
- All of the fatty acids are not always extracted which could result in erroneously low fat values for certain products. The AOAC provides recommendations for dealing with these limitations as well as precautions when using the Soxtec system.

Acid Hydrolysis. These methods are used for determination of fat in flour, baked products (including fruit-filled baked products), egg, macaroni, salad dressing, processed cheese, and various other food products. The procedure involves acid hydrolysis followed by ether extraction. The product is digested with hydrochloric acid and the released fat is then extracted with a mixture of diethyl and petroleum ether. The solvents are evaporated and the fat residue is weighed.

The limitation of this procedure is that in addition to removing fat it also extracts non-fat materials such as glycerol, low molecular weight carbo- hydrates and their polymerization products, amino acids, and urea salts. As a result, one may obtain erroneously high fat values. The AOAC provides suggestions and modifications for addressing this problem.

Alkaline Treatment. These procedures are recommended for fat determination of dairy products, milk-based products and milk-based infant formulas. The procedures require alkaline treatment of the sample, usually with ammonium hydroxide, followed by ether extraction. The ammonium hydroxide cracks the emulsion, and neutralizes and solubilizes the protein prior to ether extraction. The Rose-Gottleib procedure measures petroleum ether soluble fat. The Mojonnier method measures the total fat extracted by a mixture of diethyl ether and petroleum ether. Both methods allow for efficient separation of mono-, di-, and triglycerides as well as traces of other lipids from dairy products.

Volumetric Analysis. This is a rapid procedure that is used for many dairy products. The procedure uses strong sulfuric acid to release emulsified fat in a calibrated Babcock bottle. The low-density lipid rises and is measured volumetrically. Acceptable values (although generally slightly higher than obtained by Mojonnier) are obtained. It is suggested that this procedure be used for monitoring in the plant and that more accurate and precise methodology (e.g., Rose-Gottleib or Mojonnier methods) be used for nutrition labeling purposes.

c. Fatty Acids

There are three critical steps to fatty acid chromatographic separation: extraction, derivatization, and chromatography.

Extraction. Obtaining a total lipid extract is often difficult and as previously noted, depends on the product matrix, the solvent system selected, and the procedure used. In the case of fatty acid determination, the extraction procedure recommended by Sheppard[8] has been widely used.

Derivatization. After extraction, the AOAC procedure for methylation to derive fatty acid methyl esters is followed.

Chromatography. To date, an official AOAC procedure for determining fatty acids from C4 to C24 is not available. The Sheppard procedure uses packed-column gas chromatography (GC) and is satisfactory for fatty acids with a chain length of C8 to C24. However, shorter-chain fatty acids may also be reported under NLEA. Several unofficial gas chromatographic procedure using capillary columns have been developed and appear to be satisfactory for measuring C4 to C24 fatty acids. The American Oil Chemists Society (AOCS) has several capillary GC procedures that have received wide acceptance. These include their procedure Celc-89 that measures *cis-cis* and *trans* isomers, and has the capacity to detect short-chain fatty acids. Several procedures are presently undergoing collaborative evaluation and may become official in the near future.

d. Cholesterol

This procedure requires extraction of the food matrix with a mixed solvent system followed by saponification. The derivatized cholesterol is then analyzed by packed-column GC. The procedure has a sensitivity of 1 mg/100g.

e. Dietary Fiber

For the purposes of nutrition labeling, dietary fiber is defined as polysaccharides and lignin that are resistant to hydrolysis by human alimentary enzymes. Based on this definition, dietary fiber includes non-starch polysaccharides, resistant starch, and lignin. A distinction between soluble and insoluble dietary fiber is made based on the solubility characteristics of dietary fiber in hot aqueous buffer solutions.

While there are several approaches to dietary fiber analysis [e.g., enzymatic-gravimetric, non-enzymatic-gravimetric, and enzymatic-chemical including gas-liquid chromatography (GLC), high pressure liquid chromatography (HPLC), and colorimetric analysis] the methods currently considered most suitable for nutrition labeling are the enzymatic-gravimetric procedures developed by Prosky et al. at the FDA. They are official AOAC methods and allow for the determination of total dietary fiber (TDF), soluble dietary fiber (SDF) and insoluble dietary fiber (IDF). The methods have provisions for drying high-moisture foods, defatting foods with more than 10% fat, and removing sugar from products containing 50% or more sugar. Duplicate pretreated samples are then subjected to sequential enzymatic digestion with heat-stable amylase, protease, and amyloglucosidase to remove starch and protein. For TDF, the enzymatic digest is treated with ethanol, followed by filtration and a subsequent wash with acetone. The residue is then dried and weighed. For IDF, the digest is filtered and the residue is washed with warm water, dried, and weighed. For SDF, the filtrate and washes are combined, the SDF is precipitated with ethanol, filtered, dried, and weighed. Each residue value is corrected for protein, ash, and blank prior to final calculation.

Limitations of the method include: filtration difficulties with desugared fruits and mucilaginous products; possible errors introduced during the pretreatment procedures (drying, defatting, and desugaring) incomplete removal of protein by protease); and the amount of time and labor required.

Another TDF gravimetric procedure has been approved by the AOAC recently. It combines the neutral detergent procedure with a separate determination of autoclave-induced hot water-soluble dietary fiber to derive TDF. The procedure is not applicable to SDF or IDF determinations. This method, as well as non-enzymatic-gravimetric and enzymatic-chemical methods, are outlined in *Methods for Nutrition Labeling* (AOAC International, 1993).

f. Vitamin A

Analysis of vitamin A in foods involves the determination of natural as well as supplemental vitamin A. As with all the fat-soluble vitamins, vitamin A is not stable in its pure form and is subject to destruction by air, light, acids, and trace minerals. Consequently, vitamin A fortified foods generally contain encapsulated, coated, or otherwise protected forms of the vitamin. Oftentimes, these protected forms interfere with and complicate analysis. In the absence of any protection, vitamin A is subject to oxidation and isomerization, and appropriate precautions must be taken during extraction and analysis.

The most commonly used AOAC-approved procedure is the colorimetric Carr-Price method[9]. In this procedure, vitamin A is removed from its food matrix by saponification and extraction, and then treated with antimony trichloride to yield an intense blue color. The method may be used for a variety of matrices including baby foods[10], dairy products, fish, meats, and mixed dinners.

This method has several limitations:

- Color development is inhibited by sterols and tocopherol.
- The intense blue color fades rapidly.
- Antimony trichloride may cause turbidity and an opaque coating on the colorimeter tube in the presence of trace amounts of water.
- The procedure cannot be used when the sample contains carotenes as the main source of activity (e.g., high-potency vitamin A concentrates).
- The method does not distinguish between *cis* (low biological activity) and *trans* (high biological activity) isomers.
- Carotenoids, retinaldehyde, and dehydroretinol cause interference.

For nutrition labeling, vitamin A potency is expressed in International Units (IU). One IU of vitamin A is defined as the activity of: 0.003 μg of pure *trans* retinol; 0.344 μg of pure *trans* retinyl acetate; 0.55 μg of pure *trans* retinyl palmitate, or 0.6 μg of pure *trans* beta carotene.

The vitamin A content of foods may also be expressed in terms of retinol equivalents (RE). One RE of vitamin A is defined as 1 μg of pure *trans* retinol (3.33 IU vitamin A activity) or 6 μg of pure *trans* beta carotene (10 IU vitamin A activity).

g. Beta Carotene

Beta carotene is an important source of vitamin A activity and it is naturally present in a broad range of foods. It is usually analyzed separately from vitamin A. The official AOAC beta carotene procedure recommended for NLEA compliance is a spectrophometric method. It is suitable for determining naturally occurring beta carotene in fruits, vegetables and baby foods. The method is not applicable to the determination of beta carotene fortified foods. At present, an official AOAC procedure is not available for this purpose. However, several unofficial liquid chromatographic procedures have been developed for carotenoid determination. These are outlined in *Methods of Analysis for Nutrition Labeling* (AOAC International, 1993).

Like vitamin A, beta carotene is sensitive to air, light, acids, free halogens, and heat. Therefore, extreme care must be taken during extraction from the food matrix as well as during analysis.

h. Vitamin C

Vitamin C has two biologically active forms: ascorbic acid and dehydroascorbic acid. The official AOAC methods used for nutrition labeling are either colorimetric or fluorometric. In the colorimetric procedure, ascorbic acid is titrated with 1,2 dichloroindophenol to yield a light pink color. This procedure only measures ascorbic acid. Consequently, foods containing dehydroascorbic acid will be underestimated. Other limitations include:

- Highly colored foods may mask the titration end point.
- The presence of reducing agents and other product ingredients that are readily susceptible to oxidation may cause unrealistically high values.

In the fluorometric method, ascorbic acid is oxidized to dehydroascorbic acid which is then reacted with phenylene diamine to yield a fluorescent adduct. This is generally the preferred and most frequently used method. By eliminating the oxidation step, the analyst can directly determine the amount of dehydroascorbic acid in the product. Ascorbic acid can then be calculated by subtracting this value from the total value obtained after oxidation.

Because vitamin C is a strong antioxidant, it is very rapidly oxidized. Oxidation frequently occurs almost immediately upon extraction, the rate depending upon the concentration in solution, the presence of metals, and exposure to air and dissolved oxygen. While it is difficult to eliminate oxidation, it can be minimized by rapid and efficient handling of the sample, pH adjustment, and the use of appropriate reagents (e.g., metaphosphoric acid) to chelate metal ions.

i. Calcium

Several official AOAC procedures are available for determining the calcium content of foods. These include titration, atomic absorption spectroscopy (AAS), inductively coupled plasma (ICP), and emission spectroscopy (ES). Common to each of these methods is wet- or dry-ashing of the sample matrix, subsequent dissolution in acid followed by titration or injection into a spectrophotometric system. The atomic absorption method is most commonly used for single element analysis and ICP is used for simultaneous determination of several metals.

j. Iron

Several official AOAC procedures are available for determining the iron content of foods. These include AAS, ICP, and ES. Official spectrophotometric procedures are also available for analyzing iron. AAS and ICP are currently the most widely-used procedures. An important precaution is to control the temperature of the muffle furnace during ashing to ensure complete ashing without volatization which could cause low results. In general, ashing should be done at 550°F.

k. Sodium

Several official AOAC methods are available for the analysis of sodium in foods. These include AAS, ICP, ES, flame photometry, and specific ion electrode. To minimize analytical variation, especially when analyzing foods with very low sodium content, the samples should be dry-ashed in platinum crucibles or wet-ashed in ultra-pure acid prior to analysis.

l. Sugars

Several AOAC approved procedures are available for the determination of sugars. These include colorimetric, GLC, and HPLC methods. Table A3.4 (Appendix 3) provides a list of recommended HPLC methods and Table A3.5 (Appendix 3) provides a list of recommended GLC, colorimetric, polarimetric, infrared, and gravimetric methods. The method selected depends on the anticipated composition of the sugars in the food and potential interferences. HPLC procedures are considered most suitable and are most widely used for nutrition labeling. They have broad applicability, and allow complete separation and quantitation of mono- and disaccharides. Sample pretreatment is usually minimal and interpretation of the chromatograms is straightforward. Most HPLC systems use a refractive index (RI) detector for sugar analysis. The most frequently used system employs an amino-silica column and an acetonitrile-water (80:20) mobile phase. The elution sequence corresponds to the polarity of the sugar (e.g., monosaccharides followed by sugar alcohols, disaccharides, and higher oligosaccharides). The primary limitation of the method is interference from salt because chloride ions elute soon after glucose. The AOAC provides measures for eliminating this potential problem.

The use of GLC for sugar analysis has been very limited probably because of the low volatility of the polar carbohydrates. The AOAC-GLC procedure calls for derivatization with trimethylsilys (TMS) and use of a flame ionization detector (FID). The use of FID is more sensitive than the RI used with HPLC, and the GLC procedure may be more useful for determining the sugar content of complex samples when there is a limited amount of test product and/or its sugar content is very low.

Colorimetric procedures are nonspecific. They detect a class of sugars such as reducing sugars rather than individual sugars. They may be satisfactory for products containing a single, known sugar and when interference is not likely. They are not recommended for products containing a mixture of sugars.

Most AOAC colorimetric procedures are based on copper reduction. The Lane-Eynon volumetric method and the Munson-Walker titrimetric method are most frequently used. Both are AOAC approved. In the Lane-Eynon method, the sugar solution is titrated against a modified Fehling reagent. This method is often used for determination of glucose, lactose, and inverted sugar in various food matrices. In the Munson-Walker method, a fixed volume of solution is heated under standard conditions with Fehlings reagent and the cupric oxide that is formed is filtered off. The cupric oxide precipitate is either weighed, titrated with permanganate or thiosulphate, or measured electrolytically.

Another AOAC colorimetric procedure utilizes ferricyanide reduction. This procedure is frequently used to determine sugars in flour. It is also used in many automated reducing sugar analyses. The reduction reaction occurs in alkali and the reduction is measured iodometrically or colorimetrically.

3. Voluntary Nutrients

a. Water-Soluble Vitamins

Most of the voluntary water-soluble vitamins are analyzed microbiologically. The following organisms are typically used:

Riboflavin	*Lactobacillus casei*
Niacin	*Lactobacillus plantarum*
Pantothenic acid	*Lactobacillus plantarum*
Biotin	*Lactobacillus plantarum*
Vitamin B_6	*Saccharomyces uvarum*
Folic acid	*Streptococcus faecalis*
Vitamin B_{12}	*Lactobacillus leichmannii*

Because they occur over such a wide range of concentrations in foods and often at very low levels, it is sometimes difficult to analyze for these nutrients. Without an estimate of the amount of vitamin in a sample, multiple dilutions must be prepared and assayed over a broad concentration range. This is cumbersome, time consuming, and may compromise the sensitivity of the method, which becomes especially critical for foods containing low levels of these vitamins. Although alternative HPLC procedures are under development, they have not undergone complete collaborative study and are not officially approved.

Most of the microbiological procedures for water-soluble vitamins are similar. Because many of these vitamins are sensitive to UV light, heat, and oxidation, special precautions must be taken to avoid such exposure.

Each assay requires sample pretreatment with acid or alkali sometimes accompanied by enzymatic digestion to separate the vitamin from its matrix. This is generally followed by pH adjustments and purification steps to remove interfering materials. The digested (extracted) and purified sample is then subjected to microbiological assay. Thiamine, riboflavin, and niacin, may also be analyzed by alternate official AOAC chemical procedures. In the microbiological assay, serial dilutions of the sample are incubated with a microorganism having a requirement for the vitamin in question. Growth is measured in a medium that is void in that vitamin except for the amount provided by the sample and is compared to the growth achieved in the same medium supplemented with serial concentrations of vitamin standard. The amount of vitamin present in the test sample is then calculated from a comparison of the two growth curves. Turbidimetric or titrimetric procedures are used to measure growth. Table A3.1 (Appendix 3) lists the AOAC approved procedures for the analysis of the voluntary water-soluble vitamins.

b. Minerals

Most of the voluntary minerals (copper, magnesium, phosphorus, potassium and zinc) can be determined by instrumental analysis utilizing AAS, ICP, or ES after wet- or dry-ashing. Potassium may also be assayed by flame photometry. Official AOAC colorimetric methods are available for copper, zinc, and phosphorus, but instrumental analysis is simpler and more sensitive. Because certain minerals (e.g., copper, zinc) may occur at very low concentrations, special care must be taken to avoid contamination. The analyst should also take appropriate precautions, especially during dry-ashing to avoid losses due to volatilization. In general, the muffle furnace temperature should be kept below 525°C. Table A3.1 (Appendix 3) lists the AOAC-approved methods for the voluntary minerals.

D. Selecting a Laboratory

Selection of a well-qualified, reliable laboratory is critical to the successful and economic completion of an analytical program for nutrition labeling compliance. A number of factors should be considered when choosing a laboratory. The laboratory should have extensive experience in food analysis and, in particular, nutrient analysis. It should be staffed with qualified personnel with strong scientific credentials and solid training in analytical chemistry, food science, microbiology, computer science, and statistics. The laboratory should be able to design sampling plans for full nutrition labeling compliance. Its staff should be able to advise clients on the most economic means for achieving representative and reliable results for the product(s) undergoing analysis. An awareness of the variables that will influence the nutrient composition of foods and the values reported on the nutrition label is essential.

The methods typically employed by the laboratory should be evaluated. A laboratory providing analytical data for nutrition labeling should have an intimate working knowledge of all AOAC procedures as well as alternative procedures which are considered acceptable by FDA and USDA. Knowledge of the limits of methods, their accuracy, precision, and applicability to product-specific matrixes is critical. Many foods have unique properties and matrixes which may cause analytical difficulties and sometimes misleading results. The laboratory personnel should be able to anticipate and identify the difficulties, have strong problem solving capabilities, and be able to advise clients on any analytical limitations that may exist.

The laboratory should have ongoing internal and external quality assurance and quality control programs. Documentation should be available which reflects the laboratorys' ability to accurately reproduce its results on a regularly scheduled basis and over the course of time. Analytical results should be systematically evaluated against available AOAC-recommended reference standards. Standard procedures for check samples and recoveries should be in place for regularly evaluating accuracy and precision. As an external measure of performance, the laboratory should participate in accredited check-sample and round-robin studies (e.g., USDA, AACC, AOCS check-sample programs) and, when possible, should encourage such programs with their clients as well. Participation in methods development with other laboratories and participation in AOAC collaborative studies is an indicator of the laboratory's awareness of current methodology and willingness to compare its performance to other laboratories.

The laboratory facilities should be spacious, temperature controlled, and designed for segregated analysis of vitamins, minerals, proximates, and other nutrient food components. It should have adequate refrigerator and freezer storage space, and it should have special areas for working with nutrients sensitive to UV light.

A laboratory information management system (LIMS) should be in place to provide chain-of-custody documentation and to track samples at all points during the analytical cycle. Electronic data transfer should be available and the laboratory should have the capacity to adapt its computer system(s) to the clients' needs. The laboratory should also be able to transfer its analytical results into a client specific database and should be able to maintain, monitor, and expand it over time.

Close client communication and timely responsiveness to their needs is essential. The laboratory personnel should have a full understanding of NLEA regulations and label format requirements. Some products may not require analysis of all the mandatory nutrients, and some may benefit from declaration of voluntary nutrients. An intimate knowledge of food composition and an awareness of format requirements could save the manufacturer considerable sums of money.

The laboratory should have the capacity to construct a nutrition label from analytical data in a format that will be in full compliance with the regulations. It should have close liaison with FDA and USDA should product specific questions, particularly those related to serving size and nutrient content claims, arise.

In addition to personalized service, the laboratory should be evaluated on its ability to rapidly provide quality results at competitive prices (see Table 24.2 for estimated analytical costs). A history of turnover times (e.g., speed with which analysis are performed and reported) coupled with an assessment of work-load capacity will indicate the laboratorys' ability to satisfy client needs. In all cases, the prospective user should investigate the reputation of the laboratory and obtain references before initiating an analytical program.

IV. DATABASES

A. Introduction

Computerized database systems can be a powerful tool for developing nutrition information panels to comply with NLEA regulations. When used alone or in combination with laboratory analyses, they have the potential to facilitate the label generation process and reduce the cost. This section will:

- Briefly trace the historical evolution of databases.
- Define and describe different types of databases
- Identify sources of databases and computer systems
- Provide practical guidelines for their effective use
- Review the criteria that should be considered when selecting a database system
- Outline FDA's policies with respect to their use

B. Background

1. History

The history of database development can be traced over 100 years (1875) when Wilbur Atwater, at the Chemistry Laboratories of Weslyan University in Connecticut, established the first USDA Experiment Station for food analysis. Atwater and his staff performed proximate analysis (e.g., protein, fat moisture, ash, fiber, and calories) on a broad spectrum of foods. In 1892, his work was published in USDA Bulletin 28 entitled *The Chemical Composition of American*

Table 24.2 Estimated Price Range for Nutrition Label Analysis[a]

Analysis	High ($)	Low ($)
Mandatory		
Proximate[b]	70	34
Saturated fat	115	85
Cholesterol	100	85
Total dietary fiber	125	80
Protein	23	10
Sodium	26	9
Total sugars	90	55
Vitamin A	95	56
Vitamin C	80	32
Calcium	26	9
Iron	26	9
NLEA package	752	395
Voluntary		
Polyunsaturated fatty acid	143	95
Monounsaturated fatty acid	143	85
Soluble and insoluble fiber	175	75
Sugar alcohol	138	60
Vitamin B_1	55	32
Vitamin B_2	55	32
Niacin	65	60
Vitamin B_6	70	65
Vitamin B_{12}	110	65
Folic acid	120	65
Biotin	110	65
Pantothenic acid	120	65
Vitamin D	180	160
Vitamin E	180	70
Copper	30	14
Magnesium	26	14
Manganese	32	14
Molybdenum	46	14
Phosphorus	40	14
Potassium	24	14
Selenium	75	30
Zinc	40	14

[a] Values obtained from U.S. contract laboratory price schedules, 1993–1994.
[b] Proximate analysis includes: moisture, protein, fat, ash, carbohydrate.

Food Materials. It contained values for the proximate food components as well as caloric (fuel) values. As an outgrowth of more than 100 experiments conducted over a span of 8 years, he developed factors for calculating the calorie content of foods; today, these are known as the "Atwater" factors.

At the turn of the century, the first nutrition survey was conducted on troops drafted for World War I. The survey was initiated to identify the cause of malnutrition among draftees. Food composition data was gathered which eventually led to nutrient enrichment to correct the observed deficiencies. In 1963, USDA published *Handbook No. 8* (the "red book"), which was the first extensive book providing nutrient composition data on foods. Subsequently, USDA published *Handbook 456* and its subsection, both of which provided more information on the nutrient composition of foods. Today, the most widely used food composition source is the *USDA Handbook No. 8 Series*. A list of Agriculture *Handbook No. 8 Series* publications with availability and pricing is provided in Table A3.7 (Appendix 3). Coupled with sophisticated computer technology, the capacity to effectively use nutrient composition data has been substantially advanced.

2. Current Interest

Two major trends in the food industry have increased interest in the use of nutrient databases.

a. Healthier diets

For over a decade, there has been a trend in the United States toward healthier eating. Consumers have become more concerned about the foods they eat and aware of the impact of nutrition and diet on disease prevention and longevity. In response, the food industry has been developing healthier foods at a rate never before experienced. The competition for shelf space is intense and companies are striving to formalize and compress product development cycles. It is estimated that product development time in the food industry has been reduced by at least 30% in recent years. Nutrient database systems offer the potential to facilitate this process by enabling rapid formulation changes to satisfy specific nutrient requirements, by predicting nutrition labeling values for newly developed products, and by reducing the time to develop a label.

b. Nutrition labeling

Nutrition labeling is the second major trend that has had an impact on the use of databases. It is estimated that since the first voluntary nutrition labeling regulation in 1973, the percentage of products bearing nutrition labels increased to approximately 42% in 1978 and to approximately 66% in 1991. With the advent of NLEA, virtually all packaged foods will require nutrition labeling. As in 1973, the food industry is concerned about the cost of generating and maintaining accurate labels, and it continues to look to reduce costs by complementing laboratory analyses with database systems. As early as 1979, FDA, USDA, and FTC recognized this concern and pledged to work with industry in the development of valid databases (*Federal Register*, v.44, p. 75990, December 21, 1979). While FDA did not agree to approve databases, it offered to work with industry to resolve any compliance problems that might arise for foods labeled on the basis of a database that the Agency had reviewed. On the otherhand, FSIS endorsed the use of databases and agreed to accept nutrient declarations constructed from database values, recipe analyses using database values, and/or a combination of these approaches.

C. Definition and Type of Nutrient Databases

A nutrient database can mean different things depending on regulatory perspective. An FDA database represents a statistical analyses of single foods. A USDA database represents average values for a broad spectrum of foods. In general, a nutrient database is a collection of nutrient values on individual food products. Generally, it is stored on computer disks and/or magnetic tapes that can be accessed by computer programs of interest to suppliers of fruits, vegetables, fish, poultry, meats, frozen foods, and commodity-type products.

There are essentially two types of nutrient databases: "whole or finished food" databases (i.e., commodity-type or minimally processed foods such as canned or frozen produce) and ingredient ("recipe") databases (i.e., nutrients calculated from individual ingredients used in a food formulation) which are of great interest to food formulators.

Most current databases incorporate values from the USDA National Nutrient Data Bank (NNDB). They sometimes also include values obtained from brand name and ingredient products.

The NNDB reference database represents analytical data obtained from a large number of foods collected across seasons, geographic areas, genetic varieties, and cultivars. It is intended to reflect year-round, nationwide contributions of nutrients from foods. A number of food companies and trade associations have established, and are continuing to develop, substantial database systems that take into account these variables and also reflect numerous internally generated laboratory analyses on company products and ingredients. Brand averages are sometimes weighted by the market share of each to create a database of generic nutrient values that might be applied to identical foods. For commodity products, these combined or weighted values are often referred to as "composite values."

In addition to listing the averages of laboratory nutrient analyses, USDA provides calculated values based on laboratory measurements for cooking yields, nutrient retention during processing, and physical composition (i.e., bone, flesh skin).

D. Sources of Nutrient Databases and Computer Systems

A number of government and commercial sources of databases are available to those in the food industry who do not currently have a nutrient database and/or who do not have a computerized system for developing nutrient label claims.

The most widely used database is the *Agriculture Handbook No. 8 Series*. This series is prepared and published by the USDA Human Nutrition Information Service (HNIS). Hard copies may be purchased through the U.S. Government Printing Office, Superintendent of Documents. Track tapes and floppy disks may be purchased from the National Technical Information Service (NTIS).

The *Handbook No. 8 Database Series* contains generic values for over 5000 foods. It does not contain nutrient values for name brands, nor is it accompanied by a computer software program for constructing labels.

HNIS also maintains an electronic bulletin board which provides the latest information on its publications and computer files on food composition. The information can be viewed directly or downloaded to disk for subsequent review. The USDA nutrient data files and most NTIS files are available for downloading at no charge for use on personal computers. The USDA file provides all of the data in the *Handbook No. 8 Series* with some missing values filled in, data on the sugar content of selected foods, and data on nutrient retention during food processing. Nutrient Data Bank Electronic Bulletin Board Files available for downloading are shown in Table A3.8 (Appendix 3).

Most commercially available and industry database/computer systems build on values from the Handbook 8 Series and add brand name data and/or data from the scientific literature and other sources. Many of these programs have specific applications geared to particular product lines. Most industry developed databases are proprietary. They are usually constructed from internally generated or contracted laboratory analyses on specific products coupled with data from the *Handbook No. 8 Series* and information provided by their ingredient suppliers. In some cases companies purchase or lease commercially available nutrient database system and customize them to suit their specific needs. Some establish timesharing arrangements via a remote terminal and some, particularly smaller food companies, contract calculation services from consultants and/or vendors specializing in nutrition labeling.

There are a large number of software packages commercially available to those who do not currently have a nutrient database and who want to use a computerized system for establishing labels. These systems range in scope from less than 10 to more than 100 nutrients; from less than 100 to over 10,000 food items, and from less than $1,000 to approximately $15,000 in cost.

It should be noted that many software packages do not provide recipe calculation features for use with ingredient databases. The user should also be aware that certain software programs may not work with all computers. Also, some software packages will not accept the addition of food items and/or nutrients. Before purchasing or leasing a nutrient database computer system the potential user should request information regarding its features and limitations as well as a demonstration package (disk) for evaluation. Potential users should evaluate and select the system best suited to their needs and should consider product line, formulations, and ingredients used, anticipated changes as well as processing procedures. The 1992 Nutrient Databank Directory, provides a compilation of the important features of approximately 50 database systems and is available from the Department of Nutrition and Dietetics, University of Delaware, Alison Hall, Newark, DE, 19715-3360.

The Food and Nutrition Information Center (FNIC) at the National Agricultural Library (NAL) also has a number of computer software programs which contain nutrient analysis modules. Information about these programs can be obtained through FNIC/NAL, Room 304, 10301 Baltimore Blvd., Beltsville, MD, 20705 or by calling 301-504-5719. Information on database/computer systems is also periodically published in the *Journal of the American Dietetic Association* as well as other professional journals. A list of commercially available database/computer systems is provided in Table A3.9 (Appendix 3).

E. Attributes of Databases

1. Cost Reduction

Nutrition labels derived from nutrient databases are often considered to be less expensive than labels derived from laboratory analysis. While databases offer the potential for long-term cost savings, much depends on the objectives of the manufacturer, the product line extension, relative short- and long-term labeling needs and whether an internal nutrient database is desired or necessary to satisfy corporate strategies.

When assessing costs, one should compare the initial outlay of a computerized database system to the costs of laboratory analysis which must also include the costs of product sampling, delivery, and the cost of the product itself. As previously noted, the initial cost of a database system can vary from $1000 to $15,000. This does not take into account associated labor costs. Estimated costs for nutrition labeling analysis in the laboratory vary between approximately

$400–750 per product sample. Table 24.2 provides price ranges that contract laboratories are now charging for nutrient analyses.

In lieu of purchasing a complete computerized nutrient database system, the processor may choose to hire a consultant or database firm that has an accurate and reliable system in place. By complementing such a system with laboratory analysis, costs could be substantially reduced and food composition data can be validated. Such a choice may be particularly well suited to the small manufacturer with a limited number of products.

2. Time Savings

Nutrient database systems also have the potential for shortening the time frame required to generate nutrient information for labeling. For example, assuming no backlog, it is estimated that laboratory analysis for the mandatory nutrients for one or multiple products (the number of products being dependent on laboratory capacity) typically requires approximately five working days. This timeframe could increase if emergencies or instrumentation failures arise within the laboratory. On the other hand, once a computerized database system is set up and functioning properly, it may take as little as 15 min to generate a label. However, it may take a considerable amount of time to set up, learn, and adapt an already developed software system to satisfy individual needs and product lines; computer experience is essential. Internal development of a new or customized software system will take much longer. These factors should be considered when assessing the time savings that could accrue through database use.

3. Versatility

Nutrient databases also have the attributes of flexibility and versatility with properly developed software systems. They can be easily updated and can be used effectively when bringing new products to the marketplace. This is of paramount importance in todays world of compressed time and the intense competition for supermarket shelf space.

4. Diversity

Nutrient databases can be used for many purposes beyond nutrition labeling. Some of their current applications are listed below.

- Nutrition labeling
- Product development
- Formulation adjustments
- Research
- Nutrition surveys

- Clinical studies
- Menu planning
- Competitive product assessment

F. Practical Guidelines

The development and use of databases is complex and requires attentiveness to the accuracy, completeness, and precision of the information provided in the database as well as continuous maintenance of that information. Its accuracy and effective implementation is influenced by adjustments for processing, handling, storage, and such natural variables that could influence nutrient composition of foods as climate, season, region, variety, and cultivar.

The people developing and using databases are critical to their successful implementation. They must be knowledgeable in nutrition, food composition, and food science, and they should be thoroughly familiar with the manufacturing processes of their products. They must be able to exercise good judgement when selecting and entering database values and when evaluating the calculated data for accuracy and compliance. Database information (particularly ingredient database information) should not be taken at face value.

As noted earlier, there are two basic types of databases. One is a "whole or finished food" commodity-type database represented by average values for over 5000 foods (USDA, *Handbook No. 8 Series*). The other, about which FDA has expressed reservations (see Section C on FDA policies), is a computerized ingredient database system where nutrient values are calculated by combining the nutrient data from the individual ingredients.

Most current computerized nutrient database systems build on *Handbook No. 8 Series* values and include the mandatory NLEA nutrients. They differ primarily in the number of food products and ingredients they contain and in their quality and completeness.

A good quality, computerized nutrient database system should include several fundamental components:

- An accurate, complete, and well-documented tabulation of the nutrient composition of the foods and ingredients used by the food processor
- A validated computer software program that has the capacity to incorporate ingredient nutrient values into recipes (formulations) and to calculate a nutrition information panel for the final product after adjusting for processing changes and applying NLEA rounding rules
- A quality assurance program to ensure accuracy of the system with a mechanism for regularly evaluating and maintaining the integrity of individual ingredient nutrient values

The following are some practical guidelines for evaluating, developing, and using computerized database systems.

G. Accuracy and Quality

1. Documentation

In addition to providing a full complement of nutrient values for the foods and ingredients used in a product line, a quality database should be supported by documentation describing the protocol followed in establishing the database. This should include:

- The source of the data (USDA *Handbook No. 8 Series*, scientific literature, recipe data, internally generated values, supplier data)
- The vintage of the data; the quality assurance procedures used to validate the data (i.e., laboratory analysis, data and computation checks)
- Historical plant data with records of processing losses, yields, nutrient degradation
- The calculations employed to compensate for processing changes
- Records of adjustments and accountability for natural variation (i.e., season, environment, variety, region, and cultivar)

2. Database Accuracy

Nutrient values for the ingredients used in a database should be based on sound laboratory analysis. The quality of a database and its accuracy ultimately rests with the laboratory analyses performed on the food or ingredient product. The procedures used should be documented and assessed with respect to their appropriateness to the food or ingredient being analyzed. An officially approved analytical procedure suitable for one type of food may not be suitable for another (see Section III.B of this chapter). Knowledge of the sampling process is imperative. Documentation should be available describing the sampling design, the number of samples selected and analyzed, the source of the samples (region and location, i.e., processing plant, farm, or retail store) and the type of sample (season, variety, cultivar).

There are several practical steps towards achieving accurate (reliable) nutrition labels.

a. Start with good information.

The nutrient database component of the system is one of the most important features for achieving accuracy. At the outset, obtain accurate nutrient data. Document the source(s) used to compile the database. Many food composition tables contain extant data which have not been specifically collected for database use. This data is often not well documented with respect to sampling, analytical methodology, and may be outdated[11]. Whenever possible, newly generated data with appropriate documentation should be used. Quality assurance procedures should be in place to verify the accuracy and consistency of the nutrient values. A schedule should be established for reviewing, auditing, and confirming values through laboratory analysis. The database should be updated on a regular basis (at least annually) to incorporate the most recent data generated by USDA, suppliers, or published in the scientific literature.

The user should be intimately familiar with the ingredients used in a product formulation. If nutrient values are not available from handbooks or the scientific literature, request the data from the supplier. Be aware that prior to NLEA many suppliers did not gather nutrient information on their raw materials. Today, they are starting to supply this information as a necessary service to their customers. However, many ingredients still have missing values and inaccurate values are sometimes encountered. As a precaution, when values are missing or are questionable, the ingredient should undergo laboratory analysis. If possible, avoid using nutrient values for generic ingredients. They are variable and may not match the specific ingredients used in a formulation. In addition to regularly reviewing the entire database, institute a system for regularly validating supplier data. Insist upon unrounded values when obtaining data from a supplier. The most accurate nutrient values are derived from unrounded data. When rounded values are used errors can be magnified, so rounding should be done at the end of the calculation process.

While the size of a database is important in terms of the number of nutrients, foods and ingredients it contains, it is far more critical that the database includes all the components specific to a manufacturers product line and provides the specificity needed to calculate complex multi-ingredient formulations. The database should also allow for expansion and editing to facilitate application to new products.

b. Be sure to use an accurate formula.

Verify that the ingredients and quantities stipulated in the product formulation are correct and complete before calculating. Use exact formula quantities that are identical to the quantities used in the processing plant. Before inputting any data, recheck it to be certain that the sum of all ingredients equals 100%. A potential source of error sometimes observed, relates to the use of substitute ingredients and/or presumably minor alterations in the formula at the time of processing. When ingredients and/or foods used in a formulation (recipe) are missing from the database, substitution of similar items is an alternative that is frequently exercised. If such a substitution is intended to be carried through production, be sure it is compatible with the production process and that product quality (i.e., sensory characteristics) will not be adversely effected. If the substitution is intended to provide missing nutrient values, be aware of the potential for discrepancies in the final results. As a general rule avoid errors by avoiding substitutions. However, if substitutes are necessary, the database system should have the versatility to provide accurate nutrient values and valid labels.

c. Select or develop a database system that will make appropriate adjustments for changes in nutrient composition as the result of processing and validate the system.

The database should have built in programs to adjust for such processing effects as moisture changes due to volatilization or absorption; fat changes due to drainage or absorption; and nutrient degradation or losses due to heat, leeching, chelation, enzymatic activity, precipitation, flocculation, and physical separations. The processor should be able to predict accurately, product yields by carefully controlling all processing steps and maintaining accurate records of these steps. A well-designed sampling plan should be developed and coupled with laboratory analysis to document any nutrient changes that could alter any label declarations. Appropriate adjustment factors can then be built into the database for subsequent application. Such a plan will also provide an assessment of the effect of processing on the nutritional value of the product and its overall quality. Alternatively, the processor can use the USDA guide for estimating nutrient changes during processing if applicable. These can also be built into the database system to correct for nutrient changes.

d. Do not ignore missing nutrient values.

The handling of missing nutrient values is a major source of potential error in deriving nutrition labels from ingredient databases. They are often entered and calculated as zero and result in underestimates. In the absence of any sources of nutrient information, the ingredient or food should be submitted for laboratory analysis.

e. Perform a final review for accuracy, completeness, and reasonableness.

Before finalizing a nutrition information panel, the results of the database computer analysis should be carefully reviewed and checked for reasonableness. It is generally helpful to use a spreadsheet for this purpose. This will facilitate comparisons and help identify abnormal and missing values. A number of trouble shooting tips are listed below:

- Verify that the correct ingredients have been selected and are the same as used in the processing plant. Use common terminology to describe foods. Database ingredients, formulation ingredients and the ingredients used in processing should all be the same. Inappropriate selections that could lead to errors include interchanging raw and cooked food values, using values from a fortified rather than unfortified ingredient, using data from a defatted or dehydrated product instead of its regular form, and replacing a low-sodium product with its regular counterpart.

- Be sure that substitute ingredients are accounted for and confirm their nutrient values.
- Verify that selection of the ingredients from the database has addressed season, region, variety, and cultivar.
- Be sure numbers are properly converted and properly transcribed. Use standard and consistent units of weight and volume for all calculations. Some typical sources of error are conversion of retinol units to International Units of vitamin A, conversion of international units of vitamin D to micrograms, conversion of parts per million (ppm) to milligrams (mg), and conversion and proper transcription of grams to milligrams.
- Recheck adjustment entries to be sure that corrections for moisture or fat gain/loss and other nutrient changes were properly entered.
- Recheck laboratory reports to verify results.

If possible, compare the nutrient values of the product in question to similar products in the market place or listed in handbooks. If the discrepancies are out of tolerance or if the data seems unreasonable, the product ultimately must be submitted to laboratory analysis for reconciliation. Proximate analysis (moisture, fat, protein, fiber, ash, carbohydrate) can be performed rapidly and inexpensively and frequently provides important clues for tracing the source of discrepancies.

Knowledge of the product, its nutrient composition, and processing techniques will allow the manufacturer to establish internal precision objectives and acceptable tolerances. Thus, while discrepancies may exist between laboratory values and database values (i.e., between predicted and actual values), they still may satisfy the precision objectives of the manufacturer and be within acceptable FDA and USDA compliance standards.

H. FDA Policies

With the implementation of nutrition labeling regulations in 1973, FDA developed a manual entitled *A Guide for Developing and Using Databases.* An updated version was issued in 1993 and can be obtained from the Center for Food Safety and Nutrition (CFSAN). The manual is intended to assist industry in understanding the regulations and develop labels. It provides information on the Agency's policies on the use of nutrient databases for nutrition labeling and includes generic instructions for developing and preparing an acceptable database including procedures for sampling, design, collecting samples, statistical treatment of data, and compliance procedures.

The FDA's general policy for databases appears in the *Federal Register*, Vol. 56, No. 229, pp. 60884–5, November 27, 1991. In it, the Agency states that if a manufacturer plans to use a database for nutrition labeling, it is advantageous to have the database reviewed and accepted by FDA and to employ the procedures described in the manual. If the Agency finds that the nutrition label of a product derived from a database that has been accepted by FDA is not in compliance with the regulations, the Agency will not take immediate action if the company has followed good manufacturing practices during its production. In such cases, the Agency will work with the manufacturer to resolve the regulatory issue. On the other hand, action will be taken if non-compliance is a result of failure to follow good manufacturing practices.

The Agency points out that submission of a database for purposes of nutrition labeling is voluntary. FDA does not intend to prescribe how a company determines the nutrient content on its products. A company may use alternative procedures to those presented in the FDA manual. However, it is suggested that such procedures be discussed with the Agency before implementation to avoid the potential risk of rejection and unnecessary expenditures of time and money. The Agency emphasizes that the data used for nutrition labeling is the prerogative and responsibility of the manufacturer. However, the manufacturer must be judicious in this selection to ensure compliance with the regulations.

The current NLEA regulations [21 CFR 101.9 (g)(8)] state that compliance may be achieved

> . . . by the use of an FDA approved database that has been computed following FDA guideline procedures and where food samples have been handled in accordance with current good manufacturing practice to prevent nutrient loss. FDA approval of a database shall not be considered granted until the CFSAN has agreed to all aspects of the database in writing.

The regulations further state that "approvals will be in effect for a limited time, e.g., 10 years, and will be eligible for renewal in the absence of significant changes in agricultural or industry practices."

FDA has expressed concern about the use of existing databases with respect to their reliability in generating nutrition label values to comply with regulations. The Agency notes that a large body of nutrient content data developed primarily for determining average daily intakes is satisfactory for that purpose but may not be adequate for nutrition label compliance.

The 1993 FDA Nutrition Labeling Manual states that the Agency is not currently accepting ingredient composition databases to calculate the final composition of a mixed product. The stated reasons for this position are that such databases are usually not accompanied by: (a) indicators of data quality, (b) indicators of the methods of analysis or sampling used to obtain the data, (c) indicators of the design and execution of quality management procedures used to monitor the sampling and analysis of the components, and (d) indicators of subsequent loss of nutrients during processing and handling of the mixed product. However, the Agency has indicated that if a successful model can be developed to define the relationship between ingredient composition and final product composition that accounts for nutrient losses during processing, it might receive acceptance. Extensive analyses of ingredients and final product composition will be required to develop and validate a successful model.

Over the last year, FDA has been developing a review system to serve as a guide to those intending to submit databases for the Agency to review. The system describes the FDA review process and provides the criteria the Agency is currently using to evaluate individual food (finished food) databases. Although still under development, it also provides information on ingredient database evaluation.

The Agency's review of *individual* food (finished food) databases consist of the following:

- Data reasonableness and variables
- Analytical methods (shipping, handling, storage, quality control, sample preparation, laboratory controls)
- Sampling design (sampling objectives, target population and samples population, sampling frame, sample size, sample collection and handling, types of test samples)
- Statistical treatment of data (statistical analysis and calculations of label values)

The Agency's review of *ingredient* database systems consists of the following:

- The suitability of the ingredient database system for the particular food products
- Background information regarding the development of the system
- Variables that affect the nutrient content of the food and/or the major ingredients
- The ingredient database (number of ingredients, source of data, completeness of database, how information is collected and reviewed)
- Procedures to ensure accuracy and currency of the database
- Corrections for moisture and nutrient loss
- Validation processes, including sample collection and analytical methods for validation, and comparison of calculated and analytical values
- The software system description and validation

Since the advent of NLEA and through June of 1994, FDA had received 53 databases covering raw commodities, frozen and canned produce, ingredients, and ingredient model systems. Forty-nine of these were still under review in June 1994. It would appear that the Agency may not have the necessary resources to adequately handle the current flow of submissions and some modifications in the review process may be necessary.

Furthermore, the stringent acceptance standards set forth by the Agency are not being satisfied, for the most part, and may be too difficult to achieve. The databases and systems being submitted to the Agency have been lacking documentation, lacking information about variables that may alter nutrient content and do not have appropriate sampling designs. FDA is trying to find a workable solution that is fair to the food industry while providing consumers with valid nutrient information on product labels. It is considering the possibility of evaluating databases with less information than suggested in the manual. It is also considering label authorization based on preliminary database review (interim approval). This would give the manufacturer some assurance that FDA would not take action against products so long as the manufacturer quickly corrects any labels that are out of compliance.

Towards these ends, the Agency has requested comments about evaluation criteria, interim approvals, and guidelines for timely follow-up. It is FDA's intention to encourage database developers to continue their efforts and to improve their databases. Whatever the final outcome, databases that are accepted by FDA will undoubtedly require periodic updating. The frequency will depend upon the type of product (i.e., single or mixed), the size of the accepted database and the stability of the nutrients over time.

V. ENDNOTES

1. Details may be found in *Methods of Analysis for Nutrition Labeling* (AOAC International, 1993).
2. In recent years (1987), the International Union of Pure and Applied Chemistry (IUPAC) has developed a protocol for the design, conduct, data handling, and interpretation of interlaboratory method performance (e.g., collaborative studies). This protocol has been adopted by other methods standardizing organizations in the food area and has allowed for the assessment of the variability (Relative Standard Deviation) that can be expected for nutrient analyses by a group of laboratories.
3. AOAC Task Force Subcommittee on Definition of Foods for Analytical Progress
4. AOAC Referee, August 1992.
5. AOAC Referee, July 1993.
6. AOAC 15th ed., 1990.
7. *Methods of Analysis for Nutrition Labeling* (AOAC International, 1993).
8. Sheppard, A. J. (1992) *Lipid Manual, Methodology Suitable for Fatty Acid-Cholesterol Analysis*, Wm. C. Brown Publishers, Dubuque, IA.
9. Alternatively, triflouroacetic acid (TFA) may replace the "Carr-Price" reagent.
10. Infant formulas and medical diets may be saponified with ethanolic potassium hydroxide; partitioned with organic solvent; separated from the sample matrix and quantified by HPLC (AOAC 16th ed., 1995).
11. See Chapter 23 for a discussion of extant data.

Appendix 1

Reference Amounts, Protein Digestibility Coefficients, and FDA/USDA Regulations and Contacts

Table A1.1 Reference Amounts Customarily Consumed Per Eating Occasion: General Food Supply[1,2,3,4]

Product category	Reference amount	Label statement[5]
Bakery products		
Biscuits, croissants, bagels, tortillas, soft bread sticks, soft pretzels, corn bread, hush puppies	55 g	___ piece(s) (___ g)
Breads (excluding sweet quick type), rolls	50 g	___ piece(s) (___ g) for sliced bread and distinct pieces (e.g., rolls); 2 oz. (56 g/ in. slice) for unsliced bread
Bread sticks, *see* crackers		
Toaster pastries, *see* coffee cakes		
Brownies	40 g	___ piece(s) (___ g) for distinct pieces; ___ fractional slice (___ g) for bulk

Table A1.1 (Continued)

Product category	Reference amount	Label statement[5]
Bakery products (cont.)		
Cakes, heavy weight (cheese cake; pineapple upside-down cake; fruit, nut, and vegetable cakes with more than or equal to 35% of the finished weight as fruit, nuts, or vegetables[6]	125 g	___ piece(s) (___ g) for distinct pieces (e.g., sliced or individually packaged product(s); ___ fractional slice (___ g) for large discrete units
Cakes, medium weight (chemically leavened cake with or without icing or filling except those classified as light weight cake; fruit, nut and vegetable cake with less than 35% of the finished weight as fruit, nuts, or vegetables; light weight cake with icing; Boston cream pie; cupcake; eclair; cream puff)[7]	80 g	___ piece(s) (___ g) for distinct pieces (e.g., cupcake); ___ fractional slice (___ g) for large discrete units
Cakes, light weight (angel food, chiffon, or sponge cake without icing or filling)[8]	55 g	___ piece(s) (___ g) for distinct pieces (e.g., sliced or individually packaged products); ___ fractional slice (___ g) for large discrete units
Coffee cakes, crumb cakes, doughnuts, Danish, sweet rolls, sweet quick type breads, muffins, toaster pastries	55 g	___ piece(s) (___ g) for sliced bread and distinct pieces (e.g., doughnut); 2 oz. (56 g/visual unit of measure) for bulk products (e.g., unsliced bread)
Cookies	30 g	___ piece(s) (___ g)
Crackers that are usually not used as snack, melba toast, hard bread sticks, ice cream cones[9]	15 g	___ piece(s) (___ g)
Crackers that are usually used as snacks	30 g	___ piece(s) (___ g)

Product category	Reference amount	Label statement[5]
Croutons	7 g	___ tbsp(s) (___ g); ___ cup(s) (___ g); ___ piece(s) (___ g) for large pieces
French toast, pancakes, variety mixes	110 g prepared for french toast and pancakes; 40 g dry mix for variety mixes	___ piece(s) (___ g); ___ cup(s) (___ g) for dry mix
Grain-based bars with or without filling or coating, e.g., breakfast bars, granola bars, rice cereal bars	40 g	___ piece(s) (___ g)
Ice cream cones, *see* crackers		
Pies, cobblers, fruit crisps, turnovers, other pastries	125 g	___ piece(s) (___ g) for distinct pieces; ___ fractional slice (___ g) for large discrete units
Pie crust	1/6 of 8 in. crust; 1/8 of 9 in. crust	1/6 of 8 in. crust (___ g); 1/8 of 9 in. crust (___ g)
Pizza crust	55 g	___ fractional slice (___ g)
Taco shells, hard	30 g	shell(s) (___ g)
Waffles	85 g	___ piece(s) (___ g)
Beverages		
Carbonated and noncarbonated beverages, wine coolers, water	240 ml	8 fl oz. (240 ml)
Coffee or tea, flavored and sweetened	240 ml prepared	8 fl oz. (240 ml)
Cereals and other grain products		
Breakfast cereals (hot cereal type), hominy grits	1 cup prepared; 40 g plain dry cereal; 55 g flavored, sweetened cereal	___ cup(s) (___ g)
Breakfast cereals, ready-to-eat, weighing less than 20 g per cup, e.g., plain puffed cereal grains	15 g	___ cup(s) (___ g)

Table A1.1 (Continued)

Product category	Reference amount	Label statement[5]
Cereals and other grain products (cont.)		
Breakfast cereals, ready-to-eat, weighing 20 g or more but less than 43 g per cup; high fiber cereals containing 28 g or more of fiber per 100 g	30 g	___ cup(s) (___ g)
Breakfast cereals, ready-to-eat, weighing 43 g or more per cup; biscuit types	55 g	___ piece(s) (___ g) for large distinct pieces (e.g., biscuit type); ___ cup(s) (___ g) for all others
Bran or wheat germ	15 g	___ tbsp(s) (___ g); ___ cup(s) (___ g)
Flours or cornmeal	30 g	___ tbsp(s) (___ g); ___ cup(s) (___ g)
Grains, e.g., rice, barley, plain	140 g prepared; 45 g dry	___ cup(s) (___ g)
Pastas, plain	140 g prepared, 55 g dry	___ cup(s) (___ g); ___ piece(s) (___ g) for large pieces (e.g., large shells) or 2 oz. (56 g/visual unit of measure) for dry bulk products (e.g., lasagna or spaghetti noodles)
Pastas, dry, ready-to-eat, e.g., fried canned chow mein noodles	25 g	___ cup(s) (___ g)
Starches, e.g., cornstarch, potato starch, tapioca, etc.	10 g	1 tbsp (8 g) for cornstarch; 1 tbsp (10 g) for tapioca, 1 tbsp (___ g) for others
Stuffing	100 g	___ cup(s) (___ g)
Dairy products and substitutes		
Cheese, cottage	110 g	1/2 cup (105 g) for small curd; 1/2 cup (113 g) for large curd, low fat, or with fruit added; 1/2 cup (___ g) for others
Cheese, used primarily as ingredients, e.g., dry cottage cheese, ricotta cheese	55 g	1/3 cup (48 g) for dry curd cottage cheese; 1/4 cup (62 g) for ricotta cheese

Product category	Reference amount	Label statement[5]
Cheese, grated hard, e.g., Parmesan, Romano	5 g	1 tbsp(s) (5 g)
Cheese, all others except those listed as separate categories, includes cream cheese and cheese spread	30 g	___ piece(s) (___ g) for distinct pieces; ___ tbsp(s) (___ g) for cream cheese and cheese spread; 1 oz. (28 g/visual unit of measure) for bulk
Cheese sauce, *see* sauce category		
Cream or cream substitutes, fluid	15 ml	1 tbsp (15 ml)
Cream or cream substitutes, powder	2 g	1 tsp (2 g)
Cream, half & half	30 ml	2 tbsp (30 ml)
Eggnog	120 ml	1/2 cup (120 ml) or 4 fl oz. (120 ml)
Milk, condensed, undiluted	30 ml	2 tbsp (30 ml)
Milk, evaporated, undiluted	30 ml	2 tbsp (30 ml)
Milk, milk-based drinks, e.g., instant breakfast, meal replacement, cocoa	240 ml	1 cup (240 ml) or 8 fl oz. (240 ml)
Shakes or shake substitutes, e.g., dairy shake mixes, fruit frost mixes	240 ml	1 cup (240 ml) or 8 fl oz. (240 ml)
Sour cream	30 g	2 tbsp (30 g)
Yogurt	225 g	1 cup (___ g)
Desserts		
Ice cream, ice milk, frozen yogurt, sherbet; all types, bulk and novelties (e.g., bars, sandwiches, cones)	1/2 cup, includes the volume for coatings and wafers for the novelty type varieties	___ piece(s) (___ g) for individually wrapped or packaged products; 1/2 cup (___ g) for others
Frozen flavored and sweetened ice and pops, frozen fruit juices; all types, bulk and novelties (e.g., bars, cups)	85 g	___ piece(s) (___ g) for individually wrapped or packaged products; ___ cup(s) (___ g) for others
Sundae	1 cup	1 cup (___ g)

Table A1.1 (Continued)

Product category	Reference amount	Label statement[5]
Desserts (cont.)		
Custards, gelatin or pudding	1/2 cup	___ piece(s) (___ g) for distinct unit (e.g., individually packaged products); 1/2 cup (___ g) for bulk
Dessert toppings and fillings		
Cake frostings or icings	35 g	___ tbsp(s) (___ g)
Other dessert toppings, e.g., fruits, syrups, spreads, marshmallow cream, nuts, dairy and nondairy whipped toppings	2 tbsp	2 tbsp (___ g)
Pie fillings	85 g	___ cup(s) (___ g)
Egg and egg substitutes		
Egg mixtures, e.g., egg foo young, scrambled eggs, omelets	110g	___ piece(s) (___ g) for discrete pieces; ___ cup(s) (___ g)
Eggs (all sizes)[9]	50 g	1 large, medium, etc. (___ g)
Egg substitutes	An amount to make 1 large (50 g) egg	___ cup(s) (___ g)
Fats and oils		
Butter, margarine, oil, shortening	1 tbsp	1 tbsp (14 g) for butter, margarine, or oil; 1 tbsp (9 g) for whipped butter or margarine; 1 tbsp (13 g) for shortening
Butter replacement, powder	2 g	___ tsp(s) (___ g)
Dressings for salads	30 g	2 tbsp (___ g)
Mayonnaise, sandwich spreads, mayonnaise-type dressings	15 g	1 tbsp (14 g) for mayonnaise; 1 tbsp (15 g) for imitation mayonnaise, mayonnaise-type dressings or sandwich spread
Spray types	0.25 g	About ___ seconds spray (___ g)

Product category	Reference amount	Label statement[5]
Fish, shellfish, game meats[10], and meat or poultry substitutes		
Bacon substitutes, canned anchovies[11], anchovy pastes, caviar	15 g	___ piece(s) (___ g) for discrete pieces; ___ tbsp(s) (___ g) for others
Dried, e.g., jerky	30 g	___ piece(s) (___ g)
Entrees with sauce, e.g., fish with cream sauce, shrimp with lobster sauce	140 g cooked	___ cup(s) (___ g); 5 oz. (140g/visual unit of measure) if not measurable by cup
Entrees without sauce, e.g., plain or fried fish and shellfish, fish and shellfish cake	85 g cooked; 110 g uncooked[12]	___ piece(s) (___ g) for discrete pieces; ___ cup(s) (___ g); oz. (g/visual unit of measure) if not measurable by cup[13]
Fish, shellfish or game meat[10], canned[11]	55 g	___ piece(s) (___ g) for discrete pieces; ___ cup(s) (___ g)
Substitute for luncheon meat, meat spreads, Canadian bacon, sausages and frankfurters	55 g	___ piece(s) (___ g) for distinct pieces (e.g., slices, links); ___ cup(s) (___ g); 2 oz. (56 g/visual unit of measure) for nondiscrete bulk product
Smoked or pickled[11], fish, shellfish, or game meat[10], fish or shellfish spread	55 g	___ piece(s) (___ g) for distinct pieces (e.g., slices, links) or ___ cup(s) (___ g); 2 oz. (56 g/visual unit of measure) for nondiscrete bulk product
Substitutes for bacon bits, *see* miscellaneous category		
Fruits and fruit juices		
Candied or pickled[11]	30 g	___ piece(s) (___ g)
Dehydrated fruits, *see* snacks category		
Dried	40 g	___ piece(s) (___ g) for large pieces (e.g., dates, figs, prunes); ___ cup(s) (___ g) for small pieces (e.g., raisins)
Fruits for garnish or flavor, e.g., maraschino cherries[11]	4 g	1 cherry (___ g)
Fruit relishes, e.g., cranberry sauce, cranberry relish	70 g	___ cup(s) (___ g)

Table A1.1 (Continued)

Product category	Reference amount	Label statement[5]
Fruits and fruit juices (cont.)		
Fruits used primarily as ingredients, avocado	30 g	*See* note 13
Fruits used primarily as ingredients, others (cranberries, lemon, lime)	55 g	___ piece(s) (___ g) for large fruits; ___ cup(s) (___ g) for small fruits measurable by cup[13]
Watermelon	280 g	*See* note 13
All other fruits (except those listed as separate categories), fresh, canned or frozen	140 g	___ piece(s) (___ g) for large pieces (e.g., strawberries, prunes, apricots, etc.); ___ cup(s) (___ g) for small pieces (e.g., blueberries, raspberries, etc.)[13]
Juices, nectars, fruit drinks	240 ml	8 fl oz. (240 ml)
Juices used as ingredients, e.g., lemon juice, lime juice	5 ml	1 tsp (5 ml)
Legumes		
Bean cake (tofu)[11], tempeh	85 g	___ piece(s) (___ g) for discrete pieces; 3 oz. (84 g/visual unit of measure) for bulk products
Beans, plain or in sauce	130 g for beans in sauce or canned in liquid; 90 g for others	1/2 cup (___ g)
Miscellaneous category		
Baking powder, baking soda, pectin	1 g	1/4 tsp (___ g)
Baking decorations, e.g., colored sugars and sprinkles for cookies, cake decorations	1/4 tsp or 4 g if not measurable by teaspoon	___ piece(s) (___ g) for discrete pieces; 1/4 tsp (___ g)
Batter mixes, bread crumbs	30 g	___ tbsp(s) (___ g) or ___ cup(s) (___ g)
Cooking wine	30 ml	2 tbsp (30 ml)

Product category	Reference amount	Label statement[5]
Drink mixers (without alcohol)	Amount to make 240 ml drink (without ice)	___ fl oz. (___ ml)
Chewing gum[9]	3 g	___ piece(s) (___ g)
Meat, poultry, and fish coating mixes, dry; seasoning mixes, dry, e.g., chili seasoning mixes, pasta salad seasoning mixes	Amount to make one reference amount of final dish	___ tsp(s) (___ g) or ___ tbsp(s) (___ g)
Salad and potato toppers, e.g., salad crunchies, salad crispins, substitutes for bacon bits	7 g	___ tbsp(s) (___ g)
Salt, salt substitutes, seasoning salts (e.g., garlic salt)	1 g	___ tsp(s) (___ g); ___ piece(s) (___ g) for discrete pieces (e.g., individually packaged product(s)
Spices, herbs (other than dietary supplements).	1/4 tsp or 0.5 g if not measurable by teaspoon	1/4 tsp (___ g);___ piece(s) (___ g) if not measurable by teaspoons (e.g., bay leaf)
Mixed dishes		
Measurable with cup, e.g., casseroles, hash, macaroni and cheese, pot pies, spaghetti with sauce, stews, etc.	1 cup	1 cup (___ g)
Not measurable with cup, e.g., burritos, egg rolls, enchiladas, pizza, pizza rolls, quiche, all types of sandwiches	140 g, add 55 g for products with gravy or sauce topping, e.g., enchilada with cheese sauce, crepe with white sauce[14]	___ piece(s) (___ g) for discrete pieces; ___ fractional slice (___ g) for large discrete units

Table A1.1 (Continued)

Product category	Reference amount	Label statement[5]
Nuts and seeds		
Nuts, seeds, and mixtures, all types: sliced, chopped, slivered, and whole	30 g	___ piece(s) (___ g) for large pieces (e.g., unshelled nuts); ___ tbsp(s) (___ g) or ___ cup(s) (___ g) for small pieces (e.g., peanuts, sunflower seeds)
Nuts and seed butters, pastes, or creams	2 tbsp	2 tbsp (___ g)
Coconut, nut, and seed flours	15 g	___ tbsp(s) (___ g)
Potatoes and sweet potatoes/yams		
French fries, hash browns, skins, or pancakes	70 g prepared; 85 g for frozen unprepared french fries	___ piece(s) (___ g) for large distinct pieces (e.g., patties, skins); 2.5 oz. (70 g/___ pieces) for prepared fries; 3 oz. (84 g/___ pieces) for unprepared fries
Mashed, candied, stuffed, or with sauce	140 g	___ piece(s) (___ g) for discrete pieces (e.g., stuffed potato); ___ cup(s) (___ g)
Plain, fresh, canned, or frozen	110 g for fresh or frozen; 160 g for canned in liquid	___ piece(s) (___ g) for discrete pieces; ___ cup(s) (___ g) for sliced or chopped products
Salads		
Gelatin salad	120 g	1/2 cup (120 g)
Pasta or potato salad	140 g	___ cup(s) (___ g)
All other salads, e.g., egg, fish, shellfish, bean, fruit, or vegetable salads	100 g	___ cup(s) (___ g)
Sauces, dips, gravies, and condiments		
Barbecue sauce, hollandaise sauce, tartar sauce, other sauces for dipping (e.g., mustard sauce, sweet and sour sauce), all dips (e.g., bean dips, dairy-based dips, salsa)	2 tbsp	2 tbsp (___ g)

Product category	Reference amount	Label statement[5]
Major main entree sauces, e.g., spaghetti sauce	125 g	1/2 cup (___ g)
Minor main entree sauces (e.g., pizza sauce, pesto sauce), other sauces used as toppings (e.g., gravy, white sauce, cheese sauce), cocktail sauce	1/4 cup	1/4 cup (___ g)
Major condiments, e.g., catsup, steak sauce, soy sauce, vinegar, teriyaki sauce, marinades	1 tbsp	1 tbsp (___ g)
Minor condiments, e.g., horse-radish, hot sauces, mustards, worcestershire sauce	1 tsp	1 tsp (___ g)
Snacks		
All varieties, chips, pretzels, popcorns, extruded snacks, fruit-based snacks (e.g., fruit chips), grain-based snack mixes	30 g	___ cup(s) (___ g) for small pieces (e.g., popcorn); ___ piece(s) (___ g) for large pieces (e.g., large pretzels; pressed dried fruit sheet); 1 oz. (28 g/visual unit of measure) for bulk products (e.g., potato chips)
Soups		
All varieties	245 g	1 cup(s) (___ g)
Sugars and sweets		
Baking candies (e.g., chips)	15 g	___ piece(s) (___ g) for large pieces; ___ tbsp(s) (___ g) for small pieces; 1/2 oz. (14 g/visual unit of measure) for bulk products
Hard candies, breath mints	2 g	___ piece(s) (___ g)
Hard candies, roll-type, mini-size in dispenser packages	5 g	___ piece(s) (___ g)
Hard candies, others	15 g	___ piece(s) (___ g) for large pieces; ___ tbsp(s) (___ g) for "mini-size" candies measurable by tablespoon; 1/2 oz. (14 g/visual unit of measure) for bulk products

Table A1.1 (Continued)

Product category	Reference amount	Label statement[5]
Sugars and sweets (cont.)		
All other candies	40 g	___ piece(s) (___ g); 1 1/2 oz. (42 g/visual unit of measure) for bulk products
Confectioner's sugar	30 g	1/4 cup (30 g)
Honey, jams, jellies, fruit butter, molasses	1 tbsp	1 tbsp (___ g)
Marshmallows	30 g	___ cup(s) (___ g) for small pieces or ___ piece(s) (___ g) for large pieces
Sugar	4 g	1 tsp (___ g);___ piece(s) (___ g) for discrete pieces (e.g., sugar cubes, individually packaged products)
Sugar substitutes	An amount equivalent to one reference amount for sugar in sweetness	___ tsp(s) (___ g) for solids; ___ drop(s) for liquid; ___ piece(s) (___ g) (e.g., individually packaged products)
Syrups	30 ml for syrups used primarily as an ingredient (e.g., light or dark corn syrup); 60 ml for all others	2 tbsp (30 ml) for syrups used primarily as an ingredient; 1/4 cup (60 ml) for all others
Vegetables		
Vegetables primarily used for garnish or flavor, e.g., pimento, parsley	15 g	___ piece(s) (___ g); ___ tbsp(s) (___ g) for chopped products
Chili pepper, green onion	30 g	___ piece(s) (___ g)[13]; ___ tbsp(s) (___ g) or ___ cup(s) (___ g) for sliced or chopped products

Product category	Reference amount	Label statement[5]
All other vegetables without sauce: fresh, canned, or frozen	85 g for fresh or frozen; 95 g for vacuum canned; 130 g for canned in liquid, cream-style corn, canned or stewed tomatoes, pumpkin, or winter squash	___ piece(s) (___ g) for large pieces (e.g., brussel sprouts); ___ cup(s) (___ g) for small pieces (e.g., cut corn, green peas); 3 oz. (84 g/visual unit of measure) if not measurable by cup[13]
All other vegetables with sauce: fresh, canned or frozen	110 g	___ piece(s) (___ g) for large pieces (e.g., brussel sprouts); ___ cup(s) (___ g) for small pieces (e.g., cut corn, green peas); 4 oz. (112 g/visual unit of measure) if not measurable by cup
Vegetable juice	240 ml	8 fl oz. (240 ml)
Olives[11]	15 g	___ piece(s) (___ g); ___ tbsp(s) (___ g) for sliced products
Pickles, all types[11]	30 g	1 oz. (28 g/visual unit of measure)
Pickle relishes	15 g	1 ___ tbsp (15 g)
Vegetable pastes, e.g., tomato paste	30 g	2 ___ tbsp (33 g) for tomato paste; 2 ___ tbsp (___ g) for all others
Vegetable sauces or purees, e.g., tomato sauce, tomato puree	60 g	1/4 cup (61 g) for tomato sauce; 1/4 cup (63 g) for tomato puree; 1/4 cup (___ g) for all others

[1] These values represent the amount (edible portion) of food customarily consumed per eating occasion and were primarily derived from the 1977-1978 and the 1987-1988 Nationwide Food Consumption Surveys conducted by the U.S. Department of Agriculture

[2] Unless otherwise noted in the Reference Amount column, the reference amounts are for the ready-to-serve or almost ready-to-serve form of the product (i.e., heat and serve, brown and serve). If not listed separately, the reference amount for the unprepared form (e.g., dry mixes; concentrates; dough; batter; dry, fresh, and frozen pasta) is the amount required to make one reference amount of the prepared form. Prepared means prepared for consumption (e.g., cooked).

Table A1.1 (Continued)

[3] Manufacturers are required to convert the reference amount to the label serving size in a household measure most appropriate to their specific product using the procedures in 21 CFR 101.9(b).
[4] Copies of the list of products for each product category are available from the Divison of Nutrition (HFF-260), Center for Food Safety and Applied Nutrition, Food and Drug Administration, 200 C St. SW, Washington, DC 20204.
[5] The label statements are meant to provide guidance to manufacturers on the presentation of serving size information on the label, but they are not required. The term "piece" is used as a generic description of a discrete unit. Manufacturers should use the description of a unit that is most appropriate for the specific product (e.g., sandwich for sandwiches, cookie for cookies, and bar for frozen novelties). [The guidance provided is for the label statement of products in ready-to-serve or almost ready-to-serve form. The guidance does not apply to the products which require further preparation for consumption (e.g., dry mixes, concentrates) unless specifically stated in the product category, reference amount, or label statement column that it is for these forms of the product. For products that require further preparation, manufacturers must determine the label statement following the rules in § 101.9(b) using the reference amount determined according to § 101.12(c).]
[6] Includes cakes that weigh 10 g or more per cubic in.
[7] Includes cakes that weigh 4 g or more per cubic in. but less than 10 g per cubic in.
[8] Includes cakes that weigh less than 4 g per cubic in.
[9] Label serving size for ice cream cones and eggs of all sizes will be 1 unit. Label serving size of all chewing gums that weigh more than the reference amount that can reasonably be consumed at a single-eating occasion will be 1 unit.
[10] Animal products not covered under the Federal Meat Inspection Act or the Poultry Products Inspection Act, such as flesh products from deer, bison, rabbit, quail, wild turkey, geese, ostrich, etc.
[11] If packed or canned in liquid, the reference amount is for the drained solids, except for products in which both the solids and liquids are customarily consumed (e.g., canned chopped clam in juice).
[12] The reference amount for the uncooked form does not apply to raw fish in § 101.45 or to single-ingredient products that consist of fish or game meat as provided for in § 101.9(b)(j)(11).
[13] For raw fruit, vegetables, and fish, manufacturers should follow the label statement for the serving size specified in Appendices A and B to the regulation entitled "Food Labeling: Guidelines for Voluntary Nutrition Labeling; and Identification of the 20 Most Frequently Consumed Raw Fruits, Vegetables, and Fish; Definition of Substantial Compliance; Correction" (56 FR 60880 as amended 57 FR 8174, March 6, 1992.
[14] Pizza sauce is part of the pizza and is not considered to be sauce topping.
Reprinted from: *Federal Register* 58(3):2294–2297, January 6, 1993.

Table A1.2 True Protein Digestability Value of Common Foods[a]

Major product group	Subgroup	Product	True digestibility	Ref.
Cereals and grains	Barley	Barley	90	1
	Corn and corn products	Corn, extruded cereal	62	2
		Corn, corn flake	70	2
		Corn, degerminated opaque 2	94	2
		Corn, puffed cereal	76	2
		Corn, whole	89	2
		Corn, whole opaque 2	92	2
		Corn meal	84	2
	Millett	Millett	79	2
	Oats and oat products	Oat, sugared flakes	67	2
		Oat flakes	70	2
		Oat, extruded oat/wheat	76	2
		Cereal	89	2
		Oat, quick oatmeal	82	2
		Oat, oatmeal	90	3
		Oats, rolled		
	Rice and rice products	Rice	91	2
		Rice, germ	87	4
		Rice, brown, cooked	72	5
		Rice, high protein	85	2
		Rice, milled, cooked	86	6
		Rice, polished	87	2,5
		Rice, crisped rice cereal	77	2
		Rice, flakes	63	2

Table A1.2 (Continued)

Major product group	Subgroup	Product	True digestibility	Ref.
Cereals and grains (cont.)	Sorghum	Sorghum, cooked	73	6
	Triticale	Triticale	90	2
	Wheat and wheat products	Bread	96	2
		Bread, coarse, brown	91	2
		Bread, white	89	2
		Bread, whole wheat	92	2
		Wheat, bran	95	15
		Wheat, brown, cooked	93	6
		Wheat, endosperm (farina)	98	2
		Wheat, flour, 90% extracted	89	2
		White, 80% extracted	92	2
		Wheat, germ	81	2
		Wheat and wheat products	98	2
		Wheat, hard spring	86	2
		Wheat, meat analogue	95	2
		Wheat, puffed wheat	84	2
		Wheat, shredded	73	2
		Wheat, wheaties	80	2
		Wheat, white flour	97	2
		Wheat, whole	87	2
		Wheat, whole, hot cereal	85	2
		Wheat, 40% bran flakes	69	2
Dairy products	Casein	Acid-casein	95	7
		Casein	96	8,4

		Caseinate	95	7
		Rennet-casein	94	7
	Cheese	Cheese, chedder	99	4
		Cottage cheese	99	2
	Lactalbumin	Lactalbumin	94	2
	Milk	Milk retentate	97	7
		Milk, skim	94	8
		Milk, whole	94	2
		Milk, whole, powdered	95	2
	Whey	Whey protein	95	4
Eggs and egg products		Egg albumin	97	2
		Egg, flakes	92	2
		Egg, powdered, dried	93	4
		Egg, dried	98	2
		Egg, powdered, defatted	100	2
		Egg, scrambled	96	2
		Egg, spray dried	92	2
		Egg, whole, unprocessed	97	2,9
Legumes and oilseed products	Beans (*Mucunoa* spp.)	Beans, velvet	68	4
	Beans (*Phaseolus lunatus*)	Beans, butter	57	4
		Beans, lima	78	4
	Beans (*Phaseolus vulgaris*)	Beans, black	69	4,10
		Beans, brown, cooked	79	11
		Beans, common	82	4
		Beans, haricot	71	4
		Beans, kidney	81	8
		Beans, Natal round yellow	80	4
		Beans, pinto, canned	73	8
		Beans, red	78	4

Table A1.2 (Continued)

Major product group	Subgroup	Product	True digestibility	Ref.
Legumes and oilseed products (cont.)		Beans, seafarer	84	4
		Beans, snap, frozen	82	11
		Beans, spotted sugar	81	4
		Beans, sugar	69	4
		Beans, sugar, speckled	78	4
		Beans, white kidney	78	4
	Beans (*vicia faba*)	Beans, broad	87	4
		Beans, faba	86	10
	Cottonseed	Cottonseed	78	2
		Cottonseed meal	80	4
		Cottonseed, glandless flour	98	2
	Flaxseed	Flaxseed	85	4
	Lentils (*Lan culinaris*)	Lentils	85	4,8
	Lupine (*Lupinus albus*)	Lupine	76	4
	Peanut products	Peanut butter	95	2
		Peanut flour	93	2
		Peanuts	87	4
		Peanut meal	91	12
	Peas (*Cajanus cajan*)	Pidgeon peas	76	13
		Pidgeon peas, raw	41	13
	Peas (*Cicer arietinum*)	Chick peas, canned	88	8
	Peas (*Pisum sativum*)	Pea concentrate	94	3
		Peas	88	14
		Peas, century, autoclaved	83	10

		Peas, green, frozen	94	11
		Peas, trapper, autoclaved	84	10
		Peas, yellow, cooked	86	11
		Pea flour	88	10
		Peas, Alaskan field	88	2
	Peas (*Vigna unguiculata*)	Cowpeas	79	4
	Sesame	Sesame seed, dehulled	82	4
	Soy products	Soybean	91	4
		Soy concentrate	95	4
		Soy flour	84	2
		Soy flour, defatted	87	2
		Soy isolate	96	2
		Soy protein, spun	100	2
	Sunflower	Sunflower seed	82	2
		Sunflower seed, flour	90	2
Meat and meat products	Beef	Beef	95	2
		Beef, low fat, ground	91	2
		Beef, powdered, defatted	97	2
		Beef, salami	98	8
		Beef, stew	89	8
		Beefsteak	97	2
		Beef tenderloin, roasted	91	2
	Fish and seafood	Haddock	100	4
		Sardine	95	4
		Shark	72	4
		Tuna, canned	90	2
	Luncheon meats	Canned frankfurters	97	4
		Chicken, frankfurters	97	8
		Sausage	94	8

Table A1.2 (Continued)

Major product group	Subgroup	Product	True digestibility	Ref.
Meat and meat products (cont.)	Pork	Pork loin and tenderloin	98	2
	Poultry	Chicken	100	4
		Chicken, dark meat	92	4
		Chicken, light meat	93	4
		Turkey breast, roasted	91	2
Miscellaneous foods		Macaroni/Cheese, canned	94	3
Nuts and nut products		Cashew	85	4
		Coconut, meal (defatted)	80	4
		Pecan	71	4
Starchy roots, tubes		Potato	89	4
Vegetables		Cabbage	88	4
		Kale	85	4
		Rape	85	4
		Mustard	82	4
		Turnip leaves	86	4
		Mushrooms	90	4

Notes

True digestibility values obtained using adult subjects were considered first followed by data using the rat as the animal model. When more than one value was considered the values were averaged. Data sources:

1. B. O. Eggum, B. Pederson, and I. Jacobson, "The influence of dietary tea, coffee, and cocoa on protein and energy utilization of soybean meal and barley in rats," *Brit. J. Nutr*. 50:197–205, 1983.

2. D. T. Hopkins, "Effects of variation in protein digestibility," in *Protein Quality in Humans: Assessment and In Vitro Estimation*, C. E. Bodwell, J. S. Adkins, and D. T. Hopkins, (eds.), AVI Publishing Company INC, Westport, CT, pp. 169–193, 1981.

3. F. E. McDonough, F. H. Steinke, G. Sarwar, B. O. Eggum, R. Bressani, P. J. Huth, W. E. Barbeau, G. V. Mitchell, and J. G. Phillips, In vitro rat assay for true protein digestibility: Collaborative study, *J AOAC* 73(5):801–805, 1990.

4. FAO, *Amino Acid Content of Foods and Biological Data on Proteins*, Food Policy and Food Science Service, Nutrition Division, FAO of the United Nations, Rome, 1970.

5. H. Miyoshi, T. Okuda, K. Okuda, and H. Koishi, "Effects of brown rice on apparent digestibility and balance of nutrients in young men on low protein diets," *J. Nutr. Sci. Vitamin.* 33:207–218, 1987.

6. B. O. Eggum, B. O. Juliano, M. G. B. Ibabao, C. M. Perez, and V. R. Carangal, "Protein and energy utilization of boiled rice-legume diets and boiled cereals in growing rats," *Plant Foods for Hum. Nutr.* 37:237–245, 1987.

7. R. Fragne and J. Adrian, "Protein value of some industrial dairy products," *Internat. J. Vitam. Nutr. Res.* 56:391–393, 1986.

8. G. Sarawar, "Digestibility of protein and bioavalability of aminio acids in foods. Effects on protein quality assessment," *World Rev. Nutr. and Dietet.* 54:26–70 Karger, Basel, 1987.

9. K. Kaneko and G. Koike, Utilization and requirement of egg protein in Japanese women, *J. Nutri. Sci. Vitamin.* 31:43–52, 1985.

10. G. Sarwar and F. E. McDonough, Evaluation of protein digestibility—Corrected amino acid score method for assessing protein quality of foods, *J. AOAC* 73(3):347–355, 1990.

11. M. Lowgren and L. Hambraeus, Protein quality in legumes—Effect of soaking and heat treatment and amino acid supplementation on nitrogen retention in rats, *Nutr. Rep. Internat.* 38(4):745–756, 1988.

12. FAO, *Protein Quality Evaluation*, Report of the Joint FAO/WHO Expert Consultation, Rome, 1991.

13. A. G. Griffiths and G. P. Savage, The composition and nutritive value of pigeon pea (*Cajanus cajan* Millsp.), *Nutr. Abstr. Rev. (Ser. A)* 61(4):261–283, 1991.

14. R. Mongeau, G. Sarwar, R. W. Peace, and R. Brassard, "Relationship between dietary fiber levels and protein digestibility in selected foods as determined in rats," *Plant Foods Hum. Nutr.* 39:45–51, 1989.

15. I. Hansen, T. Larsen, K. E. Bach-Knudsen, and B. O. Eggum, "Nutrient digestibilities in ingredients fed alone or in combination," *Brit. J. Nutr.* 66:27–35, 1991.

Source: *Federal Register* 58(3):2193–2195, January 6, 1993.

Table A1.3 A Comparison of FDA and FSIS Nutrition Labeling Requirements

Item	FDA final rule	FSIS final rule
Compliance date	May 8, 1994	July 6, 1994
Mandatory labeling	Processed foods (other than meat and poultry) including fish	Most processed meat and poultry products other than single ingredient raw meat and poultry.
Nutrition information	"As packaged[b]" "As consumed" (optional)	Same
Compliance	Industry responsibility. Compliance will be verified by chemical analysis.	In addition to chemical analysis, data base values, recipe analyses based on data base values, and combinations of these methods are permitted for compliance.
Voluntary labeling	Raw product and seasonal, unpackaged or packaged at retail	Single ingredient raw meat and poultry, fresh or frozen.
Nutrition information	"As consumed" for raw fish. "As packaged" for raw fruits and vegetables	"As packaged" or "as consumed."
"As packaged"	Nutrition information based on serving size of product in package without further preparation.	Same
"As consumed"	Nutrition information provided on cooked fish and game meat, using common cooking methods without addition of components that could alter nutrient profile.	Nutrition information provided on cooked meat and poultry using common cooking methods without addition of components that could alter nutrient profile.
Point of purchase labeling	Permits placards, pamphlets, brochures	Same

Voluntary compliance	60% of retailers surveyed must comply for at least 90% of the 20 most commonly consumed raw fruits, vegetables and fish by May 1993. Otherwise, mandatory program will be developed. Stores will be surveyed every 2 years to ensure compliance	60% of stores surveyed must comply for at least 90% of major cuts by July 1994. If participation not satisfactory, FSIS will investigate course of action.
Data for nutrient values	Allows use of data provided by FDA; products using FDA data will not be subject to compliance procedures.	Allows use of USDA *Handbook No. 8* for nutrient values; products using *Handbook No. 8* will not be subject to compliance procedures if nutrient content claims are not made.
Nutrition label content	*Mandatory disclosures* Calories Calories from fat Total fat (grams + %DV) Saturated fat (grams + %DV) Cholesterol (milligrams + %DV) Sodium (milligrams + %DV) Total carbohydrate (grams + %DV includes dietary fiber) Dietary fiber (grams + %DV) Sugars (grams) Protein (grams) Vitamin A (%DV) Vitamin C (%DV) Calcium (%DV) Iron (%DV)	*Mandatory disclosures* Same
	Voluntary disclosures Calories from saturated fat	*Voluntary disclosures* Same

Table A1.3 (Continued)

Item	FDA final rule	FSIS final rule
	Polyunsaturated and monounsaturated fats (grams, unless a claim is made about fatty acid or cholesterol content)	Same, plus stearic acid
	Soluble and insoluble fiber, sugar alcohols, and other carbohydrates (grams)	Same
	Protein as %DV for foods other than foods for infants and children under 4 years of age	Same
	Potassium (milligrams + %DV)	Same
	Thiamine, riboflavin, niacin, and other vitamins and minerals (%DV)	Same
	Voluntary disclosures become mandatory if a claim is made about a nutrient, except "calories from saturated fat" and "other carbohydrates."	Same
Simplified label format	May be used when 7 or more of the mandatory nutrients are present in insignificant amounts.	May be used when any of the required nutrients; except for the 5 core nutrients: calories, total fat, total carbohydrate, protein, and sodium are present in insignificant amounts.
	Insignificant amount is that which may be declared as zero—except for total carbohydrate, dietary fiber, and protein, where it is an amount that may be labeled as "less than 1 gram."	
Exemptions from mandatory rule	*Standard exemptions*	*Standard exemptions*
	Products for export	Same
	Custom slaughter products	Same
	Products used in institutional food services (e.g., hospitals, prisons)	Restaurant menus
	Products intended for further processing	

Exemptions from mandatory rule (cont.)	*Other exemptions, unless a nutrient content claim is made*:	
	Small businesses: based on number of full time equivalent employees and units within a product line made yearly (phase in process-first year, begin with fewer than 300 employees and less than 600,000 units of product; end with fewer than 100 employees and less than 100,000 units after 3 years).	Small businesses with 500 or fewer employees and 100,000 pounds or less production of a particular product (phase in process-begin with 250,000 pounds and end with 100,000 pounds after 3 years).
	Small packages-a total surface area of less than 12 square inches available to bear labeling.	Same, in addition to small individually wrapped packages of less than 1/2 ounce net weight.
		Products intended for further processing.
	Ready-to-eat foods primarily prepared, processed, or portioned to consumer specification at retail establishments.	Ready-to-eat products that are packaged or portioned at retail, multi-ingredient products processed at retail.
	Restaurants	Products not for sale to consumers, including foods used in institutional food services.
Serving sizes	The amount customarily consumed per eating occasion by persons 4 years of age or older.	Same
	139 food product categories with reference amounts; including 11 groups of food specially formulated or processed for infants or children under 4 years of age.	23 meat, 22 poultry product categories with reference amounts.
	Serving size for raw fish is 3 oz. cooked.	Serving size for single ingredient raw meat and poultry is 3 oz. cooked.
	Household measures (cups, tablespoons, teaspoons, fl. oz., pieces, or oz.).	Same, except for fl. oz.

Table A1.3 (Continued)

Item	FDA final rule	FSIS final rule
	When household measure is oz., visual representation required.	Not required.
	Meals-weigh 10 oz. or more, contain 3 components from 2 or more food groups. Main dish items-weigh 6 or more ounces; contain 2 components from 2 or more food groups.	Meal-type products (dinners, entrees) that weigh 6-12 oz. to be labeled in their entirety.
	Units/pieces of varied sizes (e.g., seafood, pickles), labeled in the number of ounces closest to the reference amount plus the approximate number of pieces.	Units or pieces of varied sizes. Serving size expressed as a number and labeled to the average.
	Individual units that weight between 67 and 200 percent of the reference amount must use 1 unit as their serving size.	Same
Percent Daily Values (%DV)	Percent of Daily Values and daily values (represent Recommended Daily Intakes and Daily Reference Values) will appear on food labels to put food in context of a healthful daily diet.	Same
	Establishes RDIs and DRVs, which are meant to serve as reference values, to help consumers compare how nutrient levels in foods contribute toward a healthy diet.	Same
	RDIs established for 19 vitamins and minerals.	Same
	DRVs established for total fat, saturated fat, cholesterol, protein, total carbohydrate, dietary fiber, sodium and potassium.	Same

Nutrient content claims	FDA has general requirements for the use of descriptors and definitions for several nutritional components (e.g., calories, fat, saturated fat, sodium, and cholesterol). Listed below are some of the definitions.	FSIS adopts FDA's requirements for the use of descriptors and definitions for several nutritional components.
	High: Product contains 20% or more of the DRV or RDI per reference amount.	Same
	Good source of: Product contains 10–19% of the DRV or RDI per reference amount.	Same
	Fat free: Product contains less than 0.5 g of fat per reference amount and labeled serving for individual foods and per labeled serving for meal-type products, and no added ingredient that is fat or oil unless identified as trivial in the ingredient statement.	Same
	Low fat: Product contains 3 g or less of fat per Reference Amount for individual foods and per 100 g for meal-type products that have no more than 30% of calories from fat.	Same
	Light or lite: If product derives more than 50% or more of its calories from fat, its fat content is reduced by 50% or more compared to the	Same

Table A1.3 (Continued)

Item	FDA final rule	FSIS final rule
	reference food. If the product derives less than 50% of its calories from fat, the fat is reduced by 50% or more or the number of calories is reduced by at least one-third compared to the reference food.	
	Reduced/less fat: Product has reduced fat content of 25% or more.	Same
	More/added/enriched: Amount of nutrient in the product exceeds the amount in the reference food by at least 10% of the DRV or RDI per reference amount.	Same
	Percent fat-free: Only used on products that meet low fat definition.	Same; permit *percent lean* claims, synonymous with percent fat-free claims.
	"Lean" and "extra lean": Same as USDA-regulated single-ingredient raw meat and poultry product for use with a FDA-regulated seafood or game meat product. Same as USDA-regulated multi-ingredient meat or poultry meal-type product for use with all FDA-regulated meal-type products.	*"Lean" and extra lean":* Descriptors for use with all meat and poultry products, including meal-type, multi-ingredient products and raw single ingredient meat and poultry cuts.
	Same	*Lean:* Meat or poultry products with less than 10 g of fat, 4.5 or less g of saturated fat, and less than 95 milligrams cholesterol per 100 g, and per Reference Amount for individual foods, and per 100 g and labeled serving size for meal-type products.

Nutrient content claims (cont.)	Same	*Extra-lean:* Meat or poultry products with less than 5 g of fat, less than 2 g of saturated fat, and less than 95 mg cholesterol per 100 g and Reference Amount for individual foods, and per 100 g and labeled serving for meal-type products.
Health claims	Eight health claims are permitted on foods only where claims are specifically provided for by regulations and where no other nutrient is present in an amount that increases the risk of disease. The claim must be supported by publicly available scientific evidence and there must be significant scientific agreement among qualified scientists	No provisions in January 1993 rule; will publish proposal at later date.
"Healthy"	Final rule effective May 8, 1994 for foods labeled with the claim for the first time; January 1, 1996 for pre-existing products.	Final rule effective November 1995 for meat and poultry products.

[a] Single ingredient fish or game meat products may provide nutrition information as consumed in lieu of as packaged.

Source: *A Comparison of FDA and FSIS Nutrition Labeling Requirements*, USDA/FSIS Publication, May 1944.

Table A1.4 FDA Contact Persons for Food Labeling Questions

Department	Contact	Telephone
Analytical or "laboratory" methods	Dr. Martin Bueno	202-205-4455
	Mr. Mike Deutsch	202-205-4263
Available label space	Dr. Raymond Schucker	202-205-5676
Calorie labeling and calorie factors	Ms. Virginia Wilkening	202-205-5763
Corrections/technical amendments	Mr. Gerad McCowin	202-205-4561
Current labeling provisions (not NLEA)	Mr. James Summers	202-205-5229
Databases for nutrient content	Dr. May Bender	202-205-5592
Dietary supplement labeling, nutrient content claims	Ms. Camille Brewer	202-205-5966
Dietary supplement labeling, nutrition labeling	Ms. Susan Thompson	202-205-5587
Dietary supplement labeling, health claims	Mr. James Taylor	202-205-5099
Economic impact/regulatory impact analysis	Dr. Richard Williams, Jr.	202-205-5271
Enforcement strategy/priorities	Ms. Elizabeth Campbell	202-205-5229
Exemptions and special nutrition labeling provisions	Ms. Virginia Wilkening	202-205-5763
Fat labeling	Ms. Susan Thompson	202-205-5587
Fiber labeling	Dr. James Hoadley	202-205-5593
Folate or folic acid	Dr. Jeanne Rader	202-205-5375
Food labeling education	Ms. Naomi Kulakow	202-205-8682
Format for nutrition labeling	Ms. Virginia Wilkening	202-205-5763
	Dr. Raymond Schucker	202-205-5657
	Mr. Donald Plumb	202-205-5229
Fresh fruits/vegetables/fish labeling	Dr. Jean Pennington	202-205-5434
Health claims (general)	Dr. Christine Lewis	202-205-5588
"Healthy" proposal	Ms. Felicia Satchell	202-205-5099
Ingredient labeling (new regulations)	Ms. Felicia Satchell	202-205-5099
Metric labeling	Mr. James Taylor	202-205-5099
Net content inspection procedures	Dr. Margaret Cole	202-205-4291
Nutrient content claims	Ms. Connie Henry	202-205-5229
Nutrition labeling details	Ms. Virginia Wilkening	202-205-5763
Part 101.9(g) (9) exemptions	Ms. Carol Adler	202-205-5494
Percent juice labeling	Ms. Nan Rainey	202-205-5099
Petition process	Ms. Gerad McCowin	202-205-4561
Product dating	Mr. James Summers	202-205-5229
Protein	Dr. Geraldine Mitchell	301-594-5829
BST labeling	Ms. Elizabeth Campbell	202-205-5229
RDIs/DRVs/DVs	Dr. Christine Lewis	202-205-5588
Restaurant labeling	Dr. Michelle Smith	202-205-5099

Table A1.4 (Continued)

Department	Contact	Telephone
Serving size	Dr. Ellen Anderson	202-205-5662
Slack fill	Dr. Michelle Smith	202-205-5099
Slides	Mr. John Tisler	202-205-5215
Small business	Mr. Gerad McCowin	202-205-4561
Standards of identity	Ms. Nan Rainey	202-205-5099
State exemption petitions	Dr. Michelle Smith	202-205-5099
State enforcement	Mr. Gerad McCowin	202-205-4561
Sugars/sugar alcohols labeling	Dr. James Hoadley	202-205-5593
USDA/FDA regulations comparisons	Ms. Nancy Crane	202-205-5615

Updated lists may be obtained from: Center for Food Safety and Applied Nutrition, Office of Food Labeling (HFS-150), 200 C Street, S.W., Washington, D.C. 20204. Telephone: 202-205-4561, Fax: 202-205-4594.

Table A1.5 USDA Contact Persons for Food Labeling Questions

Department	Contact	Telephone
Analytical or "laboratory" methods	David Soderberg	202-205-0063
Compliance and data bases	Linda Posati	202-254-2576
Current label provisions (not 1/6/93 regulations)	Food Labeling Division	202-254-2569
Education program	Patricia White	202-254-2576
Enforcement directive	Lynn Dickey	202-254-2576
Exemptions/exclusions from nutrition labeling	Linda Singletary	202-254-2576
Format for nutrition label	Linda Singletary	202-254-2576
Ground beef and hamburger labeling	Barbara Dwyer	202-254-2576
Health claims	Carol Fletcher	202-254-2576
"Healthy"	Carol Fletcher	202-254-2576
Ingredient labeling	Robert Post	202-254-2588
Listing nutrients on the nutrition label	Linda Singletary	202-254-2576
Nutrient content claims	Barbara Dwyer	202-254-2576
Placement of nutrition labeling: proposed rule	Linda Posati	202-254-2576
Prior label approval: proposed rule	Kathleen Leddy	202-254-2596
Reference amounts (RACC's)	Tara Kuspis	202-254-2576
Safe handling instructions	Sally Jones	202-254-2596
Serving size	Tara Kuspis	202-254-2576
Small business exemption	Carol Fletcher	202-254-2576

Updated lists may be obtained from: Center for Food Safety and Applied Nutrition, Office of Food Labeling (HFS-150), 200 C Street, S.W., Washington, D.C. 20204. Telephone: 202-205-4561, Fax: 202-205-4594.

Appendix 2

Enforcement Policy Statement on Food Advertising

I. INTRODUCTION

The Federal Trade Commission (FTC) is issuing this statement to provide guidance regarding its enforcement policy with respect to the use of nutrient content and health claims in food advertising. The Commission believes the statement is appropriate in light of the passage of the Nutrition Labeling and Education Act of 1990 (NLEA),[1] and the Food and Drug Administration's (FDA) January 6, 1993, issuance of food labeling regulations implementing the NLEA.[2]

The FTC, FDA, and USDA share jurisdiction over claims made by manufacturers of food products pursuant to a regulatory scheme established by Congress through complementary statutes. Section 5 of the Federal Trade Commission Act (FTC Act) (hereinafter "Section 5") "prohibits unfair or deceptive acts or practices," and, in the case of food products, Sections 12 and 15 of the FTC Act prohibit "any false advertisement" that is "misleading in a material respect."[3] FDA's authority is embodied in part in Section 403(a) of the Federal Food, Drug, and Cosmetic Act (FDCA) which prohibits "labeling [that] is false or misleading in any particular."[4] Since 1954, the FTC and the FDA have operated under a Memorandum of Understanding,[5] under which the Commission has assumed primary responsibility for regulating food advertising, while FDA has taken primary responsibility for regulating food labeling.[6]

The NLEA amended Section 403 of the FDCA and effected broad changes in the regulation of nutrition claims on food labels. In addition to requiring nutrition information on virtually all food products, the NLEA directed FDA to standardize and limit the terms permitted on labels, and allows only FDA-approved nutrient content claims and health claims to appear on food labels.[7] While the NLEA is designed in part to prevent deceptive and misleading claims on labels, Congress also intended that nutrient content and health claims educate consumers in order to assist them in maintaining healthy dietary practices.[8] The NLEA also mandated that FDA undertake a consumer education effort to educate consumers about the new food label and the importance of diet to health.[9] Therefore, in keeping with its recently expanded and unique jurisdictional mandate, the requirements set forth in FDA's regulations have a broader purpose than preventing false and misleading claims in food labeling.

The NLEA applies only to labeling and did not change the FTC's statutory authority to prohibit deceptive acts or practices under Section 5 of the FTC Act. Nevertheless, in light of the comprehensive regulatory scheme established for food labeling claims by the NLEA, the Commission is issuing this statement to clarify how its own authority relates to issues raised by FDA's food labeling regulations.

The Commission recognizes the importance of consistent treatment of nutrient content and health claims in food advertising and labeling and seeks to harmonize its advertising enforcement program with FDA's food labeling regulations to the fullest extent possible under the statutory authority of the FTC Act. The Commission also recognizes the scientific expertise of FDA in this area. The Commission has traditionally accorded great weight to FDA's scientific determinations in matters of nutrition and health and will continue to do so. In addition, as a general matter, it is unlikely that the Commission will take action under Sections 5 and 12 of the FTC Act regarding nutrient content and health claims if they comply with FDA's regulations.[10]

The principal elements of the Commission's authority to regulate nutrient content and health claims in food advertising are set forth below in the discussion of the Commission's legal framework in Part II of this statement. Part III of the statement addresses the Commission's approach to harmonization with the NLEA and FDA's regulations in the area of nutrient content claims in food advertising. Part IV of the statement addresses the Commission's approach to health claims in food advertising. Claims made in food advertising may raise issues addressed in more than one section of this statement. Advertisers, therefore, should comply with all relevant provisions of the statement and not simply the provision that seems most directly applicable.

In issuing this statement, the Commission recognizes that the FDA intends its regulatory approach to be dynamic, designed to respond to changes in science and consumer understanding of nutrition and diet-disease issues. Therefore, while the Commission's purpose in issuing this statement is to provide guidance on how it will enforce Sections 5 and 12 in the food advertising area, the statement is not intended to provide a comprehensive analysis of how each of FDA's regulations relates to the Commission's enforcement policy. Instead, this statement focuses on the general issues that are likely to remain relevant to the Commission's regulation of food advertising over time, as specific provisions in the FDA regulations are amended.

II. LEGAL FRAMEWORK FOR COMMISSION ACTION

As noted above, the FTC regulates food advertising under its statutory authority to prohibit deceptive acts or practices under Section 5 of the FTC Act. The Commission has set forth its interpretations of this authority in its Deception Policy Statement[11] and its Statement on Advertising Substantiation.[12] FTC food cases, applying the principles articulated in these statements, have also established a growing body of precedent against which food advertisers can assess the lawfulness of their claims.[13]

As set out in the Deception Statement, the Commission will find an advertisement deceptive under Section 5 and, therefore, unlawful, if it contains a representation or omission of fact that is likely to mislead consumers acting reasonably under the circumstances, and that representation or omission is material.14

The first step in a deception analysis is to identify representations made by an advertisement. A representation may be made by express or implied claims. An express claim directly makes a representation. The identification of an implied claim requires an examination of both the representation and the overall context of the ad,[15] including the juxtaposition of phrases, images, and the nature of the claim and the transaction.[16] In other words, in ascertaining the meaning of an advertisement, the Commission will focus on the ad's overall net impression.[17]

In addition to deception arising from affirmative representations in an advertisement, the omission of material information may also be deceptive in certain circumstances. First, deception can occur through omission of information that is necessary to prevent an affirmative representation from being misleading.[18] Second, "it can also be deceptive for a seller to simply remain silent, if he does so under circumstances that constitute an implied but false representation."[19] However, "[n]ot all omissions are deceptive, even if providing the information would benefit consumers."[20] As with advertisements that contain affirmative representations, the test for whether an omission is deceptive is whether the overall impression created by the ad is deceptive.[21]

The next step in identifying deception in an ad requires the Commission to consider the representation from the perspective of a consumer acting reasonably under the circumstances.[22] Finally, a representation must be material, i.e., likely to affect a consumer's choice or use of a product or service.[23] Express claims and claims involving health or safety are presumptively material.[24]

In addition, objective claims carry with them the implication that they are supported by valid evidence. It is deceptive, therefore, to make an express or implied nutrition or health benefit claim for a food unless, at the time the claim is made, the advertiser possesses and relies upon a reasonable basis substantiating the claim.[25] A reasonable basis consists of competent and reliable evidence. In the context of nutrient content or health claims, substantiation will usually require competent and reliable scientific evidence sufficient to support the claim that is made.[26] Commission orders generally require that scientific evidence consist of tests, analyses, research, studies or other evidence conducted and evaluated in an objective manner by persons qualified to do so, using procedures generally accepted in the relevant profession to yield accurate and reliable results.[27] The substantiation must also be examined in the context of the entire body of relevant evidence, particularly if it produces results that are contrary to that body of evidence.

III. CONTENT CLAIMS

A. Describing the Absolute and Comparative Nutrient Content of Foods

As mandated by the NLEA, FDA's regulations define certain absolute and comparative terms that can be used to characterize the level of a nutrient in a food. "Absolute" terms (e.g., "low," "high," "lean") describe the amount of nutrient in one serving of a food. "Relative" or comparative terms (e.g., "less," "reduced," "more") compare the amount of a nutrient in one food with the amount of the same nutrient in another food. With very few exceptions, only these specific terms, and certain approved synonyms, may be used on food labels to characterize the level of a nutrient, although interested parties may petition FDA to authorize new nutrient content terms and synonyms.[28]

1. Absolute Nutrient Content Claims

Prior to the finalization of FDA's regulations, there was no comprehensive set of standardized definitions for absolute terms such as "low" and "high" to describe the level of a nutrient in a food. Now that FDA has established a standard metric to describe the nutrient content of foods, the Commission will apply FDA's definitions for absolute nutrient content terms when those terms are used in the same context in advertising. In general, the Commission will use FDA's serving size or reference amounts customarily consumed, as set forth in FDA's regulations, in its analysis of a claim. If, however, an advertiser chooses to depict a non-standard serving size in an advertisement, the Commission will require the advertiser to meet the FDA's standard both for the reference amount customarily consumed and for the serving size depicted.[29]

The Commission has previously indicated that where a claim is subject to the joint jurisdiction of the FTC and the FDA, it will accord significant deference to the FDA's standards.[30] Consumer understanding will be improved if the agencies responsible for regulating the use of express or implied absolute nutrient content descriptors have consistent requirements for use of these terms. Multiple governmental definitions for the same terms would have the potential to mislead consumers.[31]

Similarly, the use in advertising of FDA-defined terms in a manner inconsistent with FDA's definitions is likely to mislead consumers. The uniform and detailed nutrient content information required on food labels, as well as the NLEA-mandated educational effort, are likely to familiarize consumers with both the FDA-defined terms and their definitions, further reinforcing consumer expectation that nutrient content terms are consistently applied.

Furthermore, the principle that certain claims may be deceptive unless they are based on a common standard of measurement or testing is well founded under Section 5.[32] At the same time, statements that a food is "high" or "low" in a particular nutrient are objective product claims that imply support by a reasonable basis.[33] The Commission generally determines what level of substantiation constitutes a reasonable basis by weighing the six factors set forth in *Pfizer, Inc.* and subsequent cases.[34] Applying those factors here leads the Commission to conclude that to avoid deception, advertisers should meet FDA's definitions for absolute nutrient content claims.

Where FDA has not established any standard metric, such as "low" or "high," for a specific nutrient, the Commission will closely review claims in food advertising that characterize the level of that nutrient.[35] The Commission has traditionally deferred to FDA's scientific and public health determinations, and will consult with FDA and other government and public health authorities regarding the significance of the nutrient for which such a claim is made.

2. Comparative Nutrient Content Claims

FDA's regulations also establish definitions for comparative terms that characterize the nutrient content of a labeled food relative to that of a comparison or "reference" food. These definitions require that a food bearing a comparative term meet specified minimum percentage differences in the relevant nutrient. For example, the regulations permit use of the terms "less" and "reduced" only where there is a minimum 25 percent difference in the relevant nutrient. In addition, comparative claims must disclose the reference food, the percentage difference in the nutrient between the labeled and reference food (e.g., "50 percent less fat than our regular cheese"), and quantitative information regarding the absolute amount of the nutrient in the labeled and reference foods (e.g., "fat reduced from 6 g. to 3 g. per serving").

Comparative nutrient content claims that comply with FDA's regulations will generally comply with Section 5.[36] The Commission will scrutinize carefully comparative nutrient content claims that characterize nutrient differences in ways that do not comply with FDA's regulations. However, a comparative advertising claim that is accurately qualified to identify the nature of a nutrient difference and to eliminate misleading implications[37] may comply with Section 5, even if the nutrient difference does not meet FDA's prescribed differences for purposes of labeling.[38]

In examining comparative claims, several principles are likely to be applied by the Commission. First, comparative claims should make clear the basis for the comparison.[39] Claims should identify the reference food to which the product is being compared to so that the appropriate comparison is clear to consumers. Second, consistent with the position it has taken on the use of descriptors, the Commission believes that advertisers using unqualified comparative terms must meet FDA's minimum percentage difference requirements for those claims. For example, if an ad represents that a food has "less fat than Brand X," without indicating the percentage or absolute difference in fat, the Commission will rely on FDA's 25% minimum difference requirement in determining whether the claim is deceptive.

Third, comparative claims should not overstate the significance of a nutrient difference.[40] For this reason, some comparative claims may need to be qualified in a manner sufficient to ensure that consumers are not misled regarding the significance of the nutrient difference. For example, a simple statement of percentage difference for a food that contains only a small amount of a nutrient, such as "our crackers have one-third less fat than Brand X," may suggest that the nutrient difference is greater in an absolute sense than it actually is. This type of claim may need further qualification to prevent the claim from creating a misleading impression (e.g., "one third less fat than Brand X—theirs has 3 g., ours has 2 g.").

Even where nutrient differences are substantial in an absolute sense, careful qualification may be necessary for products that despite such absolute reductions, still contain appreciable amounts of a nutrient, to ensure that consumers are not misled regarding the absolute level of the nutrient. Thus, a claim such as "20% less fat in our frozen entree compared to Brand X," regarding a product that nevertheless contains a significant amount of fat, may need to identify the quantitative amount of fat in the advertised food and the reference food (e.g.,

"20% less fat than Brand X—Brand X has 25 g. fat, ours has 20 g. fat"), particularly in situations where consumers are not likely to be aware that the item is generally high in fat.

In summary, the Commission ordinarily will not challenge comparative nutrient content claims that comply with FDA's regulations, and will carefully scrutinize comparative nutrient content claims that characterize nutrient differences in ways that do not comply with FDA's regulations.[41]

3. Synonyms for Nutrient Content Claims

In addition to authorizing the use of only a limited set of defined nutrient content terms on food labels, FDA's regulations authorize the use of only certain synonyms for these defined terms.[42] The impetus behind Congress's requirement that FDA limit defined terms and synonyms may be found in the educational and public health goals of the NLEA—to promote consumer understanding of the meaning of the terms through a limited lexicon that will allow consumers to make informed dietary choices.[43]

The Commission will examine advertising to ensure that claims that characterize the level of a nutrient, including those using synonyms that are not provided for in FDA's regulations, are consistent with FDA definitions. Commission precedent establishes that an advertisement that can reasonably be interpreted in a misleading way is deceptive, even though other, nonmisleading interpretations may be equally possible.[44] Thus, when express or implied claims suggest that a food product meets the standard for use of an FDA-defined term, advertisers should ensure that the food actually meets the relevant FDA standard. For example, depending on the context of an ad, use of the phrases "packed with" or "lots of" to describe the level of fiber in a food could convey to some reasonable consumers that the food is "high" in fiber. Because FDA's regulations define the terms "good source" and "high" with respect to fiber,[45] consumers are likely to be misled if a "high fiber" claim is implied by an ad for a food that is only a "good source" of fiber.

4. Implied Nutrient Content Claims

As defined in FDA's regulations, an implied nutrient content claim is a claim that:

> (i) Describes the food or an ingredient therein in a manner that suggests that a nutrient is absent or present in a certain amount (e.g., "high in oat bran"); or
> (ii) Suggests that the food, because of its nutrient content, may be useful in maintaining healthy dietary practices and is made in association with an explicit claim or statement about a nutrient (e.g., "healthy, contains 3 grams (g) of fat").[46]

Under this definition, statements about ingredients may or may not be nutrient content claims.[47] FDA has generally adopted a case-by-case approach to statements about ingredients that depends on the overall context of the label. The regulations also provide, however, that certain ingredient statements will be treated as nutrient content claims whenever they appear on labels.[48]

The Commission's approach to implied claims also relies on an analysis of the overall context in which a claim appears. As explained above, the Commission evaluates the overall impression created by an ad, including the ad itself, the arrangement of phrases and images in the ad, and the nature of the claim being made, in order to determine whether a representation is likely to mislead reasonable consumers.[49] If the net impression produced by an ad is likely to mislead reasonable consumers, the ad is deceptive and violates Section 5.

FTC food cases and consent agreements also demonstrate the principle that statements regarding ingredients may have nutrient content implications. For example, advertising may implicitly characterize the amount of a nutrient in a product through representations regarding the ingredients with which the product is made.[50] An ad may imply that a food is free of a particular nutrient by suggesting that the product is free of ingredients that are essentially the same from the consumer's perspective.[51]

Consistent with its statutory authority and its commitment to harmonization, the Commission will look closely at advertisements that may implicitly characterize the level of a nutrient. The Commission will give great weight to any FDA determinations concerning ingredient statements in analyzing the net impression conveyed by an ad.

B. Nutrient Content Claim Disclosures

As mandated by the NLEA, FDA's nutrient content labeling regulations require a number of disclosures. These mandated disclosures include, but are not limited to: (1) a referral statement to the nutrition panel, required whenever a nutrient content claim is made;[52] (2) disclosure of nutrients (fat, saturated fat, cholesterol, and sodium) present in a food at a level that FDA has concluded increases the risk of diet-related disease, required whenever a nutrient content claim is made;[53] and (3) "triggered" disclosures of the amount of certain related nutrients when claims concerning fiber, saturated fat, and cholesterol appear.[54]

As set forth in Part II above, disclosure of material information that is necessary to prevent deception may be required under Section 5 of the FTC Act.[55] For example, it is misleading to fail to disclose qualifying information necessary to prevent an affirmative statement from creating a misleading impression.[56] However, a seller's silence in circumstances that do not give a particular meaning to the silence is not deceptive.[57] The failure to provide nutrition information that consumers may find useful in improving their diet, while subject to challenge under the NLEA with respect to labels, therefore, is not necessarily subject to challenge as deceptive under Section 5.[58] In the context of advertising that makes affirmative nutrient content claims, the Commission's analysis of deception by omission will be based on a consideration of whether a nutrient content claim gives rise to a misleading impression absent disclosure of other nutrition information.

Some of FDA's disclosures appear designed to fulfill the educational goals of the NLEA, which are beyond the scope of the Commission's law enforcement mandate. For example, all nutrient content claims on a label must be accompanied by a statement referring the consumer to the nutrition panel, where complete nutrition information regarding the product is found.[59] While a complete nutrition portrait of a food may be useful to consumers, it is unlikely that the absence of this referral statement from an advertisement would render the ad deceptive to consumers.

In contrast, other disclosures mandated for food labels may also appropriately be required under certain circumstances to prevent deception in advertising under Section 5. In determining whether such disclosures are necessary to prevent deception, the Commission will consider several factors. First, the Commission will carefully evaluate nutrient content claims for foods that contain a nutrient at a level considered by FDA to increase the risk of a diet-related disease.[60] When the context of an ad as a whole conveys to consumers the net impression that the food makes only positive contributions to a diet, or does not contain any nutrients at levels that raise the risk of diet-related disease, the failure to disclose the presence of risk-increasing nutrients is likely to be deceptive.[61]

Second, the Commission will also scrutinize nutrient content claims for cholesterol, saturated fat, and fiber. Congress enacted "special rules"[62] requiring that claims for these nutrients trigger disclosure of other nutrients.[63] Consumers often may infer that certain nutrient claims imply a characterization of the amount of another nutrient. Similarly, where different nutrients are linked to the same health issue (for example, cholesterol and saturated fat, or dietary fiber and total fat), a claim regarding one of these nutrients is likely to give rise to a misleading impression regarding the benefit of the food absent disclosure of the presence of the other nutrient. Under these circumstances, the failure to correct these misimpressions through adequate disclosures is likely to be deceptive.

IV. HEALTH CLAIMS[64]

FDA's regulations for health claims in food labeling establish general standards for the use of claims that characterize the relationship of a substance in a food to a disease or health-related condition.[65] These general standards include, among other things: (1) limiting authorization of health claims only to those categories for which there is "significant scientific agreement" that the relevant diet-disease relationship is supported by the scientific evidence;[66] (2) establishing disqualifying levels for total fat, saturated fat, cholesterol, and sodium, above which foods are disqualified from bearing any health claims;[67] (3) for the specific substance that is the subject of a health claim, setting a threshold level for the amount of such substance in the food, that is either sufficiently low or sufficiently high to support the health claim;[68] (4) requiring that foods bearing health claims have some minimal nutritional value;[69] and (5) requiring that health claims identify those factors, other than dietary intake of the substance, that affect the diet-disease relationship.[70] In addition, as required by the NLEA, FDA's regulations provide a petition process for interested persons to seek FDA authorization of additional health claims.[71]

The Commission shares the concerns underlying the NLEA, and embodied in FDA's regulations, that health claims be adequately substantiated and presented in a manner that is truthful and not misleading. These same principles form the foundation of the Commission's well-established deception and advertising substantiation doctrines, described in Part II above. The Commission's approach to the regulation of health claims in food advertising and FDA's approach to such claims in labeling therefore share many basic elements.

A. Standard for Substantiation of Health Claims

The NLEA directed FDA to promulgate regulations authorizing claims about diet-disease relationships only if FDA determined,

> based on the totality of the publicly available scientific evidence (including evidence from well-designed studies conducted in a manner which is consistent with generally recognized scientific procedures and principles), that there is significant scientific agreement, among experts qualified by scientific training and experience to evaluate such claims, that the claim is supported by such evidence.[72]

The NLEA directed FDA to apply this "significant scientific agreement" standard in determining whether there was adequate substantiation to permit health claims for ten specific diet-disease relationships.[73] After reviewing the scientific literature, FDA issued regulations authorizing a number of specific categories of health claims.

The Commission's standard for substantiation of health claims in food advertising shares many elements with FDA's approach to such claims in labeling. Like FDA, the Commission imposes a rigorous substantiation standard for claims relating to the health or safety of a product, including health claims for food products.[74] The Commission's standard that such claims be supported by "competent and reliable scientific evidence" has been more specifically defined in Commission orders addressing health claims for food products to mean:

> tests, analyses, research, studies or other evidence based on the expertise of professionals in the relevant area, that have been conducted and evaluated in an objective manner by persons qualified to do so, using procedures generally accepted in the profession to yield accurate and reliable results.[75]

Thus, both the Commission and FDA look to well-designed studies, including clinical research and other forms of reliable and probative scientific evidence, in evaluating health claims for foods.

In addition, the Commission, like FDA, evaluates substantiation for health claims in the context of the surrounding body of evidence, and does not look to isolated studies, especially if those studies are unrepresentative of the larger body of evidence. However, the Commission does not require food advertisers to establish that there is scientific consensus in support of their claims. Similarly, FDA has clearly indicated that its "significant scientific agreement" standard does not require that such agreement represent a "full consensus among scientists."[76]

In evaluating health claims, the Commission looks to a number of factors to determine the specific level of scientific support necessary to substantiate the claim.[77] Central to this analysis is an assessment of the amount of substantiation that experts in the field would consider to be adequate. The Commission regards the "significant scientific agreement" standard, as set forth in the NLEA and FDA's regulations, to be the principal guide to what experts in the field of diet-disease relationships would consider reasonable substantiation for an unqualified[78] health claim.[79] Thus, it is likely that the Commission will reach the same conclusion as FDA as to whether an unqualified claim about the relationship between a nutrient or substance in a food and a disease or health-related condition is adequately supported by the scientific evidence.

The Commission also recognizes the importance of the petition process, established under the NLEA and FDA's regulations, as a mechanism for authorizing health claims in food labeling. The Commission will look with particular care at any health claims not specifically considered by the FDA in this process. The absence of an FDA determination that a health claim is scientifically valid will be a significant factor in the Commission's assessment of the adequacy of substantiation for the claim.[80]

While the Commission's approach to evaluation of unqualified health claims will generally parallel FDA's assessment of whether there is significant scientific agreement supporting the relevant diet-disease relationship, the Commission recognizes that there may be certain limited instances in which carefully qualified health claims may be permitted under Section 5 although not yet authorized by the FDA, if the claims are expressly qualified to convey clearly and fully the extent of the scientific support. At the same time, however, the Commission believes that qualified claims based on evidence that is inconsistent with the larger body of evidence have the potential to mislead consumers, and, therefore, are likely to violate Section 5.

The Commission recognizes the need to scrutinize closely qualified claims to maintain the credibility of health claims in food advertising and labeling. The Commission will therefore be especially vigilant in examining whether qualified claims are presented in a manner that ensures that consumers understand both the extent of the support for the claim and the existence of any significant contrary view within the scientific community.[81] In the absence of adequate qualification, the Commission will find such claims deceptive.[82]

B. Health Claims for Foods That Contain a Nutrient at a Level That Increases the Risk of a Disease

FDA's health claim regulations identify four nutrients—total fat, saturated fat, cholesterol, and sodium—the consumption of which has been associated with increased risk of certain diseases or health-related conditions, particularly cancer, cardiovascular disease, and hypertension. For each of these nutrients, the regulations establish levels above which foods containing the nutrient are disqualified from bearing health claims.[83] The disqualifying levels set by FDA were based on an analysis of what level of these nutrients in a food would increase, "to persons in the general population, the risk of a diet-related disease, taking into account the significance of the food in the total daily diet."[84]

The Commission will rely heavily on FDA's scientific determination as to what levels of total fat, saturated fat, cholesterol, and sodium may increase the risk of a diet-related disease or other health condition[85] and, while not necessarily prohibiting all health claims in advertising for such foods that contain such levels, will carefully scrutinize health claims for such foods to ensure that the claims are truthful and adequately qualified.[86] Situations involving risk-increasing levels established by FDA should not be interpreted as an exhaustive list of instances in which a broad, unqualified health claim for a food may be found deceptive by the Commission.

Unqualified health claims in advertising for such foods are likely to be deceptive when the risk-increasing nutrient is closely related to the subject health claim. Often the presence and significance of such a nutrient will have to be disclosed. Without such disclosures, consumers could infer from the health message that the food does not present any related health risks.[87] The failure to disclose the presence and significance of risk-increasing nutrients that are closely related to the health claim for such foods is likely to constitute an omission of a material fact and render the health claim deceptive.[88]

For example, a claim that a food will reduce the risk of one specified disease is likely to convey to reasonable consumers that the food will not increase the risk of some other health condition closely related to that disease. Thus, an unqualified claim that a food is low in saturated fat and cholesterol, and therefore compatible with a diet designed to reduce the risk of cardiovascular disease, would be deceptive if the food contained so much sodium that it might increase the risk of hypertension and thus, cardiovascular disease.[89] To prevent deception, a health claim for such a food is likely to need a disclosure that clearly conveys both the presence and significance of the risk-increasing nutrient.[90]

Even when the risk-increasing nutrient does not bear directly on the health condition that is the subject of the health claim, it may be necessary to disclose the presence of a risk-increasing nutrient. Depending on context, a specific health claim may convey to consumers a broader message that the food is healthful in all respects. For example, a health claim describing the benefits of calcium in reducing the risk of osteoporosis, when made in advertising for a dairy product that is high in saturated fat, may create the deceptive impression among reasonable consumers that consuming the dairy product will reduce the risk of osteoporosis without increasing the risk of any other health-related condition or disease, for example, heart disease. To prevent deception, a health claim for such a food may need to include a disclosure that conveys the presence and significance of the risk-increasing nutrient.[91]

In those instances, as outlined above, where disclosure of a risk-increasing nutrient level is necessary to prevent deception, the Commission will carefully scrutinize the disclosure to ensure that it is adequate to convey clearly the limited nature of the health claim being asserted.

C. Nutrient/Substance Levels Sufficient to Ensure Meaningful Health Benefits

In addition to establishing levels of total fat, saturated fat, cholesterol, and sodium, above which foods are disqualified from bearing health claims, FDA's regulations also establish threshold levels for the specific nutrients that are the subject of particular health claims made in food labeling. If a health claim is about the effects of consuming a substance at decreased dietary levels (e.g., lowering saturated fat and cholesterol intake to reduce the risk of coronary heart disease), FDA sets the threshold at a level that it determines is "sufficiently low to justify the claim."[92] If a claim relates to the effects of consuming the substance at other than decreased dietary levels (e.g., increasing calcium intake to reduce the risk of osteoporosis), FDA sets the threshold at a level that it determines is "sufficiently high to justify the claim."[93] In establishing these "high" and "low" thresholds, FDA specifically considered both whether these levels were sufficient to advance the public health policy of assisting consumers in maintaining healthy dietary practices,[94] and whether health claims for foods not meeting such thresholds would be "misleading because the nutrient levels [were] not low enough, or not high enough, to really contribute to the claimed effect."[95]

The Commission shares FDA's view that health claims should not be asserted for foods that do not significantly contribute to the claimed benefit. A claim about the benefit of a product carries with it the implication that the benefit is significant.[96] Thus, consistent with its position on the use of absolute nutrient content descriptors and unqualified comparative nutrient content claims, the Commission will ordinarily apply FDA's thresholds for specific nutrient levels in examining unqualified health claims for the specific nutrient levels that are the subject of the particular health claim.

The Commission recognizes, however, that there may be certain limited instances in which it is possible to craft a qualified, truthful, and nonmisleading claim comparing the relative health benefits of a food product to other products for which the food can be substituted, even if the nutrient level does not meet FDA's prescribed threshold for the food. Such comparative claims, encouraging consumers to substitute a food that is significantly lower or higher in the relevant nutrient than other foods in the same category, will be unlikely to mislead consumers if the claimed benefit from the substitution will contribute significantly to the claimed health effect.

In addition, such comparative claims must be sufficiently qualified to make clear to consumers that the benefit derives only from the substitution of the advertised food for a significantly less healthful alternative and that the subject product does not otherwise offer an overall health benefit. It may be necessary to disclose the actual level of the nutrient that is the basis for the claim and its significance to prevent deception.[97]

D. Minimum Nutritional Value for Foods Bearing Health Claims

Under FDA's regulations, any food bearing a health claim must not only meet the threshold level for the specific substance or nutrient that is the subject of the health claim, as discussed in Part IV, Section C., *supra*, but also must contain a sufficient amount of at least one of six nutrients and substances specified by FDA.[98] For example, a food that is sufficiently low in total fat to meet FDA's threshold level for a health claim about dietary fat and cancer would also need to contain one or more of the six specific nutrients or substances at a sufficient quantity to ensure that the food contributed significantly to a healthful diet. Like FDA's threshold levels, this rule ensures that health claims are reserved for foods that contribute significantly to a healthy diet.[99]

The Commission shares FDA's view that health claims may be misleading to the extent that they encourage consumers to choose foods that provide calories but have little or no nutritional value, under the mistaken belief that their choices will contribute to a healthy diet. The Commission believes that, like claims for foods that fail to meet FDA's threshold levels, health claims for foods with little or no positive nutritional value have the potential to be deceptive since they imply that the health benefit being asserted is significant.[100] Therefore, the Commission will generally give great deference to FDA's standards for minimum nutritional value for foods bearing unqualified health claims.

The Commission recognizes, however, that there may be some instances in which it is possible to craft a qualified, truthful, and nonmisleading claim comparing the relative health benefits of a food product to other products for which the food can be substituted, even if the food does not meet FDA's minimum nutritional value standards. While the food bearing such a qualified comparative health claim may not contribute in any absolute sense to a healthful diet, the substitution of such food for a less healthful food in the same category could result in a meaningful contribution toward the claimed health effect without detracting from the healthfulness of the overall diet.[101]

As noted in Part IV, Section C., *supra*, such comparative claims must be sufficiently qualified to convey clearly that the claimed health benefit derives only from the substitution of the advertised food for a significantly less healthful alternative.

E. Relevance of Dietary Factors to Claimed Health Benefit

For each category of health claims approved by FDA, the regulations present model health claim language that places the health benefits to be derived from consuming a nutrient in the context of other factors that bear on the relevant disease or health-related condition.[102] For example, in authorizing claims about calcium/osteoporosis, FDA developed model language explaining how other factors like gender, age, ethnicity, and exercise bear on the relationship between calcium consumption and osteoporosis.[103] FDA's model health claims are intended to ensure that health claims are complete, truthful and not misleading. The model statements therefore include reference to the fact that factors other than consumption of the food also bear on the claimed health effect.[104]

The Commission shares FDA's concern that health claims for food products may mislead consumers if they oversimplify the diet-disease relationship or otherwise overstate the relative significance of dietary factors in achieving certain health effects. Health claims in food advertising should therefore be sufficiently qualified to avoid implying to reasonable consumers that consumers can achieve the claimed effect simply by consuming the food and without regard to other factors, such as overall diet, exercise, age, or family history, that may either contribute or detract from the claimed effect.

However, while the Commission recognizes the desirability of educating consumers about the role of other factors that bear on the risk of disease and how such factors interact with diet, the Commission must evaluate whether the failure to disclose such qualifying information in a claim about the health effects of a food would mislead consumers. As explained above, not all omissions of information are deceptive in violation of Section 5. In assessing whether an omission is deceptive, the Commission examines whether the omitted information would be necessary to prevent an affirmative claim from creating a misleading impression.[105]

The Commission will not require food advertisers to include in advertising containing health claims all potentially relevant information about the specific diet-related disease, or affirmatively to disclose that the risk of the disease depends on many factors, unless such disclosure is necessary to prevent consumers from being misled about the significance of diet as one of those factors. Indeed, in many forms of advertising it would not be feasible to include all nutritional information that may be of interest to consumers. While the additional dietary and nondietary factors associated with a health condition may be of interest to consumers, in most cases Section 5 would not require full disclosure of such information to prevent consumers from being misled by statements about the contribution of a particular food to a health effect.

V. ENDNOTES

1. Nutrition Labeling and Education Act of 1990, Pub. L. No. 101-535, 104 Stat. 2353 (codified in part at 21 U.S.C. § 343(i), (q) and (r)).
2. Simultaneously, the Food Safety and Inspection Service (FSIS) of the United States Department of Agriculture (USDA) issued its own nutrition labeling regulations relating to meat and poultry products. While FSIS's regulations were not mandated by the NLEA, these regulations were intended to implement the NLEA's goals for products regulated by USDA. Although the principles in this statement relate to FDA's regulations, the Commission intends to apply similar principles to consideration of claims for products regulated by USDA.
3. 15 U.S.C. §§ 45, 52, 55 (1980).
4. 21 U.S.C. § 343(a). USDA's authority is derived from the Federal Meat Inspection Act, 21 U.S.C. § 601(n)(1) (prohibiting labeling of meat or meat products that is "false or misleading in any particular"), and the Poultry Products Inspection Act, 21 U.S.C. § 453(h)(1) (prohibiting labeling of poultry products that is "false or misleading in any particular").

5. Working Agreement Between FTC and Food and Drug Administration, 4 Trade Reg. Rep. (CCH) ¶ 9,850.01 (1971) (hereinafter "Memorandum of Understanding").
6. The Memorandum of Understanding also reaffirms the agencies' shared commitment to prevent deception of the public, to coordinate their work to eliminate duplication of effort, and to promote consistency in handling matters of mutual concern.
7. The NLEA defines a "nutrient content claim" as any claim that expressly or by implication "characterizes the level of any nutrient." 21 U.S.C. § 343(r)(1)(A) (Supp. 1990). A "health claim" is defined as any claim that characterizes the relationship of any nutrient to a "disease or health related condition." 21 U.S.C. § 343(r)(1)(B) (Supp. 1990).
8. "Health claims supported by a [sic] significant scientific agreement can reinforce the Surgeon General recommendations and help Americans to maintain a balanced and healthful diet. Similarly, statements regarding the level of these nutrients in foods will assist Americans in following the Surgeon General's guidelines." House Committee on Energy and Commerce, Nutrition Labeling and Education Act of 1990, H.R. Doc. No. 538, 101st Cong., 2d Sess. 9-10 (1990).
9. NLEA, § 2(c).
10. The Commission notes that the manner in which such information is conveyed in advertising may differ from the way it would be presented in labeling. The Commission cautions advertisers to consider carefully the importance of the context in which they make claims. Some claims that would technically comply with FDA's labeling regulations might be deceptive in advertising if the context of the ad renders the express message of the claim misleading.
11. *See Cliffdale Associates, Inc.*, 103 F.T.C. 110, 176 (1984), reprinting as appendix letter dated Oct. 14, 1983, from the Commission to The Honorable John D. Dingell, Chairman, Committee on Energy and Commerce, U.S. House of Representatives ("Deception Statement").
12. FTC Policy Statement on Advertising Substantiation, 48 Fed. Reg. 10,471 (1984), reprinted in *Thompson Medical Co.*, 104 F.T.C. 648, 839 (1984), *aff'd*, 791 F.2d 189 (D.C. Cir. 1986), *cert. denied*, 479 U.S. 1086 (1987) ("Substantiation Statement").
13. *See*, e.g., cases cited *infra* notes 26, 29, 32, 36, 50, 51, 74, 75, 81, 87, 96.
14. Deception Statement, 103 F.T.C. at 183.
15. *Kraft, Inc.*, FTC Dkt. No. 9208, slip op. at 7 (Jan. 30, 1991), *aff'd*, 970 F.2d 311 (7th Cir. 1992), *cert. denied*, 113 S. Ct. 1254 (1993) (citing *Thompson Medical Co.*, 104 F.T.C. at 789, 799; *Cliffdale Associates*, 103 F.T.C. at 164; Deception Statement, 103 F.T.C. at 176).
16. Deception Statement, 103 F.T.C. at 176. The Commission may rely on its own expertise in finding claims that are reasonably clear from the face of an advertisement. *Kraft*, 970 F.2d at 319, and cases cited therein. If the Commission is unable to conclude that an implied claim is conveyed based on a review of the ad itself, the Commission may rely on extrinsic evidence demonstrating that the ad implies a claim. *Kraft*, slip op. at 7; *Thompson Medical*, 104 F.T.C. at 789.

17. *Kraft*, slip op. at 7-8; *Removatron Int'l Corp.*, 111 F.T.C. 206, 292 (1988), *aff'd*, 884 F.2d 1849 (1st Cir. 1989); *Thompson Medical*, 104 F.T.C. at 790.
18. Deception Statement, 103 F.T.C. at 175 n.4; *see also International Harvester Co.*, 104 F.T.C. 949, 1057 (1984); *Campbell Soup Co.*, FTC Dkt. No. 9223 (Aug. 18, 1992) (consent order).
19. *International Harvester*, 104 F.T.C. at 1058.
20. Deception Statement, 103 F.T.C. at 175 n.4; *International Harvester*, 104 F.T.C. at 1059.
21. Deception Statement, 103 F.T.C. at 175 n.4.
22. Deception Statement, 103 F.T.C. at 177.
23. *Id.* at 182.
24. *Kraft*, slip op. at 22-23, *Thompson Medical*, 104 F.T.C. at 816-17; Deception Statement, 103 F.T.C. at 182-83.
25. Substantiation Statement, 104 F.T.C. at 839.
26. *See*, e.g., *Kraft*, slip op. at 2 (scientific evidence required to substantiate calcium content claims and comparative calcium content claims); *Bertolli, Inc.*, FTC Dkt. No. C-3396 (Aug. 17, 1992) (consent order) (scientific evidence required to substantiate claims regarding edible oil's impact on any physiologic function or risk factor for disease or other health benefit); *Pacific Rice Prods.*, FTC Dkt. No. C-3395 (Aug. 17, 1992) (consent order) (scientific evidence required to substantiate claims regarding health benefits derived from consumption of products); *see also Thompson Medical*, 104 F.T.C. at 822.
27. *See Bertolli*; *Pacific Rice*.
28. 21 C.F.R. § 101.69(b) (1993).
29. *See*, e.g., *Nestle Food Co.*, FTC Dkt. No. C-2265 (Jan. 21, 1992) (consent order) and *Presto Food Prods., Inc.*, FTC Dkt. No. C-3480 (Feb. 23, 1994) (consent order) (resolving allegations that low fat claims based on the small serving of nondairy creamers that might be used in coffee were deceptive when made with respect to a larger serving that might be used over cereal or fruit or in cooking).
30. *See Thompson Medical*, 104 F.T.C. at 826.
31. In the past, courts have upheld the Commission's position that inconsistent meanings for the same terms have the potential to mislead consumers. In *FTC v. Brown & Williamson Tobacco Corp.*, 778 F.2d 35 (D.C. Cir. 1985), the court held that Brown & Williamson had deceptively advertised its Barclay cigarettes as "1 mg. tar." The 1 mg. tar rating was a result of the cigarettes' different design, which caused the amount of tar that Barclay cigarettes delivered to smokers to be disproportionately greater than that delivered by cigarettes that were similarly rated under the FTC rating system. Considering the claim against the background of the Commission's tar and nicotine rating system, the court affirmed the Commission's position that the claim misled consumers who had come to rely on the FTC rating system to make comparative assessments regarding cigarettes.
32. e.g., *Presto Food Prods., Inc.*, FTC Dkt. No. C-3480 (Feb. 23, 1994) (consent order); *Clorox Co.*, FTC Dkt. No. C-3427 (May 17, 1993) (consent order); *Isaly*

Klondike Co., FTC Dkt. No. C-3412 (Jan. 28, 1993) (consent order); *Nestle Food Co.*, FTC Dkt. No. C-2265 (Jan. 21, 1992) (consent order).

33. Substantiation Statement, 104 F.T.C. at 839.
34. 81 F.T.C. 23, 64 (1972); *Thompson Medical*, 104 F.T.C. at 813, 821; *Bristol-Myers*, 102 F.T.C. at 321. These are: (1) the type of product advertised, (2) the type of claim, (3) the benefits of a truthful claim, (4) the ease of developing substantiation for the claim, (5) the consequences of a false claim, and (6) the amount of substantiation that experts in the field believe is reasonable.
35. Under FDA's regulations, a label claim characterizing the level of a nutrient (i.e., a nutrient content claim) is prohibited unless made in accordance with the regulations. 21 C.F.R. § 101.13(b) (1993). However, the label of a product may contain a statement of the amount of a nutrient, such as "1 g. of omega-3 fatty acids" if it does not explicitly or implicitly characterize the level of the nutrient. 21 C.F.R. § 101.13(i)(3) (1993). Thus, statements that merely note the amount of a nutrient without characterizing the level are permitted even for nutrients not approved to appear on the nutrition panel.
36. This principle is already apparent from recent Commission consent orders, which provide safe harbors for those claims specifically permitted in labeling. *See*, e.g., *Nestle Food Co.*, FTC Dkt. No. C-2265 (Jan. 21, 1992) (consent order) (providing that nothing in the relevant portions of the order shall prohibit certain representations regarding total fat, saturated fat or cholesterol if such representations are specifically permitted in labeling, for the serving size advertised or promoted, by FDA regulation); *Isaly Klondike Co.*, FTC Dkt. No. C-3412 (Jan. 28, 1993) (consent order) (providing that nothing in the Order shall prevent respondent from making representations specifically permitted in labeling for food by the NLEA regulations).
37. As it has in the past, the Commission emphasizes that truthful comparisons may need to be sufficiently qualified to remove deceptive implications. *See* Policy Statement in Regard to Comparative Advertising, 16 C.F.R. § 14.15 (1979) (comparative advertising regarding objective measurable attributes must have sufficient clarity or disclosures to ensure that such comparisons are not deceptive).
38. For example, a small nutrient difference that appears as part of a claim touting the multidimensional nutritional differences offered by a product is less likely to overstate the significance of that difference than would such a claim standing alone. Thus, an advertiser may seek to signal to consumers that, while it has reduced total fat and saturated fat in its product by 25%, it has also achieved a small reduction in sodium compared with other products in the category. In these circumstances, a truthful claim that makes clear that the sodium reduction is less than the 25% reduction in other nutrients and does not overstate the significance of this incidental reduction is unlikely to mislead consumers.
39. *See* Policy Statement in Regard to Comparative Advertising, 16 C.F.R. § 14.15 (1979). The Commission's Guides for the Use of Environmental Marketing Claims also include this requirement. 16 C.F.R. § 260.6(d) (1993).
40. *See P. Lorillard Co. v. FTC*, 186 F.2d 52, 57 (4th Cir. 1950) (advertising claiming that cigarette was lowest in nicotine, tar and resins challenged in part because the

difference was, in fact, insignificant); *Sun Co.*, FTC Dkt. No. C-3381 (May 6, 1992) (consent order) (challenging advertising for octane gasoline that represented gas would provide superior power that would be significant to consumers).

41. Although the term "light" is defined in FDA's regulations as a comparative descriptor, the term also has been used to describe the food itself, much like an absolute descriptor such as "low." As reflected in FDA's preamble and regulations, the term also is associated chiefly with substantial reductions in fat or calories. *See* 58 Fed. Reg. 2351-2358. Given the unique characteristics of the term "light" as reflected in FDA's regulations, it is unlikely that the term can be used in advertising without undue confusion unless the food meets FDA's definitions. Accordingly, the Commission will apply FDA's definition for "light" in determining whether advertising using the term is deceptive.
42. 21 C.F.R. § 101.13(b) (1993). Interested parties may petition FDA to authorize additional synonyms. 21 C.F.R. § 101.69(b)(2) (1993).
43. 58 Fed. Reg. 2319-20 (1993). *See* Nutrition Labeling and Education Act of 1990, § 403(4)(2)(A)(i).
44. *Chrysler Corp. v. FTC*, 561 F.2d 357, 363 (D.C. Cir. 1977); *Kraft*, slip. op. at 6 n.8. *see also* Deception Statement, 103 F.T.C at 178 n.21 ("A secondary message understood by reasonable consumers is actionable if deceptive even though the primary message is accurate").
45. 21 C.F.R. § 101.54(b) and (c) (1993).
46. 21 C.F.R. § 101.13(b)(2) (1993).
47. 58 Fed. Reg. 2371 (1993).
48. For example, the regulations state that "a claim that a food contains oat bran is a claim that it is a good source of dietary fiber; that a food is made only with vegetable oil is a claim that it is low in saturated fat; and that a food contains no oil is a claim that it is fat free." 21 C.F.R. § 101.65(c)(3) (1993).
49. *Kraft*, slip op. at 7-8; *Removatron*, 111 F.T.C. at 292; *Thompson Medical*, 104 F.T.C. at 790. *See also FTC v. Sterling Drug*, 317 F.2d 669, 674 (2d Cir. 1963) (the Commission examines "the entire mosaic ... rather than each tile separately").
50. *Kraft*, 970 F.2d at 322 (upholding Commission's finding that claims about the amount of milk in processed cheese slices were, in context, implied claims about calcium content).
51. *See Estee Corp.*, 102 F.T.C. 1804 (1983) (consent order) (advertisements that claimed that foods sweetened with high-fructose corn syrup did not contain sugar and were accepted by the American Diabetes Association implied (falsely) that the foods were appropriate for people who needed to avoid sugar).
52. 21 C.F.R. § 101.13(g) (1993).
53. 21 C.F.R. § 101.13(h) (1993). As discussed in Part IV, *infra*, these same levels of nutrients serve to disqualify foods from bearing health claims. *See* 21 C.F.R. § 101.14(a)(5) (1993).
54. *See* 21 C.F.R. § 101.54(d) (requirements for fiber claims); 21 C.F.R. § 101.62(c) (requirements for saturated fat claims); 21 C.F.R. § 101.62(d) (requirements for cholesterol claims).

55. Deception Statement, 103 F.T.C. at 176.
56. *International Harvester*, 104 F.T.C. at 1057.
57. *Id.* at 1059.
58. *Id.* at 1058 ("[n]ot all omissions are deceptive, even if providing the information would benefit consumers").
59. 21 C.F.R. § 101.13(g) (1993).
60. *See North American Philips Corp.*, 111 F.T.C. 139, 177-84 (1988) (Initial Decision) (according great weight to other government agencies' determinations regarding the significance of a chemical added to drinking water by the water filter and thus whether the failure to disclose this fact was material).
61. *Id.* at 175 (Commission's complaint alleged, and the Administrative Law Judge found, that failure to disclose that water filter device introduced a potentially hazardous chemical into drinking water was misleading in light of representations that device would remove organic chemicals and clean the water).
62. House Committee on Energy and Commerce, Nutrition Labeling and Education Act of 1990, H.R. Rep. No. 538, 101st Cong., 2d Sess. 20 (1990).
63. 21 U.S.C. § 343(r)(2)(A)(iii)-(v).
64. FDA's definition of a health claim includes two basic elements: (1) a substance or nutrient; and (2) the relationship of that substance or nutrient to a disease or health-related condition. 21 C.F.R. § 101.14(a)(1) (1993). Thus, claims on food labels are not governed by FDA's health claims regulations unless they include either express or implied references to both a substance and a disease. FDA's approach to implied health claims is similar to the Commission's in that this definition includes claims in which the disease element is implied through symbols or by other means, looking at the context of the entire label. *Id.*; *see also* discussion of FDA's definition of implied health claims, 58 Fed. Reg. 2483 (1993). Like FDA, the Commission examines food claims in the context of the entire advertisement to determine whether an implied health claim is being made. Therefore, the Commission may determine in certain instances, based on its review of the entire context of an advertisement, that a nutrient content claim, even in the absence of any express reference to a disease or health-related condition, conveys an implied health message to consumers.
65. 21 C.F.R. § 101.14 *et seq.* (1993).
66. 21 C.F.R. § 101.14(c) (1993).
67. 21 C.F.R. § 101.14(a)(5) (1993).
68. 21 C.F.R. § 101.14(d)(2)(vi)-(vii) (1993).
69. 21 C.F.R. § 101.14(e)(6) (1993).
70. 21 C.F.R. § 101.14(d)(2)(iii) (1993).
71. 21 C.F.R. § 101.70 (1993). This regulation requires that FDA take final action within 190 days of the receipt of a petition, either to deny the petition or to publish a proposal to amend the regulations to allow the use of the requested health claim.
72. 21 U.S.C. § 343(r)(3)(B)(i). This standard is also set forth in FDA's regulations at 21 C.F.R. § 101.14(c) (1993).
73. NLEA, § 3(b).

74. *See*, e.g., *Pacific Rice*, FTC Dkt. No. C-3395 (Aug. 17, 1992) (consent order) (claims about health benefits of consuming rice bran cereal challenged as unsubstantiated); *see also Thompson Medical*, 104 F.T.C. at 822 (claims involving health or safety issues require a "relatively high level of substantiation, typically scientific tests").
75. *Gracewood Fruit Co.*, FTC Dkt. No. C-3470 (Oct. 29, 1993) (consent order); *see also Pompeian, Inc.*, FTC Dkt. No. C-3402 (Oct. 27, 1992) (consent order).
76. 58 Fed. Reg. 2505 (1993).
77. *See Pfizer, Inc., supra* note 34. *see also* Substantiation Statement, 104 F.T.C. at 840; *Thompson Medical*, 104 F.T.C. at 821.
78. Unqualified as used in this discussion of substantiation refers to health claims that do not include specific disclosures concerning the extent of supporting scientific evidence.
79. This approach is consistent with the Commission's approach to evaluating the substantiation for claims made for drug products and medical devices regulated by FDA. *See*, e.g., *Removatron*, 111 F.T.C. at 305 (FDA's determination of efficacy of hair removal device given substantial weight); *Thompson Medical*, 104 F.T.C. at 826 (recognizing importance of applying standard consistent with FDA's in evaluating safety and efficacy of a drug product subject to jurisdiction of both agencies).
80. Food marketers should not expect to circumvent FDA's petition process for health claims simply by limiting the assertion of unapproved or unreviewed claims to advertising.
81. *See*, e.g., *National Comm'n on Egg Nutrition (NCEN)*, 517 F.2d 485 (7th Cir. 1975), *appeal after remand*, 570 F.2d 157 (7th Cir. 1977), *cert. denied*, 483 U.S. 921 (1978). The final Commission order in *NCEN*, as modified by the court, required that the advertiser, if it made any claims regarding the relationship between dietary cholesterol and heart disease, disclose that there was a controversy among experts about the scientific basis for the link between egg consumption and heart disease, and that NCEN was presenting its side of that controversy. Where NCEN characterized the level of scientific evidence, the order further required a disclosure that many medical experts believed that increasing egg consumption might increase the risk of heart disease.
82. In order to be effective, qualifications or disclosures should be sufficiently clear and prominent to prevent deception. *See* Deception Statement, 103 F.T.C. at 180; *Thompson Medical*, 104 F.T.C. at 789 n.9, 842-43; *see also* Guides for the Use of Environmental Marketing Claims, 16 C.F.R. § 260.6(a) (1993). Clarity of language, relative type size and proximity to the claim being qualified, and an absence of contrary claims that could undercut effectiveness, will maximize the likelihood that the qualifications and disclosures are appropriately clear and prominent. *See*, e.g., *Figgie Int'l, Inc.*, 107 F.T.C. 313, 401 (1986), *aff'd*, 817 F.2d 102 (4th Cir. 1987). For example, the Commission is unlikely to find a video superscript, without accompanying audio, to be an effective method of disclosure in a television ad. *See*, e.g., *Kraft*, slip. op. at 10. As always, the Commission will also consider any extrinsic evidence of the effectiveness of qualifications and disclosures in its

determination of whether a claim is deceptive. In making this determination, the Commission will consider all reasonable interpretations of the advertisement. The Commission will find an advertisement to be deceptive if it can reasonably be interpreted in a misleading way, even though other, nonmisleading interpretations may be equally possible. *See Kraft*, slip. op. at 6 n.8.

83. These specific disqualifying levels are set forth at 21 C.F.R. § 101.14(a)(5) (1993).
84. 58 Fed. Reg. 2489 (1993).
85. The Commission has routinely accorded great weight to FDA determinations of the safety and efficacy of food and drug products. *See*, e.g., *Removatron*, 111 F.T.C. at 305; *Thompson Medical*, 104 F.T.C. at 826; *see also Sterling Drug, Inc.*, 102 F.T.C. 395, 768-69, *aff'd*, 741 F.2d 1146 (9th Cir. 1984), *cert. denied*, 470 U.S. 1084 (1985).
86. For example, USDA has stated its "intention to publish a proposed rule on health claims in line with FDA's proposal." *See* 58 Fed. Reg. 632, 664 (Jan. 6, 1993). If so, the regulation's disqualifying level for cholesterol will preclude health claims on the labels of virtually all meat and poultry products. Notwithstanding the regulations, however, the Commission would not prohibit a truthful advertising claim that explains in a nondeceptive manner the health advantages of substituting meat or poultry items that are relatively low in fat and saturated fat for higher fat alternatives (e.g., a claim suggesting the merit of substituting skinless breast of turkey for hamburger). Such claims would assist consumers who are trying to improve their diets but who are unwilling to forgo all meat and poultry.
87. *See*, e.g., *Campbell*, FTC Dkt. No. 9223 (Aug. 18, 1992) (consent order required disclosure of sodium content and recommended maximum daily sodium intake in advertisements making claims about heart disease for soups with more than 500 mg. of sodium per 8-oz. serving).
88. The Commission has traditionally required that material information be disclosed if its absence could mislead reasonable consumers. *See* Deception Statement, 103 F.T.C. at 182; *see also International Harvester*, 104 F.T.C. at 1057; *North American Philips*, 111 F.T.C. at 175, 195 (failure to disclose the fact that a water filter could introduce a harmful chemical into the water was misleading).
89. In *Campbell*, the Commission charged that claims that the company's soups contained little fat or cholesterol, and were heart-healthy, were deceptive because the company had failed to disclose that the soups were high in sodium. Specifically, the complaint alleged that the high level of sodium was a material fact given that a diet high in sodium can contribute to hypertension, a risk factor associated with heart disease. FTC Dkt. No. 9223 (Aug. 18, 1992) (consent order).
90. A statement indicating both the amount of the risk-increasing nutrient and the recommended maximum daily intake of that nutrient, as determined by FDA, would be one example of an acceptable disclosure, provided such information adequately conveys the health implications of the risk-increasing nutrient. *See*, e.g., *Campbell*, *supra*.
91. Further, the FDA's treatment of health claims in labeling for any food containing a risk-increasing level of a nutrient, as well as the NLEA-mandated educational effort, could well increase consumers' expectations concerning the scope of un-

qualified health claims, including expectations that the foods do not present any significant health risks.

92. 21 C.F.R. § 101.14(d)(2)(vi) (1993).
93. 21 C.F.R. § 101.14(d)(2)(vii) (1993).
94. 58 Fed. Reg. 2514 (1993).
95. 56 Fed. Reg. 60,553 (1992) (discussion of proposed regulations).
96. *See*, e.g., *Gracewood Fruit Co.*, FTC Dkt. No. C-3470 (Oct. 29, 1993) (consent order). The complaint accompanying the *Gracewood* consent agreement challenged claims that eating grapefruit could reduce serum cholesterol levels, in part because there was no evidence that the small amount of pectin (the relevant nutrient) in grapefruit was sufficient to cause any meaningful reduction in serum cholesterol. *See also Lorillard*, 186 F.2d at 57 (advertising claiming that cigarettes were lowest in nicotine, tars, and resins challenged in part because the difference was so small as to be insignificant). Similarly, the Commission's Guides for the Use of Environmental Marketing Claims include the general principle that claims should not be presented in a manner that overstates the attribute or benefit of a product, and that "[m]arketers should avoid implications of significant environmental benefits if the benefit is in fact negligible." 16 C.F.R. § 260.6(c) (1993).
97. *See* discussion *supra* at Part III, Section A.2., (comparative nutrient claims).
98. 21 C.F.R. § 101.14(e)(6) (1993).
99. 58 Fed. Reg. 2522 (1994).
100. *See* discussion *supra* at Part IV, Section C.
101. For example, a qualified comparative health claim suggesting that consumers switch from a high fat to a fat-free salad dressing, and indicating that diets low in total fat may contribute to a reduced risk of some forms of cancer, could encourage a dietary choice resulting in a significant health benefit, even if the fat-free salad dressing did not contain sufficient levels of any of the six nutrients or substances specified by FDA.
102. FDA has stated that model health claim language can be paraphrased as long as all mandatory elements of the model statements are addressed. 58 Fed. Reg. 2510 (1993).
103. 21 C.F.R. § 101.72(e) (1993). In authorizing other health claims, FDA provides alternative approaches of either expressly enumerating the relevant factors, or stating more simply that the development of the disease depends on many factors. *See*, e.g., 21 C.F.R. § 101.73 (1993) (governing claims about dietary fat and cancer).
104. 58 Fed. Reg. 2511 (1993); *see also* 21 U.S.C. § 343(r)(3)(B)(iii).
105. Deception Statement, 103 F.T.C. at 176. In *J.B. Williams Co. v. FTC*, for example, the Commission challenged as deceptive advertising claims that a vitamin and iron supplement would reduce tiredness because the advertiser failed to disclose that those symptoms are usually caused by factors other than vitamin and iron deficiency. 381 F.2d 884, 890 (6th Cir. 1967). *See also Keele Hair & Scalp Specialists*, 55 F.T.C. 1840 (1959), *aff'd*, 275 F.2d 18 (5th Cir. 1960) (baldness cure claims challenged for failure to disclose significance of male heredity as cause of baldness, for which cure was ineffective).

Appendix 3

Analytical Methods and Databases

Table A3.1 Acceptable Official Methods for Nutrition Labeling

AOAC No.	Method	Matrix[a]
Ash		
940.26A[b]	Ash of fruits and fruit products	A,J
920.153[b]	Ash of meat	B,L,M
925.51A[b]	Ash of canned vegetables	C,P,T
950.14A[b]	Ash of nonalcoholic beverages	D
920.100[b]	Ash in tea	D
925.49[b]	Ash of confectionery	E
972.15	Ash of cocoa products	E
945.18	Ash of cereal adjuncts, direct method	F
930.22	Ash of bread, direct method	F
935.42	Ash of cheese, gravimetric method	G
945.46	Ash of milk, gravimetric method	H
920.117	Ash of butter	H
930.30	Ash of dried milk	H
920.108	Ash of cream, gravimetric method	H
945.48E	Ash of evaporated milk	H
920.115E[b]	Ash of sweetened condensed milk	H
938.08[b]	Ash of seafood	I,Q
986.25	Ash of milk-based infant formula, gravimetric method	K

Table A3.1 (Continued)

AOAC No.	Method	Matrix[a]
950.49[b]	Ash of nuts and nut products	N
923.03	Ash of flour, direct method	R
935.39	Ash of baked products, direct method (without fruit)	R
941.12	Ash of spices, gravimetric method	S
Beta carotene		
941.15	Carotene in fresh plant materials and silages, spectrophotometric method (naturally occurring carotene only)	A,C,J,P,T
Calcium		
968.08	Minerals in animal feed, AAS method	All
975.03	Metals in plants, AAS method	All
980.03	Metals in plants, spectrographic method	All
965.09	Minor nutrients in fertilizers, AAS method	All
953.01	Metals in plants, emission spectrographic method	A,C,F,J,P,T
985.01	Metals and other elements in plants, ICP method	A,C,F,J,P,T
921.01	Ca in plants, microtitrimetric method	A,C,F,J,P,T
951.01	Ca in plants, nitrocreosol method	A,C,F,J,P,T
983.19	Ca in mechanically separated poultry and beef titrimetric method	B,L
968.31	Ca in canned vegetables, titrimetric method	C,T
945.41	Ca in bread, titrimetric method	F
944.03	Ca in flour, titrimetric method	F
991.25	Ca, Mg, and P in cheese, AAS and colorimetric methods	G
984.27	Ca, Cu, Fe, Mg, Mn, P, K, Na, and Zn in infant formula, ICP method	K
985.35	Minerals in ready-to-feed milk-based infant formula, AAS method	K
Carbohydrates, sugars (mono- and di-)		
982.14	Glucose, fructose, sucrose, and maltose in presweetened cereals, LC method	All
980.13	Fructose, glucose, lactose, maltose and sucrose in milk chocolate, LC method	All
984.17	Sugars in licorice extract, LC method	All
977.20	Separation of sugars in honey, LC method	All
971.18	Carbohydrates in fruit juices, GC method	All
954.11	Separation of sugars in honey, charcoal column chromatographic method I	All
979.21	Separation of sugars in honey, charcoal column chromatographic method II	All

AOAC No.	Method	Matrix[a]
925.52	Sugars in canned vegetables	C,T
939.03	Sugars (reducing and nonreducing) in flour, titrimetric method	F,P,R
925.42	Sugars (reducing) before inversion in food dressings	O
925.43	Sugars (reducing) after inversion in food dressings	O
925.44	Sucrose in food dressings	O
950.30	Sugars (reducing) in nonalcoholic beverages	D
950.50	Sugars (reducing) in nuts and nut products, Munson-Walker method	N
950.51	Sucrose in nuts and nut products	N
931.02	Sugars in plants, sample preparation	A,C,J,T
906.01	Sugars (reducing) in plants, Munson-Walker general method	A,C,J,T
930.07	Sucrose in plants, inversion method	A,C,J,T
925.15	Sugars in roasted coffee	D
925.35	Sucrose in fruits and fruit products	A,D,J,T
925.36	Sugars (reducing) in fruits and fruit products, inversion method	A,J
975.14	Sugars in bread	F,R
935.39G	Sugars in baked products	F,R
Cholesterol		
976.26	Cholesterol in multicomponent foods, GC method (only > 10 mg/100 g)	All
Copper		
985.35	Minerals in ready-to-feed milk-based infant formula, AAS method	All
984.27	Ca, Cu, Fe, Mg, Mn, P, K, Na and Zn in infant formula, ICP method	All
960.40	Copper in food, colorimetric method	All
968.08	Minerals in animal feed, AAS method	All
975.03	Metals in plants, AAS method	All
980.03	Metals in plants, spectrographic method	All
953.01	Metals in plants, emission spectrographic method	A,C,F,J,P,T
985.01	Metals and other elements in plants, ICP method	A,C,F,J,P,T
953.03	Cu in plants, colorimetric method	A,C,F,J,P,T
985.40	Cu in liver, AAS method	L
990.05	Cu, Fe, Ni in edible oils and fats, AAS method	O
Cyanocobalamine		
960.46	Vitamin assays, microbiological method with 952.20 cobalamin (vitamin B_{12} activity) in vitamin preparations, microbiological method	All

Table A3.1 (Continued)

AOAC No.	Method	Matrix[a]
986.23	Cobalamin (vitamin B_{12} activity) in milk-based infant formula, turbidimetric method	K
Dietary fiber, insoluble		
991.42	Insoluble dietary fiber in food and food products, enzymatic-gravimetric method	A,C-F,J,K, N,P,R-T
991.43	Total, insoluble and soluble dietary fiber in foods, enzymatic-gravimetric method	A,C-F,J,K ,N,P,R-T
Dietary fiber, soluble		
991.42	Insoluble dietary fiber in food and food products, enzymatic-gravimetric method (by calculation; direct method is under AOAC review)	A,C-F,J,K, N,P,R-T
991.43	Total, insoluble and soluble dietary fiber in foods, enzymatic-gravimetric method	A,C-F,J,K, N,P,R-T
Dietary fiber, total		
985.29	Total dietary fiber in foods, enzymatic-gravimetric method	A,C-F,J,K, N,P,R-T
991.43	Total, insoluble and soluble dietary fiber in foods, enzymatic-gravimetric method	A,C-F,J,K, N,P,R-T
Fat, polyunsaturated		
979.19	*cis-cis* Methylene interrupted polyunsaturated fatty acids in oils, spectrophotometric method (suitable after isolation of fat)	All
957.13	Polyunsaturated acids in oils and fats, spectrophotometric method	O
Fat, total		
Total fat has been concisely defined under NLEA as the sum of lipid fatty acids expressed as triglycerides. Currently, no AOAC Official Methods meet that definition per se. See the February 1993 issue of *The Referee* for the task force report on fat that contains a number of methods that approximate the definition and may be useful on an interim basis until methods can be collaboratively studied.		
Folacin		
960.46	Vitamin assays, microbiological method with 944.12 folic acid in vitamin preparations, microbiological method (suitable for free form only) (*L. casei* commonly preferred, but not collaboratively studied.)	All

AOAC No.	Method	Matrix[a]
992.05	Folic acid in infant formulas, LC method	K
Iron		
968.08	Minerals in animal feed, AAS method	All
975.03	Metals in plants, AAS method	All
980.03	Metals in plants, spectrographic method	All
965.09	Minor nutrients in fertilizers, AAS method	All
953.01	Metals in plants, emission spectrographic method	A,C,F,J,P,T
945.40	Iron in bread, spectrophotometric method	F
950.39	Iron in macaroni products, spectrophotometric method	F
944.02B	Iron in flour, spectrophotometric method	F
984.27	Ca, Cu, Fe, Mg, Mn, P, K, Na, and Zn in infant formula, ICP method	K
985.35	Minerals in ready-to-feed milk-based infant formula, AAS method	K
990.05	Cu, Fe, and Ni in edible oils and fats, AAS method	O
Magnesium		
975.03	Metals in plants, AAS method	All
980.03	Metals in plants, spectrographic method	All
965.09	Minor nutrients in fertilizers, AAS method	All
953.01	Metals in plants, emission spectrographic method	A,C,F,J,P,T
985.01	Metals and other elements in plants, ICP method	A,C,F,J,P,T
991.25	Ca, Mg, and P in cheese, AAS and colorimetric methods	G
984.27	Ca, Cu, Fe, Mg, Mn, P, K, Na, and Zn in infant formula, ICP method	K
985.35	Minerals in ready-to-feed milk-based infant formula, AAS method	K

Moisture

See the task force report on moisture in the January 1993 issue of *The Referee*.

AOAC No.	Method	Matrix[a]
Niacin		
960.46	Vitamin assays, microbiological method with 944.13 niacin and niacinamide in vitamin preparations, microbiological method	All
961.14	Niacin and niacinamide in drugs, foods and feeds, colorimetric method	All
981.16	Niacin and niacinamide in foods, drugs and feeds, automated method	All
975.41	Niacin and niacinamide in cereal products, automated method	F
985.34	Niacin and niacinamide in ready-to-feed milk-based infant formula, microbiological-turbidimetric method	K

Table A3.1 (Continued)

AOAC No.	Method	Matrix[a]
Pantothenic acid		
960.46	Vitamin assays, microbiological method with 945.74 pantothenic acid in vitamin preparations, microbiological method (suitable for free form only)	All
Phosphorus		
958.01	Total phosphorus in fertilizers, spectrophotometric molybdovanadophosphate method	All
957.18	Microchemical determination of phosphorus, Kjeldahl digestion method	All
964.06	Phosphorus in animal feed, alk ammonium molybdophosphate method	All
965.17	Phosphorus in animal feed, photometric method	All
980.03	Metals in plants, spectrographic method	A,C,F,J,P,T
953.01	Metals in plants, emission spectrographic method	A,C,F,J,P,T
985.01	Metals and other elements in plants, ICP method	A,C,F,J,P,T
966.01	Phosphorus in plants, gravity quinolinium molybdophosphate method	A,C,F,J,P,T
931.01	Phosphorus in plants, microbiological method	A,C,F,J,P,T
970.39	Phosphorus in fruits and fruit products, spectrophotometric molybdovanadate method	A,J
969.31	Phosphorus in meat, alk ammonium molybdophosphate method	B,L
972.22	Phosphorus in meats, automated method	B,L
991.27	Phosphorus in meat and meat products, spectrophotometric method	B,L
948.09	Phosphorus in flour	F
990.24	Total phosphorus in cheese and processed cheese, photometric method	G
991.25	Ca, Mg, and P in cheese, AAS and colorimetric methods	G
984.27	Ca, Cu, Fe, Mg, Mn, P, K, Na, and Zn in Infant formula, ICP method	K
986.24	Phosphorus in milk-based infant formula, spectrophotometric method	K
935.59	Total phosphorus in food dressings, alk ammonium molybdophosphate method	O
930.35G, H,I	Phosphorus in vinegars, alk ammonium molybdophosphate method	S

AOAC No.	Method	Matrix[a]
Potassium		
985.35	Minerals in ready-to-feed milk-based infant formula, AAS method	All
956.01	K/Na in plants, flame photometric method	A,C,F,J,P,T
953.01	Metals in plants, emission spectrographic method	A,C,F,J,P,T
985.01	Metals and other elements in plants, ICP method	A,C,F,J,P,T
975.03	Metals in plants, AAS method	A,C,F,J,P,T
980.03	Metals in plants, spectrographic method	A,C,F,J,P,T
965.30	K in fruits and fruit products, flame photometric method	A,J
990.23	Na and K in dried milk, flame emission spectrophotometric method	H
969.23	Na and K in seafood, flame photometric method	I,Q
984.27	Ca, Cu, Fe, Mg, Mn, P, K, Na, and Zn in infant formula, ICP method	K
Protein		
981.10	Crude protein in meat, block digestion method	All
955.04	Total N in fertilizers, Kjeldahl method	All
977.02	Total N (crude protein) in plants, automated and semiautomated methods	A,C,F,J,P,T
978.04	Total N (crude protein) in plants, Kjeldahl method	A,C,F,J,P,T
920.152	Protein in fruit products, Kjeldahl method	A,J
992.15	Crude protein in meat and meat products, combustion method	B,L
939.02	Milk protein in milk chocolate, Kjeldahl method	E
990.03	Crude protein in animal feed, combustion method	F
990.02	Crude protein in animal feed, semiautomated method	F
920.87	Total protein in flour	F
945.18B	Protein in cereal adjuncts, Kjeldahl method	F
930.25	Protein in macaroni products	F
950.36	Protein in bread	F
991.20	Total nitrogen content of milk, Kjeldahl method	H
991.22	Protein nitrogen content of milk, Kjeldahl method (direct)	H
991.23	Protein nitrogen content of milk, Kjeldahl method (indirect)	H
930.33	Protein in ice cream and frozen desserts, Kjeldahl and dye binding methods	H
930.29	Protein in dried milk, Kjeldahl method	H
967.12	Protein in milk, dye binding method I	H
975.17	Protein in milk, dye binding method II	H
975.18	Protein in milk, mid-infrared spectrophotometric method	H

Table A3.1 (Continued)

AOAC No.	Method	Matrix[a]
972.16	Fat, lactose, protein, and solids in milk, mid-infrared spectrophotometric method	K
986.25C	Protein in milk-based infant formula, Kjeldahl method	K
928.08	Nitrogen in meat, Kjeldahl method	L
977.14	Nitrogen in meat, automated Kjeldahl method	L
981.10	Nitrogen in meat, block digestion method	L
950.48	Crude protein in nuts and nut products, Kjeldahl method	N
935.39C	Protein in baked products	R
Protein quality		
991.29	True protein digestibility of foods and food ingredients, rat bioassay method with 982.30 protein efficiency ratio, calculation method	All
960.48	Protein efficiency ratio, rat bioassay method	All
Pyridoxine		
960.46	Vitamin assays, microbiological method with 961.15 vitamin B_6 in food extracts, microbiological method	All
985.32	Vitamin B_6 in ready-to-feed milk-based infant formula, microbiological method	K
Riboflavin		
960.46	Vitamin assays, microbiological method with 960.33 riboflavin (vitamin B_2) in vitamin preparations, microbiological method	All
970.65	Riboflavin (vitamin B_2) in foods and vitamin preparations, fluorometric method	All
981.15	Riboflavin in foods and vitamin preparations, automated method	All
985.31	Riboflavin in ready-to-feed milk-based infant formula, fluorometric method	K
Sodium		
985.35	Minerals in ready-to-feed milk-based infant formula, AAS method	All
956.01	K/Na in plants, flame photometric method	A,C,F,J,P,T
953.01	Metals in plants, emission spectrographic method	A,C,F,J,P,T
980.03	Metals in plants, spectrographic method	A,C,F,J,P,T
966.16	Na in fruits and fruit products, flame spectrophotometric method	A,J
990.23	Na and K in dried milk, flame emission spectrophotometric method	H

AOAC No.	Method	Matrix[a]
969.23	Na and K in seafood, flame photometric method	I,Q
976.25	Na in foods for special dietary use, ISE method	K
984.27	Ca, Cu, Fe, Mg, Mn, P, K, Na, and Zn in infant formula, ICP method	K
Thiamine		
942.23	Thiamine (B_1) in foods, fluorometric method	All
957.17	Thiamine (B_1) in bread, fluorometric method	F,R
953.17	Thiamine (B_1) in grain products, fluorometric method	F,R
986.27	Thiamine (B_1) in milk-based infant formula, fluorometric method	K
Tryptophan		
960.46	Vitamin assays, microbiological method with 988.15 tryptophan in foods and food and feed ingredients, ion exchange chromatographic method	All
Vitamin A		
974.29	Vitamin A in mixed feeds, premixes, and foods, colorimetric method	B,G,H,I,K, L,M,Q
992.04	Vitamin A (retinol isomers) in milk and milk-based infant formula, LC method	H,K
992.06	Vitamin A (retinol) in milk-based infant formula, LC method	K
Vitamin C		
984.26	Total vitamin C in food, semi-automated fluorometric method	A,C-H,J,K, P,R,T
967.22	Vitamin C (ascorbic acid) in vitamin preparations, microfluorometric method	A,C-H,J,K, P,R,T
967.21	Vitamin C (ascorbic acid) in vitamin preparations and juices, titrimetric method (colorless juices only)	D
985.33	Vitamin C (reduced ascorbic acid) in ready-to-feed milk-based infant formula, titrimetric method	K
Vitamin D		
982.29	Vitamin D in mixed feeds, premixes, and pet foods, LC method	All
936.14	Vitamin D in milk, vitamin preparations, and feed concentrates, rat bioassay method	All
981.17	Vitamin D in Fortified milk and milk powder, LC method	G,H,K

Table A3.1 (Continued)

AOAC No.	Method	Matrix[a]
Vitamin E		
971.30	Alpha-tocopherol and alpha-tocopheryl acetate in foods and feeds, colorimetric method	All
992.03	Vitamin E activity in milk-based infant formula, LC method	K
Zinc		
968.08	Minerals in animal feed, AAS method	All
975.03	Metals in plants, AAS method	All
980.03	Metals in plants, Spectrographic method	All
965.09	Minor nutrients in fertilizers, AAS method	All
986.15	As, Cd, Pd, Se, and Zn in food, multielement method	All
944.09	Zn in food, colorimetric method	All
969.32	Zn in food, AAS method	All
953.01	Metals in plants, emission spectrographic method	A,C,F,J,P,T
985.01	Metals and other elements in plants, ICP method	A,C,F,J,P,T
941.03	Zn in plants, mixed color method	A,C,F,J,P,T
953.04	Zn in plants, single color method	A,C,F,J,P,T
984.27	Ca, Cu, Fe, Mg, Mn, P, K, Na, and Zn in infant formula, ICP method	K
985.35	Minerals in ready-to-feed milk-based infant formula, AAS method	K

[a] Key to matrixes: A, Baby Foods-Fruits; B, Baby Foods-Meats; C, Baby Foods-Vegetables; D, Beverages and Juices; E, Candy; F, Cereals and Products; G, Cheese; H, Dairy Products; I, Fish; J, Fruits; K, Infant Formula/Medical Diet; L, Meat (Beef/Pork/Fowl); M, Mixed Dinners (TV Dinners); N, Nuts; O, Oils/Fats (Dressings); P, Potatoes and Products; Q, Shellfish; R, Sweet Mixes (Cake/Pie, etc.); S, Spices; T, Vegetables. Note: These matrixes were considered as representative of all food types.

[b] These methods refer to 900.02A, Ash of Sugar, in surplus status, no longer in use with sugar. Method 923.03 may be substituted.

Source: *Methods of Analysis for Nutrition Labeling*, Proceedings and Final Report, AOAC International, 1993.

Table A3.2 Moisture Analysis for Nutrition Labeling, AOAC Official Methods

AOAC No.	Title	Brief description	Applicable matrixes	Comments
925.04	Moisture in animal feed	Distillation with toluene	Spices	Method measures water as moisture in sample.
986.21	Moisture in spices	Distillation with toluene or hexane	Spices	Method measures water as moisture in samples.
972.20	Moisture in prunes and raisins	Moisture meter method	Dried fruit	Measures conductivity and equates this to moisture.
977.10	Moisture in cacao products	Karl Fischer, electro-chemical method	Milk, chocolate, and confectionery coatings	Method measures water as moisture in samples.
984.20	Moisture in oils and fats	Karl Fischer, electrochemical method	Oils and fats except for alk. or oxidized samples	Method measures water as moisture in samples.
920.116	Moisture in butter	Pressure = atmospheric temperature = bp (H_2O) on sand to constant weight	Butter	Method measures all 100°C volatile components, residual solids method.
925.10	Solids (total) and moisture in flour	Pressure = atmospheric temperature = 130°C – 100°C air oven for 3 h	Flour	Method measures all 100°C volatile components, residual solids method.
925.23	Solids (total) in milk	Pressure = atmospheric temperature = 98°C – 100°C air oven for 3 h	Milk	Method measures all 100°C volatile components, residual solids method.
930.15	Moisture in animal feed	Pressure = atmospheric temperature = 135°C for 2 h	Animal feeds	Method measures all 130°C volatile components, residual solids method.

Table A3.2 (Continued)

AOAC No.	Title	Brief description	Applicable matrixes	Comments
931.04	Moisture in cacao products	Pressure = atmospheric temperature = 100°C air oven to constant weight	Cacao products	Method measures all 100°C volatile components, residual solids method.
935.29	Moisture in malt	Pressure = atmospheric temperature = 103°C – 106°C air oven for 3 h	Malt	Method measures all 103°C – 106°C volatile components, residual solids method.
941.08	Total solids in ice cream and frozen desserts	Pressure = atmospheric temperature = 100°C air oven for 3.5 h	Ice cream and other frozen desserts	Method measures all 100°C volatile components, residual solids method.
948.12	Moisture in cheese (rapid screening method)	Pressure = atmospheric temperature = 130°C air oven for 1.25 h	Cheese	Method measures all 130°C volatile components, residual solids method.
952.08	Solids (total) in seafood	Pressure = atmospheric temperature = 100°C air oven 4 h or temperature = 100°C draft oven for 1 h	All marine products except raw oysters	Method measures all 100°C volatile components, residual solids method.
984.25	Moisture (loss of mass on drying) in frozen french-fried potatoes	Pressure = atmospheric temperature = 103°C air oven for 16 h	Quick-frozen french-fried potatoes	Method measures all 130°C volatile components, residual solids method.
985.14	Moisture in meat and poultry products	Pressure = atmospheric temperature = 130°C + microwave oven	Meat and poultry	Method measures all 130°C volatile components, residual solids method.

950.46B(a)	Moisture in meat	Pressure = atmospheric temperature = 100°C – 102°C air oven for 16–18 h	Meat and meat products	Method measures all 100°C volatile components, residual solids method.
950.46B(b)	Moisture in meat	Pressure = atmospheric temperature = 125°C air oven for 2–4 h	Meat and meat products	Method measures all 125°C volatile components, residual solids method.
920.151	Solids (total) in fruits and fruit products	Pressure = < 100 mm Hg; temperature = 70°C 2 h wts to < 3 mg change	Fruits and fruit products	Method measures moisture in sample.
925.40	Moisture in nuts and nut products	Pressure < 100 mm Hg temperature = 95°C – 100°C constant weight (ca 5 h)	Nuts and nut products excluding high sugars or glycerols	Method measures moisture in samples.
926.08	Moisture in cheese	Pressure < 100 mm Hg temperature = 100°C constant weight (ca 4 h)	Cheese	Method measures all 100°C volatile components, residual solids method.
926.12	Moisture and volatile matter in oils and fats	Pressure = < 100 mm Hg; temperature = bp (H_2O) + 25°C to constant weight	Fats and oils	Measures all volatile components in sample, residual solids method.
927.05	Moisture in dried milk	Pressure = < 100 mm Hg; temperature = 100°C constant weight (ca 5 h)	Dried milk	Method measures all 100°C volatile components, residual solids method.
934.01	Moisture in animal feed	Pressure = < 100 mm Hg; temperature = 95°C – 100°C (P = < 50 T = < 70 w/i sugar)	Animal feeds	Method measures moisture, better at lower T and P.

Table A3.2 (Continued)

AOAC No.	Title	Brief description	Applicable matrixes	Comments
934.06	Moisture in dried fruits	Pressure < 100 mm Hg temperature = 70°C constant weight (ca 6 h)	Dried fruit	Method measures moisture in samples.
920.115(d)	Sweetened condensed milk, total solids	Pressure < 100 mm Hg temperature = 100°C constant weight (ca 6 h)	Sweetened condensed milk	Method measures all 100°C volatile components, residual solids method.
925.30(b)	Solids (total) in eggs	Pressure < 25 mm Hg temperature = 98°C – 100°C constant weight (ca 5 h)	Liquid or dried eggs	Method measures all 98°C – 100°C volatile components, residual solids method.
925.09	Solids (total) and moisture in flour	Pressure < 25 mm Hg temperature = 98°C – 100°C constant weight (ca 5 h)	Flour	Method measures moisture in samples.
945.43	Moisture in fig bars and raisin-filled crackers.	Pressure < 50 mm Hg temperature = 98°C – 70°C 16 h with sand	Fig bars and raisin-filled crackers	Method measures moisture in samples.
964.22	Solids (total) in canned vegetables	Pressure < 50 mm Hg temperature = 69°C – 71°C multistep steam-air oven-vacuum oven drying, final = 2 h in vacuum	Canned vegetables	Method measure moisture in samples.

925.45(a)	Moisture in sugars, vacuum drying	Pressure < 50 mm Hg temperature = 70°C (preferably 60°C) 2 h + until weight loss < 2 mg	Cane and beet, raw and refined sugars	Method measures moisture in samples.
925.45(b)	Moisture in sugars, drying at atmospheric pressure	Pressure = atmospheric temperature = 100°C 3 h + until weight loss < 2 mg	Cane and beet, raw and refined sugar	Method measures all 100°C volatile components, residual solids method.
925.45(d)	Moisture in sugars, drying on quartz sand	Pressure < 50 mm Hg temperature = 70°C (preferably 60°C) until wt loss < 2 mg (ca 18 h)	Massecuites, molasses, and other liquid and semiliquid products	Method measures moisture in samples.
985.26	Total solids in processed tomato products by microwave moisture analyzer	Pressure = atmospheric temperature = 125°C + to constant weight	Tomato products	Method measures volatile in samples.
972.16	Solids in milk by NIR	Reflectance spectroscopic method	Milk products	Method measures moisture in samples.

Source: *Methods of Analysis for Nutrition Labeling*, Proceedings and Final Report, AOAC International, 1993.

Table A3.3 Fat Analysis for Nutrition Labeling, AOAC Official Methods

Method No.	Title	Brief description	Applicable matrixes	Comments
960.39	Crude fat in meat	Petroleum or diethyl ether extraction	Baby food-meats, meats	Mono-, di-, and triglycerides and most of the sterols, glycolipids, phospholipids, and waxes
976.21	Crude fat in meat	Rapid Sp gr method; tetrachloroethylene extraction	Baby food-meats, meats	Mono-, di-, and triglycerides and most of the sterols, glycolipids, phospholipids, and waxes
985.15	Crude fat in meat and poultry products	Rapid microwave-methylene chloride solvent extraction	Baby food-meats, meats	Mono-, di-, and triglycerides, glycolipids, phospholipids, and waxes; sterol yield may be reduced
920.39B	Crude fat in animal feed	Diethyl ether extraction	Cereals and products; not adequate if heat processed, extruded, or has sugar	Mono-, di-, and triglycerides and traces of other lipid components
920.39C	Crude fat in animal feed	Diethyl ether extraction; H_2O prewash if high in sugar	Cereals and products sweet mixes (cakes and pies); not adequate if heat processed	Mono-, di-, and triglycerides; may not quant. extract total lipids; recommended further review or study of method
945.18A	Fat in cereal adjuncts	Petroleum ether extraction	Cereals and products sweet mixes (cakes and pies); not adequate if heat processed	Mono-, di-, and triglycerides; may not quant. extract total lipids; recommended further review or study of method

945.38F	Fat in grains	Refers to 920.39C	Cereals and products; adequate if not heat processed	Mono-, di-, and triglycerides and traces of other lipid components
933.05	Fat in cheese	Acid hydrolysis, petroleum and diethyl ether extraction and re-extraction	Cheese	Mono-, di-, and triglycerides and traces of other lipid components; sterol yield greatly reduced
905.02	Fat in milk	Alk treatment, petroleum and diethyl ether extraction and re-extraction	Dairy	Mono-, di-, and triglycerides and traces of other lipid components
989.05	Fat in milk	Alk treatment, petroleum and diethyl ether extraction	Dairy	Mono-, di-, and triglycerides and traces of other lipid components
938.06	Fat in butter	Petroleum or diethyl ether extraction	Butter	Mono-, di-, and triglycerides and traces of other lipid components
920.111A	Fat in cream	Alk treatment, petroleum and diethyl ether extraction and re-extraction	Dairy	Mono-, di-, and triglycerides and traces of other lipid components
920.111B	Fat in cream	Babcock, acid hydrolysis, volumetric analysis	Dairy	Mono-, di-, and triglycerides and phospholipids; sterol yield reduced
952.06	Fat in ice cream and frozen desserts	Alk treatment, petroleum and diethyl ether extraction and re-extraction	Dairy	Mono-, di-, and triglycerides and traces of other lipid components
945.48G	Fat in evaporated milk	Alk treatment, petroleum and diethyl ether extraction and re-extraction (refers to 905.02)	Dairy	Mono-, di-, and triglycerides and traces of other lipid components

Table A3.3 (Continued)

Method No.	Title	Brief description	Applicable matrixes	Comments
932.06	Fat in dried milk	Alk treatment, petroleum and diethyl ether extraction	Dairy	Mono-, di-, and triglycerides and traces of other lipid components
948.15	Crude fat in seafood	Acid hydrolysis, petroleum and diethyl ether extraction	Fish, shellfish	Mono-, di-, and triglycerides, fatty acid portion of phospholipids and glycolipids; in some products with high sugar content, may overestimate fat; sterol yield may be reduced 964.12
986.25	Fat in milk-based infant formula	Alk treatment, petroleum and diethyl ether extraction and re-extraction (refers to 945.48G)	Infant formula/medical	Mono-, di-, and triglycerides and traces of other lipid components
925.32	Fat in eggs	Acid hydrolysis, petroleum and diethyl ether extraction	Eggs/egg products; may not be adequate for some egg products containing sugar	Mono-, di-, and triglycerides, fatty acid portion of phospholipids and glycolipids; sterol yield may be reduced
948.22	Crude fat in nuts and nut products	Diethyl ether extraction	Nuts; not adequate for nuts containing sugar	Mono-, di-, and triglycerides and traces of other lipid components
950.54	Total fat in food dressings	Acid hydrolysis, petroleum and diethyl ether extraction	Oils/fats (dressings)	Mono-, di-, and triglycerides, fatty acid portion of phospholipids and glycolipids

945.44	Fat in fig bars and raisin filled cookies	Acid hydrolysis, petroleum and diethyl ether extraction and re-extraction	Sweet mixes (cakes and pies) and baked cereal products	Mono-, di-, and triglycerides and fatty acid portion of phospholipids and glycolipids; sterol yield may be reduced; re-extraction may not remove all sugars; recommend further review or study of method
963.15	Fat in cacao products	Petroleum ether extraction	Chocolate products	Mono-, di-, and triglycerides and traces of other lipid components
925.07	Fat in cacao products	Petroleum and diethyl ether extraction	Candy	Mono-, di-, and triglycerides and traces of other lipid components
920.177	Ether extract of confectionery	Petroleum and diethyl ether extraction and re-extraction	Candy	Mono-, di-, and triglycerides and traces of other lipid components
920.172	Ether extract of prepared mustard	Diethyl ether extraction	Mustard	Mono-, di-, and triglycerides and traces of other lipid components
963.22[b]	Methyl esters of fatty acids in oils and fats	GC method following preparation of methyl esters (according to 969.33)	All	Fatty acid profile; recommend further review or study of method to make method adequate for quantitative analysis of fatty acids

Table A3.3 (Continued)

Method No.	Title	Brief description	Applicable matrixes	Comments
979.19[b]	*cis, cis*-Methylene interrupted polyunsaturated fatty acids in oils	Enzymatic-spectrophotometric method	All	

[a] This list has been prepared thanks to the efforts of S. Bailey, D. Carpenter, H. Chin, J. DeVries, N. Fraley, W. Hummer, A. Kistler, S. Lee, J. Ngeh-Ngwainbi, P. Oles and D. Sullivan

[b] Fatty acid methodology and *cis,cis*-methylene interrupted procedure.

Source: *Methods of Analysis for Nutrition Labeling*, Proceedings and Final Report, AOAC International, 1993.

Table A3.4 HPLC Methods for Carbohydrate Analysis

Method	Sugars	Matrix	Traction	Clean-up	Analytical column/detector	Mobile phase
Analysis on amino silica columns[a,b,c]						
982.14	Glucose, fructose, sucrose, maltose	Presweetened cereals (defatted)	50% ethanol, 85–90°C, water bath, 25 min	Centrifuge; C_{18}; Cartridge 0.45 µm filter	Amino bonded silica 300 × 4 mm/RI	Acetonitrile H_2O (80:20)
980.13	Fructose, glucose, lactose, maltose, sucrose	Milk chocolate (defatted)	H_2O, 85–90°C water bath, 25 min	Centrifuge; 0.45 µm filter	Amino bonded silica 300 × 4 mm/RI	Acetonitrile H_2O (80:20)
984.17	Fructose, glucose, sucrose, maltose	Licorice extracts	H_2O, 0–90°C, 25 min	0.45 µm filter C_{18} cartridge	Amino bonded silica 300 × 4 mm/RI	CH_3CN-H_2O (83 + 17)
977.20	Fructose, glucose, sucrose	Honey	Aqueous CH_3CN	0.45 µm filter	Amino bonded silica 300 × 4 mm/RI	CH_3CN-H_2O CH_3CN-H_2O (83 + 17)
Analysis on ion exchange resin columns						
983.22	Minor saccharides (dextrose total DP_2 total DP_3)	Corn sugar	H_2O	5% (w/w) mixed exchange resin in sample solution.	610 × 7 mm ion exchange Ca^{2+} 78°C	H_2O
979.23	Major saccharides (glucose, fructose, maltose, psicose)	Corn syrup	H_2O	5% (w/w) mixed exchange resin in sample solution	610 × 7 mm ion exchange 50 W-x 4, Ca^{2+} 78°C	H_2O

[a] Methods 982.14 and 980.13 should have broad applicability, because samples are defatted before extraction.
[b] May need to eliminate the NaCl interference as indicated in 982.14.
[c] Recommended collaborative study to extend to other matrixes (i.e., dairy products).
Source: *Methods of Analysis for Nutrition Labeling*, Proceedings and Final Report, AOAC International, 1993.

Table A3.5 Other Methods for Sugar Analysis

AOAC No.	Method	Brief description	Applicable matrixes
Gas chromatography			
971.18	Carbohydrates in fruit juices	Sample clarified with Pb acetate, derivitize to TMS derivitive and separated on silicone phase; determine mono- and disaccharides[a]	All
973.28	Sorbitol in foods	Acetate derivitive of sorbitol are determined	Fruits
Charcoal column chromatography			
954.11	Separation of sugars in honey	Charcoal column chromatography followed by determination of fractions[a]	All
979.21	Separation of sugars in honey	Charcoal column chromatography followed by determination of fractions[b]	All
Enzymatic methods			
969.39	Glucose in corn syrups and sugars	Glucose oxidase method (colorimetric determination using –dianisidine)	Corn syrups
984.15	Lactose in milk	Hydrolyzed lactose with beta-galactosidase, released galactose measured by NAD reduced method (UV)	Milk
Reducing sugar methods, copper reduction			
906.03	Invert sugar in sugars and syrups	Munson-Walker method	Sugars and syrups[b]
923.09	Invert sugar in sugars and syrups	Lane-Eynon general volume method using Soxhlet reagent (modified Fehling solution)[d]	Sugars and syrups[c,d]
929.09	Invert sugar in sugars and syrups		Sugars and syrups
945.59	Invert sugar in sugars and syrups[e]		Sugars and syrups[f]
945.60	Invert sugar in sugars and syrups	Meissl-Hiller method	Sugars and syrups[e]

950.56	Invert sugar in sugars and syrups[e]	Quisumbing-Thomas method	Sugars and syrups
955.36	Invert sugar in sugars and syrups	Berlin Institute method with Muller's solution	Sugars and syrups
970.58	Invert sugar in molasses	Lane-Eynon constant volume method	Molasses
931.07	Glucose and sucrose in eggs	Dissolve, precipitation of protein, quantified by titration	Eggs
936.06	Glucose in cacao products	Defat, dissolve in H_2O product clarified with Pb acetate and quantified by modified Munson-Walker method (using Sichert-Bleyer table for glucose)	Chocolate
938.18	Glucose in cacao products	Defat, dissolve in H_2O, quantified using Soxhlet reagent, clarify with Pb acetate, Zerban-Sattler method	Chocolate
938.02	Glucose in cacao products	Refer to 938.18 or 936.06	
925.37	Glucose (commercial) in fruits and fruit products	Refer to 930.37	Fruits and fruit product
933.02	Glucose in plants	Shaffer-Somogyi microbiological method	
935.62	Glucose in sugars and syrups	Lane-Eynon and Munson-Walker methods	
945.67	Glucose in corn syrups and sugars[e]	Steinhoff method	
959.11	Glucose in sugars and syrups	Shaffer-Somogyi microbiological method	
935.63	Fructose in sugars and syrups	Lane-Eynon and Munson-Walker method	
932.15	Fructose in sugars and syrups	Jackson-Matthews, modification of Nyns selective method (similar to Shaffer-Somogyi)	
960.06	Fructose in plants	Shaffer-Somogyi microbiological or Munson-Walker method	
950.57	Arabinose, galactose, and xylose and other sugars in sugars and syrups	Shaffer-Somogyi microbiological and other methods	
920.139	Sucrose in lemon, orange, and lime extracts	Refer to 945.27, 925.48, 930.36	
902.02	Sucrose in vanilla extract	Refer to 925.47, 925.48, 930.36	
920.184	Sucrose in honey	Refer to 920.183	
920.189	Sucrose in maple products	Refer to 930.36	

Table A3.5 (Continued)

AOAC No.	Method	Brief description	Applicable matrixes
925.35	Sucrose in fruits and fruit products	Invert with HCl or invertase then Munson-Walker method	
925.44	Sucrose in food dressings	Munson-Walker % invert sugar before/after inversion	
930.07	Sucrose in plants	Invert sugar by HCl or invertase, then Lane-Eynon method	
930.36	Sucrose in sugars and syrups	Invert sugar with HCl or invertase, Munson-Walker method	
950.51	Sucrose in nuts and nut products	Refer to 920.40[e]	
927.07	Lactose in meat	Benedict solution method	
930.32	Lactose in process cheese	Invert sugar with HCl, then Munson-Walker method	
933.04	Lactose in milk chocolate		
935.65	Lactose in sugars and syrups	Munson-Walker or Lane-Eynon method	
945.48	Evaporated milk-lactose	Refer to 896.01, 930.28	
952.05	Lactose in bread	Titration method with Somogyi reagent	
984.15	Lactose in milk	Enzymatic method	
935.64	Maltose in sugars and syrups	Munson-Walker or Lane Eynon methods	
906.01	Sugars (reducing) in plants	Munson-Walker method	
920.190	Sugars (reducing) in maple products as invert sugar		
920.183	Sugars (reducing) in honey	Munson-Walker and Lane-Eynon methods	
921.03	Sugars (reducing) in plants[e]	Quisumbing-Thomas method	
925.15	Sugars in roasted coffee	Munson-Walker or Lane-Eynon methods	
925.36	Sugars (reducing) in fruits and fruit products	Refer to 925.35	
925.42	Sugars (reducing) before inversion in food dressings	Munson-Walker method	

925.43	Sugars (reducing) after inversion in food dressings	Invert sugars by HCl then Munson-Walker method	
925.52	Sugars in canned vegetables	Munson-Walker method on reduced sugars before/after inversion with HCl	
931.02	Sugars in plants	Clarification with Pb acetate or ion-exchange resins	
935.39G	Sugars in baked products	Refer to 975.14	
945.29	Sugars (total reducing) in brewing sugars and syrups	Munson-Walker or Lane-Eynon methods	
945.66	Total reducing sugars	Refer to 923.09 and 906.03	
950.50	Sugars (reducing) in nuts and nut products	Munson-Walker method	
968.28	Total sugars in molasses as invert sugar	Lane-Eynon constant volume method	
975.14	Sugars in bread	Munson-Walker method for reduced sugars, for sucrose refer to 925.05	
Reducing sugar method, ferricyanide reduction			
939.03	Sugars (reducing and nonreducing) in flour	Clarification with Na tungstate followed by titration with ferricyanide and thiosulfate	Cereals, potatoes, sweet mixes
Polarimetric methods			
896.01	Lactose in milk		
896.02	Sucrose in sugars and syrups	Polarizing before/after inversion with HCl or invertase	Reducing sugars
920.189	Sucrose in maple products	Direct and invert polarizations before/after inversion with HCl	
920.115	Sweetened condensed milk-sucrose	Polarizing before/after inversion with HCl	
925.35	Sucrose in fruits and fruit products	Polarizing before/after inversion with HCl	
925.47	Sucrose in sugars and syrups	Polarizing before/after inversion with invertase	

Table A3.5 (Continued)

AOAC No.	Method	Brief description	Applicable matrixes
925.46	Sucrose in sugars and syrups	Saccharimeter reading, sample solution clarified with Pb acetate or alumina cream, ICUMA method[c]	
925.48	Sucrose in sugars and syrups	Polarizing before/after inversion with HCl	
926.13	Sucrose in raffinose in sugars and syrups	Polarizing before/after inversion with invertase-melibiase	
926.14	Sucrose and raffinose in sugars and syrups	Polarizing before/after inversion with HCl	
970.57	Sucrose in molasses	Polarizing before/after inversion with invertase	
942.20	Sucrose in sugar beets	Hot H_2O digestion then polarization	
945.55	Sucrose in gelatin	Polarizing before/after inversion with HCl	
950.30	Sugars (reducing) in nonalcoholic beverages	Refer to 950.29	
950.29	Sucrose in nonalcoholic beverages	Polarizing before/after inversion with HCl or invertase	
950.31	Glucose (commercial) in nonalcoholic beverages	Refer to 930.37	
930.37	Glucose (commercial) in sugars and syrups		

Infrared methods			
972.16H,I	Lactose in milk	Mid-infrared[g]	
975.19	Lactose in milk	Refer to 972.16H,I	
Gravimetric methods			
930.28	Lactose in milk	Gravimetric method	Dairy product
920.110	Lactose in cream	Refer to 930.28	Dairy product

[a] Applicable to broad range of matrixes if samples are defatted and extracted before analysis.
[b] International Commission for Uniform Methods of Sugar Analysis (ICUMSA) approved method.
[c] Lane-Eynon method may require filtration of reagents. Recommend not using asbestos as a filtering aid.
[d] Be cautious of potential interferences from oligosacch and lactic acid.
[e] Surplus method.
[f] For low levels of invert in sucrose.
[g] Must be calibrated against a primary method.
Source: *Methods of Analysis for Nutrition Labeling*, Proceedings and Final Report, AOAC International, 1993.

Table A3.6 Matrixes Needing Methods

Recommendations	Matrixes[a]
Ash (D. Soderberg)	O
Recommend replace references to 900.02 with 923.03	
Beta carotene (M. Deutsch)	B, D-I, K-O, Q-S
Need method/cs[b] and source of standards Need method to separate *cis* and *trans* isomers Need method for encapsulated products	
Biotin (M. Deutsch)	All
Need method/cs Modification of 960.46 + FDA SOP 305 to use *L. plantarum* recommended	
Carbohydrates, sugar alcohols (M. Clarke)	All
Need method/cs	
Carbohydrates, sugars (M. Clarke)	
Cs of HPLC method with wide matrix applicability recommended	
Cholesterol (M. Deutsch)	All < 10 mg/100
Cs underway on direct saponification method	
Copper (Open[c])	
Need cs for Cu as analyte in 968.08, 975.03, 965.09, 984.27, 985.35, 970.18, 980.03 Add note to 968.08 that Pt dishes are preferable, that losses may occur using porcelain dishes	
Cyanocobalamine (M. Deutsch)	
Cs recommended for 960.46 + 952.20 to use *L. leichmanui* for all matrixes	
Dietary fiber, insoluble (L. Prosky)	
Cs recommended for sample preparation of high-fat samples	
Dietary fiber, soluble (L. Prosky)	
Cs recommended for sample preparation of high-fat samples	
Dietary fiber, total (L. Prosky)	
Cs recommended for sample preparation of high-fat samples	

Table A3.6 (Continued)

Recommendations	Matrixes[a]
Drained weight (Open[c])	
Need procedure for pickled vegetables, fruits, and tuna	
Fat, monounsaturated (D. Firestone)	All
Recommend adoption of AOCS method where applicable	
Fat, polyunsaturated (D. Firestone)	
Recommend revision of extraction procedure	
Fat, saturated (D. Firestone)	A, C, J, T
Recommend revision of extraction procedure	
Modification of 969.33 to extend matrixes and quantitation of fatty acids in all foods recommended	
Recommend cs for samples < 350 mg fat content	
Need method for C-4 to C-6 fatty acids	
Fat, total (D. Firestone)	All
Need quantitation method for fatty acids	
Folacin (M. Deutsch)	
Cs recommended for 960.46 + 944.12 + FDA SOP 315 to use *L. casei* for free form	
Cs recommended for 960.46 + 944.12 + FDA SOP 320 to use *S. faecalei* for products containing free form	
Cs recommended for 960.46 and FDA SOP 330 to use *K. apriculata* for bound form	
Iodine (Open[c])	All
Need method/cs; consideration of ceric sulfate reduction recommended	
Iron (Open[c])	
Recommend extension of applicability statements to foods for 965.09, 968.08, 985.35, 975.03, 980.03	
Magnesium (Open[c])	
Recommend extension of applicability statements to foods for 965.09, 985.35, 975.03, 985.35, 980.03	
Recommend extension of applicability statement to A-C, F-H, K-M, P, Q, T for 984.27	

Table A3.6 (Continued)

Recommendations	Matrixes[a]
Moisture (Open[c])	B, F, I, J, L-O, Q, R for high sugar products
Recommend modification of 952.08 to eliminate asbestos use	
See Moisture Report in the January 1993 issue of *The Referee*	
Niacin (M. Deutsch)	
Recommend extension of applicability statement to B, D, F-H, J, P for 985.24	
Recommend extension of applicability statement to all foods for 944.13	
Recommend cs on niacin in fortified foods using HPLC method in *J. AOAC* (in press)	
Cs recommended for 960.46 and 944.13 to use *L. plantarum* for all matrixes	
Pantothenic acid (M. Deutsch)	All
Cs recommended for 960.46 and 945.74 to use *L. plantarum* for free form	
Cs recommended for 960.46 + FDA SOP 350 + USDA 97 to use L. plantarum for total pantothenic acid for all matrixes	
Cs recommended for 960.46 and FDA SOP 340 to use *A. suboxydaus* for panthenol	
Phosphorus (Open[c])	
Recommend extension of applicability statements to foods for 958.01, 957.18, 964.06, 965.17	
Recommend extension of applicability statement to A, E-M, O-R, T for 986.24	
Potassium (Open[c])	
Recommend extension of applicability statement to include food for 985.35	
Recommend extension of applicability statement to C, K, O for 984.27	
Protein (D. Soderberg)	
Recommend modification of Kjeldahl method to include spices	
Recommend cs of Dumas for all matrixes	
Pyridoxine (M. Deutsch)	
Recommend cs based on *J. AOAC* 39:157 (1956)	
Cs recommended for 960.46 and 961.15 to use *S. uvarum* for total B_6 all matrixes	

Table A3.6 (Continued)

Recommendations	Matrixes[a]
Riboflavin (M. Deutsch)	
Recommend cs for HPLC method	
Recommend extension of applicability statement to foods for 940.33	
Recommend modification of 970.65 and 981.15 to incorporate standard addition	
Cs recommended for 960.46 and 940.33 to use *L. casei* for all matrixes	
Sampling/sample prep (Open[c])	
Recommend completion of AOAC book on sampling	
Recommend inclusion of sampling and sample preparation in future AOAC methods	
Sodium (Open[c])	
Recommend extension of applicability statement to food for 985.35	
Recommend extension of applicability statement to C, K, O for 984.27	
Stearic acid (D. Firestone)	
Need quantitative method	
Thiamine (M. Deutsch)	
Recommend extension of applicability statement to all food for 986.27	
Recommend extension of applicability statement to D, H-J, Q for 957.17	
Cs recommended for simultaneous determination of B_1, B_2, and B_6 using fluorescence detection, JAOAC 75:561 (1992)	
Cs recommended for 960.46 and FDA SOP 360 to use *L. viridescens* for all matrixes	
Tryptophan (M. Deutsch)	
Cs recommended for FDA SOP 500, 501, and 502, HPLC for free form	
Cs recommended for 960.46 and FDA SOP 365 to use *L. plantarum* for bound and free forms	

Table A3.6 (Continued)

Recommendations	Matrixes[a]
Vitamin A (M. Deutsch)	A, C-F, J, N-P, R-T
Cs recommended for new HPLC method based on 992.04, for other than infant formula to apply to variety of matrixes. Modify sample size and replace ambient 18 h with high temperature/short time saponification	
Vitamin C (M. Deutsch)	B, I, L-O, Q, S
Need method/cs using HPLC to separate erythorbate, ascorbate, and dehydroascorbate	
Recommend modification of 984.26 to include meats, nuts, seafood, spices, and oils/fats	
Recommend adoption of ACNA HPLC method, *J. AOAC* 75:887 (1992), for vitamin C in variety of foods	
Vitamin D (M. Deutsch)	
Recommend extension of applicability statements to foods for 982.29, 936.14	
Recommend cs on GPC/HPLC method, *NIST-Anal. Chem.* 60:1929 (1988)	
Recommend cs using HPLC method, reversed or normal phase	
Vitamin E (M. Deutsch)	
Recommend cs of HPLC method	
Recommend modification of current methods with solvent substitution for chloroform	
Need further investigation into standards, methodology, and standard conversion factors	
Zinc (Open[c])	
Recommend extension of applicability statements to foods for 985.35, 968.08, 975.03, 980.03, 965.09	
Recommend extension of applicability statement to A-F, H-M, O, P, R, T for 984.27	

[a] Those analytes for which no matrix is specified can be analyzed using AOAC Official Methods listed in the March 1993 issue of *The Referee*. The methods in that list are deemed to be adequate on the basis of collaborative studies and/or common use for the analyte/matrix combinations listed although not all analyte/matrix combinations listed have been fully collaboratively studied. See Table A2.1 (Appendix 2), for matrix identification.

[b] cs, collaborative study.

Table A3.6 (Continued)

[c] No current general referee (GR) for these specific analytes/topics. Contact AOAC International, Technical Services, 2200 Wilson Blvd., Suite 400, Arlington, VA 22201, (703) 522-3032.

AOAC general referees and their topics:
M. Bueno—Infant Formula/Medical Diets—FDA HFF-266, 200 C St., SW, Washington, DC 20204, (202) 205-4445.
D. Soderberg—Meat, Poultry, Meat and Poultry Products—USDA, FSIS, 300 12th St, SW, Washington, DC 20205, (202) 205-0039
M. Deutsch—Vitamins and Other Nutrients —FDA, HFF-264, 200 C St., SW, Washington, DC 20204, (202) 205-4263
M. Clarke—Sugars and Sugar Products—Sugar Processing Research, Inc., 1100 Robert E. Lee Blvd, New Orleans, LA 70124, (504) 286-4542
L. Prosky—Dietary Fiber—FDA, HFF-268, 8301 Muirkirk Rd, Laurel, MD 20708, (301) 344-6005
D. Firestone—Fats and Oils—FDA, HFF-426, 200 C St, SW, Washington, DC 20204, (202) 205-4381.
Thanks for the efforts of M. Bueno, D. Carpenter, H. Chin, M. Deutsch, J. DeVries, N. Fraley, W. Hummer, W. Ikins, W. Landen, S. Lee, J. Morawski, P. Oles, L. Prosky, D. Soderberg, A. Soliman (deceased), D. Sullivan, J. Tanner, W. Wolf
Reprinted from *Methods of Analysis for Nutrition Labeling*, Proceedings and Final Report, AOAC International, 1993.

Table A3.7 USDA Food Composition Publications

Series	Title	Date	GPO Stock No.	Price ($)
Agriculture Handbook No. 8 Series				
AH-8-1	Dairy and egg products	1976	001-000-03635-1	9.00
AH-8-2	Spices and herbs	1977	001-000-03646-7	3.75
AH-8-3	Baby foods	1978	001-000-03900-8	12.00
AH-8-4	Fats and oils	1979	001-000-03984-9	7.50
AH-8-5	Poultry products	1979	001-000-04008-1	17.00
AH-8-6	Soups, sauces, and gravies	1980	001-000-04114-2	12.00
AH-8-7	Sausages and luncheon meats	1980	001-000-04183-5	6.00
AH-8-8	Breakfast cereals	1982	001-000-04283-1	9.00
AH-8-9	Fruits and fruit juices	1982	001-000-04287-4	14.00
AH-8-10	Pork products	1992	001-000-04593-8	14.00
AH-8-11	Vegetable and vegetable products	1984	001-000-04427-3	16.00
AH-8-12	Nut and seed products	1984	001-000-04429-0	7.50
AH-8-13	Beef products	1990	001-000-04555-5	21.00
AH-8-14	Beverages	1986	001-000-04468-1	9.50
AH-8-15	Finfish and shellfish products	1987	001-000-04497-4	10.00
AH-8-16	Legumes and legume products	1986	001-000-04488-5	8.50
AH-8-17	Lamb, veal, and game products	1989	001-000-04541-5	13.00
AH-8-18	Baked products	1992	001-000-04584-9	28.00
AH-8-19	Snacks and sweets	1991	001-000-04577-6	19.00
AH-8-20	Cereal grains and pasta	1989	001-000-04549-1	8.50
AH-8-21	Fast foods	1988	001-000-04524-5	11.00
AH-8	Composition of foods: raw, processed, prepared. Supplements update selected items, add data for new food items, and add tables of data on selected nutrients.			
	1989 Supplement	1990	001-000-04554-7	11.00
	1990 Supplement	1991	001-000-04571-7	17.00
	1991 Supplement	1992	001-000-04585-7	12.00
Other				
HERR-48	Sugar content of selected foods: individual and total sugars	1987	001-000-04514-8	2.00

Order from the Government Printing Office (GPO) as follows:

By mail: Make check or money order payable to "Superintendent of Documents", or use Government Printing Office Deposit Account. Mail to: Superintendent of Documents, U.S. Government Printing Office, Washington,, DC 20402.

By telephone: Use VISA or Mastercard credit cards or Government Printing Office Deposit Account. Call (202) 783-3238.

By telefax: Call (202) 275-0019 anytime.

Table A3.7 (Continued)

By visiting or calling a GPO bookstore as follows:

State	Address	Telephone
Alabama	O'Neill Bldg. 2021 Third Ave. North Birmingham, AL 35203	205-731-1056
California	ARCO Plaza-Level C, 505 S. Flower St. Los Angeles, CA 90071	213-239-9844
	Room 1023 Federal Office Bldg. 450 Golden Gate Ave. San Francisco, CA 94102	415-252-5334
Colorado	Room 117, Federal Bldg. 1961 Stout St. Denver, CO 80294	303-844-3964
	World Savings Bldg. 720 N. Main St. Pueblo, CO 81003	714-544-3142
District of Columbia	710 North Capitol St. NW Washington, D.C. 20402	202-512-0132
	1717 H. St., NW Washington, D.C. 20005	202-653-5075
Florida	P. O. Box 35089 Federal Bldg. 400 W. Bay St. Jacksonville, FL 32202	904-353-0569
Georgia	P. O. Box 56445 275 Peachtree St. NE Atlanta, GA 30343	404-331-6947
Illinois	Room 1365, Dirksen Bldg. 219 S. Dearborn St. Chicago, IL 60604	312-353-5133
Maryland	Laurel Bookstore 8660 Cherry Lane Laurel, MD 20707	301-953-7974
Massachusetts	Room 169 O'Neill Federal Bldg. 10 Causeway St. Boston, MA 02222	617-720-4180

Table A3.7 (Continued)

State	Address	Telephone
Michigan	Suite 160 McNamara Federal Bldg. 477 Michigan Ave. Detroit, MI 48226	313-226-7816
Missouri	120 Bannister Mall 5600 E. Bannister Rd. Kansas City, MO 64137	816-765-2256
New York	Room 110 26 Federal Plaza New York City, NY 10278	212-264-3825
Ohio	Room 1653 Federal Office Bldg. 1240 E. 9th St. Cleveland, OH 44199	216-522-4922
	Room 207, Federal Bldg. 200 N. High St. Columbus, OH 43215	614-469-6956
Oregon	1305 S.W. First Ave. Portland, OR 97201	503-221-6217
Pennsylvania	Morris Bldg. 100 N. 17th St. Philadelphia, PA 19103	215-597-0677
	Room 118, Federal Bldg. 1000 Liberty Ave. Pittsburgh, PA 15222	412-644-2721
Texas	Suite 1C46, Federal Bldg. 1100 Commerce St. Dallas, TX 75242	214-767-0076
	Texas Crude Bldg. 801 Travis St. Houston, TX 77002	713-228-1187
Washington	Room 194 Federal Office Bldg. 915 Second Ave. Seattle, WA 98174	206-442-4270
Wisconsin	Room 190, Federal Bldg. 517 E. Wisconsin Ave. Milwaukee, WI 53202	414-297-1304

Source: *FSIS Manual on Use of Data Bases for Nutrition Labeling*, February, 1993.

Table A3.8 Nutrient Data Bank Electronic Bulletin Board Files for Downloading

---

USDA Nutrient Data Base for Standard Reference, Release 10: Release 10 contains all of the nutrient data on all foods published in the complete 21 volumes of the AH-8 series, including the Pork Section published in February 1993. Some, but not all, of the missing values contained in the published versions have been filled in by HNIS staff. The files include raw and cooked forms of food and contain about 5,500 entries.

Survey Nutrient Data Base, Release 5: Release 5 contains nutrient data used in the 1989 USDA Continuing Survey of Food Intakes by Individuals. Files contain many items that are calculated from recipes and do not correspond to any published food composition table. Values for meat and poultry products are for cooked forms of the food. The data base contains about 6,500 foods.

Primary Data Set for Standard Reference, Release 5: Release 5 contains nutrient values for foods used to create the Survey Nutrient Data Base, including raw foods used in calculating the recipe items and certain ingredients such as baking powder, which do not appear in AH-8. The file does not contain all foods in AH-8 but holds about 3,000 items.

USDA Table of Nutrient Retention Factors, Release 3: Release 3 contains retention factors for 18 vitamins and minerals for 260 foods used in recipe calculations for the survey data base. It does not contain many of the specific retention factors shown for the individual cuts of meat and poultry products as published in the hard copy versions of the AH-8 series.

Recipes for the Survey Data Base, Release 5: This file contains the component records of the recipes used to create the survey data base, e.g., yields for recipes, moisture and fat changes, and retention factor codes.

Sugar: The sugar file contains data from the Home Economic Research Report No. 48, which was published in 1987. It includes data on mono- and disaccharides, other sugars, and total sugars for over 500 foods. More information on sugar content of foods may be obtained from HNIS at 301-436-8491.

The board also contains a file of nutrient data on about 900 foods from Home and Garden Bulletin No. 72, a Dietary Analysis Program for the PC, data on vitamins K and D, and data on the nutrient content of the U.S. food supply.

Source: *FSIS Manual on Use of Data Bases for Nutrition Labeling*, February, 1993.

Table A3.9 Commercial Sources of Databases[1]

Organizations	System names	Contacts
CAMDE Corporation Suite 331 4435 S. Rural Road Tempe, AZ 85282	Nutri-Calc HD and Plus	Julie McDonald 602-821-2310
Case Western Reserve University, School of Medicine 2109 Adelbert Road Cleveland, OH 44106-4945	HVH-CWRU Nutrient Data Base	Harold Houser 216-368-3195
The CBORD Group, Inc. 61 Brown Road Ithaca, NY 14850	Diet Analyzer System	Nancy Vergara 607-257-2410
Computrition, Inc. P.O. Box 4689 Chatsworth, CA 91313-4689	Nutritional Software Library	Bridgett Harvey-Elliott 818-701-5544
DDA Software P.O. Box 26 Hamburg, NJ 07419	Nutrient Analysis System 2	Carl Bredbenner, Jr. 201-764-6677
DINE Systems, Inc. 586 N. French Road, Suite 2, Amherst, NY 14228	The DINE System: Nutrient Analysis (DINE Windows, Macdine II)	Kathryn Dennison 716-688-2492
ESHA Research P.O. Box 13028 Salem, OR 97309-9730	Food Processor II and Genesis	Patricia Bishop 503-585-6242
Formulas/Now! Software 2755 Chelsea Drive, Suite A Oakland, CA 94611-2505	Formulas Now	Jerry Salzman 510-482-2705
Hopkins Technology 421 Hazel Lane Hopkins, MN 55343-7116	Food/Analyst Plus	Phillip Dunn 612-931-9376
N-Squared Computing Analytic Software, Suite 240 3040 Commercial St., SE Salem, OR 97302	Nutritionist III and IV	Laurie North 503-364-9118

Table A3.9 (Continued)

Organizations	System names	Contacts
Nutrition and Food Associates P.O. Box 47007 Minneapolis, MN 55447	NutriForm	Patricia Godfrey 612-550-9475
Palatino Products P.O. Box 20452 Castro Valley, CA 94546	Kwik-answer	George Shaw 510-538-0132
Practorcare, Inc. Suite 2F 10951 Sorrento Valley Road San Diego, CA 92121	Nutriplanner	Barbara Harris 800-421-9073
University of Minnesota, Nutrition Coordinating Center Suite 310 2221 University Ave., SE Minneapolis, MN 55414-3076	Minnesota Nutrition Data System	Keren Price 612-627-4869
University of Texas Health Science Center School of Public Health P.O. Box 20186 Houston, TX 77225	Food Intake Analysis System	Deidre Douglass 713-792-4533
Wellsource, Inc. P.O. Box 569 Clackamas, OR 97015	Professional Dietitian	Donald Hall 503-656-7446
WholeGrain Software 1427 Bancroft Way Berkeley, CA 94702	NutriLabeler	David Stone 415-848-5903

[1] Mention of commercial products is solely for identification purposes to report factually on available information and does not constitute endorsement by USDA. USDA neither evaluates nor approves nutrient calculation systems and neither guarantees nor warrants the quality of the listed products. Use of brand names by USDA implies no approval of the products to the exclusion of others which may also be suitable.

Source: *FSIS Manual on Use of Data Bases for Nutrition Labeling*, February, 1993. This table is for information purposes only and shall not be considered a complete listing of commercial sources of databases or an endorsement of any of the organizations listed.

Index